ARL Reader Planungstheorie Band 2

Thorsten Wiechmann

(Hrsg.)

ARL Reader
Planungstheorie Band 2

Strategische Planung – Planungskultur

Hrsg.
Thorsten Wiechmann
Fakultät Raumplanung, TU Dortmund
Dortmund, Deutschland

ISBN 978-3-662-57623-6   ISBN 978-3-662-57624-3   (eBook)
https://doi.org/10.1007/978-3-662-57624-3

Die Deutsche Nationalbibliothek verzeichnet diese Publikation in der Deutschen Nationalbibliografie; detaillierte bibliografische Daten sind im Internet über http://dnb.d-nb.de abrufbar.

Springer Spektrum
© Springer-Verlag GmbH Deutschland, ein Teil von Springer Nature 2019

Einbandabbildung: Parco Dora © Thorsten Wiechmann
Covergestaltung: deblik, Berlin
Planung/Lektorat: Sarah Koch

Springer Spektrum ist ein Imprint der eingetragenen Gesellschaft Springer-Verlag GmbH, DE und ist ein Teil von Springer Nature.
Die Anschrift der Gesellschaft ist: Heidelberger Platz 3, 14197 Berlin, Germany

Vorwort

Die erste Idee zu diesem Reader entstand in einem Gespräch mit Prof. Dr. Hans Heinrich Blotevogel im März 2011 in Dortmund. Das Präsidium der Akademie für Raumforschung und Landesplanung (ARL) wollte damals eine neue Initiative zur Planungstheorie starten. Dafür sprachen mehrere Gründe: Zum einen bedürfen die Planungswissenschaften, wie jede akademische Disziplin, einer soliden theoretischen Fundierung. Sie müssen einen leistungsfähigen Theoriekern ausbilden, der kumulativ auszubauen ist. Zum anderen werden in Deutschland die lebhaften internationalen Diskussionen zur Planungstheorie traditionell nur wenig rezipiert und erreichen selten die Planungspraxis. Es zeigt sich eine ausgeprägte Kommunikationslücke zwischen der Planungswissenschaft und der Planungspraxis.

Vor diesem Hintergrund sollte die Initiative der ARL den aktuellen Stand planungstheoretischer Diskurse aufarbeiten und reflektieren. In einem Expertengespräch im November desselben Jahres wurden die Ideen konkretisiert und die Einrichtung eines ARL-Arbeitskreises beschlossen. Der hier vorliegende ARL Reader Planungstheorie ist eines der zentralen Ergebnisse dieses Arbeitskreises. Er will eine umfassende und doch pointierte Bestandsaufnahme des planungstheoretischen Diskussionsstandes leisten.

Aus der Fülle möglicher Theorien und Methodologien konzentriert sich der Reader auf vier Themenkomplexe:
1. Kommunikative Planung (Band 1)
2. Neoinstitutionalismus und Governance (Band 1)
3. Strategische Planung (Band 2)
4. Planungskultur (Band 2)

In Form eines Sammelwerkes präsentiert er Debatten bestimmende Originaltexte national und international bekannter Autoren. Diese werden durch namhafte Planungswissenschaftler eingeordnet und kritisch diskutiert. Damit bietet der Band einen so bisher nie dagewesenen Einstieg und Überblick über die Grundlagen der aktuellen planungstheoretischen Debatten für Studierende der Raum- und Planungswissenschaften sowie fachlich interessierte Wissenschaftler und Planungspraktiker.

All jenen, die bei der Entstehung des Readers geholfen haben, sei an dieser Stelle ganz herzlich gedankt. Hierzu zählen zuvorderst die Mitglieder des ARL-Arbeitskreises „Planungstheorien – Stand und Perspektiven", die in den Jahren 2013 bis 2015 in sieben zweitägen Sitzungen Konzept und Inhalt des Readers gemeinsam entwickelt haben:
Dipl.-Ing. Judith Marie Böttcher (geb. Bornhorst), ARL/HafenCity Universität Hamburg
Prof. Dr. Rainer Danielzyk, ARL/Leibniz Universität Hannover
Dr. Ludger Gailing, Leibniz-Institut für Raumbezogene Sozialforschung
Dr. Marian Günzel, TU Dortmund (AK-Geschäftsführer)
Dr. Alexander Hamedinger, TU Wien
Dr. Gérard Hutter, Leibniz-Institut für ökologische Raumentwicklung
Prof. Dr. Thomas Krüger, HafenCity Universität Hamburg
Prof. Dr. Frank Othengrafen, TU Dortmund
Dr. Mario Reimer, Institut für Landes- und Stadtentwicklungsforschung
Prof. Dr. Walter Schönwandt, Universität Stuttgart
Prof. Dr. Thorsten Wiechmann, TU -Dortmund (Leiter des Arbeitskreises)
Prof. Dr. Karsten Zimmermann, TU Dortmund

An den Beratungen des AK haben darüber hinaus eine Reihe weiterer Personen in Form von thematischen Inputs und Statements mitgewirkt. Auch sie haben wesentlichen Anteil am Zustandekommen des Readers: Prof. Dr. Uwe Altrock, Prof. Dr. Hans Heinrich Blotevogel, Prof. Dr. Christian Diller, Klaus Einig, Prof. Dr. Huib Ernste, Prof. Dr. Dietrich Fürst, Prof. Dr. Jean-David Gerber, Prof. Dr. Panos Getimis, Prof. Dr. Enrico Gualini, Dr. Christoph Hemberger, Antje Herbst, Prof. Dr. Oliver Ibert, Dr. Meike Levin-Keitel, Marlies Meijer, Andreas Putlitz, Prof. Dr. Wolf Reuter, Prof. Dr. Klaus Selle und Dr. Martin Sondermann (in alphabetischer Reihenfolge). Als zuständige Wissenschaftlerin in der Geschäftsstelle der ARL hat Dr. Evelyn Gustedt die Lenkungsgruppe des Arbeitskreises in allen administrativen und organisatorischen Fragen tatkräftig unterstützt. Als Geschäftsführer des Arbeitskreises liefen viele Fäden bei Dr. Marian Günzel zusammen, der maßgeblichen Anteil am Erfolg des Vorhabens hatte.

Ein großes Dankeschön gilt auch dem Präsidium der Akademie für Raumforschung und Landesplanung, das nicht nur den Arbeitskreis in allen Phasen wohlwollend unterstützt, sondern auch den Reader durch die Bereitstellung der erforderlichen Mittel erst ermöglicht hat.

Last but not least gilt ein besonderer Dank auch Dr. Simone Jordan und Sarah Koch (Associate Editors) sowie Anja Dochnal (Projektmanagerin)

vom Springer Spektrum Verlag, die die Entstehung des Werkes professionell und mit viel Engagement begleitet haben. Beide Bände wurden zudem durch den Springer Spektrum Verlag beim Erwerb der Abdruckrechte für die Originalartikel maßgeblich unterstützt.

Thorsten Wiechmann
Dortmund
8. Februar 2019

Inhaltsverzeichnis

Autorenverzeichnis

Prof. Dr. Rainer Danielzyk
Institut für Umweltplanung
Universität Hannover
Hannover, Deutschland

Prof. Dr. Frank Othengrafen
Fakultät Raumplanung
TU Dortmund
Dortmund, Deutschland

Dr. Gérard Hutter
Leibniz-Institut für ökologische Raumentwicklung, IÖR
Dresden, Deutschland

Dr. Mario Reimer
ILS – Institut für Landes- und Stadtentwicklungsforschung
Dortmund, Deutschland

Prof. Dr. Thomas Krüger
Projektentwicklung und Projektmanagement in der
Stadtplanung
Hafencity Universität Hamburg
Hamburg, Deutschland

Prof. Dr. Thorsten Wiechmann
Fakultät Raumplanung
TU Dortmund
Dortmund, Deutschland

Einleitung – Zum Stand der deutschsprachigen Planungstheorie

Thorsten Wiechmann

© Springer-Verlag GmbH Deutschland, ein Teil von Springer Nature 2019
T. Wiechmann (Hrsg.), *ARL Reader Planungstheorie Band 2*, https://doi.org/10.1007/978-3-662-57624-3_1

1

» *There is no planning practice without a theory about how it ought to be practiced. That theory may or may not be named or present in consciousness, but it is there all the same.* (Friedmann 2003, S. 8)

» *In a changing and globalizing world, planning theory is core to understanding how planning and its practices both function and evolve.* (Gunder et al. 2018, S. 1)

1.1 Der ARL Reader Planungstheorie

Die Planungsdisziplin hat sich seit ihren Anfängen in den 1950er und 1960er Jahren zu einer in Forschung, Lehre und Praxis relevanten Wissenschaft entwickelt. Parallel zur Entwicklung der Stadt- und Raumplanung entwickelte sich auch das Nachdenken über Planung, sowohl innerhalb der räumlichen Planung als auch in anderen gesellschaftlichen Bereichen. Dies führte zur Herausbildung von Planungstheorien, deren Vermittlung heute im Curriculum aller Planungsstudiengänge verankert ist. Allerdings fehlte es bislang im deutschsprachigen Raum an einer aktuellen Bestandsaufnahme des internationalen planungstheoretischen Diskussionsstandes, die eine Grundlage bilden könnte für weiterführende Fachdiskurse.

Der von der Akademie für Raumforschung und Landesplanung (ARL) – Leibniz-Forum für Raumwissenschaften 2013 eingesetzte Arbeitskreis „Planungstheorien – Stand und Perspektiven" hatte sich zur Aufgabe gesetzt, eine derartige Bestandsaufnahme in Form einer Reflexion ausgewählter planungstheoretischer Diskurse zu leisten, Defizite in der bisherigen Auseinandersetzung sowie offene Forschungsfragen zu identifizieren und damit einen Beitrag zur konzeptionellen Debatte in den Planungswissenschaften beizusteuern. Entstehung und Rezeption wichtiger Diskurse und Metaerzählungen sollten in einem möglichst kohärenten Rahmen nachgezeichnet und interpretiert werden. Dabei sollte es auch darum gehen, die meist unbewussten Prämissen hinter den bestehenden theoretischen Zugängen deutlich zu machen und den Umgang mit auftauchenden Widersprüchen zu thematisieren.

Reader zur Planungstheorie gibt es 50 Jahre nach Gründung der ersten deutschen Planungsfakultät an der Universität Dortmund im Jahr 1968 bislang nur in englischer Sprache: von dem ersten Kompendium *A Reader in Planning Theory*, herausgegeben von Andreas Faludi (1973), über das in vier Auflagen erschienene Standardwerk *Readings in Planning Theory* (Campbell und Fainstein 1996; Fainstein und Campbell 2003, 2012; Fainstein und DeFilippis 2016) bis hin zu neueren Textsammlungen wie dem dreibändigen Werk *Critical Essays in Planning Theory* (Hillier und Healey 2008) sowie dem aktuellen *Routledge Handbook of Planning Theory* (Gunder et al. 2018).

Kompendien für den deutschsprachigen Raum wurden bisher bestenfalls als vorlesungsbegleitende Handapparate an Planungsschulen zusammengestellt, nicht aber einer

breiteren Leserschaft zur Verfügung gestellt. Dies ist im Grunde erstaunlich, denn die englischsprachigen Werke sind weder in der Lage, deutschsprachige Debattenbeiträge zu vermitteln, noch richten sie sich in ihrer Zusammenstellung und Aufbereitung an ein deutschsprachiges Publikum. Aufgabe derartiger Reader ist ja nicht nur eine reine Anordnung bzw. Sammlung von Debattenbeiträgen, sondern die planungstheoretischen Debatten zusammenzufassen, einzuordnen und kritisch zu diskutieren. Die einzelnen Originalbeiträge stehen nicht solitär, sondern werden zueinander in Bezug gesetzt und in einen inhaltlichen Gesamtzusammenhang gestellt. Es geht nicht nur darum, einzelne, besonders lesenswerte Beiträge der planungstheoretischen Debatte pointiert aufzugreifen, sondern auch einen Gesamtüberblick und gegenseitige Verknüpfungen der Debatte zu vermitteln. Dies erfordert aber eine kontextspezifische Betrachtung: „Planning theories, therefore, need to be contextualized and localized, as they are narratives developed in the context of particular circumstances and in response to certain concerns. To map out the current state of planning theory, therefore, these theories need to be located in the context of their historical evolution, and in reference to the changing conditions of the societies in which they emerge" (Gunder et al. 2018, S. 2).

Ziel des *ARL Reader Planungstheorie* ist es, eine umfassende und dennoch pointierte Bestandsaufnahme des internationalen planungstheoretischen Diskussionsstandes aus Sicht der deutschsprachigen Planungswissenschaft zu leisten. In Form eines Sammelbandes präsentiert der Reader hierfür debattenbestimmende Originaltexte national und international bekannter Autoren, die sichtbare Spuren in den jeweiligen Debatten der letzten drei Jahrzehnte hinterlassen haben. Diese Originaltexte werden durch die Autoren des Readers in eigenen Beiträgen zusammengefasst, kontextualisiert und diskutiert. Damit wird ein niedrigschwelliger Zugang zu den Debatten ermöglicht und mit weiteren Literaturhinweisen versehen, um je nach Interessenlage die jeweilige Thematik weiter zu vertiefen. Aus der Fülle möglicher Theorien und Methodologien konzentriert sich der Reader auf folgende vier Themenkomplexe:

1. Kommunikative Planung (Zimmermann; Bd. 1)
2. Neoinstitutionalismus/Governance (Gailing, Hamedinger; Bd. 1)
3. Strategische Planung (Hutter, Wiechmann, Krüger; Bd. 2)
4. Planungskultur (Othengrafen, Reimer, Danielzyk; Bd. 2)

Als erster Reader dieser Art im deutschsprachigen Raum bieten die beiden Bände einen Einstieg und Überblick über ausgewählte Grundlagen der aktuellen planungstheoretischen Debatte für Studierende der Raum- und Planungswissenschaften sowie fachlich interessierte Wissenschaftlerinnen und Wissenschaftler. Aber auch den planungspraktisch Tätigen bietet das Werk einen Rahmen zur theoretischen Reflexion und analytischen Einordnung des eigenen Handelns bzw. des Berufsalltags in der Planungspraxis. Damit bietet es mittelbar auch Erklärungshilfen und Anregungen für die Lösung planungspraktischer Problemstellungen.

1.2 Was ist Planungstheorie?[1]

Je nach wissenschaftstheoretischem Standpunkt werden unterschiedliche Anforderungen an den Theoriebegriff geknüpft. Im Allgemeinen entwirft eine Theorie ein System von Aussagen, um damit einen spezifischen Ausschnitt der Realität modellhaft zu beschreiben, zu erklären oder auch Vorhersagen zu treffen. Soweit daraus Handlungsempfehlungen abgeleitet werden, bilden Theorien die Grundlage für die Praxis. In der Planungsdisziplin führten Anwendungsnähe, verbunden mit einem traditionell eher geringen Theorieinteresse, disziplinäre Vielfalt an theoretischen Zugängen und mangelnde eigenständige Paradigmatisierung in der Vergangenheit jedoch wiederholt zum Vorwurf des Eklektizismus und eines fehlenden gemeinsamen Theoriekerns. Bislang wurde weder über den Gegenstandsbereich der Planung noch über die relevanten Denkschulen ein disziplinärer Konsens hergestellt. Es ist nicht einmal klar, was unter Planung zu verstehen ist, denn das Begriffsverständnis hängt maßgeblich vom planungstheoretischen Standpunkt des Betrachters ab.

Planungstheorien können wahlweise empirisch-analytisch arbeiten und auf ein besseres Verständnis der Planungspraxis abzielen (deskriptive oder explanative Planungstheorien) oder auch dezidiert Handlungsanleitungen geben, wie geplant werden sollte (normative Planungstheorien): „Planning theory is […] divided into those who understand planning through analyzing existing practices and those who theorize in an effort to transform planning practices" (Fainstein und DeFilippis 2016, S. 2).

Traditionell ist das Verhältnis von Planungstheorie und Planungspraxis problematisch. Die meisten praktisch Planenden schenken der Planungstheorie wenig Beachtung und erinnern sich auch nur ungern an das Planungstheorieseminar, das sie im Studium absolvieren mussten. Raumplanung ist ein angewandtes Fach, und die meisten Planungsstudierenden werden zu Praktikern und nicht zu akademischen Forschern. Absolventen der Geographie, der Wirtschafts- oder Sozialwissenschaften werden für theoretische Beiträge in der Regel honoriert. In der Stadt- und Regionalplanung geht es hingegen in erster Linie um die praktische Anwendung. Auch dies verlangt vorausschauendes Denken. Die Abschätzung künftiger Auswirkungen von Planungsinterventionen erfordert ein theoretisches Verständnis der Prozesse, die Räume und Orte prägen. Daher sind Theorien für Planerinnen und Planer unentbehrlich, auch wenn diese möglicherweise intuitiv, implizit und nicht hinterfragt sind. Planungspraxis basiert oftmals eher auf Intuition als auf expliziten Theorien: „Yet this intuition may in fact be assimilated theory. In this light, theory represents cumulative professional knowledge" (Fainstein und Campbell 2012, S. 3). Theorien, ob nun als verdichtete Praxis oder explizit formuliert, sind letztlich Voraussetzung für jedwedes intentionale Handeln in der Praxis. Es spricht allerdings viel dafür, gerade auch die impliziten Theorien, die die Planungspraxis beeinflussen, explizit zu machen und kritisch zu reflektieren.

Planungstheorie ist ferner inter- und transdisziplinär angelegt. Sie greift auf theoretische Vorarbeiten aus ganz unterschiedlichen disziplinären Kontexten zurück. Die Planungswissenschaft ist daher konfrontiert mit einer unaufhebbaren Pluralität und Konkurrenz divergenter Paradigmen. Jeder Versuch, eine umfassende und einheitliche Theorie der Planung aufzustellen, ist unter diesen Umständen zum Scheitern verurteilt.

Den in den vergangenen Jahrzehnten entstandenen planungstheoretischen Ansätzen, die sich immer auch im Spiegel des Zeitgeistes und disziplinfremder Theorieeinflüsse entwickelten, kommt letztlich eine wichtige Funktion zu: die Verständigung der Profession über sich selbst. Wie jede Wissenschaft bedarf auch die Planungswissenschaft der kritischen Selbstreflexion auf Basis von Theorien. Die enge Verknüpfung der vergleichsweise jungen Planungswissenschaft mit dem politisch-administrativen System der Stadt- und Raumplanung macht die Aufgabe der Selbstreflexion besonders dringlich, zumal beide – Wissenschaft und Praxis der Planung – seit ihrer Entstehung unter erheblichem Legitimationsdruck stehen: „Planning theory is one of the few means we have at our disposal to hold us together as a family of practitioners" (Friedmann 2011, S. 130). Zugleich sollte eine praxisbezogene Theorie verstärkt Beiträge zur Lösung von Praxisproblemen leisten (Selle 2006).

1.2.1 Dimensionen der Planungstheorie

Planungstheoretische Ansätze befassen sich mit drei Grundfragen, die eng mit den unterschiedlichen Dimensionen des Politikbegriffs in den Politikwissenschaften – Polity, Policy und Politics – sowie des Strategiebegriffs in den Organisationswissenschaften – Kontext, Inhalt und Prozess – korrelieren (◻ Tab. 1.1): Warum wird geplant? Was wird geplant? Wie wird geplant?

Der Kontext adressiert die institutionelle Dimension von Planung, die von Strukturen, Organisationen, Regeln und Normen bestimmt wird und einerseits Handlungen ermöglicht, aber andererseits den Handlungsspielraum der Akteure auch begrenzt. Die inhaltliche Dimension steht für die normative Substanz von Planung, bei der der materielle Gehalt von Plänen zum Gegenstand der Analyse wird. Es geht um problembezogene Themenbearbeitung und Aufgabenerfüllung, um planerische Leitbilder und Ziele. Die Prozess-Dimension meint schließlich den prozeduralen Verlauf von Planung und stellt auf formelle und informelle Willensbildungs-, Entscheidungs- und Implementations-

---

1 Die nachfolgenden Ausführungen basieren auf Wiechmann (2018).

◻ Tab. 1.1 Planungstheoretische Grundfragen. (Wiechmann 2018)

Frage	*Warum* wird geplant?	*Was* wird geplant?	*Wie* wird geplant?
	Legitimität von Planung	Substanz von Planung	Rationalität von Planung
Dimension	Kontext	Inhalt	Prozess
	Polity	Policy	Politics
Fokus	Planung als öffentliche Aufgabe	Planerische Leitbilder und Inhalte	Planung als Handlungssystem
Themen	Strukturen, Organisationen, Normen, Institutionen	Probleme, Aufgaben, Ziele, Werte, Issues	Konflikt, Konsens, Macht, Instrumente, Akteure

prozesse sowie die Durchsetzung von Interessen durch Macht, Konflikt und Konsens ab. Ähnlich wie beim Policy-Zyklus der Politikwissenschaft werden in der Planungswissenschaft Prozesse häufig mithilfe von Phasenmodellen beschrieben.

So wie Inhalt und Prozess untrennbar miteinander verbunden sind, so sind alle Planungsprozesse in spezifische Kontexte eingebettet und können letztlich nur kontextbezogen interpretiert werden. Variationen in Kontext oder Prozess führen ebenso wie auch Verschiebungen im Zeitablauf zu veränderten Ergebnissen. Die drei Dimensionen dürfen dabei nicht als eigenständige Komponenten missverstanden werden. Es ist heute anerkannt, dass unterschiedliche Ansätze der Stadt- und Raumplanung nur mit Blick auf alle drei Dimensionen – Kontext, Inhalt und Prozess – und ihre Wechselwirkungen umfassend analysiert und erklärt werden können.

Aus analytischen Gründen, zur Fokussierung der Argumentation und zur Reduktion von Komplexität ist bei der theoretischen Auseinandersetzung mit Planung eine Konzentration auf eine der Dimensionen aber sinnvoll, solange die anderen berücksichtigt werden. Aus planungstheoretischer Sicht kommt der Prozessdimension herausragende Bedeutung zu, da Planung die gedankliche Vorwegnahme eines Ablaufs von Handlungsschritten beinhaltet. Spätestens seit den einflussreichen Arbeiten Faludis (1969) mit der Unterteilung in prozessuale „Theories of Planning" und substanzielle „Theories in Planning" ist die Suche nach „allgemeinen Planungstheorien" (Selle 2005), die den Vorgang des Planens unabhängig von konkreten gesellschaftlichen Aufgabenfeldern thematisieren, verbreitet.

1.2.2 Historische Wurzeln der Planungstheorie

Die Entwicklung der Planungstheorie lässt sich nicht losgelöst von der Entwicklung der Planungspraxis verstehen, da sich die Theorie immer auch über die Auseinandersetzung mit der Praxis und den Erkenntniswert für die Praxis definiert hat: „The first question of theory is one of identity, which in turn leads to history" (Fainstein und Campbell 2012, S. 6).

Die Geschichte der modernen Stadt- und Raumplanung als öffentlicher Aufgabe beginnt in der zweiten Hälfte des 19. Jahrhunderts mit dem Bemühen weitsichtiger Stadtplaner, die Folgen der Industrialisierung, die wohnungshygienischen und sozialen Missstände der Gründerzeit zu überwinden. Räumliche Entwicklung galt in dieser Phase der „Anpassungsplanung" (Albers 1992) aber weder als prognostizier- noch steuerbar. Planung beschränkte sich auf „Regulierungsbemühungen" (Düwel und Gutschow 2001, S. 37), auf Gefahrenabwehr, die Behebung konkreter Missstände und stadthygienische Maßnahmen. Der planerische Anspruch blieb bescheiden: „The 19th century revolution town is an example of piecemeal (and bad) planning" (Keeble 1969, S. 1).

In der folgenden Phase der „Auffangplanung" in der ersten Hälfte des 20. Jahrhunderts ging es dagegen schon um ein vorausschauendes Steuerungsverständnis: „Planning emerged as the 20th century response to the 19th century industrial city" (Fainstein und Campbell 2012, S. 6; vgl. Hall 2002). Eine rationale und wissenschaftlich begründete staatliche Planung wurde als Möglichkeit betrachtet, die beste Alternative zur Erreichung eines vorgegebenen Zieles auszuwählen. In den divergierenden politischen Systemen dieser Epoche wurde die Planung zu einem technischen Hilfsmittel deklariert, ohne dass diese Auffassung explizit als Planungstheorie formuliert worden wäre. Einen Meilenstein stellte das Planungskonzept von Patrick Geddes (1915) dar, dessen Diktum „survey before plan" maßgeblichen Einfluss auf die Planung des 20. Jahrhunderts hatte. Geddes befürwortete auf Basis eines systematischen und ganzheitlichen Verständnisses von Stadtregionen die gezielte Beeinflussung sozialer Prozesse durch die Gestaltung der räumlichen Umwelt. Er war zugleich der Erste, der soziologische Ansätze in die Stadtplanung einführte. Der Mythos der rationalistischen Planung, das Gott-Vater-Modell der Planung, blieb jedoch lange der eigentliche Kern des Selbstverständnisses von Planern. Siebel (2006) hat in diesem Zusammenhang auf die Kontinuität der autoritären Planung hingewiesen: Planung wurde systemübergreifend als Akt der Herrschaft zur Reduktion von Komplexität verstanden. Die Ordnung der Gesellschaft sollte durch die Ordnung des Raumes hergestellt werden. Kritik an solch totalitären Ordnungsversuchen, wie sie z. B. Karl Popper (1945) oder Friedrich von Hayek (1945) schon früh formuliert hatten, wurde in der Stadt- und Raumplanung erst sehr spät rezipiert.

Die Geburtsstunde der (expliziten) Planungstheorie lässt sich Mitte des 20. Jahrhunderts in der unmittelbaren Nachkriegszeit verorten. Im Herbst 1947 wurde an der Universität Chicago die erste von der Architektur losgelöste, sozialwissenschaftliche Planungsfakultät eingerichtet. John Friedmann nahm dort ein Jahr später als Masterstudent an einem Seminar des jungen Politologen Edward Banfield teil, das er rückblickend als „the first ever seminar in planning theory" (Friedmann 1998, S. 245) bezeichnete. Das hier entwickelte Planungsmodell sah vor, dass rational handelnde Planer die politisch vorgegebenen Ziele in einen effektiven Plan übersetzen, der von der Verwaltung schließlich umgesetzt wird: „Planning is designing a course of action to achieve ends" (Meyerson und Banfield 1955, S. 314). Doch bereits die 1949 bis 1952 durchgeführte berühmte Fallstudie von Meyerson und Banfield über Public Housing in Chicago zeigte eindrücklich, dass das an der Chicago School erstmals umfassend beschriebene rationalistische Planungsmodell naiv und grob vereinfachend war, die Planungspraxis hingegen durchweg politisch: „Our standard of good planning – rational decision-making – is an ideal one; the standard is, we think, useful for analysis, but real organizations (like real people), if the truth is told, do not make decisions in a substantially rational manner" (Meyerson und Banfield 1955, S. 15).

Auch wenn die Chicago School bereits 1955 aus Kostengründen wieder geschlossen wurde, war ihre Ausrichtung bahnbrechend für die Etablierung einer expliziten Planungstheorie. Sie erlaubte es, die Planungspraxis kritisch zu reflektieren, und war zugleich anschlussfähig an entscheidungstheoretische Konzepte der Sozialwissenschaften von Mannheim über Simon, von Hayek und Lindblom bis Dewey. Für Faludi (1987, S. 27) kann die Wirkung der Chicago School auf die Planungstheorie daher gar nicht hoch genug eingeschätzt werden. Sie sei der „mainspring of modern planning thought".

In Europa kam es erst in Zeiten der Planungseuphorie ab Mitte der 1960er Jahre zur Herausbildung eigenständiger Planungsfakultäten. Parallel dazu entwickelte sich auch hier ein planungstheoretischer Diskurs (vgl. Luhmann 1966; Albers 1969; Faludi 1969). Bis dahin wurde Planungstheorie oftmals lediglich als ein an praktischen Problemen ausgerichteter, mit methodischen und verfahrensbezogenen Fragen befasster Bestandteil einer ingenieurwissenschaftlichen Planung verstanden: „The planning tradition itself has generally been ‚trapped' inside a modernist instrumental rationalism for many years" (Healey 1997, S. 7). Erst in der Aufbruchsstimmung der späten 1960er Jahre setzten kontroverse Debatten zum Verhältnis von Planung und Politik, zum Werteverständnis der Planung und zur Legitimation planerischer Aussagen ein (Fürst 2004, S. 240).

Als besonders einflussreich für die europäische Debatte erwiesen sich die frühen Arbeiten von Faludi (1969). Seine maßgeblich von Popper inspirierte prozessuale Planungstheorie war am Ideal rationaler Planung orientiert. Er begründete die Notwendigkeit einer Theorie der Planung sowohl mit der erforderlichen Untermauerung des Berufs-

stands als auch mit der nach Einführung von Planungsstudiengängen erforderlichen Abgrenzung gegenüber anderen Disziplinen. Als die Planungseuphorie Mitte der 1970er Jahre rasch abebbte, ging zeitgleich auch diese erste Phase planungstheoretischer Diskussionen abrupt zu Ende.

1.2.3 Entwicklungslinien der Planungstheorie

In dem halben Jahrhundert seit ihren Anfängen hat sich eine schwer überschaubare Vielfalt planungstheoretischer Ansätze entwickelt. Die einzelnen Entwicklungslinien der Planungstheorie haben wenige Überschneidungen und folgen widersprüchlichen Rationalitäten. Dabei ist es gerade der Anspruch auf besondere Rationalität, der Planung von anderen Formen sozialen Handelns unterscheidet. Für Siebel (2006) muss Planung die Widersprüche zwischen den Rationalitäten aushalten und sich in diesem Spannungsfeld bewegen. Die „eigentliche Rationalität der Planung liegt in ihrer Fähigkeit, zwischen widersprüchlichen Aufgaben zu lavieren, sich gleichsam in der Schwebe zu halten im Spannungsfeld verschiedener Rationalitäten" (Siebel 2006, S. 209).

Alle Versuche, das Feld der Planungstheorie zu kodifizieren und einzugrenzen, scheitern bereits daran, das genuine Theoriefeld der Planung zu bestimmen (Fürst 2004, S. 239): „No two of us could agree on the nature of the beast we wanted to theorize [...] We were riding off on different horses, each galloping into the sunset in a different direction" (Friedmann 1998, S. 246). Gleichwohl lassen sich unter Inkaufnahme einer weitgehenden Simplifizierung grobe Entwicklungslinien und im Zeitverlauf variierende Strömungen im planungstheoretischen Denken nachzeichnen.

1.2.3.1 1950er bis 1970er Jahre

Bis in die 1960er Jahre war in Übereinstimmung mit dem rationalistischen Planungsmodell die Auffassung verbreitet, moderne Planung sei ein leistungsfähiges Instrument zur Entscheidungsvorbereitung, wodurch auf möglichst rationale Weise komplexe gesellschaftliche Vorgänge gesteuert werden können. Aus theoretischer Perspektive hat die rationalistische Planungstheorie auch heute noch ihren Wert, da sie den Blick auf die Abweichungen vom postulierten Rationalitätsideal des informierten Nutzenoptimierers lenkt und damit die Analyse dieser Differenzen ermöglicht. Dieser Rationalitätsanspruch darf aber nicht als Verhaltensdeskription missverstanden werden. Er stellt ein Ideal dar, das in der Realität weder von Individuen noch von Organisationen erfüllt wird. Faludi (1986) folgend handelt es sich vielmehr um ein „methodologisches Prinzip", das einen Maßstab für die Bewertung von Entscheidungen bietet.

Rationalistische Planungstheorie war gleichwohl seit ihren Anfängen umstritten als „wirklichkeitsfremdes Konzept" (Selle 2005, S. 65), das weder theoretisch noch

1

praktisch einzulösen sei. Bereits in den 1950er Jahren formulierte Lindblom (1959) ein alternatives Planungsmodell, bei dem er sich auf Poppers Kritik an holistischer Systemplanung bezog. Statt des aussichtslosen Unterfangens, große Verbesserungen mit großen Plänen und zentraler Steuerung zu erreichen, strebt der fragmentierte Inkrementalismus schrittweise, aber stete Verbesserungen in einem dezentral organisierten sozialen Prozess an. Der Planer konzentriert sich auf eine begrenzte Zahl an Handlungsalternativen und vermeintlich „wichtigen" Konsequenzen und nimmt in Kauf, dass er auch wesentliche Konsequenzen ausspart. Im Vergleich zum rationalistischen Planungsmodell ist damit eine drastische Reduzierung der Anforderungen an den Planer verbunden. Konsequenterweise muss aber auch die Erwartung an die Planungsergebnisse reduziert und der Anspruch aufgegeben werden, ein Problem „endgültig" zu lösen. An die Stelle eines einmaligen kräftigen Zubeißens tritt beständiges Nagen (Lindblom 1968, S. 25).

Der fragmentierte Inkrementalismus wurde in der planungswissenschaftlichen Literatur häufig pauschal kritisiert. Dabei verkürzen viele Kritiker den Ansatz auf ein richtungsloses „Durchwursteln", dem jede strategische Komponente fehle. Diese Gleichsetzung, die durch den Titel des bekanntesten Artikels von Lindblom (1959), „The Science of ‚Muddling Through'", nahegelegt wurde, ist jedoch falsch. Lindblom beschreibt viele Aspekte politischer Entscheidungsfindung realistischer, als es die Ansätze des rationalistischen Planungsmodells vermögen. Trotzdem finden sich bei ihm kaum Hinweise, wie der Verzicht auf langfristige Zielformulierungen und die Verbesserung durch kleine Schritte zu kollektiv wünschenswerten Ergebnissen führen können.

Der Streit zwischen Rationalisten und Inkrementalisten in der Frühphase der Planungstheorie führte zu wiederholten Versuchen, Mittelwege zwischen dem geschlossenen Modell einer synoptischen Entwicklungsplanung und dem offenen Modell der Stückwerkstechnik zu beschreiten (Wiechmann 2008, S. 38 ff.). Zu den prominentesten Beispielen zählen das Konzept des „Mixed Scanning" von Etzioni (1967) und der „Strategic Choice Approach" (Friend und Jessop 1969; Friend und Hickling 1987).

1.2.3.2 1980er bis 2000er Jahre

Ab den 1970er Jahren vollzog sich ein fundamentaler planungstheoretischer Wandel. Insbesondere die planungstheoretischen Arbeiten der 1980er und 1990er Jahre betonten den reflexiven und kommunikativen Charakter von Planung. Vertreter der Planungstheorie wie Schön (1983), Forester (1989), Innes (1995) und Healey (1997) sahen den Fokus von Planung nicht mehr auf der technischen Rationalität, sondern auf der Funktion von Planung als kommunikativer Handlung und Lerninstrument. Nicht mehr Kontrolle stand im Mittelpunkt, sondern das Erzeugen von Handlungen und Innovationen (Friedmann 2003, S. 8).

Der „communicative turn in planning theory" (Healey 1992) basiert wesentlich auf den Vorstellungen „kommunikativer Rationalität" von Habermas. Gefragt wird nach normativen Prinzipien, wie strategische Konsensbildung in fragmentierten Gesellschaften gelingen kann. Planung soll dabei durch die Macht des besseren Arguments in hierarchiefreien Verhandlungssituationen demokratischer werden, der Planer selbst zum „Ermöglicher" von Kommunikationsprozessen. Kritisiert wurden die Ansätze **kommunikativer Planung** sowohl hinsichtlich ihrer mangelnden Legitimationsbasis und ihrer begrenzten Konfliktregelungskapazität als auch hinsichtlich ihrer dominant präskriptiven Natur. Die Realität der Planung sei dagegen weit entfernt von den normativen Idealen herrschaftsfreier Kommunikation (vgl. Selle 2004; Allmendinger 2009).

Spätere Ansätze kommunikativer Planung basieren daher eher auf den Arbeiten des französischen Philosophen Foucault, der davon ausgeht, dass Macht allen Diskursen immanent ist, durch sie manifestiert und reproduziert wird. Deswegen sind das darin entstehende Wissen und die „Lösungen" durch die Machtverhältnisse determiniert. Flyvbjerg (1998) hat in einer viel beachteten Fallstudie zum „Aalborg-Projekt" auch empirisch aufgezeigt, dass reale Planungsprozesse stärker von der „Rationality of Power" als von der „Power of Rationality" bestimmt werden.

Die US-amerikanische Planungsdebatte hat sich im Zuge des „argumentative turn" (Fischer und Forester 1993) kritisch mit dem planungspraktischen Spannungsfeld von Ideal und Wirklichkeit auseinandergesetzt. In der Tradition der pragmatischen Planungstheorie wird der durch Handlung geschaffenen Realität ein faktischer Geltungsanspruch zugesprochen. Sie baut auf der in Nordamerika verbreiteten philosophischen Strömung des Pragmatismus auf und betont die Gleichberechtigung von Wissen und Praxis (Dewey 1925). Nach dem Scheitern der meisten planungstheoretischen Ansätze wurde pragmatische Planung als „antitheoretischer" Getting-Things-Done-Ansatz verstanden, in dem Theorie und Praxis keine getrennten Sphären sind, sondern sich gemeinsam entwickeln (Healey 2008). Bei Forester (1989) rückt die Thematisierung von Macht in Planungsprozessen in den Mittelpunkt des Interesses. Im „Critical Pragmatism" (Forester 1993) setzt er sich kritisch mit der politischen Rolle des Planers und den realweltlichen Hindernissen von Planung auseinander. Die Kernidee des Ansatzes ist es, Planung als die Restrukturierung der Kommunikation zwischen Stakeholdern mit divergierenden und widerstreitenden Interessen und großen Ungleichheiten in Bezug auf Macht und Einfluss zu verstehen. Der Planer wird hier nicht als durch sein Fachwissen überlegener Entscheider oder als neutraler Moderator gesehen, sondern als ein pragmatischer Spezialist, der inklusive und partizipative Formen kollektiven Handelns unterstützt.

Wichtige Impulse erhielt die Planungstheorie in den 1990er Jahren auch durch unterschiedliche Theorien des

Neoinstitutionalismus und der **Governance-Forschung.** Der Neoinstitutionalismus stellt aus sozialwissenschaftlicher Perspektive eine Gegenbewegung zu herkömmlichen behavioristischen Theorieansätzen und zum Rational-Choice-Ansatz dar. Er betrachtet neben formellen Institutionen zur besseren Abbildung der Realität auch informelle Regeln und zusätzliche Ordnungsprinzipien. Besondere Verbreitung in der Planungswissenschaft fanden das „Institutional Analysis and Development Framework" von Ostrom (1990) sowie der Ansatz des „akteurzentrierten Institutionalismus" nach Mayntz und Scharpf (1995).

Die Governance-Perspektive lenkt den Blick auf die Bedeutung kollektiven Handelns. Mit dem Governance-Begriff verbindet sich jedoch keine bestimmte Theorie, vielmehr versprechen eine Reihe von theoretischen Bezügen – Systemtheorie, Spieltheorie, ökonomische und soziologische Institutionentheorien, Urban-Regime-Theorie und Netzwerktheorien – fruchtbare Verknüpfungsoptionen. Anders als die klassischen Ansätze der politikwissenschaftlichen Steuerungstheorie, die einem stärker akteursorientierten Ansatz folgen, nehmen die Ansätze der Governance-Theorie eine stärker institutionalistische Perspektive ein (Mayntz 2004). Sie fragen nach intermediären Regelungsstrukturen, also dem institutionellen Rahmen, der das Handeln der Akteure in Staat, Wirtschaft und Zivilgesellschaft lenkt. Die Akteure konstituieren die Regelungsstrukturen und werden zugleich von ihnen gelenkt. Benz und Fürst (2003, S. 12) verwenden den Rahmenbegriff „Regional Governance" zur „Bezeichnung einer komplexen Steuerungsstruktur in Regionen". Im Zentrum steht die Koordination kollektiven Handelns auf der regionalen Ebene.

1.2.3.3 Neuere Ansätze

Seit den 1990er Jahren ist in der Planungspraxis auch eine Rückbesinnung auf die Notwendigkeit eines planvollen, integrativen Vorgehens beobachtbar. Die unübersehbaren Nachteile projektorientierter Planung führten sowohl in der angelsächsisch geprägten internationalen als auch in der deutschsprachigen Planungsforschung zu einer Debatte über eine Renaissance **strategischer Planung** (Healey et al. 1997; Salet und Faludi 2000; Wiechmann 2008). Der „Turn to Strategy" (Healey 2007, S. 183) ist als eine Antwort auf die Defizite inkrementeller Planung durch Projekte zu verstehen. Theoretisch-konzeptionell orientierte wie auch empirische Arbeiten zur strategischen Planung befassen sich damit, inwieweit leistungsfähige Strategien zu einer effektiveren Planungspraxis führen. In Abhängigkeit vom Planungskontext, den theoretischen Zugängen und den Erkenntnisabsichten sind in den Arbeiten unterschiedliche Verständnisse von strategischer Planung festzustellen. Mit Bezug auf Ansätze der Management-Theorie werden auch emergente Strategien in den Strategiebegriff einbezogen (Wiechmann 2008). Es geht nicht mehr nur darum, die zur Umsetzung eines Zieles notwendigen Mittel einzusetzen. Vielmehr entstehen Strategien auch „planlos" aus alltäglichen Handlungsroutinen und durch spontane

Entscheidungen. Neben die formale Planung treten andere Möglichkeiten, eine Strategie zu entwickeln.

Die europäische Debatte über „Strategic Spatial Planning" (Albrechts und Balducci 2013) unterscheidet sich von der amerikanischen Debatte über „Strategic Planning" (Bryson 2004) insbesondere dadurch, dass in Europa strategische Planung als integrative und entwicklungsorientierte Form der Planung diskutiert wird, in den USA hingegen in enger Anlehnung an das „Corporate Planning" als planungsbasierte Form der Strategieentwicklung zur Herbeiführung fundamentaler Entscheidungen.

Mit der **Planungskulturforschung** hat sich in den letzten Jahren ein weiterer Strang planungstheoretischer Ansätze etabliert, der die kulturelle Einbettung und Gebundenheit von Planungspraktiken in den Blick nimmt (Othengrafen und Reimer 2013). Planungskultur meint hier das raumzeitlich gebundene, ortsspezifische Planungsverständnis und die dazugehörigen formellen und informellen Planungsroutinen. Es geht um die Art und Weise, wie die jeweiligen Akteure ihre Rollen und Aufgaben verstehen, wie sie Probleme wahrnehmen, damit umgehen und dabei bestimmte Regeln, Verfahren und Instrumente anwenden. Diese Ansätze bauen auf dem „Cultural Turn" in den Sozialwissenschaften auf und versuchen, die große Varianz an Planungspraktiken weltweit zu erklären. Kritik an Planungskulturforschung macht sich vor allem an dem vagen Kulturbegriff fest. Fürst (2007) spricht von einem „slippery concept", das für empirische Arbeiten ungeeignet sei, weil es zu viele Variablen und wechselseitige Abhängigkeiten berücksichtige, um kausale Zusammenhänge valide ermitteln zu können.

Mit dem Fokus auf kulturelle Phänomene stehen Teile der Planungskulturforschung in der Tradition des Strukturalismus. Andere folgen stärker praxeologischen Ansätzen. Dem stehen jüngere planungstheoretische Ansätze gegenüber, die sich in der Tradition des französischen Post-Strukturalismus sehen. So bezieht sich Gunder (2011) auf Foucault, Derrida und Lacan, während sich Hillier (2008) dezidiert auf Deleuze und Guattari beruft. Poststrukturalistische Planungstheorie knüpft eher an planungskritische Theorien wie den Inkrementalismus und den Pragmatismus an: „So while post-structuralist approaches are part of the contemporary face of planning theory, they actually echo more traditional concerns with ‚non-planning'" (Allmendinger 2009, S. 189).

Poststrukturalistische Planungstheorie geht wie der Pragmatismus davon aus, dass Kommunikation von Macht durchzogen ist. Sprache konstituiert Identifikation und Auffassungen über die Gesellschaft im Allgemeinen und Planung im Speziellen: „We act as planners in and through language" (Gunder 2011, S. 201). Durch Sprache vermittelte Planung versuche, die Realität zu ordnen: „Ideas in planning, such as the role of green belts, can and do have a powerful permanence outside of formal planning policy or plans" (Allmendinger 2009, S. 189). Im Poststrukturalismus werden „Master Signifikanten", wie z. B. Green Belt, als in einem Wort vereinfachte Ordnungen von Wissen verstanden. Die Sprache gilt aber als unvollständig. Symbolische Ankerpunkte für Gruppenidentitäten mit bestenfalls vagem Bedeutungskern werden nach Laclau und Mouffe (2001) „leere Signifikanten" genannt. Für Gunder und Hillier (2009) handelt es sich bei Planung selbst um einen solchen leeren Signifikanten. Dies gelte aber auch für planerische Schlüsselbegriffe wie „Nachhaltigkeit", „Rationalität" und „Verantwortung". Das Unbewusste und die Unmöglichkeit eindeutiger Sinnzuschreibungen, das Verschwimmen von Kategorien wie menschlich vs. nichtmenschlich sind wichtige Bestandteile poststrukturalistischen Denkens. Handlungsfähigkeit wird als relationale Auswirkung von in Netzwerken Handelnden, Macht selbst als relationaler Prozess verstanden. Ziel dieser planungstheoretischen Ansätze ist es letztlich, tieferliegende Gründe und Kräfte für das Entstehen von Planungspraktiken zu verstehen (Balducci et al. 2011, S. 487).

1.2.4 Wandel im planungstheoretischen Denken

Die hier vorgenommene Darstellung von einzelnen Strömungen im planungstheoretischen Denken stellt zwangsläufig eine grobe Simplifizierung dar und kann daher auch keine vollständige Auflistung sein. So fehlen Ausführungen zur marxistischen Planung, zu „Advocacy Planning", zu systemtheoretischen Ansätzen oder auch zum „Evidence Based Planning". Letztlich vermögen aber auch umfassende Darstellungen, wie sie Friedmann (1987) aus amerikanischer Sicht oder Allmendinger (2009) aus britischer Sicht vorgelegt haben, keinen kompletten Überblick zu geben. Alle Versuche zur Systematisierung von Planungstheorien heben zwangsläufig spezifische Aspekte als strukturbestimmende Momente hervor. Ziel der Ausführungen hier war es, die aus Sicht der deutschsprachigen Planungswissenschaft einflussreichen Entwicklungslinien im planungstheoretischen Denken in der gebotenen Kürze nachzuzeichnen.

Einen knappen Überblick über das planungstheoretische Denken zu geben, gleicht nichtsdestotrotz dem Versuch, einem Zuhörer ein diffus auslaufendes Mosaik aus unzähligen Einzelteilen, dem augenscheinlich jedes Muster zu fehlen scheint, in wenigen Worten am Telefon präzise zu beschreiben. Es fehlt an etablierten planungstheoretischen Denkschulen und von der Scientific Community gemeinsam geteilten Grundlagen. Ein Blick in die vorliegenden englischsprachigen Reader zur Planungstheorie belegt dies. Von den Texten in Faludis *Reader in Planning Theory* aus dem Jahr 1973 fanden sich nur zwei (Davidoff 1965; Lindblom 1959) in dem von Campbell und Fainstein 1996 erstmals herausgegebenen Standardwerk *Readings in Planning Theory*. Die drei jüngsten Auflagen die-

ses Readers (Fainstein und Campbell 2003, 2012; Fainstein und DeFilippis 2016) enthalten insgesamt 62 Originaltexte, von denen sich lediglich sechs durchgängig in allen drei Auflagen wiederfinden[2]. Immerhin sind John Friedmann, Patsy Healey, Frank Fischer und Susan Fainstein jeweils mit drei unterschiedlichen Texten vertreten, Robert Beauregard, Heather Campbell, John Forester, Leonie Sandercock, Bent Flyvbjerg, Iris Young und June Thomas mit zwei verschiedenen Texten. Von einem etablierten Kanon planungstheoretischer Literatur kann auch ein halbes Jahrhundert nach Gründung eigenständiger Planungsfakultäten keine Rede sein.

Gleichwohl ist ein genereller Wandel im planungstheoretischen Denken unverkennbar. In Anlehnung an Friedmann (2011), Fainstein und Campbell (2012) sowie Fürst (2005) lassen sich unabhängig von einzelnen Theorieansätzen vier „big shifts in planning theory" hervorheben:

- vom administrativ-technischen Plänemachen zur gesamtgesellschaftlichen Aufgabe,
- von der verwissenschaftlichen Suche nach optimalen Lösungen zu kollektiven Lernprozessen,
- vom interventionistischen Steuerungsanspruch zu kommunikativem Handeln und
- vom planenden Erfüllungsgehilfen zu politisch agierenden Planungsakteuren.

Dieser Wandel darf nicht darüber hinwegtäuschen, dass das Feld der Planungstheorie auch heute noch heterogen und fragmentiert ist: „Planning theory, like planning practice, is an eclectic or, put it more elegantly, an interdisciplinary, even transdisciplinary field" (Friedmann 2011, S. 222). Trotz der großen Anwendungsnähe eine eigenständige Paradigmatisierung der Planungswissenschaften als universitäre Disziplin voranzutreiben und der Vielfalt an theoretischen Zugängen einen gemeinsamen Grundstock an Denkansätzen und planungstheoretischen Schulen gegenüberzustellen, bleibt Aufgabe künftiger Generationen von Planungswissenschaftlerinnen und -wissenschaftlern.

1.3 Zur Zusammenstellung des Readers

Das Zusammenstellen eines Readers zur Planungstheorie ist alles andere als trivial. Das Feld der Planungstheorie ist nicht klar umrissen. Was gehört dazu, was nicht? Sollte sich der Reader auf originäre Planungstheorien – prozessuale „Theories of Planning" im Sinne Faludis (1969) – beschränken oder auch substanzielle „Theories in Planning" aus Nachbardisziplinen wie der Ökonomie und der Geographie umfassen? Sind Letztere als disziplinär zuordenbare Ansätze nicht nur in Bezug auf die jeweilige „Mutterdisziplin" interpretierbar? Sind mit den zugrunde liegenden fachwissenschaftlichen Diskursen nicht

---

2   Die sechs Texte sind: Jacobs (1961), Davidoff (1965), Fishman (1982), Foglesong (1986), Campbell (1996) und Scott (1998).

axiomatische Vorentscheidungen verbunden, die nur paradigmeninternen Kontrollmechanismen unterworfen werden können? Wie können divergierende, ja miteinander unvereinbare theoretische Grundlagen aus verwandten Disziplinen in einem kohärenten Überblick dargestellt werden? Erfordert nicht alleine die Bestimmung des Planungsbegriffs a priori eine Positionierung in einem Theoriegebäude, und führt dies nicht zwangsläufig zur Inkommensurabilität unterschiedlicher Begriffe und Positionen? Welche Debattenbeiträge werden aufgenommen, welche weggelassen? Wie viele Texte sind für den angestrebten Überblick erforderlich? Was kann den Leserinnen und Lesern, was dem Verlag zugemutet werden? Zudem stellt sich auch immer die Frage, ob vorwiegend Klassiker der Planungsliteratur aufgenommen werden oder ob aktuellen Diskursen der Vorzug gegeben wird.

Der vorliegende Reader verfolgt ein Konzept, das sich zum einen auf den Kern der Planungstheorie und damit die prozessualen „Theories of Planning" fokussiert. Zum anderen unterliegt er nicht der Illusion einer umfassenden Theorie der räumlichen Planung. Vielmehr werden bedeutsame und – theoretisch wie praktisch – einflussreiche Planungstheorien in jeweils eigenständigen Kapiteln nachgezeichnet und interpretiert. Dabei werden sowohl ältere, einflussreiche Originaltexte als auch neuere Debattenbeiträge aufgenommen und kommentiert.

Der ARL-Arbeitskreis „Planungstheorien – Stand und Perspektiven" hat vor dem Hintergrund der dargestellten Entwicklungslinien der Planungstheorie vier große Diskurse herausgegriffen. In Band 1 werden die beiden wohl einflussreichsten planungstheoretischen Metaerzählungen des späten 20. Jahrhunderts behandelt: Mit verschiedenen Ansätzen der kommunikativen Planung sowie des Neoinstitutionalismus und der Governance-Forschung verbindet sich die erfolgreiche Überwindung einer technokratisch-rationalistischen Vorstellung von Planung. Kommunikation, Macht und Konflikt werden in der Planungswissenschaft zu zentralen Kategorien der Auseinandersetzung mit der Planungsrealität.

Band 2 beleuchtet neuere Diskursstränge des frühen 21. Jahrhunderts: strategische Planung und Planungskultur. Während der organisationstheoretisch untersetzte „Turn to Strategy" als eine Antwort auf die Defizite inkrementeller Planung durch Projekte zu verstehen ist, zielen Ansätze die Planen als kulturelle Praxis begreifen, auf ein tieferes Verständnis lokaler und regionaler Praktiken.

Mit der Fokussierung auf diese vier großen Diskurse wird ein wichtiger, keinesfalls jedoch vollständiger Überblick über planungstheoretische Debatten gegeben. Der *ARL Reader Planungstheorie* ist daher grundsätzlich offen angelegt und könnte künftig um weitere Bände erweitert werden. Diese könnten sowohl hochaktuelle Dynamiken im planungstheoretischen Diskurs, wie z. B. den Poststrukturalismus, aufgreifen als auch die Grundlagen der modernen Planung im Widerstreit zwischen Rationalisten und Inkrementalisten in der Frühphase der Planungstheorie zum Gegenstand haben.

1.3.1 Kriterien für die Zusammenstellung des Readers

Durch die Auswahl der Texte des Readers wird ein Beitrag zu einer Kanonisierung von ausgewählten Feldern der Planungstheorie geleistet. Der Reader soll den Mainstream ausgewählter Teilbereiche der Planungstheorie widerspiegeln, nicht aber einer wissenschaftlichen Schulenbildung Vorschub leisten. Es geht vielmehr um eine kritische Aufbereitung und Reflexion vorhandener Debattenansätze. Die Zusammenstellung des Readers soll es den Leserinnen und Lesern ermöglichen, planungswissenschaftliche Ansätze und Untersuchungsgegenstände in vielfältigen Zusammenhängen zu verorten und dadurch auch neue Perspektiven zu gewinnen.

Die nachfolgend dargelegten Kriterien zu Impact, Originalität, Qualität, Form und Struktur der Texte dienten als Richtschnur zu Auswahl der Texte in den vier Themenfeldern. Allerdings konnte es keine Mechanik bei der Textauswahl geben. Die Abwägung, welche Texte aufzunehmen waren und welche Texte in den rahmensetzenden Beiträgen erwähnt, zitiert oder auch in längeren Exzerpten rezipiert werden, musste fallweise entschieden werden. In den kommentierenden Begleittexten wird auch auf weitere relevante Diskussionsstränge eingegangen, die sich in den ausgewählten Artikeln nicht unmittelbar widerspiegeln.

Impact In den Reader sollten vorrangig wirkmächtige Texte aus dem behandelten planungstheoretischen Gebiet aufgenommen werden. Hat der Text Einfluss auf andere Arbeiten gehabt? Die Zitationshäufigkeit kann hier als Indiz dienen, ist aber alleine nicht entscheidend. Nicht bei jeder Debatte war zwangsläufig der bekannteste Text aufzunehmen, sondern der für den Zweck des Readers interessanteste aus dem betrachteten planungstheoretischen Gebiet. Die Auswahl erfolgte primär nach wissenschaftlicher Relevanz, wobei im Zweifelsfall zugunsten des aktuellen Forschungsstands und gegen die fachhistorische Bedeutung eines Textes entschieden wurde. Nicht das Alter der Texte, sondern ihre Aktualität war ausschlaggebend. Neben „Leuchtturmtexten" von Vätern und Müttern der Debatte sollte auch der aktuelle Debattenstand repräsentiert sein.

Originalität Aufgenommen wurden insbesondere Texte, die eine eigene, zum Zeitpunkt ihrer Veröffentlichung originelle planungstheoretische Position beziehen oder bestehende Ansätze auf eine eigene, innovative Art und Weise verknüpfen bzw. verorten. Dabei ging es nicht darum, eine Debatte möglichst vollständig abzubilden. Stattdessen sollten in diesem Reader besonders lesenswerte und lehrreiche Texte gebündelt werden. Das können sowohl spezifische Positionierungen als auch Review-Artikel sein. Hinsichtlich der räumlichen Bezüge war die Auswahl nicht auf jene Texte beschränkt, die sich auf den deutschsprachigen Raum beziehen. Auch Beiträge, die sich mit einschlägigen planungstheoretischen Fragen in

1

der westlichen Hemisphäre befassen, wurden als relevant erachtet, sofern sie bedeutend sind für die deutschsprachige Debatte.

Qualität Die Textauswahl orientierte sich an den üblichen Qualitätsstandards wissenschaftlicher Publikationen. Darüber hinaus erforderte das Format eines Readers, dass die Texte sich im besonderen Maße durch Substanz und Dichte auszeichnen. Die Texte sollten unabhängig voneinander verständlich sein, zugleich aber in der vergleichenden Betrachtung einen Mehrwert bieten, um zu einem stimmigen Gesamtbild beizutragen.

Form und Struktur Für die behandelten planungstheoretischen Diskurse galt es, jeweils circa zehn Texte auszuwählen. Diese Auswahl orientierte sich auch an formalen Kriterien. So wurde auf die Klarheit des Ausdrucks in den ausgewählten Artikeln und damit die Verständlichkeit für eine breite Leserschaft geachtet. Die Länge der ausgewählten Beiträge hatte grundsätzlich Reader-geeignet sein. Hinsichtlich der Sprache der Texte sollten deutsch- und englischsprachige Texte in einer angemessenen Balance ausgewählt werden. Ziel war es, die deutschsprachige Perspektive zu wahren, in der Gesamtschau aber anschlussfähig zu bleiben an englischsprachige Debatten. Auch sollten theoretische und empirische Anteile in den Texten in einem ausgewogenen Verhältnis stehen.

Intensiv wurde im Arbeitskreis schließlich über die Frage diskutiert, ob eigene Artikel der Arbeitskreismitglieder von der Aufnahme in den Reader auszuschließen seien. Dabei wurde das Dilemma angesprochen, dass einerseits der Arbeitskreis Mitglieder berufen hatte, die auf den relevanten Gebieten einschlägig arbeiten, andererseits Rezeption und Akzeptanz des Readers beeinträchtigt werden könnten, wenn der Eindruck entstünde, einzelne Texte seien aus persönlichem Interesse in dem Kanon platziert worden. Da die Beteiligten die Möglichkeit hatten, sich inhaltlich umfassend in den kommentierenden Rahmentexten einzubringen, wurde auf die Aufnahme von Originalartikeln der Beteiligten gänzlich verzichtet. Wo immer es zielführend erschien, finden sich freilich Verweise auch zu jenen Quellen.

Literatur

Albers, G. (1969). Über das Wesen räumlicher Planung. *Stadtbauwelt, 21*, 10–14.

Albers, G. (1992). *Stadtplanung. Eine praxisorientierte Einführung.* Darmstadt: Primus.

Albrechts, L., & Balducci, A. (2013). Practicing strategic planning: In search of critical features to explain the strategic character of plans. *disP – The Planning Review, 49*(3), 16–27.

Allmendinger, P. (2009). *Planning theory.* Palgrave: Basingstoke.

Balducci, A., Boelens, L., Hillier, J., Nyseth, T., & Wilkinson, C. (2011). Strategic spatial planning in uncertainty: Theory and exploratory practice. *Town Planning Review, 82*(5), 481–501.

Benz, A., & Fürst, D. (2003). Region – ,Regional Governance' – Regionalentwicklung. In B. Adamaschek & M. Pröhl (Hrsg.), *Regionen erfolgreich steuern. Regional Governance – Von der kommunalen zur regionalen Strategie* (S. 11–66). Gütersloh: Bertelsmann-Stiftung.

Bryson, J. M. (2004). *Strategic planning for public and nonprofit organizations. A guide to strengthening and sustaining organizational achievement.* San Francisco: Jossey-Bass.

Campbell, S. (1996). Green cities, growing cities, just cities? Urban planning and the contradictions of sustainable development. *Journal of the American Planning Association, 62*(3), 296–312.

Campbell, S., & Fainstein, S. S. (1996). *Readings in planning theory.* Cambridge: Blackwell.

Davidoff, P. (1965). Advocacy and pluralism in planning. *Journal of the American Institute of Planners, 31*, 331–338.

Dewey, J. (1925). The development of American pragmatism. In J. Dewey (Hrsg.), *Philosophy and civilization* (S. 13–25). New York: Kessinger Publishing.

Düwel, J., & Gutschow, N. (2001). *Städtebau in Deutschland im 20. Jahrhundert. Ideen – Projekte – Akteure.* Studienbücher der Geographie. Stuttgart: Borntraeger.

Etzioni, A. (1967). Mixed-Scanning: A „third" approach to decision-making. *Public Administration Review, 27*(5), 385–392.

Fainstein, S. S., & Campbell, S. (2003). *Readings in planning theory.* Oxford: Blackwell.

Fainstein, S. S., & Campbell, S. (2012). *Readings in planning theory.* Oxford: Blackwell.

Fainstein, S. S., & DeFilippis, J. (2016). *Readings in planning theory.* Oxford: Blackwell.

Faludi, A. (1969). Planungstheorie oder Theorie des Planens? *Stadtbauwelt, 23*(38–39), 216–220.

Faludi, A. (1973). *A reader in planning theory.* Oxford: Pergamon.

Faludi, A. (1986). *Critical rationalism and planning methodology.* London: Pion.

Faludi, A. (1987). *A decision-centred view of environmental planning.* Oxford: Pergamon.

Fischer, F., & Forester, J. (Hrsg.). (1993). *Argumentative turn in policy analysis and planning.* Durham: Duke University Press.

Fishman, R. (1982). *Urban Utopias in the twentieth century: Ebenezer Howard, Frank Lloyd Wright, and Le Corbusier.* London: Cambrige.

Flyvbjerg, B. (1998). *Rationality and power. Democracy in practice.* Chicago: University of Chicago Press.

Foglesong, R. E. (1986). Planning the capitalist city. In R. E. Foglesong (Hrsg.), *Planning the capitalist city: The colonial era to the 1920s* (S. 18–24). Princeton: Princeton University Press.

Forester, J. (1989). *Planning in the face of power.* Berkeley: University of California Press.

Forester, J. (1993). *Critical theory, public policy, and planning practice: Toward a critical pragmatism.* Albany: State University of New York Press.

Friedmann, J. (1987). *Planning in the public domain: From knowledge to action.* Princeton: Princeton University Press.

Friedmann, J. (1998). Planning theory revisited. *European Planning Studies, 6*(3), 245–254.

Friedmann, J. (2003). Why do planning theory? *Planning Theory, 2*(1), 7–10.

Friedmann, J. (2011). *Insurgencies: Essays in planning theory.* London: Routledge.

Friend, J., & Hickling, A. (1987). *Planning under pressure. The strategic choice approach.* Oxford: Pergamon.

Friend, J., & Jessop, N. (1969). *Local government and strategic choice – An operational research approach to the processes of public planning.* London: Routledge.

Fürst, D. (2004). Planungstheorie – Die offenen Stellen. In U. Altrock, S. Günter, S. Huning, & D. Peters (Hrsg.), *Perspektiven der Planungstheorie* (S. 238–255). Berlin: Leue.

Fürst, D. (2005). Entwicklung und Stand des Steuerungsverständnisses in der Raumplanung. *disP – The Planning Review, 163*(4), 16–27.

Fürst, D. (2007). Planungskultur. Auf dem Weg zu einem besseren Verständnis von Planungsprozessen? PNDonline III/2007. ▶ http://www.planung-neu-denken.de/images/stories/pnd/dokumente/pndonline3-2007-fuerst.pdf. Zugegriffen: 22. Apr. 2015.

Geddes, P. (1915). *Cities in evolution: An introduction to the town planning movement and to the study of civics.* London: Williams & Norgate.

Gunder, M. (2011). Fake it until you make it, and then .... *Planning Theory, 10*(3), 201–212.

Gunder, M., & Hillier, J. (2009). *Planning in ten words or less: A Lacanian entanglement with spatial planning.* Farnham: Routledge.

Gunder, M., Madanipour, A., & Watson, V. (Hrsg.). (2018). *The Routledge handbook of planning theory.* New York: Routledge.

Hall, P. (2002). *Cities of tomorrow.* Oxford: Blackwell.

Healey, P. (1992). Planning through debate: The communicative turn in planning theory. *Town Planning Review, 20*(1), 9–20.

Healey, P. (1997). *Collaborative planning: Shaping places in fragmented societies.* London: Macmillan.

Healey, P. (2007). *Urban complexity and spatial strategies. Towards a relational planning for our times.* London: Routledge.

Healey, P. (2008). The pragmatic tradition in planning thought. *Journal of Planning Education and Research, 28*(3), 277–292.

Healey, P., Khakee, A., Motte, A., & Needham, B. (Hrsg.). (1997). *Making strategic spatial plans: Innovation in Europe.* London: UCL Press.

Hillier, J. (2008). Plan(e) speaking: A multiplanar theory of spatial planning. *Planning Theory, 7*(1), 24–50.

Hillier, J., & Healey, P. (2008). *Critical essays in planning theory* (Bd. 3). Burlington: Ashgate.

Innes, J. E. (1995). Planning theory's emerging paradigm: Communicative action and interactive practice. *Journal of Planning Education and Research, 14*(3), 183–189.

Jacobs, J. (1961). *The death and life of great American cities.* New York: Macat International Limited.

Keeble, L. B. (1969). *Principles and practice of town and country planning.* London: Estates Gazette.

Laclau, E., & Mouffe, C. (2001). *Hegemony and socialist strategy: Towards a radical democratic politics.* London: Verso.

Lindblom, C. E. (1959). The science of „muddling through". *Public Administration Review, 19*(2), 79–88.

Lindblom, C. E. (1968). *The policy-making process.* Englewood Cliffs: Prentice-Hall.

Luhmann, N. (1966). Politische Planung. *Jahrbuch für Sozialwissenschaft, 3,* 271–296.

Mayntz, R. (2004). *Governance Theory als fortentwickelte Steuerungstheorie?* Köln: Max-Planck-Institut für Gesellschaftsforschung.

Mayntz, R., & Scharpf, F. W. (Hrsg.). (1995). *Gesellschaftliche Selbstregelung und politische Steuerung.* Schriften des Max-Planck-Instituts für Gesellschaftsforschung Köln. Frankfurt a. M.: Campus.

Meyerson, M., & Banfield, E. C. (1955). *Politics, planning, and the public interest: The case of public housing in Chicago.* London: Free Press.

Ostrom, E. (1990). *Governing the commons: The evolution of institutions for collective action.* Cambridge: Cambridge University Press.

Othengrafen, F., & Reimer, M. (2013). The embeddedness of planning in cultural contexts: Theoretical foundations for the analysis of dynamic planning cultures. *Environment and Planning A, 45*(6), 1269–1284.

Popper, K. (1945). *The open society and its enemies.* London: Routledge.

Salet, W., & Faludi, A. (Hrsg.). (2000). *The revival of strategic spatial planning.* Amsterdam: Elsevier.

Schön, D. A. (1983). *The reflective practitioner. How professionals think in action.* New York: Basic Books.

Scott, J. (1998). Authoritarian High Modernism. In J. Scott (Hrsg.), *Seeing like a state: How certain schemes to improve the human condition have failed* (S. 87–102). New Haven: Yale University Press.

Selle, K. (2004). Kommunikation in der Kritik? In B. Müller, S. Löb, & K. Zimmermann (Hrsg.), *Steuerung und Planung im Wandel* (S. 229–256). Wiesbaden: VS Verlag.

Selle, K. (2005). *Planen. Steuern. Entwickeln. Über den Beitrag öffentlicher Akteure zur Entwicklung von Stadt und Land.* Dortmund: Dortmunder Vertrieb für Bau- und Planungsliteratur.

Selle, K. (Hrsg.). (2006). *Planung Neu Denken. Band 1: Zur räumlichen Entwicklung beitragen. Konzepte, Theorien, Impulse.* Dortmund: Rohn.

Siebel, W. (2006). Wandel, Rationalität und Dilemmata der Planung. In K. Selle (Hrsg.), *Planung Neu Denken. Band 1: Zur räumlichen Entwicklung beitragen. Konzepte, Theorien, Impulse* (S. 195–209). Dortmund: Rohn.

von Hayek, F. A. (1945). The use of knowledge in society. *The American Economic Review, 35*(4), 519–530.

Wiechmann, T. (2008). *Planung und Adaption. Strategieentwicklung in Regionen, Organisationen und Netzwerken.* Dortmund: Rohn.

Wiechmann, T. (2018). Planungstheorie. In Akademie für Raumforschung und Landesplanung (Hrsg.), *Handwörterbuch der Stadt- und Raumentwicklung* (S. 1771–1784). Hannover: Akademie für Raumforschung und Landesplanung.

Strategische Planung

Gérard Hutter, Thorsten Wiechmann und Thomas Krüger

© Springer-Verlag GmbH Deutschland, ein Teil von Springer Nature 2019
T. Wiechmann (Hrsg.), *ARL Reader Planungstheorie Band 2*, https://doi.org/10.1007/978-3-662-57624-3_2

2.1 Einführung

Räumliche Planung leistet Beiträge zur Koordination und Steuerung der Nutzung des Raumes durch die Gesellschaft. Sie ist damit Teil der politischen Steuerung gesellschaftlicher Ressourcen (zur Stadtentwicklung vgl. z. B. Selle 2005; zur Regionalentwicklung vgl. Fürst 2005). In den 1960er Jahren setzte sich in Deutschland allerdings ein Verständnis räumlicher Planung als sachlicher und „technischer" Prozess zweckrationaler Entscheidungen auf der Basis möglichst weitreichender wissenschaftlicher Erkenntnisse durch. Planungstheoretische Arbeiten der 1980er und 1990er Jahre betonten demgegenüber den kommunikativen Charakter räumlicher Planung sowie deren institutionelle Einbettung. Der Fokus lag nicht mehr auf zweckrationaler Planung bzw. auf „technischer" Rationalität, sondern auf der Funktion von Planung als kommunikativer Handlung, als sozialer Lernprozess und als Teil politisch-institutioneller Prozesse. Zwecke und Ziele waren nicht mehr einfach der Planung vorgegeben, sondern selbst Gegenstand politischer Auseinandersetzungen und des sozialen Lernens in der Planung. Planungsforscher reagierten damit einerseits auf wissenschaftlich-theoretische Entwicklungen (z. B. Kritiken positivistischer Forschung, neue Sozialtheorien, z. B. Theorien sozialer Praktiken, Aktor-Netzwerk-Theorie; vgl. z. B. Healey 2017). Sie reagierten andererseits auf gesellschaftliche Konfliktlagen, Unsicherheiten und Wandelungsprozesse, wie sie nach dem „Goldenen Zeitalter" der Nachkriegszeit in westlichen Gesellschaften offensichtlich geworden waren (zum „Goldenen Zeitalter" der 1950 und 1960er Jahre und zu den danach aufkommenden Unsicherheiten, insbesondere nach der „Ölkrise" in der ersten Hälfte der 1970er Jahre vgl. Hobsbawm 1995).

Vor diesem Hintergrund verwenden Beiträge zur räumlichen Planungsforschung seit ca. Mitte der 1990er Jahre vermehrt die Begriffe der Strategie und der strategischen Planung. Dies gilt sowohl für die angelsächsisch geprägte internationale Planungsforschung (z. B. Healey et al. 1997; Salet und Faludi 2000; Albrechts 2004, 2017; Friedmann 2004; Sartorio 2005; Newman 2008; Davoudi und Strange 2009; Healey 2007, 2009; Albrechts und Balducci 2013) als auch für Arbeiten im deutschsprachigen Raum (z. B. Klotz und Frey 2005; Wiechmann 2008a; Hutter und Wiechmann 2010; Kühn und Fischer 2010; Vallée 2012; Lamker 2016). Zahlreiche planerische Strategiestudien greifen auf Beiträge der Management- und Organisationsforschung zurück, um ihr Verständnis strategischer Planung zu klären. Die Managementforschung dient dabei als Quelle der Inspiration und Zitation sowie auch zur Abgrenzung zwischen Arbeiten der räumlichen Planung einerseits und Analysen andererseits, die auf das Management von Unternehmen fokussieren. Es bietet sich daher an, zentrale Beiträge zur strategischen Planung in drei Bereiche zu untergliedern:
- Strategische Planung in der internationalen Planungsforschung
- Rezeption der strategischen Planung im deutschsprachigen Raum
- Strategien und strategische Planung in der Managementforschung

Es gibt durchaus Gemeinsamkeiten im Verständnis strategischer Planung zwischen diesen drei Bereichen und es gibt zugleich deutliche Unterschiede, die das Folgende durch die Auswahl zentraler Beiträge zur Planungs- und Managementforschung erhellen möchte.

2.2 Strategische Planung in der internationalen Planungsforschung

In der Unternehmens- bzw. Managementforschung hatte strategische Planung ihre erste Blütezeit in den 1960er Jahren und Anfang der 1970er Jahre (Mintzberg et al. 1999; Bresser 2010). Seit den 1980er Jahren befassten sich nordamerikanische Planungs-, Verwaltungs- und Politikwissenschaftler dann intensiv mit der Übertragung strategischer Planungsansätze aus dem Unternehmensbereich in den öffentlichen Sektor (▶ Kap. 3, Bd. 2). Die europäische Debatte zur strategischen Planung für Städte und Regionen setzte rund zehn Jahre später ein und stand unter anderen Vorzeichen (Salet und Faludi 2000; Wiechmann 2008a). Strategische Planung meinte in Großbritannien zunächst „Strukturplanung" und „Regionalplanung". Mit den wegweisenden Arbeiten von Autoren wie Louis Albrechts, Andreas Faludi, Patsy Healey und Hans Mastop kristallisierte sich in der europäischen Diskussion seit ungefähr Mitte der 1990er Jahre ein eigenständiges normativ orientiertes, relativ breit gefasstes und situativ orientiertes Verständnis strategischer räumlicher Planung heraus:
- Strategische Planung ist *normativ* orientiert, da sie die Kommunikation und Verständigung über soziale, ökologische und ökonomische Ziele umfasst. Autoren wie Albrechts betonen in der Rückschau die hohe Bedeutung „progressiver Politik" und von Zielen wie Gerechtigkeit und Fairness (Albrechts 2017; vgl. Healey 2017 mit allerdings etwas mehr normativer Zurückhaltung). Im Mittelpunkt ihrer Arbeiten stehen deshalb die Möglichkeiten strukturellen bzw. grundlegenden Wandels von Städten und Regionen (vgl. Healey 2013).
- Strategische Planung ist *breit* gefasst. Sie erschöpft sich keineswegs in vorrangig formal geprägten und „technischen" Planungsprozessen, wie sie die Unternehmensforschung in den 1960er Jahren vor Augen hatte. Strategische Planung ist ein vielschichtiger, von Unsicherheiten geprägter sozialer Prozess, in dem Akteure aus Politik, Verwaltung, Wirtschaft, Zivilgesellschaft usw. eine Rolle spielen und den vielfältige sozialökonomische Strukturen und Koordinationsmechanismen prägen (vgl. Healey 2007, 2017).
- Zahlreiche Autoren der europäischen Planungsdiskussion verstehen strategische Planung nicht als

„Blaupause" und allgemeingültiges „Prozessschema" für die Formulierung und Umsetzung von Strategien, sondern als *situativ und wertorientiert zu bestimmenden Prozess,* der in aller Regel iterativ abläuft (also mit zahlreichen Vor- und Rückbezügen, Wiederholungen usw.). Damit ergeben sich Herausforderungen für Versuche wissenschaftlicher Generalisierung (Yin 2014), die Planungsforscherinnen wie Patsy Healy (2017) durch vergleichende Fallstudien, konzeptionelle Exploration und Erfahrungswissen zu bewältigen suchen.

Die „Renaissance" strategischer räumlicher Planung seit den 1990er Jahren ist auch eine Reaktion auf die Defizite unkoordinierter Fachplanungen und neoliberaler Politiken in den 1980er Jahren, geprägt von Ronald Reagan, Margaret Thatcher und Helmut Kohl, sowie eine Reaktion auf die Grenzen projektorientierter Planungsansätze („Planung durch Projekte"). Strategische Planung soll sowohl den sektoralen Planungen als auch den einzelnen Projekten eine integrative Orientierung und einen Rahmen setzen. Viele Autoren verwenden den Ausdruck „Bezugsrahmen" (Frame of Reference), um die erwünschten Wirkungen strategischer Planungen im Kontext von Governance in Städten und Regionen zu benennen (vgl. Mastop und Faludi 1997; Albrechts 2004; Healey 2009). Wie die ausgewählten Aufsätze zeigen, ist strategische Planung zugleich ziel- und handlungsorientiert und unterscheidet sich daher systematisch von integrativen Entwicklungsplanungen der 1970er Jahre.

Nach über 20 Jahren intensiver Forschung und Diskussion zur strategischen räumlichen Planung hat die internationale Planungsforschung einen differenzierten Entwicklungsstand erreicht, der durch die drei nachfolgend ausgewählten Aufsätze nur selektiv erfasst werden kann. Folgende Gründe für die Wahl der drei Aufsätze sollen hervorgehoben werden:

- Es handelt sich um viel beachtete Aufsätze, die zentrale Autoren der internationalen Planungsforschung verfassten. Sie spielten nicht zuletzt bei der Gründung der Association of European Schools of Planning (AESOP) eine entscheidende Rolle.
- Insbesondere der Beitrag von Albrechts (2017), aber auch der Beitrag von Healey (2017), verdeutlicht die normative Orientierung strategischer räumlicher Planung. Es ergeben sich Anschlussmöglichkeiten für Theorien kommunikativer Planung.
- Die Aufsätze zeigen, dass strategische räumliche Planung über den Bezug zu einzelnen Organisationen, wie sie in der Managementforschung vorherrschen, hinausgeht. Strategische Planung steht in einem engen Zusammenhang mit Theorien zu Governance und Institutionen.
- Das situative Argument findet sich in allen drei Aufsätzen. Der Beitrag von Mastop und Faludi (1997) zeigt dabei, dass und wie strategische Pläne auch evaluationsbezogen untersucht werden können, um ihre Konzeption und Anwendung weiterzuentwickeln.

Das Lesen der drei ausgewählten Aufsätze und insbesondere das „Verdauen" und Interpretieren der darin enthaltenen Argumente liefern, so unsere Hoffnung, eine geeignete Ausgangbasis für das Verständnis strategischer räumlicher Planung. Um den Einstieg in dieses Verständnis zu erleichtern, geht das Folgende näher auf die ausgewählten Aufsätze ein. Dabei steht zunächst der Aufsatz von Albrechts (2004) im Vordergrund, da er in besonderer Weise den Zusammenhang zwischen der normativen, inhaltlichen und situativen Orientierung strategischer Planung verdeutlicht.

Louis Albrechts: Strategic (Spatial) Planning Reexamined (Albrechts 2004)

Der Belgier Louis Albrechts, einer der Gründer von AESOP und Herausgeber der Zeitschrift *European Planning Studies,* zählt zu den profiliertesten Planungswissenschaftlern in Europa. Er hat sich seit den 1990er Jahren intensiv mit Fragen der strategischen Planung befasst und seinen eigenen Ansatz entwickelt. In dem Artikel „Strategic (Spatial) Planning Reexamined" gibt er einen Überblick über die einschlägige Debatte seit den 1980er Jahren und beschreibt von ihm sog. Bausteine für eine neuartige strategische Planung, die den heutigen Herausforderungen in einer postmodernen Welt gerecht werde. Er grenzt dieses neue, von der Managementtheorie beeinflusste Planungsverständnis gegen die traditionelle Flächennutzungsplanung ab, die zu starr und thematisch zu eingeschränkt sei, um heutzutage adäquate Lösungen für komplexe gesellschaftliche Probleme bereitzustellen. Albrechts versteht sich selbst als Autor, der Planungstheorie und Planungspraxis aus der Perspektive des „reflective practitioner" verknüpft. Er entwickelt seinen Ansatz einerseits aus der planungs- und managementtheoretischen Literatur – hier bezieht er sich insbesondere auf Autoren wie Mintzberg, Healey, Kunzmann und Bryson, berücksichtigt aber auch ältere planungswissenschaftliche Werke aus den USA und England der 1980er Jahre – sowie andererseits auf Basis der beobachteten Praxis strategischer Planung in Europa. Seinen komplexen, „viergleisigen" Ansatz strategischer Planung beschreibt er als einen demokratischen, offenen, selektiven und dynamischen Prozess. Ziel einer solchen Planung sei das Erzeugen einer Vision, um der Bestimmung von Problemen, Herausforderungen und kurzfristigen Maßnahmen einen Rahmen zu geben.

Der Aufsatz von Albrechts verdeutlicht das weit gespannte Spektrum von Ansprüchen an strategische räumliche Planung (vgl. die komplexe Abbildung in Albrechts 2004, S. 753) und skizziert die Vielfalt an Elementen, die potenziell für strategische Planungen von Bedeutung sind. Sie umfassen inhaltliche, prozessuale und kontextbezogene Elemente. Dieses Verständnis strategischer Planung unterläuft die bekannte Unterscheidung zwischen „Theorien in der Planung (Theories in Planning)" einerseits und „Theorien der Planung (Theories of Planning)" andererseits (Faludi 1969), was Albrechts (vgl. 2004, S. 749) z. B. durch die Betonung der Bedeutung von Werten und Visionen hervorhebt. Andere Arbeiten der internationalen Planungsforschung richten ihr Hauptaugenmerk auf prozessuale Fragen strategischer räumlicher Planung, so der hier ausgewählte Beitrag von Healey (2009). Sie bezeichnet den Aufsatz als eine Art „Essay", der auf der Grundlage ihrer zahlreichen konzeptionellen und empirischen Forschungen auf „normative Abstraktion" zielt (Healey 2009, S. 441). Diese dient der Bewusstwerdung und Reflexion über vor allem auch wertbezogene und ethische Fragen strategischer räumlicher Planung. Prozessorientierung zeigt sich insbesondere in der Wahl der Sprache und Gliederung des Beitrags, bei der sprachliche Prozessformen im Vordergrund stehen, z. B. „creating frames", „selecting actions" (Healey 2009, S. 442) anstelle von „frame creation", „action selection". Dies korrespondiert mit Arbeiten der Managementforschung, die allgemein für Prozessorientierung im sozialen Handeln argumentieren (vgl. z. B. Weick 2001).

Patsy Healey: In Search of the „Strategic" in Spatial Strategy Making (Healey 2009)

Patsy Healey zählt, wie Louis Albrechts, zu den profiliertesten Planungswissenschaftlern in Europa. Der von ihr herausgegebene Sammelband *Making Strategic Spatial Plans: Innovation in Europe* ist ein Meilenstein in der planungswissenschaftlichen Befassung mit strategischen Fragen (vgl. Healey et al. 1997). Mit Monografien wie *Collaborative Planning: Shaping Places in Fragmented Societies* (Healey 1997) hat sie zudem wichtige Beiträge zu weiteren planungstheoretischen Diskussionen geleistet (vgl. z. B. die Beiträge zur kommunikativen Planung). Eine Reihe von ihren Arbeiten befasst sich vertiefend mit Fragen der Strategieentwicklung in Städten und Regionen (z. B. Healey 2005, 2007, 2009). Der Reader wählt den Aufsatz „In Search of the ‚Strategic' in Spatial Strategy Making". Den Strategiebegriff versteht Healey im Anschluss an Albrechts im Kern als einen „selektiven Fokus", durch den Akteure in Städten und Regionen Orientierung angesichts komplexer und potenziell verwirrender Themen, Ideen, Ansprüche und Argumente gewinnen. Healey betont damit, ähnlich wie Henry Mintzberg, den synthetischen Charakter von Strategien. Strategien in diesem Sinne können sehr unterschiedliche Ausprägungen annehmen (z. B. Konzepte, Bilder, Prinzipien). Sie sind nicht zwangsläufig das Resultat analytisch-systematischer Arbeit. Imagination, Kreativität und die Verfolgung von Partikularinteressen durch einzelne Gruppen können – um nur einige Beispiele zu nennen – ebenfalls zu Strategien führen. Das in dem gewählten Aufsatz erwähnte Konzept „framing selectively" als Kennzeichen von Strategien in Prozessen der räumlichen Entwicklung macht es anschlussfähig an Arbeiten, die sich mit „kognitiven Rahmen" von Akteuren und Prozessen befassen (z. B. Analysen von sozialen Bewegungen). Dieses Verständnis strategischer Planung entwickelt sich aus theoretischen Bezügen von Healey im „soziologischen Institutionalismus", in der Governance-Forschung und auch Arbeiten in der Tradition der interpretativen und selbst ethnographischen Forschung geprägt (vgl. Healey 2003, 2017).

Räumliche Planer sehen sich in einer wertkomplexen, veränderlichen und dynamischen Welt mit der Herausforderung konfrontiert, über Planaussagen hinaus Prozesse der Planumsetzung zu beachten und aus Unterschieden zwischen Plänen, Umsetzung und „realer Welt" zu lernen. Umsetzungsanalysen und Evaluationen gewinnen heute erheblich an Bedeutung – zumindest theoretisch. Strategische räumliche Planung sieht sich dabei spezifischen Anforderungen gegenüber. Versteht man strategische räumliche Planung als einen sozialen Prozess, der auch einer politischen Logik folgt, kann der Wert strategischer Pläne nicht einfach an der Übereinstimmung zwischen Planfestlegungen und Handlungen der Planadressaten festgemacht werden. Es ist durchaus möglich, wenn nicht sogar die Regel, dass die Planadressaten im Laufe eines Planungsprozesses die Ziele und Maßnahmen hinterfragen und korrigieren. Sofern dies Ausdruck eines Lernprozesses ist, wäre es unangemessen, automatisch auf ein Scheitern der Planung zu schließen. Nicht nur „conformance", sondern auch „performance", also Leistungsfähigkeit, sollte der Maßstab zur Bewertung strategischer Planung sein (vgl. Mastop und Faludi 1997). Faludi (2000) fordert statt einer mechanischen Planumsetzung eine Plananwendung, d. h. die Beeinflussung der Einsichten und Haltungen der Akteure der Raumentwicklung, und zwar unabhängig davon, ob die Ergebnisse dieses Einflusses mit den im Plan formulierten Zielen und Maßnahmen übereinstimmen. Die Wirkung der strategischen Planung ist hier also indirekt. Die handelnden Akteure setzen den Plan nicht einfach um, sondern handeln mehr oder weniger autonom, nach eigenen Prämissen und unter Ausnutzung von gegebenen Handlungsspielräumen. Die Leistungsfähigkeit strategischer Pläne bemisst sich hier vor allem daran, ob sie dabei helfen, Handlungsalternativen zu verdeutlichen und Entscheidungssituationen zu definieren. Dies kann auch dann der Fall sein, wenn die operative Entscheidung vom Plan abweicht, die Begründung dieser Abweichung jedoch auf den Plan Bezug nimmt (vgl. Mastop und Faludi 1997).

Hans Mastop und Andreas Faludi: Evaluation of Strategic Plans: The Performance Principle (Mastop und Faludi 1997)

Mastop und Faludi prägen seit den 1980er Jahren insbesondere die angelsächsische Diskussion zum Wesen von strategischer Planung und ihrer Evaluation. Mastop und Faludi verstehen Pläne nicht nur als Instrumente zur Lösung von Problemen, sondern vor allem als gedankliche Konstrukte und Elemente von sozial-interaktiven Prozessen. Aus diesem Grund sei insbesondere die Evaluation von strategischen Plänen komplex. Im Gegensatz zu den Projektplänen, deren Wirkungen an den realen Veränderungen entsprechend einer Ziel-Mittel-Logik gemessen werden können, schließen bei einem strategischen Plan die Handlungen nicht automatisch an dessen Beschluss an. Mastop und Faludi gehen davon aus, dass der strategische Plan auch dann eine Wirkung entfaltet, wenn seine direkte Umsetzung misslingt. In Anlehnung an Healeys Verständnis von strategischer Planung als sozialen Prozess, richten Mastop und Faludi den Blick auf die soziale Interaktion zwischen den „Planmachern" und den „Planadressaten". Damit der strategische Plan Wirkung entfalten kann, müssen die Adressaten den Plan als einen Teil ihres Handlungskontextes ansehen. Der Plan wird zu einem Bezugsrahmen für ihr Handeln. Bei der Evaluation von strategischer Planung sei deshalb die Anwendung der strategischen Pläne bzw. die „Performance" (Leistungsfähigkeit), als weitere Evaluationsdimension neben der Umsetzung der Pläne bzw. der „Conformance" (Konformität), von entscheidender Bedeutung. Der Artikel von Mastop und Faludi fördert das Verständnis zu dem Wesen und der Wirkungsweise strategischer Planung und liefert Hinweise zur Evaluation von strategischen Plänen. Es werden erste empirische Erkenntnisse des methodischen Ansatzes dargestellt.

Die drei ausgewählten Aufsätze können als zentrale Ausgangspunkte für Arbeiten zur strategischen räumlichen Planung gelten. Der Beitrag von Albrechts gibt einen Überblick und betont die normative Dimension der Planung. Der Beitrag von Healey zeigt prozessuale Möglichkeiten strategischer Urteilsbildung auf. Der Beitrag von Mastop und Faludi argumentiert, dass strategische Pläne – anders als Projekte – auch spezifischer Evaluationskriterien bedürfen (das Kriterium der „Performanz" eines strategischen Plans). Strategische räumliche Planung als normatives, breit gefasstes und situativ anzuwendendes Konzept hat – wenig überraschend – auch Kritik der internationalen Planungsforschung auf sich gezogen (z. B. Newman 2008 mit Betonung politischer Prozesse und ihrer unzureichenden

Beachtung in der strategischen Planung). Der folgende Abschnitt widmet sich allerdings einem anderen Strang der Planungsforschung. Er geht der Frage nach, welche Rezeption strategische räumliche Planung im deutschsprachigen Raum erfahren hat und welche Beiträge diese Rezeption verdeutlichen.

2.3 Rezeption strategischer Planung im deutschsprachigen Raum

Der Begriff „strategische Planung" hat in der deutschsprachigen Diskussion über räumliche Planung bis zur Jahrtausendwende kaum eine Rolle gespielt, auch wenn die damit umrissenen Inhalte, Prozesse und Rahmenbedingungen natürlich auch hier thematisiert worden sind. Diese wurden aber unter anderen Begriffen gefasst, von der integrierten Entwicklungsplanung über die Leitbildentwicklung („Leitbilder der Raumentwicklung") und das Regionalmanagement bis hin zu regionalen Entwicklungskonzepten. Vor allem seit den 2000er Jahren findet der Terminus „strategische Planung" auch im deutschen Sprachraum Widerhall (vgl. z. B. Fassbinder 1993; Brake 2000; Altrock 2004; Klotz und Frey 2005; Hutter 2006; Krüger 2007; Hamedinger et al. 2008; Kühn 2008, 2013; Wiechmann 2008a, b; Hutter und Wiechmann 2010; Kühn und Fischer 2010; ARL 2011; Vallée 2012; Böttcher 2017).

Die terminologischen Verschiebungen durch die vermehrte Verwendung des Begriffs „strategische Planung" in der deutschsprachigen Planungsdebatte spiegeln eine veränderte Planungswirklichkeit wider. Konstatiert werden ein Bedürfnis nach Wiedergewinnung der strategischen Dimension und eine Rückkehr der „großen Pläne" (Klotz und Frey 2005). Öffentliche Verwaltungen stehen unter einem zunehmenden Druck, zur Steigerung der Wettbewerbsfähigkeit von Städten und Regionen auch auf Methoden der Unternehmenssteuerung zurückzugreifen. Exemplarisch steht hierfür das Steuerungsmodell des New Public Management. Auch die Praxis der Raumplanung verwendet vermehrt Methoden aus der Managementpraxis, wie Planungsmanagement, Szenarioplanung und Evaluation (Sinning 2006). Die geforderte stärkere Orientierung des Planungssystems auf seine Entwicklungsfunktion erfordert neue integrative Strategien, die neben den Trägern öffentlicher Belange auch gezielt Akteure aus Wirtschaft und Zivilgesellschaft ansprechen (Wiechmann 2008a; Vallée 2012).

Das Verständnis strategischer Planung ist auch im deutschsprachigen Raum durch Vielfalt gekennzeichnet: Vereinzelt beziehen sich Autoren auf die historisch-militärischen Wurzeln der Strategieforschung (z. B. Scholl 2005). Andere Autoren verstehen strategische Planung primär als Versuch der Übertragung eines betriebswirtschaftlichen Konzepts für Unternehmen in die Stadt- und Regionalentwicklung (vgl. Wiechmann 2008a; Fürst

2012). Wiederholt finden sich Anknüpfungen an den im Rahmen der IBA Emscher Park entwickelten perspektivischen Inkrementalismus von Karl Ganser und Kollegen (Kühn 2008; Reimer 2012, S. 37). Allerdings besteht kein Konsens, ob dieser primär konzeptionelle Ansatz eine Weiterentwicklung der klassischen Entwicklungsplanung ist, die Anforderungen an Flexibilität, Kooperation und Umsetzungsorientierung aufgreift (vgl. Ritter 2007), oder der Versuch eines „dritten Weges", der die fundamentalen Gegensätze von pragmatischer, von politischen Opportunitäten geprägter Praxis und an langfristigen Zielen orientierten integrativen Strategien der Planung aber nicht überwinden kann (Kühn 2008).

Die ausgewählten Aufsätze repräsentieren sicherlich nicht insgesamt die deutschsprachige Planungsforschung zur strategischen räumlichen Planung (vgl. Wiechmann 2008a; Hutter und Wiechmann 2010). Sie verdeutlichen allerdings (quasi im Sinne von Fallbeispielen), wie die Rezeption strategischer Planung im deutschsprachigen Raum erfolgt ist und welche Unterschiede zur internationalen Diskussion dabei auffallen:

- Die Autoren verstehen strategische Planung zwar als normatives Konzept. So entwirft Kühn (2008, S. 241) ein normatives Strukturmodell für die Stadt- und Regionalplanung; auch Fürst (2012) orientiert sich an strategischer Planung als normatives Konzept. Kühn und Fürst argumentieren jedoch nicht selbst vorrangig normativ. Kühn verwendet das normative Strukturmodell dazu, um Abweichungen der Realität von diesem Modell und Problemfelder strategischer Planung zu beschreiben (zu den empirischen Ergebnissen ihres strategischen Planungsprojekts vgl. auch Kühn und Fischer 2010). Der Beitrag von Fürst (2012) dient eher dazu, die internationale und nationale Planungsdiskussion zu berücksichtigen und zu interpretieren, als sie um weitere normative Argumente zu bereichern. Der Beitrag von Ritter (2007) ist verhalten normativ im Hinblick auf die Stadtentwicklungsplanung.
- Auch die drei deutschsprachigen Beiträge verfolgen ein breites Verständnis strategischer Planung. Sie beziehen sie auf komplexe Konstellationen von Akteuren aus Politik, Verwaltung, Zivilgesellschaft, Öffentlichkeit, Wirtschaft usw. Strategische Planung als Prozess umfasst eine Vielzahl von Aktivitäten (Analyse, Zielformulierung, Entwurf von Optionen und Alternativen, Bewertung, Controlling, Lernen, Öffentlichkeitsarbeit usw.). Im Vergleich zu den englischsprachigen Beiträgen fällt die häufige Nennung der Stärken-Schwächen-Risiken-Chancen-Analyse auf (SWOT = Strengths, Weaknesses, Opportunities, Threats).
- Die ausgewählten Aufsätze betonen, dass strategische Planung situativ zu bestimmen ist und dass sie in aller Regel iterativ abläuft. Sie sind durch die jeweils im Vordergrund stehenden inhaltlichen Problemlagen und Rahmenbedingungen gekennzeichnet. So argumentiert

Kühn vor dem Hintergrund der Problemlagen schrumpfender Städte, Fürst mit Blick auf formale Planungsinstrumente und Prozesse der Regionalplanung und Ritter vor dem Hintergrund der Geschichte der Stadtentwicklung und Stadtentwicklungsplanung in Deutschland. Im Vergleich zu den englischsprachigen Aufsätzen fällt die geringere Betonung wertbezogener und ethischer Fragestellungen auf.

Das Folgende geht zunächst vertiefend auf die Aufsätze von Kühn (2008) und Fürst (2012) ein. Beide Aufsätze berücksichtigen in überzeugender Weise den damaligen internationalen und nationalen Forschungsstand zur strategischen räumlichen Planung. Sie ziehen daraus allerdings, wie oben angedeutet, andere Schlussfolgerungen als in den drei englischsprachigen Aufsätzen.

Manfred Kühn: Strategische Stadt-und Regionalplanung (Kühn 2008)

Der Aufsatz von Manfred Kühn ist im Rahmen eines konzeptionell-empirischen Forschungsprojekts zur strategischen Stadtplanung entstanden, wobei das Projekt vergleichende Fallstudien in einer Reihe schrumpfender Städte beinhaltete und seitens der Deutschen Forschungsgemeinschaft (DFG) gefördert wurde. Kühn folgt in dem konzeptionellen Aufsatz einem analytisch orientierten Verständnis von Strategie. Strategie wird verstanden als planvolles Handeln von Akteuren und Institutionen, das die Bestimmung langfristiger Ziele aus der Analyse externer und interner Rahmenbedingungen ableitet und mit der Auswahl kurzfristiger und flexibler Schritte zur Realisierung dieser Ziele kombiniert. Kühn betont die theoretisch-konzeptionelle Einheit der beiden Steuerungsfunktionen Orientierung und Umsetzung von strategischer Planung und möchte deshalb anhand vergleichender Fallstudien empirisch untersuchen, inwieweit diese Einheit angesichts zahlreicher, unsicherer und dynamischer Rahmenbedingungen der Stadtentwicklung von Akteuren der Stadtplanung tatsächlich durchgehalten werden kann. Dieses Strategieverständnis weist interessante Unterschiede und Gemeinsamkeiten mit dem Verständnis von Healey auf. Letzteres zeigt sich u. a. in der Betonung des Zusammenhangs von Orientierung und Handeln, von Strategien und Projekten, von Strategieformulierung und Umsetzung. Unterschiede zeigen sich vor allem in der unterschiedlichen Beurteilung der Bedeutung von „systematisch-methodischen Analysen" für die Strategieentwicklung. Healey sieht, wie oben erwähnt, Situations-, Ziel- und Mittelanalysen als einen möglichen Weg zur Strategieentwicklung. Das Strategieverständnis von Kühn hingegen ähnelt eher der strategischen Planung im Sinne einer „klassischen" Unternehmensplanung (Mintzberg 1994).

Der Aufsatz von Fürst zur strategischen Regionalplanung ist im Kontext eines Arbeitskreises der Akademie für Raumforschung und Landesplanung (ARL), Hannover, entstanden (vgl. die Beiträge in Vallée 2012). Er steht damit in einem engen Zusammenhang zu den weiteren Beiträgen dieses Arbeitskreises. Dieser fasste seine Überlegungen anhand der Formulierung eines normativen Strukturmodells strategischer Regionalplanung zusammen (vgl. Vallée et al. 2012). Damit zeige sich, dass strategische Planung die Erfüllung von Ordnungs- und Entwicklungsaufgaben sowie den Einsatz formeller und informeller Planungsinstrumente umfasst (vgl. oben zum breiten Verständnis strategischer Planung). Strategische Planung ist also nur in besonderen Fällen[1] und unter spezifischen Rahmenbedingungen vorrangig ein informeller Planungsprozess, der den Einsatz formaler Instrumente vorbereitet, parallel zu diesen erfolgt oder der Kommunikation bereits vorgenommener Planfestlegungen dient (vgl. Danielzyk und Knieling 2011).

Dietrich Fürst: Internationales Verständnis von „Strategischer Regionalplanung" (Fürst 2012)
In dem vorliegenden Artikel resümiert Dietrich Fürst die unterschiedlichen Motivationen, Verständnisse und Ansätze von strategischer Planung in Praxis und Theorie und bietet dem Leser eine Kontextualisierung und eigene Einschätzung jener. Die vor allem länderspezifischen Unterschiede im Begriffsverständnis und die Kontroversen in der Literatur führen zu dem Schluss, dass es „keine ideale Konzeption der strategischen Planung geben kann, sondern diese stark kontextgebunden organisiert und inhaltlich gestaltet werden muss" (Fürst 2012, S. 23). Für den deutschsprachigen Raum erkennt Fürst jedoch eine „Art gemeinsamen Nenner", bei dem sich strategische Planung durch eine klare Zielorientierung, eine diskursive Methodik, ein mehrstufiges Verfahren sowie ein lernorientiertes Controlling auszeichnet und die Öffentlichkeitsbeteiligung als Ressource einsetzt. Der Begriff „strategische Planung" ist „Ausdruck einer gewünschten Neuausrichtung" (Fürst 2012, S. 18), demnach ein stark wertbeladenes Konzept. Die wissenschaftliche Diskussion hält Fürst für normativ und nicht methodisch-instrumentell ausgerichtet.

---

1   Diese Aussage bedarf in künftigen Arbeiten zur strategischen Planung der vertieften Bearbeitung. So argumentiert z. B. Kunzmann (2013, S. 29) tentenziell für ein Verständnis strategischer Planung als reine Ergänzung des formellen Planungsinstrumentariums der Stadt- und Regionalentwicklung. Mäntysalo (2013) hingegen argumentiert für strategische Planung als einen Bezugsrahmen, in dem sowohl formale Instrumente als auch informelle Handlungsansätze eine Rolle spielen. Damit könnten die verschiedenen Bewertungskriterien der räumlichen Planung (wie z. B. Legitimität, Flexibilität) besser in Einklang gebracht und Paradoxien berücksichtigt werden. Autoren wie Wiechmann (2008a) argumentieren in eine ähnliche Richtung.

In der Planungspraxis existiert strategische Planung nicht in seiner reinen Form. Eine Gemeinsamkeit aller strategischen Planungsansätze besteht jedoch nach Fürst darin, dass sie wesentliche Auswirkungen auf das Feld der Akteure und den Institutionenrahmen besitzen. Aufgrund des Stellenwerts von Kooperationen bringt strategische Planung eine stärkere Verknüpfung von Planung und den neuen Governance-Formen mit sich. Nach Fürst geht mit der Aktualität von strategischer Planung jedoch kein Paradigmenwechsel einher.

Die Beiträge von Kühn und Fürst befassen sich allgemein mit strategischer Stadt- und Regionalplanung. Zahlreiche weitere Publikationen fokussieren auf spezifische Planungsinstrumente, z. B. die Stadtentwicklungsplanung, wenn es um die Verwendung des Strategiebegriffs geht. Dies zeigt sich im Positionspapier „Integrierte Stadtentwicklungsplanung und Stadtentwicklungsmanagement – Strategien und Instrumente nachhaltiger Stadtentwicklung" des Deutschen Städtetages (DST 2013). Es liegt durchaus nahe, dass strategische Planung am Beispiel der Stadtentwicklungsplanung konkretisiert wird. Aufgabe der Stadtentwicklungsplanung ist es, übergeordnete Ziele (Leitbilder), Handlungsfelder und Maßnahmen für die nachhaltige Entwicklung der Stadt sektoral oder teilräumlich differenziert für einen mittelfristigen Planungshorizont aufzuzeigen. Aktuelle Analysen zur Stadtentwicklungsplanung werfen allerdings Fragen zu ihrer Wirksamkeit nicht nur als Steuerungsinstrument, sondern sogar dahingehend auf, ob bzw. wann mittelfristige Planungen in der Praxis als Referenz- und Diskursmaterial von Bedeutung sind (vgl. Böttcher 2017). Der Aufsatz von Ritter (2007) widmet sich demgegenüber der konzeptionellen Begründung einer „zeitgemäßen" strategischen Stadtentwicklungsplanung.

Ernst-Hasso Ritter: Strategieentwicklung heute: Zum integrativen Management konzeptioneller Politik (am Beispiel der Stadtentwicklungsplanung) (Ritter 2007)
Der Beitrag „Strategieentwicklung heute" von Ernst-Hasso Ritter kann als exemplarisch für das Selbstverständnis der Stadtentwicklungsplanung in der jüngeren kommunalen Praxis in Deutschland gelten, die als strategische Steuerung verstanden wird. Dabei soll eine „integrative" konzeptionelle Planung mit der untergeordneten „operativen Ebene" der Umsetzung verknüpft werden. Das Vorgehen soll rational strukturiert, im Unterschied zur „Entwicklungsplanung" der 1970er Jahre angesichts vieler Ungewissheiten und Akteursinteressen allerdings offen, kommunikativ, nur rahmensetzend, kooperativ und rückkoppelnd bzw. im „Gegenstromprinzip" gestaltet werden. Stadtentwicklung wird als ein strategisch an längerfristigen

Grundorientierungen ausgerichteter, multilateraler Managementprozess konzipiert, in dem auch die neuen quantitativen Steuerungsmodelle der Verwaltung unterstützend eingesetzt werden können. Ritter hält eine so konzipierte strategische Stadtentwicklungsplanung sachlich aufgrund wachsender Komplexität und Koordinationsanforderungen für zunehmend erforderlich sowie zur Gemeinwohlbestimmung und Legitimation geboten. Er konstatiert allerdings eine sehr unterschiedliche Praxis in den Kommunen.

Die drei deutschsprachigen Aufsätze sind wichtige Ausgangspunkte für Arbeiten zur strategischen räumlichen Planung. Der Beitrag von Kühn zeigt, wie strategische Planung als Konzept in problemorientierten empirischen Analysen zur Stadtentwicklung in Deutschland Anwendung finden kann. Der Aufsatz von Fürst ermuntert dazu, den Neuigkeitswert strategischer Regionalplanung angesichts zahlreicher bereits laufender Aktivitäten in Forschung und Praxis nicht zu überschätzen bzw. überzeugender theoretisch zu begründen und empirisch aufzuzeigen. Der Beitrag von Ritter leistet Ähnliches mit Blick auf die spezifischen Herausforderungen der Stadtentwicklungsplanung. Auffällig ist der geringere normative Anspruch der ausgewählten Arbeiten im Vergleich zur Emphase, wie sie bei Autoren wie Albrechts und Healey zu finden ist. Planungstheoretische Arbeiten mit dem Ziel der vertieften Begründung strategischer Planung könnten sich – neben deskriptiv-erklärenden Zielen (vgl. z. B. das deskriptive Prozessmodell von Wiechmann 2008a) – auch verstärkt normativ-planungskulturellen Fragestellungen widmen.

2.4 Strategien und strategische Planung in der Managementforschung

Die Managementforschung befasst sich vorrangig mit Organisationen und Netzwerken. Traditionell stehen gewinnorientierte Privatunternehmen im Vordergrund, die an Märkten in den Branchen bzw. Wirtschaftszweigen der Gesellschaft agieren. Die Managementforschung hat sich allerdings auch mit Fragen des Managements von Organisationen und Netzwerken im öffentlichen Sektor und im sog. intermediären Sektor beschäftigt (bzw. im „dritten Sektor" zwischen Markt und Staat). Dieser Blick über Privatunternehmen und Markt hinaus spiegelt sich in der Managementforschung auch darin wieder, dass neben typisch wirtschaftswissenschaftlichen und organisationstheoretischen Ansätzen (z. B. Transaktionskosten, dynamische Fähigkeiten) zunehmend auch weitere

sozialwissenschaftliche Konzepte und Theorien Beachtung finden (z. B. „Praktiken", „Institution" und „Diskurs"; vgl. Bryson 2010; Clegg et al. 2011).

Die Managementforschung speziell zu Strategien und zur strategischen Planung zeigt eine ähnliche Entwicklung. Im Anschluss an die im Militär geschaffenen Grundlagen (z. B. Strategieverständnis des Carl von Clausewitz) entwickelte sich die strategische Planung zu einem eigenständigen Gebiet als Teil des strategischen Managements von Unternehmen bereits in den 1950er Jahren (vgl. Mintzberg et al. 1999; De Wit und Meyer 2010). Einen Popularitätsschub erfuhr strategisches Management in den 1960er und 1970er Jahren, als der Glaube verbreitet war, formalisierte strategische Planung böte die Antwort auf wesentliche Probleme der Unternehmensführung und Organisationsentwicklung. Dem Boom folgten die Ernüchterung und eine Phase der Planungsskepsis in den 1980er Jahren. Erst in den 1990er Jahren erlebte strategische Planung eine Art Renaissance, wenn auch in anderer Form. Die neuen Managementmodelle fokussierten auf die Fähigkeit des Unternehmens, sich veränderten Umwelten anzupassen, und auf organisationales Lernen. Strategische Flexibilität wurde wichtiger als die Strategie selbst. Emergente Phänomene, organisatorischer und institutioneller Wandel entwickelten sich zu Schlüsselthemen der unternehmerischen Strategieforschung. Nach wie vor gibt es Lehrstühle für strategisches Management weltweit nahezu an jeder Universität oder Business School.

Die Managementforschung zu Strategien und zur strategischen Planung ist sehr differenziert und schwer zu überblicken (ein weit verbreitetes Lehrbuch ist das von Johnson et al. 2011; vgl. auch Clegg et al. 2011). Bei aller gebotenen Zurückhaltung von pauschalen Urteilen über *die* Managementforschung bietet sich im Vergleich mit der internationalen und deutschsprachigen räumlichen Planungsforschung folgende Charakterisierung an:

- In der Managementforschung spielen genuin ökonomische Ziele von Unternehmen nach wie vor oftmals eine zentrale Rolle. Es gibt zwar auch Unternehmensforschungen, die das Prinzip der Nachhaltigkeit thematisieren, die speziell mit Strategieentwicklung und strategische Planung befasste Literatur räumt wertbezogenen und ethischen Fragen allerdings deutlich weniger Platz ein, als dies in der Planungsforschung der Fall ist. Damit einher geht eine größere Gewichtung deskriptiver und explanativer Ziele der Managementforschung (vgl. Bresser 2010).
- Die Managementforschung verfolgt ein breites Verständnis von Strategie und strategischem Management. Dies zeigt z. B. der von Bresser (2010) herausgegebene Reader, in dem u. a. Fragen kollektiver Strategien und des strategischen Wandels angesprochen werden. Auch die

einschlägigen Lehrbücher zum strategischen Management zeugen von einem breiten Steuerungsverständnis, in dem strategische Planung allerdings eng als kommunikativer Prozess innerhalb von einzelnen Organisationen gefasst wird (Grant 2003). Management ist also der Oberbegriff, dem Planung als Teil eingeordnet wird.

— Zahlreiche Managementforscher sind mit situativen Argumenten wenig zufrieden. Die Managementforschung zur strategischen Planung befindet sich auf dem Weg in eine empirisch auch mit großen Fallzahlen arbeitende Wissenschaft (Bresser 2010). Grant (2003) und Bresser (2010) sehen mittlerweile die Zeit von Grundsatzdiskussionen zwischen Befürwortern und Kritikern formalisierter strategischer Planung als überwunden. Empirische Analysen mit hohen Fallzahlen zeigten die Vorteilhaftigkeit strategischer Planung auch bei zunehmend komplexen, unsicheren und dynamischen Marktbedingungen. Managementforscher, die sich stärker an Konzepten wie Praktiken, Diskurs, Macht und „Mikropolitik" orientieren, thematisieren hingegen eher die „Details" des Strategiemachens im gesellschaftlichen Kontext anhand von Einzelfallstudien oder vergleichenden Fallstudien (z. B. Clegg et al. 2011).

Für den Einstieg in die Managementforschung speziell zu Strategien und zur strategischen Planung eignen sich die Arbeiten von Henry Mintzberg. Dieser hat sich in kritischer Weise mit formalisierter strategischer Planung in Organisationen, insbesondere Unternehmen, und dem „rationalistischen Paradigma" menschlichen und organisationalen Verhaltens auseinandergesetzt. Dieses Paradigma ist nach wie vor für weite Bereiche der Wirtschaftswissenschaften und der Managementforschung von hoher Bedeutung. Der ausgewählte, „klassisch" zu nennende Aufsatz von Mintzberg (1987) verdeutlicht vor allem folgendes Argument: Das Wort „Strategie" hat nicht nur eine Bedeutung, sondern mehrere. Mintzberg unterscheidet fünf Bedeutungen: Strategie als Plan, Handlungsmuster, kulturelle Perspektive, Marktposition und List.

Dieses Verständnis von Strategie verdeutlicht Anknüpfungspunkte an andere Bereiche des Readers (z. B. Theorien zu Planungskulturen). Entscheidend ist, dass Mintzberg vorrangig auf das Management in einzelnen Organisationen fokussiert, Unternehmen an Märkten. Strategie als kulturelle Perspektive ist damit ein Element der Organisationskultur einer Organisation, z. B. eines Unternehmens, die wiederum Relationen zu weiter gespannten Bedingungen sozialen Handelns aufweist (z. B. nationale Regulierungen, regionale Innovationssysteme und Kulturen).

Henry Mintzberg: The Strategy Concept I: Five Ps For Strategy (Mintzberg 1987)
Die Beiträge von Henry Mintzberg gehören international zu den prägenden im akademischen Diskurs zu Management und Strategien von Organisationen. Er entwickelte seine Erklärungsansätze zu *The Nature of Managerial Work* (Mintzberg 1973) aus empirischen Analysen über die Tätigkeiten und Rollen von Managern. In „Five Ps For Strategy" (Mintzberg 1987) wird sein Verständnis von Strategien, das wertvolle Anregungen für die räumliche Planung gibt, pointiert erläutert. Strategien sind nach Mintzberg das Ergebnis der Wirkungen von verschiedenen Kräften, von denen schließlich die Entwicklung von Organisationen bestimmt wird. Dazu gehören neben einer expliziten Planung, der deliberativen Strategie, insbesondere prägende Verhaltensmuster, die sich im laufenden Handeln herausschälen, die emergente Strategie, die sich von der geplanten oft unterscheidet. Die tatsächliche Strategie wird erst durch das Zusammenwirken der beabsichtigten Handlungen („plan") mit den im Prozess auftretenden Verhaltensmustern („pattern") bestimmt. Prägende Wirkungen haben dabei die Stellung der Organisation im jeweiligen Handlungskontext („position") sowie die in ihr vorherrschenden Sichtweisen und Werthaltungen („perspective"). Das zukunftsgerichtete Verhalten von Organisationen ist also nicht allein das Ergebnis einer rational nachvollziehbaren strategischen Planung, sondern vielmehr das Ergebnis eines laufenden Prozesses unterschiedlicher, teilweise widerstrebender und teilweise komplementärer, nur zum Teil kalkulierbarer Kräfte.

Mintzberg versteht damit Planung und Pläne nicht als „Königsweg" und schon gar nicht als alleinigen Weg zur Strategie. Absichtsvoll, vorausschauend und explizit formulierte Strategien und die entsprechenden „Strategen" sind nicht unter allen Umständen das Zentrum organisatorischer Strategieentwicklung. Organisationen können Strategien auch auf anderen Wegen entwickeln. Akteure beurteilen ggf. im Rückblick weitgehend ungeplant ablaufende Einzelerfahrungen in bestimmten Situationen und deren „Impulse" für neue Problemlösungen als von „strategischer Bedeutung" und initiieren weitere Prozesse des strategischen Wandels. Mintzberg hat dies auf die Formel gebracht, dass Strategien nicht nur deliberativ, sondern auch emergent entstehen ("deliberate versus emergent strategy"). Managementforscher wie Karl E. Weick (1987) gehen noch einen Schritt weiter. Sie betonen Fähigkeiten und Ressourcen zur Improvisation, ohne dabei die Möglichkeit von Strategien und strategischer Planung grundlegend infrage zu stellen. Pläne können Akteuren Orientierung verschaffen und handlungsmotivierend wirken, nicht allein durch die Planinhalte selbst, sondern bereits durch das Vertrauen darin, über einen „Plan" für den Umgang mit einer unsicheren Zukunft zu verfügen.

Karl E. Weick: Substitutes for Strategy (Weick 1987)
Karl E. Weick hat zwei Monografien vorgelegt, die als Meilensteine der Organisations- und Managementforschung gelten. Mit dem Buch *Prozeß des Organisierens* (Weick 1979) verfolgt er die Absicht, ein neues, prozessorientiertes Verständnis von Organisationen und des Organisierens auf der Grundlage eines für soziale Handlungen geeigneten Evolutionsverständnisses zu entwickeln. Die Monografie *Sensemaking in Organizations* (Weick 1995) interpretiert dieses Verständnis von organisationalen Prozessen neu aus Sicht einer interpretativen Forschungsperspektive. Vor diesem Hintergrund ist es nicht verwunderlich, dass sich die Ausdrücke „Strategie" und „Planung" nur selten in seinen Schriften finden. Der ausgewählte Aufsatz ist gleichwohl für einen Reader zu Planungstheorien im Allgemeinen und zur strategischen Planung im Besonderen von Interesse: Wie andere evolutionstheoretisch orientierten Forscher versteht Weick „Strategie" als Element zur Erhaltung („Retention") von ausgewählten Variationsmöglichkeiten. Strategien in diesem Verständnis dienen nicht vorrangig der Planung von künftigen Entscheidungen und Handlungen, sondern für die rückblickende Interpretation bereits erfolgter Handlungen. Sie können z. B. als zusammenfassende „Praxistheorie" zur Erfolgserklärung verstanden werden. Weick grenzt sich damit deutlich vom „klassischen" Strategieverständnis von Alfred Chandler und auch von „Strategie als planvolles Handeln" ab. Gewisse Übereinstimmung ergibt sich auch mit dem Verständnis von Evaluation strategischer Planung nach Mastop und Faludi, weil sich diese von einem einfachen, linearen, vorausschauenden Verständnis der Erfolgsmöglichkeiten von Planungen abheben. Interessant ist zudem, dass sich Weick weniger direkt mit Strategie oder Planung befasst, sondern vielmehr damit, was diese ersetzen kann, in dem Akteure Orientierung erlangen und durch Handeln Bedeutungen generieren, die sie dann rückblickend als Erfolg interpretieren. Er thematisiert hierfür „confidence as strategy" und „improvisation as strategy" und schafft auf diese Weise mögliche Verbindungslinien zu weiteren Bereichen des planungstheoretischen Readers (z. B. zu Theorien zu Planungskulturen).

Die Anwendbarkeit von Ergebnissen der Managementforschung auf die Planungspraxis im Handlungsfeld der Stadt- und Regionalentwicklung ist bereits seit Langem Gegenstand von Debatten (z. B. Kritik an strategischer Planung als „wirtschaftsorientierte" Planung, Managementlösungen als Ausdruck der „Neoliberalisierung" der Stadtpolitik unter „postfordistischen" Rahmenbedingungen). Dies ist nicht verwunderlich. Denn das Handlungsfeld des Managements von Organisationen, insbesondere von Unternehmen, ist deutlich anders konstituiert als das der räumlichen Planung. Dies gilt nicht nur für die grundsätzlich hierarchische Struktur einer Organisation, ihre Abgrenzung nach „außen", selbst im Falle von Unternehmensnetzwerken, und die Zweckbestimmung der Sicherung oder Vermehrung des Kapitals und von Renditen. Spätestens mittelgroße Unternehmen können nicht mehr allein durch Intuition und spontane Maßnahmen von Unternehmenslenkern geführt und schon gar nicht sinnvoll weiterentwickelt werden. Wettbewerb, Wandel der Märkte, die Einführung neuer Technologien usw. erfordern Planung zumindest des Einsatzes der Ressourcen und eine Orientierung, wohin die Entwicklung der Organisation und ihrer Leistungen gehen soll.

Dagegen sind Städte und Regionen in ihrem baulichen Bestand weitgehend determiniert, und das diesbezügliche „Delta" an Veränderungsmöglichkeiten, die vor Ort gestaltet werden können, ist gering. Sehr stark sind dagegen die Kräfte der Fortschreibung der bestehenden Verteilung von Ressourcen und Strukturen. So können hoch entwickelte Regionen zumindest eine Weile scheinbar gut mit punktuellen und inkrementellen Einzelmaßnahmen auskommen und die Entwicklung den wirtschaftlichen und politischen „Märkten" überlassen. Dafür gibt es in Ländern mit starkem Städtewachstum viele Beispiele. Unternehmen würden bei solchen Prozessen schnell ruiniert sein und vom Markt verschwinden.

Mit der Übertragung strategischer Planungsansätze aus dem Unternehmensbereich in den öffentlichen Sektor haben sich insbesondere US-amerikanische Planungs-, Verwaltungs- und Politikwissenschaftler seit den 1980er Jahren intensiv befasst. Als frühe Anwendungsfälle strategischer Planung in der kommunalen Praxis gelten die Strategiepläne von Dallas (1982) und San Francisco (1983). Kritische Stimmen zur Übertragung privatwirtschaftlicher Steuerungsansätze auf den öffentlichen Bereich im Allgemeinen und Städte und Regionen im Speziellen finden sich in der US-amerikanischen Literatur allerdings nur selten. Strategische Planung wird als praktikables Set von Standards für die erfolgreiche Steuerung von privatwirtschaftlichen, gemeinnützigen und öffentlichen Organisationen und die Anpassung der eigenen Maßnahmen an eine komplexe und dynamische Umwelt betrachtet (vgl. Bryson 2010). Ein viel beachteter Aufsatz, der dieses Planungsverständnis zum Ausdruck bringt, stammt von Bryson aus dem Jahr 1988.

John M. Bryson: A Strategic Planning Process for Public and Non-profit Organizations (Bryson 1988)

John M. Bryson hat ab Mitte der 1980er Jahre wesentliche Grundlagen der amerikanischen Debatte über strategische Planung gelegt (z. B. Bryson 2004). Er versteht unter strategischer Planung in enger Anlehnung an das „Corporate Planning" eine disziplinierte Anstrengung zur Herbeiführung fundamentaler Entscheidungen. Es umfasse eine Reihe unterschiedlicher Ansätze und Instrumente. Mit einigen Adaptionen, insbesondere hinsichtlich der Beteiligungsprozesse, wird das im Unternehmensbereich entwickelte Instrumentarium strategischer Planung vom SWOT-Modell über das „Strategic Issue Management" bis hin zu Portfolio-modellen auch für den öffentlichen Sektor als geeignet angesehen. Sein präskriptives Modell eines zyklischen strategischen Planungsprozesses für öffentliche Einrichtungen und Non-Profit-Organisationen, das er später im „Strategy Change Cycle" zusammenfasste, ist in den USA weit verbreitet. Es wurde grundsätzlich auch für die Anwendung in Kommunen, Regionen und Staaten sowie in interorganisatorischen Netzwerken konzipiert. Das Modell baut wie die klassischen Prozessmodelle auf einer linearen Abfolge von Planungs- und Umsetzungs-schritten auf, beschreibt aber in Abgrenzung zu diesen einen dynamischen, iterativen Managementprozess mit zahlreichen Rückkopplungsschleifen. Der abgedruckte Text aus dem Jahr 1988 steht stellvertretend für die frühe amerikanische Debatte. Bereits hier hat Bryson die Grundidee seines pragmatischen Ansatzes eines strategischen Planungsprozesses im öffentlichen Bereich dargelegt und illustriert diese an zwei Beispielen, an denen der Autor beratend mitwirkte. In dem Beitrag macht Bryson zugleich die für ihn enge Verbindung von Planen und Entscheiden deutlich: So sei es hilfreich, Entscheider als strategische Planer und strategische Planer als „Facilitator" der Entscheidungsfindung über Ebenen und Funktionen hinweg zu sehen.

Die ausgewählten Beiträge der Managementforschung können in der planungstheoretischen Forschung als Ausgangspunkte für die Klärung des Planungsverständnisses dienen. So motiviert der Aufsatz von Mintzberg dazu, alternative Definitionen von Strategie zu berücksichtigen, die das „Strategische" in der räumlichen Planung fundieren. Der Beitrag von Weick erhellt, dass die erwünschten Wirkungen von strategischer Planung (z. B. Rahmensetzung und Handlungsmotivation) auch durch andere Prozesse zu erreichen sind als durch soziale Prozesse, in denen Analysen und Plandokumente eine gewisse Rolle spielen. Der Beitrag von Bryson macht verständlich, dass zwischen Planungskulturen und strategischer Planung wichtige Zusammenhänge bestehen können. So unterscheidet sich die europäische Debatte über „Strategic Spatial Planning"

von der amerikanischen Debatte über „Strategic Planning" im öffentlichen Sektor insbesondere dadurch, dass in Europa strategische Planung als integrative und entwicklungsorientierte Form der Planung diskutiert wird, in den USA hingegen als planungsbasierte Form der Strategie-entwicklung. Dementsprechend wird strategische Planung in Europa primär von anderen Formen der Planung abgegrenzt, etwa dem traditionellen Flächennutzungsplan (vgl. Albrechts 2004). In Amerika steht dagegen in Anlehnung an Diskurse der Managementforschung die Frage im Vordergrund, ob Städte und Regionen überhaupt integrativ planen sollten und ob nicht andere Modi der Prioritätensetzung, wie etwa Aushandlungsprozesse, leistungsfähiger sind.

2.5 Fazit und Ausblick

Komplexe Konzepte wie solche der strategischen räumlichen Planung sind am besten durch Rückblicke auf ihre Genese zu verstehen (vgl. oben die ▶ Abschn. 2.1 und 2.2). Die deutschsprachige Planungsforschung hat dabei bisher insbesondere zur deskriptiv-analytischen (z. B. das Prozessmodell von Wiechmann 2008a) und empirischen strategischen Planungsforschung beigetragen (z. B. Hutter 2007; Kühn und Fischer 2010; Böttcher 2017). Beiträge der Managementforschung dienten vor allem als Inspirationsquellen und Grundlagen für die Entwicklung eines „zeitgemäßen" Verständnisses strategischer Planung in Städten und Regionen.

Bryson (2010) argumentiert, dass strategische Planung künftig eher noch an Bedeutung gewinnen wird. Auch aktuelle Beiträge aus der räumlichen Planungsforschung lassen vermuten, dass strategische Planung lokale und regionale Akteure im Umgang mit einer wertkomplexen, unsicheren und von Wandel gekennzeichneten Welt unterstützt (z. B. Albrechts 2017; Healey 2017). Für die Planungsforschung liegt es nahe, normative, deskriptive und explanative Argumente weiterzuentwickeln und insbesondere deren Wechselwirkungen verstärkt in den Blick zu nehmen. Damit könnte ein weiteres Problem angegangen werden: die Beziehung von Theorie und Praxis bzw. Wissenschaft und Planungspraxis (Corley und Gioia 2011; Silva et al. 2015; Hutter und Otto 2017; Weick 2005). Die Erwartungen, unmittelbar praktisch anwendbare bzw. umsetzbare Befunde und Beiträge zum „realen" gesellschaftlichen Wandel zu erarbeiten, steigen (z. B. Van de Ven 2007; Schneidewind 2014). Arbeiten zur strategischen räumlichen Planung sind damit aufgerufen, sich ihrer historischen Einbettung, aktuellen theoretischen Orientierung und Perspektive zu versichern. Hierfür dürfte – über methodologische und methodische Überlegungen hinaus – insbesondere auch die Stärkung des Dreiklangs zwischen normativer, deskriptiver und explanativer Orientierung von hoher Bedeutung sein, der für die strategische räumliche Planung genuin ist. Es ist unsere Hoffnung, dass dieser Dreiklang künftig öfters zu hören ist.

Literatur

Albrechts, L. (2004). Strategic (spatial) planning re-examined. *Environment and Planning B: Planning and Design, 31*, 743–758.

Albrechts, L. (2017). Strategic planning as a catalyst for transformative practices. In B. Haselsberger (Hrsg.), *Encounters in planning thought: 16 autobiographical essays from key thinkers in spatial planning* (S. 184–201). New York: Routledge.

Albrechts, L., & Balducci, A. (2013). Practicing strategic planning: In search of critical features to explain the strategic character of plans. *DISP, 49*(3), 16–27.

Altrock, U. (2004). Anzeichen für eine „Renaissance" der strategischen Planung? In U. Altrock, S. Günther, S. Huning, & D. Peters (Hrsg.), *Perspektiven der Planungstheorie* (S. 221–238). Berlin: Leue Verlag.

ARL (Akademie für Raumforschung und Landesplanung). (2011). *Strategische Regionalplanung.* Hannover: Verlag der ARL.

Böttcher, J. M. (2017). *Wie wirkt Planung? Theorie und Praxis der strategischen Stadtentwicklungsplanung am Beispiel Wohnen in wachsenden Großstädten.* Lemgo: Verlag Dorothea Rohn.

Brake, K. (2000). Strategische Entwicklungskonzepte für Großstädte – Mehr als eine Renaissance der „Stadtentwicklungspläne"? *Deutsche Zeitschrift für Kommunalwissenschaften, 2*, 269–288.

Bresser, R. K. F. (2010). *Strategische Managementtheorie.* Stuttgart: de Gruyter.

Bryson, J. M. (1988). A strategic planning process for public and non-profit organizations. *Long Range Planning, 21*(1), 73–81.

Bryson, J. M. (2004). *Strategic Planning for Public and Nonprofit Organizations. A Guide to Strengthening and Sustaining Organizational Achievement.* San Francisco: Wiley.

Bryson, J. M. (2010). The Future of Public and Nonprofit Strategic Planning in the United States. *Public Administration Review, 70*, S255–S267.

Clegg, St. R., Carter, Ch., Kornberger, M., & Schweitzer, J. (2011). *Strategy. Theory and practice.* London: SAGE.

Corley, K. G., & Gioia, D. A. (2011). Building theory about theory building: What constitutes a theoretical contribution. *Academy of Management Review, 36*(1), 12–32.

Danielzyk, R., & Knieling, J. (2011). Informelle Planungsansätze. In Akademie für Raumforschung und Landesplanung (ARL) (Hrsg.), *Grundriss der Raumordnung und Raumentwicklung* (S. 473–498). Hannover: Akademie für Raumforschung und Landesplanung (ARL).

Davoudi, S., & Strange, I. (Hrsg.). (2009). *Conceptions of space and place in strategic spatial planning.* New York: Routledge.

De Wit, B., & Meyer, R. (2010). *Strategy. Process, Content, Context. An International Perspective.* London: Thomson Learning.

DST (Deutscher Städtetag). (2013). *Integrierte Stadtentwicklungsplanung und Stadtentwicklungsmanagement – Strategien und Instrumente nachhaltiger Stadtentwicklung. Positionspapier des Deutschen Städtetages.* Berlin: Deutscher Städtetag.

Faludi, A. (1969). Planungstheorie oder Theorie des Planens? *Stadtbauwelt, 23*(38/39), 216–220.

Faludi, A. (2000). The performance of spatial planning. *Planning Practice and Research, 15*(4), 299–318.

Fassbinder, H. (1993). Zum Begriff der strategischen Planung – Planungsmethodischer Durchbruch oder Legitimation notgedrungener Praxis? In H. Fassbinder (Hrsg.), *Strategien der Stadtentwicklung in europäischen Metropolen* (S. 9–16). Hamburg: Titelzusatz.

Friedmann, J. (2004). Honk Kong, Vancouver and Beyond: Strategic spatial planning and the longer range. *Planning Theory & Practice, 5*(1), 50–56.

Fürst, D. (2005). Entwicklung und Stand des Steuerungsverständnisses in der Raumplanung. *Disp – The Planning Review, 163*(4), 16–27.

Fürst, D. (2012). Internationales Verständnis von „Strategischer Regionalplanung". In D. Vallée (Hrsg.), *Strategische Regionalplanung* (S. 18–30). Hannover: Verlag der ARL.

Grant, R. (2003). Strategic Planning in a Turbulent Environment: Evidence from the Oil Majors. *Strategic Management Journal, 24*(6), 491–517.

Hamedinger, A., Frey, O., Dangschat, J. S., & Breitfuss, A. (Hrsg.). (2008). *Strategieorientierte Planung im kooperativen Staat.* Wiesbaden: VS Verlag.

Healey, P. (1997). *Collaborative planning. Shaping places in fragmented societies.* Basingstoke: Macmillan Publishers Limited.

Healey, P. (2003). Collaborative planning in perspective. *Planning Theory, 2*(2), 101–123.

Healey, P. (2005). Network complexity and the imaginative power of strategic spatial planning. In L. Albrechts & S. J. Mandelbaum (Hrsg.), *The network society. A new context for planning* (S. 146–160). London: Taylor and Francis.

Healey, P. (2007). *Urban complexity and spatial strategies. Towards a relational planning for our times.* London: Routledge.

Healey, P. (2009). In search of the „Strategic" in spatial strategy making. *Planning Theory & Practice, 10*(4), 439–457.

Healey, P. (2013). Comment on Albrechts and Balducci „Practicing Strategic Planning". *Disp – The Planning Review, 49*(3), 48–50.

Healey, P. (2017). Finding my way: A life of inquiry into planning, urban development processes and place governance. In B. Haselsberger (Hrsg.), *Encounters in planning thought: 16 autobiographical essays from key thinkers in spatial planning* (S. 107–125). New York: Routledge.

Healey, P., Khakee, A., Motte, A., & Needham, B. (Hrsg.). (1997). *Making strategic spatial plans. Innovation in Europe.* London: UCL Press.

Hobsbawm, E. (1995). *Das Zeitalter der Extreme – Weltgeschichte des 20. Jahrhunderts.* München: Hanser.

Hutter, G. (2006). Strategische Planung. Ein wiederentdeckter Planungsansatz zur Bestandsentwicklung von Städten. *RaumPlanung, 210*, 210–214.

Hutter, G. (2007). Strategic planning for long-term flood risk management – Some suggestions for learning how to make strategy at regional and local level. *International Planning Studies, 12*(3), 273–289.

Hutter, G., & Otto, A. (2017). Raumwissenschaft und Politikberatung – Am Beispiel von Projekten zur Klimaanpassung in Städten und Regionen. *Disp – The Planning Review, 53*(4), 42–54.

Hutter, G., & Wiechmann, Th. (2010). *Strategische Planung – Zur Rolle der Planung in der Strategieentwicklung für Städte und Regionen.* Berlin: Altrock.

Johnson, G., Scholes, K., & Whittington, R. (2011). *Strategisches Management. Eine Einführung. Analyse, Entscheidung und Umsetzung.* München: Pearson.

Klotz, A., & Frey, O. (Hrsg.). (2005). *Verständigungsversuche zum Wandel der Stadtplanung.* Wien: Springer.

Krüger, Th. (2007). Alles Governance? Anregungen aus der Management-Forschung für die Planungstheorie. *RaumPlanung, 132*(133), 125–130.

Kühn, M. (2008). Strategische Stadt-und Regionalplanung. *Raumforschung und Raumordnung, 66*(3), 230–243.

Kühn, M. (2013). Strategiefähigkeit – Chancen und Hemmnisse lokaler Politik in schrumpfenden Städten. In M. Haus & S. Kuhlmann (Hrsg.), *Lokale Politik und Verwaltung im Zeichen der Krise?* (S. 274–289). Wiesbaden: Springer VS.

Kühn, M., & Fischer, S. (2010). *Strategische Stadtplanung. Strategiebildung in schrumpfenden Städten aus planungs- und politikwissenschaftlicher Perspektive.* Detmold: Rohn.

Kunzmann, K. R. (2013). Strategic planning: A chance for spatial innovation and creativity. *Disp – The Planning Review, 49*(3), 28–31.

Lamker, C. W. (2016). *Unsicherheit und Komplexität in Planungsprozessen. Planungstheoretische Perspektiven auf Regionalplanung und Klimaanpassung.* Lemgo: Verlag Dorothea Rohn.

Mäntysalo, R. (2013). Coping with the paradox of strategic spatial planning. *Disp – The Planning Review, 49*(3), 51–52.

Mastop, H., & Faludi, A. (1997). Evaluation of strategic plans: The performance principle. *Environment and Planning B: Planning and Design, 24*, 815–832.

Mintzberg, H. (1973). *The nature of managerial work*. New York: Harper & Row.

Mintzberg, H. (1987). The strategy concept I: Five Ps for strategy. *California Management Review, 30*(1), 11–24.

Mintzberg, H. (1994). *The Rise and Fall of Strategic Planning*. Edinburgh Gate: Free Press.

Mintzberg, H., Ahlstrand, B., & Lampel, J. (1999). *Strategy Safari. Eine Reise durch die Wildnis des strategischen Managements*. Wien: Redline Wirtschaft.

Newman, P. (2008). Strategic spatial planning: Collective action and moments of opportunity. *European Planning Studies, 16*(10), 1371–1383.

Reimer, M. (2012). *Planungskultur im Wandel. Das Beispiel der REGIONALE 2010*. Detmold: Rohn.

Ritter, E.-H. (2007). Strategieentwicklung heute. Zum integrativen Management konzeptioneller Politik (am Beispiel der Stadtentwicklungsplanung). *PNDonline, 1*, 1–12.

Salet, W., & Faludi, A. (Hrsg.). (2000). *The revival of strategic spatial planning*. Amsterdam: Koninklijke Nedelandse Akademie van Wetenschappen.

Sartorio, F. S. (2005). Strategic spatial planning. A historical review of approaches, its recent revival and an overview of the state of art in Italy. *Disp – The Planning Review, 162*(3), 26–40.

Schneidewind, U. (2014). Urbane Reallabore – Ein Blick in die aktuelle Forschungswerkstatt. *PNDonline, 3*, 1–7.

Scholl, B. (2005). Strategische Planung. In Akademie für Raumforschung und Landesplanung (ARL) (Hrsg.), *Handwörterbuch der Raumordnung* (S. 1122–1129). Hannover: ARL.

Selle, K. (2005). *Planen. Steuern. Entwickeln. Über den Beitrag öffentlicher Akteure zur Entwicklung von Stadt und Land*. Dortmund: Dortmunder Vertrieb für Bau- und Planungsliteratur.

Silva, E. A., Healey, P., Harris, N., & Van den Broeck, P. (Hrsg.). (2015). *The Routledge handbook of planning research methods*. New York: Routledge Taylor & Francis Group.

Sinning, H. (Hrsg.). (2006). *Stadtmanagement. Strategien zur Modernisierung der Stadt(-Region)*. Dortmund: Rohn.

Vallée, D. (2012). *Strategische Regionalplanung*. Hannover: ARL.

Vallée, D., Brandt, T., Fürst, D., Konze, H., Priebs, A., Schmidt, P. I., Scholich, D., & Tönnies, G. (2012). Modell einer Strategischen Regionalplanung in Deutschland. In D. Vallée (Hrsg.), *Strategische Regionalplanung* (S. 170–190). Hannover: ARL.

Van de Ven, A. H. (2007). *Engaged scholarship. A guide for organizational and social research*. Oxford: Oxford University Press.

Weick, K. E. (1979). *The social-psychology of organizing*. Reading: Wiley-Blackwell.

Weick, K. E. (1987). Substitutes for strategy. In D. J. Teece (Hrsg.), *The competitive challenge* (S. 221–233). Cambridge: Ballinger.

Weick, K. E. (1995). *Sensemaking in organizations*. Thousand Oaks: Sage.

Weick, K. E. (2001). *Making sense of the organization*. Malden: Blackwell.

Weick, K. E. (2005). Theory and practice in the real world. In Ch. Knudsen & H. Tsoukas (Hrsg.), *The Oxford handbook of organization theory* (S. 453–475). Oxford: Oxford University Press.

Wiechmann, Th. (2008a). *Planung und Adaption – Strategieentwicklung in Regionen, Organisationen und Netzwerken*. Dortmund: Rohn.

Wiechmann, Th. (2008b). Strategische Planung. In D. Fürst & F. Scholles (Hrsg.), *Handbuch Theorien und Methoden der Raum- und Umweltplanung* (S. 265–275). Dortmund: Rohn.

Yin, R. K. (2014). *Case study research. Design and methods*. Thousand Oaks: Sage.

Originaltexte

Originaltext Albrechts 2004

2

Environment and Planning B: Planning and Design 2004, volume 31, pages 743 – 758

DOI:10.1068/b3065

Strategic (spatial) planning reexamined

Louis Albrechts
Catholic University of Leuven, Institute of Urban and Regional Planning, Kasteelpark Arenberg 51,
B 3001 Leuven, Belgium; e-mail: Louis.Albrechts@isro.kuleuven.ac.be
Received 27 August 2003; in revised form 20 January 2004

Abstract. In the 1990s a strategic approach to the organization of space at different levels of scale became more prevalent. Increasingly, it is being assumed that the solutions to complex problems depend on the ability to combine the creation of strategic visions with short-term actions. The creation of strategic visions implies the design of shared futures, and the development and promotion of common assets. Moreover, all of this requires accountability within a time and budgetary frame-work and the creation of awareness for the systems of power. Delivering on these new demands implies the development of an adapted strategic planning capacity and a shift in planning style in which the stakeholders are becoming more actively involved in the planning process on the basis of a joint definition of the action situation and of the sharing of interests, aims, and relevant knowledge. In this paper I aim to provide building blocks for such an 'alternative' strategic (spatial) planning approach. It is based on two different sources. The first source is critical planning literature and strategic thinking in business, which will be used to broaden the scope of the concept. The second source consists of European strategic planning practices.

Introduction: revival of strategic planning

In the 1960s and 1970s strategic spatial planning in a number of Western countries evolved towards a system of comprehensive planning at different administrative levels. In the 1980s we witnessed a retreat from strategic planning, fuelled not only by the neoconservative disdain for planning, but also by postmodernist skepticism, both of which tend to view progress as something which, if it happens, cannot be planned (Healey, 1997a). Instead the focus of urban and regional planning practices was on projects (Motte, 1994; Secchi, 1986), especially for the revival of rundown parts of cities and regions, and on land-use regulations.

The growing complexity, an increasing concern about rapid and apparently random development (Breheny, 1991), the problems of fragmentation, the dramatic increase in interest (at all scales, from local to global) in environmental issues (Breheny, 1991), the growing strength of the environmental movement, a reemphasis on the need for long-term thinking (Friedmann, 2004; Newman and Thornley, 1996) and the aim to return to a more realistic and effective method all served to expand the agenda. In response, more strategic approaches, frameworks, and perspectives for cities, city-regions, and regions had again become fashionable in Europe by the end of the millennium (Albrechts, 1999; Albrechts et al, 2003; CEC, 1997; Healey et al, 1997; Pascual and Esteve, 1997; Pugliese and Spaziante, 2003; Salet and Faludi, 2000). As there is no 'one best or one single way' to do strategic planning the purpose of this paper is to add a new dimension in terms of values, approach, and process. It therefore reexamines strategic (spatial) planning by highlighting differences with traditional land-use plan-ning, by using views from strategic thinking in business and shifts in the overall planning approach. This is done by combining theory with practical experience.

744   L Albrechts

Land-use planning

Aims

Land-use planning is basically concerned—in an integrated and qualitative way—with the location, intensity, form, amount, and harmonization of land development required for the various space-using functions: housing, industry, recreation, transport, education, nature, agriculture, cultural activities (see also Chapin, 1965; Cullingworth, 1972). In this way a land-use plan embodies a proposal as to how land should be used—in accordance with a considered policy—as expansion and restructuring proceed in the future.

A classification of EU land-use planning systems

The emergence of land-use planning systems across Europe has some common roots. In many EU member states the first planning legislation was produced in the early 20th century as a response to increasing development pressure and the consequent problems that arose from dense and disorganized development. Cultural, institutional, and legal differences, but also the specificity of the purposes for which formal spatial planning systems were originally introduced, produced a wide variety of planning systems and traditions in the EU. *The EU Compendium of Spatial Planning Systems and Policies* (CEC, 1997) draws a line between strategic planning at the regional or national level and land-use planning at the level of the municipality and the functional urban region. Recent practice (Albrechts et al, 2001; 2003; Healey et al, 1997; Pascual and Esteve, 1997; Pugliese and Spaziante, 2003; Salet and Faludi, 2000), however, illustrates that a lot of strategic planning is going on at the level of the city and the urban agglomeration.

Land-use planning at EU level clearly focuses on the municipality or functional urban region (mainly Greece, France, Italy, and Sweden) with framework (master plan) instruments and on specific areas within the municipality with regulatory instruments. In Belgium land-use zoning also exists for subregions. The framework plans cover at least the whole of the area of the local authority and set out the broad land-use and infrastructure patterns across the area through zoning or land-allocation maps. The regulatory plan covers the whole or part of the local authority's area and indicates detailed site-specific zonings for building, land use and infrastructure.

All the EU member states, except the United Kingdom and the Republic of Ireland, use detailed planning instruments (regulatory zoning instruments, building control instruments, and implementation instruments) which play a determining role in guiding the location of development and physical infrastructure, and the form and size development takes.

The framework plans and the regulatory plans are mostly legally binding documents—with implied legal certainty and rigidity—once approved (the major exception is the United Kingdom); they are generally of no fixed duration and can be replaced only by new plans. This often relates to the extent to which a system is binding or discretionary (see CEC, 1997). In a binding system the relationship between policy and control is expected to be determined through a binding detailed land-use plan. Effects are compared with intentions like a blueprint for a house. This is the 'conformance view' (Barrett and Fudge, 1981). In a discretionary system each decision is subject to administrative and political discretion, with the plan providing general guidance. This relates to the 'performance view' (see Mastop and Faludi, 1997). The United Kingdom is the primary example of a discretionary system. Whereas in the binding system the focus is on legal certainty there is a notable absence of certainty in the discretionary system.

Local governments are generally responsible for the production of these plans and for most EU countries (Greece is an exception) the responsibility for approval is also delegated to this level. Sometimes the local plans have to be formally approved by another tier (regional or central) of government.

Consultation with other tiers of government, administrations, and official agencies are everywhere an essential part of the daily routine in plan making. There is also a general commitment to consultation with the public (in the broadest sense). The method and depth of public involvement vary considerably. The Scandinavian countries have a long-standing tradition, for others it remains very formal and restricted (Greece, Italy, Portugal).

Criticism of land-use planning

Traditional land-use planning, as more passive, pragmatic, and localized planning, aims at controlling land use through a zoning system. Land-use regulation, which in Europe has been typically plan led (Davies et al, 1989), helps to steer developments in a certain direction. Indeed, building permits are granted (or refused) if a project or development proposal is in line (or not) with the approved land-use plan and regulations. In this way the land-use plan ensures that undesirable developments do not occur but it is not able to ensure that desirable developments actually take place where and when they are needed. Cullingworth (1993) contrasts this European tradition with US local zoning systems, which are driven by concern for the specification of land rights rather than for managing the location of development (see also Healey, 1997b). The approach to planning via a single policy field (that is, spatial planning) met fierce opposition from other and usually more powerful policy fields. Although land-use plans had formal status and served as official guidelines for implementation, when it came down to the actual implementation, other policy fields—which, because of their budgetary and technical resources, were needed for the implementation—were easily able to sabotage the spatial plans if they wanted (Kreukels, 2000; Scharpf and Schnabel, 1978). Moreover, it became increasingly clear that a number of different planning concepts—such as the coherent, convenient, and compact city long advocated by planners—cannot be achieved solely through physical hard planning (see Hart, 1976).

A major emphasis on the legally binding nature of most EU land-use plans provides legal certainty but makes the plans far more rigid and inflexible and less responsive to changing circumstances. The mainly comprehensive nature of land-use plans is at odds with increasingly limited resources. Moreover, most land-use plans have a predominant focus on 'physical' aspects, providing 'physical' solutions to social or economic problems. In this way they often abstract from real historically determinate parameters of human activity and gratuitously assume the existence of transcendent operational norms. Coproduction of plans with the major stakeholders and the involvement of 'weak' groups in the land-use planning process are (with the exception of some experiences such as in Finland) nonexistent. The whole apparatus of adverse bargaining, negotiation, compromise, and deadlock, which normally surrounds the planning process, must be questioned.

In all EU systems there is evidence (CEC, 1997) to suggest that systems should become more open and less prescriptive in determining precise land uses in favor of a more flexible system to respond more quickly and adequately to changing social and economic circumstances and to the agenda of government reorganization.

Strategy and strategic planning in the literature
There are no single universally definitions for strategy and strategic planning. Various authors and practitioners use the term differently. In my selective reading of the literature I look for building blocks to broaden the perspective.

Historical roots
The word 'strategy' originated within a military context. Webster's dictionary (Webster, 1970, page 867) defines strategy as 'strategia', "science and art of employing the political, the economic, psychological and military forces of a nation or group of nations to afford the maximum support to adopted policies in peace or war". The current paper is not about the military sense of strategy.

According to Kaufman and Jacobs (1987), strategic planning originated in the 1950s in the private sector. Its roots were tied to the need for rapidly changing and growing corporations to plan effectively for and manage their futures when the future itself seemed to be increasingly uncertain (that is, strategic planning carried out by an organization for its own activities). In the early 1970s, government leaders in the USA became increasingly interested in strategic planning as a result of wrenching changes— oil crisis, demographic shifts, changing values, volatile economy, etc—(Eadie, 1983; cited in Bryson and Roering, 1988, page 995). In the early 1980s, a series of articles in the USA called on state and local governments to use the strategic planning approach developed in the corporate world (Kaufman and Jacobs, 1987).

For Mastop (1998), the first traces of strategic spatial planning in northwestern Europe date back to the 1920s and 1930s. He links strategic spatial planning closely to the idea of the modern nation-state. Strategic planning is used here to direct the activities of others (different authorities, different sectors, private actors). The differences in origin and tradition between US and European traditions reflect the historical 'statist' traditions of many postwar European states, which are linked to a battery of welfare state policies (Batley and Stoker, 1991; Esping-Anderson, 1990; cited by Healey, 1997a).

Looking for building blocks in business and planning literature
There is a large amount of literature in the USA about the use of strategy and strategic planning in business and nonprofit organizations and a growing literature in Europe about strategic planning. Mintzberg et al (1998) conclude their historical survey of strategy making by emphasizing the fact that it should be concerned with process [for Healey (1997a) it is clearly a social process, and for Kunzmann (2000) it is public sector led] and content, statics and dynamics, constraint and inspiration, the cognitive and the collective, the planned and the learned, and the economic and the political. Quinn (1980) cites a few studies that have suggested some initial criteria for evaluating a strategy: clarity, motivational impact, internal consistency, compatibility with the environment, appropriateness in light of the resources available, degree of risk, the extent to which it matches the often contradictory personal values of key figures, time horizon, flexibility, workability, focus on key concepts, and thrust and committed leadership (see also Poister and Streib, 1999). Others (Bryson, 1995; Bryson and Roering, 1988; Poister and Streib, 1999) stress the need to gather the key (internal and external) stakeholders (preferably key decisionmakers), the importance of external trends and forces, the active involvement of senior level managers, to construct a longer term vision (Kunzmann, 2000; Mintzberg, 1994; Mintzberg et al, 1998), the need to focus on implementation, to build commitment to plans (see also Albrechts, 1995; Granados Cabezas, 1995; Van den Broeck, 1996), and to be politically realistic. Faludi and Korthals Altes (1994) and Faludi and Van der Valk (1994, page 3) make a distinction between project plans and strategic plans (table 1). They define project planning as the opposite

2

Table 1. Project plans and strategic plans (source: Faludi and Van der Valk, 1994, page 3).

	Project plans	Strategic plans
Object	Material	Decisions
Interaction	Until adoption	Continuous
Future	Closed	Open
Time element	Limited to phasing	Central to problems
Form	Blueprint	Minutes of last meeting
Effect	Determinate	Frames of reference

of strategic planning. Strategic plans are defined as frameworks for action. They need to be analyzed for their performance in helping with subsequent decisions. Project plans are blueprint plans and form an unambiguous guide to action. For Granados Cabezas (1995) strategic planning anticipates new tendencies, discontinuities, and surprises; it concentrates on openings and ways of taking advantage of new opportunities.

Towards a workable normative viewpoint on strategic planning
Reflecting on the challenges spatial planning is facing and relying on the experience accumulated from business, planning practice, and a study of the planning literature leads us to the following viewpoint on the 'what' of strategic spatial planning: strategic spatial planning is a public-sector-led (Kunzmann, 2000) sociospatial (see Healey, 1997a for the emphasis on the social) process through which a vision, actions, and means for implementation are produced that shape and frame what a place is and may become. A combination of characteristics related to the 'how' of strategic planning gives a specific coloring to the concept. A first characteristic is that strategic planning has to focus on a limited number of strategic key issue areas (Bryson and Roering, 1988; Poister and Streib, 1999; Quinn, 1980); it has to take a critical view of the environment in terms of determining strengths and weaknesses in the context of opportunities and threats (Kaufman and Jacobs, 1987); it studies the external trends, forces (Poister and Streib, 1999) and resources available (Quinn, 1980); it identifies and gathers major stakeholders (public and private) (Bryson and Roering, 1988; Granados Cabezas, 1995); it allows for a broad (multilevel governance) and diverse (public, economic, civil society) involvement during the planning process; it develops a (realistic) long-term vision or perspective and strategies (Healey, 1997a; 1997b; Kunzmann, 2000; see also Mintzberg, 1994) at different levels (Albrechts et al, 2003; Quinn, 1980), taking into account the power structures (Albrechts, 2003a; Poister and Streib, 1999; Sager, 1994), uncertainties (Friend and Hickling, 1987; Quinn, 1980) and competing values; it designs plan-making structures and develops content (Mintzberg et al, 1998) images, and decision frameworks (Faludi and Van der Valk, 1994) for influencing and managing spatial change (Healey, 1997b); it is about building new ideas (Mintzberg et al, 1998) and processes that can carry them forward (Mintzberg, 2002), thus generating ways of understanding, ways of building agreements, and ways of organizing and mobilizing for the purpose of exerting influence in different arenas (Healey, 1997a); and finally it (both in the short and the long term) is focused on decisions (Bryson, 1995), actions (Faludi and Korthals Altes, 1994; Mintzberg, 1994), results (Poister and Streib, 1999), and implementation (Bryson, 1995; Bryson and Roering, 1988), and incorporates monitoring, feedback, and revision.

This may seem to some people (see Mintzberg, 1994) too broad a view of strategic planning. However, the many experiences documented in the planning literature (Albrechts et al, 2001; 2003; Healey et al, 1997; Pascual and Esteve, 1997; Pugliese and Spaziante, 2003) back up this broader view. This view also implies that strategic (spatial) planning is not a single concept, procedure, or tool. In fact, it is a set of

748   L Albrechts

concepts, procedures, and tools that must be tailored carefully to whatever situation is
at hand if desirable outcomes are to be achieved (Bryson and Roering, 1996). Strategic
planning is as much about process, institutional design, and mobilization as about
the development of substantive theories. Content is related to the strategic issues
selected in the process. In Europe the environmental agenda is linked in part to the
environmental movement's emphasis on sustainable resource use and in part to citizen
movements concerned with the quality of life in specific places (see Hall and Pfeiffer,
2000) and the appreciation of their diversity (Healey, 1997a). 'Sustainable development'
has become a widely used concept expressing the potential for creating a positive-
sum strategy combining economic, environmental, and social objectives in their spatial
manifestation. In this way it supersedes the mere focus on land use. The 'place focus' in
turn is linked to a political–cultural momentum to reassert the importance of regional
or local identity and image in the face of European 'integration' and globalization.

The term 'spatial' brings into focus the 'where of things', whether static or dynamic;
the creation and management of special 'places' and sites; the interrelations between
different activities in an area, and significant intersections and nodes within an area
which are physically colocated (see also Healey, 2004). The focus on the spatial relations

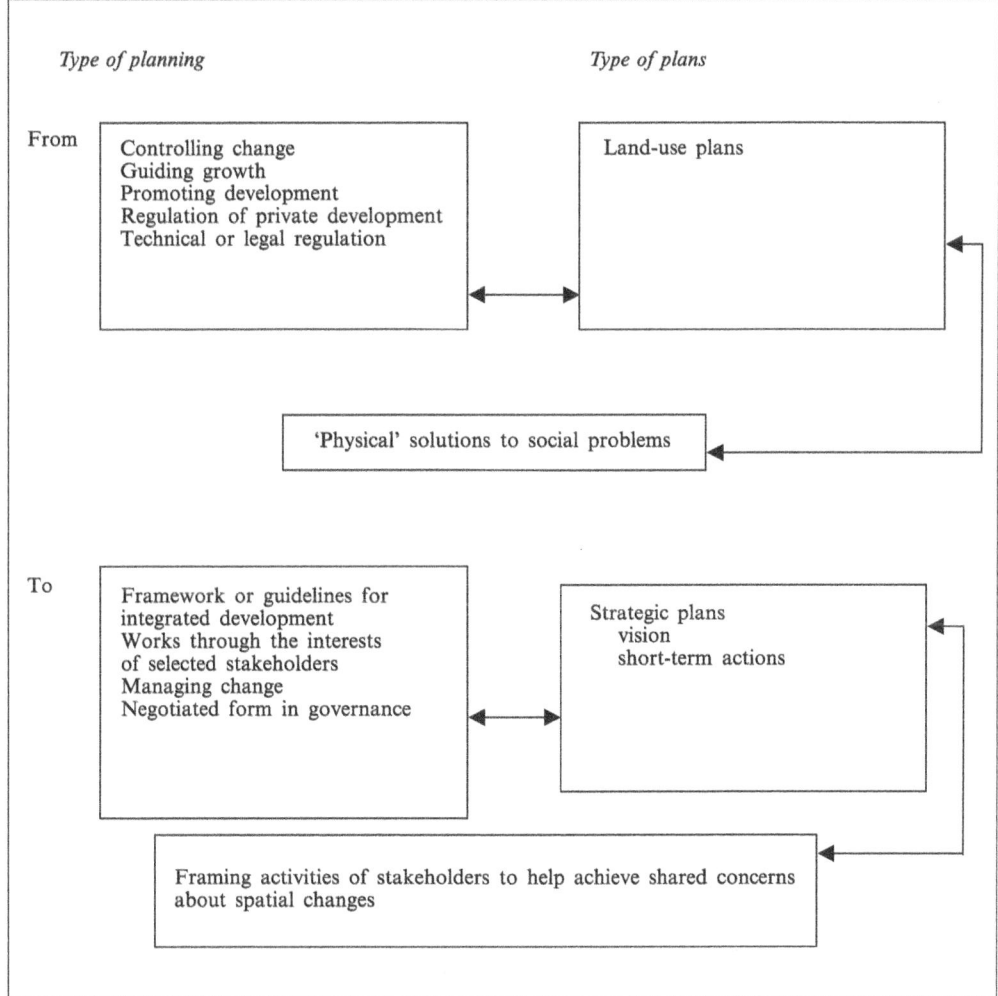

Figure 1. From traditional land-use planning to strategic planning.

2

of territories allows for a more effective way of integrating different agendas (economic, environmental, cultural, and social policy agendas) as these agendas impact on localities and of translating territorial development into specific investment programmes and regulatory practices (Albrechts et al, 2003; see also Wilkinson and Appelbee, 1999). Strategic frameworks and visions for territorial development, with an emphasis on place qualities and the spatial impacts and integration of investments, complement and provide a context for specific development projects. They also carry a potential for a 'rescaling' of issue agendas down from the national or state level and up from the municipal level. This search for new scales of policy articulation and new policy concepts is also linked to attempts to widen the range of actors involved in policy processes, with new alliances, stakeholder partnerships, and consultative processes (Albrechts et al, 2001; 2003; Healey et al, 1997). A territorial focus also provides a promising basis for encouraging levels of government to work together (multilevel governance) and in partnership with actors in diverse positions in the economy and civil society (Albrechts, 1999; Fürst, 2001; Kunzmann, 2001a).

The capacity of strategic spatial planning systems to deliver the desired outcome is dependent not only on the system itself, but also on the conditions underlying it (see also Mintzberg, 1994). These conditions—including public and professional attitudes towards spatial planning (in terms of planning content and process) and the political will on the part of the institutions involved in setting the process in motion (Granados Cabezas, 1995)—affect the ability of planning systems to implement the chosen strategies. The steps required to deliver and to implement the desired spatial outcome vary according to the underlying structure.

This normative viewpoint produces a quite different picture from traditional land-use planning in terms of plans (strategic plans), type of planning (providing a framework), and governance type (negotiated form of governance). Its rationale is to frame the activities of stakeholders to help achieve shared concerns about spatial changes (see figure 1).

Principles and points of departure for change
The efforts underway in many parts of Europe to produce strategies for cities, city-regions, and regions (Albrechts et al, 2001; 2003; Healey et al, 1997; Pascual and Esteve, 1997; Pugliese and Spaziante, 2003; Salet and Faludi, 2000) reorient spatial planning in the direction of a more strategic type of planning. Often these efforts involve the construction of new institutional arenas within structures of government, which are themselves changing. Their motivations are varied, but the objectives have typically been to articulate a more coherent spatial logic for land-use regulation, for resource protection, for action orientation, for a more open type of governance, for introducing sustainability, and for investments in regeneration and infrastructure.

Just as in general planning, there are different traditions of strategic planning. I therefore exemplify some principles of my 'new dimension' of strategic planning in terms of values, approach, and process.

Values and visioning
To keep planning from becoming more concerned with how to plan rather than with the content of planning, substantive rationality (Mannheim) or value rationality (Weber) are (re)introduced (De Jouvenel, 1964; Ozbekhan, 1969). This is needed to counteract the pure instrumental rationality that encourages an analysis of trends and extrapolates them in order to arrive at conceptions of social and economic futures. Speaking of values (for instance, spatial quality and subsidiarity) is a way of describing the sort of place we want to live in, or think we should live in. The values and the

images of what a society wants to achieve are defined in the planning process. Values and images are not generated in isolation but are—within a specific context—created and given meaning. They are validated by traditions of belief and practice, and are reviewed, reconstructed, and invented through collective experience (see Ozbekhan, 1969; but also Elchardus et al, 2000, page 24; Foucault, 1980, page 11). The introduction of value rationality is thus a clear reaction against a future that extrapolates the past and maintains the status quo. It means that time flows from the 'invented' future, which challenges conventional wisdom, toward and into the experienced present. This means inventing a world that would not otherwise be (Mintzberg et al, 1998; Ozbekhan, 1969). In this way strategic planning 'creates' a vision for a future environment, but all decisions are made in the present. This means that over time the strategic planning process must stay abreast of changes in order to make the best decisions it can at any given point. It must manage, as well as plan, strategically. This 'created future' has to be placed within a specific context (economic, social, political, and power), place, time, and scale with regard to specific issues and a particular combination of actors. It provides the setting for the process but also takes form, undergoes changes in the process. All this must be rooted in an understanding of the past.

Inclusive and accountable
Just as there are many traditions and collectivities, so also there are many images of what communities want to achieve (see Weeks, 1993). The opportunities for implementing these images are not equal. Some individuals and groups have more resources and power, which allow them to pursue their images. To give power to the range of images in a planning process requires a capacity to listen, not just for the expression of material interest, but also for what people care about, including the rage felt by many who have grown up in a world of prejudice and exclusion, of being the outsider, 'the other' (Forester, 1989; Healey, 1997b). At the core of this process is a democratic struggle for inclusiveness in democratic procedures, for transparency in government transactions, for accountability of the state and planners to the citizens they work for, for the right of citizens to be heard and to have a creative input in matters affecting their interests and concerns at different scale levels, and for reducing or eliminating unequal power structures between social groups and classes (see also Friedmann and Douglas, 1998). Forester (1989) and Albrechts (1999) stress in this respect that the planners are not only instrumental, and their implicit responsibility can no longer simply be to 'be efficient', to function smoothly as neutral means to given (and presumably well-defined) ends. Planners must be more than navigators who keep their ship on course. They must be deeply involved in formulating that course (Forester, 1989) and they must become an active force in providing direction for change by using the power available to them to anticipate and to counter the efforts of interests that threaten to make a mockery of the democratic planning process by misusing their power.

Open dialogue, accountability, collaboration, and consensus building have become key concepts. An open dialogue may have advantages as new people, new alliances, new networks, and new ideas are brought together, and as new arenas (see Bryson and Crosby, 1992) are provided in which strategy articulation takes place. The institutional capacity of a place or community is built up by means of these arenas, taking into account the very unequal balances of power. Moreover, these arenas may enable processes and may provide some initial form of a social basis reflecting plurality and diversity (see Sandercock, 1998). Developments toward more direct forms of democracy and the focus on debate, public involvement, and accountability—even with the best intentions—contain the danger of making democratic public involvement more and

2

more dependent on knowledge and skills that only the more highly educated possess
(see Benveniste, 1989, page 67). These developments may contribute to the conversion
of socioeconomic inequality into political inequality. Research on public involvement in
a local referendum (Elchardus et al, 2000) illustrates that the more highly educated
were twelve times overrepresented. Therefore empowerment (see Friedmann, 1992) is
needed for ordinary citizens and deprived groups to overcome the structural elements
of unequal access to and distribution of resources, and to overcome inequalities in
social position, class, skills, status, gender, and financial means.

Emerging alternative process
Macrostructure
The alternative strategic (spatial) planning I propose moves away from the idea of
government as the mobilizer of the public sector and the provider of solutions to
problems, towards an idea of governance as the capacity to substantiate the search
for creative and territorially differentiated solutions to problems or challenges and for a
more desirable future situation through the mobilization of a plurality of actors with
different and even competing interests, goals and strategies (see also Balducci and
Fareri, 1999). This implies a degree of selectivity (means are limited) and the mutual
dependency of actors, which means that problems cannot be solved and challenges
cannot be met by just one actor. They require the prospect of a win–win situation and
the involvement of actors on an equal basis.

Delivering on the principles discussed implies the development of a different
strategic planning capacity (see figure 2) that has the ability, in an open and creative
way, to construct alternative futures and to respond to the growing complexity,
to new demands on traditional representative democracy, and to prevailing power
structures.

Strategic planning is centered on the elaboration of a mutually beneficial dialectic
between top-down structural developments and bottom-up local uniqueness. Indeed, a
mere top-down and centrally organized approach runs the danger of overshooting the
local, historically evolved, and accumulated knowledge and qualification potential,
whereas a one-dimensional emphasis on a bottom-up approach tends to deny—or at
least to underestimate—the importance of linking local traditions with structural
macrotendencies (Albrechts and Swyngedouw, 1989). Strategic planning is selective
and oriented to issues that really matter. As it is impossible to do everything that
needs to be done, 'strategic' implies that some decisions and actions are considered

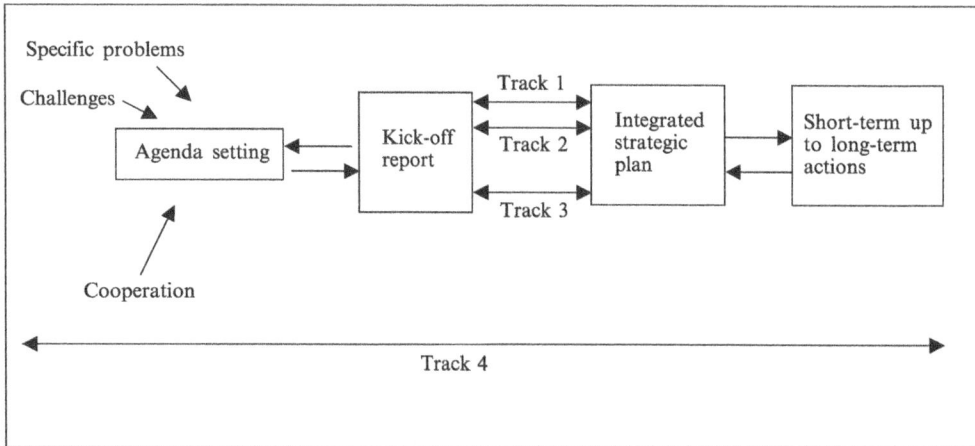

Figure 2. Possible macrostructure for the overall strategic planning process.

752   L Albrechts

more important than others and that much of the process lies in making the tough
decisions about what is most important for the purpose of producing fair, structural
responses to problems, challenges, aspirations, and diversity. Strategic planning relates
to implementation. Things must get done! This is seen as the pattern of purposes,
policy statements, plans, programs, actions (short, medium, and long term), decisions,
and resource allocation that defines what a policy is in practice, what it does, and why
it does it—from the points of view of various affected publics (Bryson and Crosby,
1992, page 296). This stresses the need to find effective connections between political
authorities and implementation actors (officers, individual citizens, community organizations, private corporations, and public departments (see Albrechts, 2003b; Hillier,
2002). Strategic spatial planning is not just a contingent response to wider forces, but
is also an active force in enabling change. This strategic planning cannot be theorized
as though its approaches and practices were neutral with respect to class, gender, age,
race, and ethnicity (Albrechts, 2002; Sandercock, 1998).

All of these factors, of course, have an impact on the role, the position, and the
skills of strategic planners. Strategic planners have on several occasions acted as
catalysts (Albrechts, 1999; Mintzberg, 1994), as counterweights, and as initiators of
change (Albrechts, 1999; Krumholz, 1982). They mobilize and build alliances. They
present real political opportunities, learning from action not only what works but
also what matters. They substantiate change and refuse to function smoothly as neutral
means to given and presumably well-defined ends.

Strategic planning does not flow smoothly from one phase to the next. It is a
dynamic and creative process. New points of view and facts that come to light today
might very well alter certain decisions made yesterday.

Four-track approach
The four-track approach is based on interrelating four types of rationality: value
rationality (the design of alternative futures), communicative rationality (involving a
growing number of actors—private and public—in the process), instrumental rationality (looking for the best way to solve the problems and achieve the desired future), and
strategic rationality (a clear and explicit strategy for dealing with power relationships)
(see Albrechts, 2003a).

The four tracks (Albrechts et al, 1999; see also Van den Broeck, 1987; 2001) can be
seen as working tracks: one for the vision, a second for the short-term and long-term
actions, a third for the involvement of the key actors, and finally a fourth track for a
more permanent process (mainly at the local level) involving the broader public in
major decisions (see figure 3). The proposed tracks may not be viewed in a purely linear
way. The context forms the setting of the planning process but also takes form and
undergoes changes in the process (see Dyrberg, 1997). In the first track the emphasis
is on the long-term vision. The vision is constructed (communicative rationality) in
relation to the social values (value rationality) to which a particular environment is
historically committed (see Ozbekhan, 1969). The creation of a vision is a conscious and
purposive action to represent values and meanings for the future. Power is at the heart
of these values and meanings (strategic rationality). To avoid pure utopian thinking, the
views of social critics such as Harvey, Friedmann, and Krumholz have been integrated
into the first track.

In track 2 the focus is on creating trust by solving problems through actions on a
very short term. It concerns acting in such a way as to make the future conform to the
vision constructed in track 1 and to tackle problems in view of this vision. In this track
the four types of rationality interrelate.

2

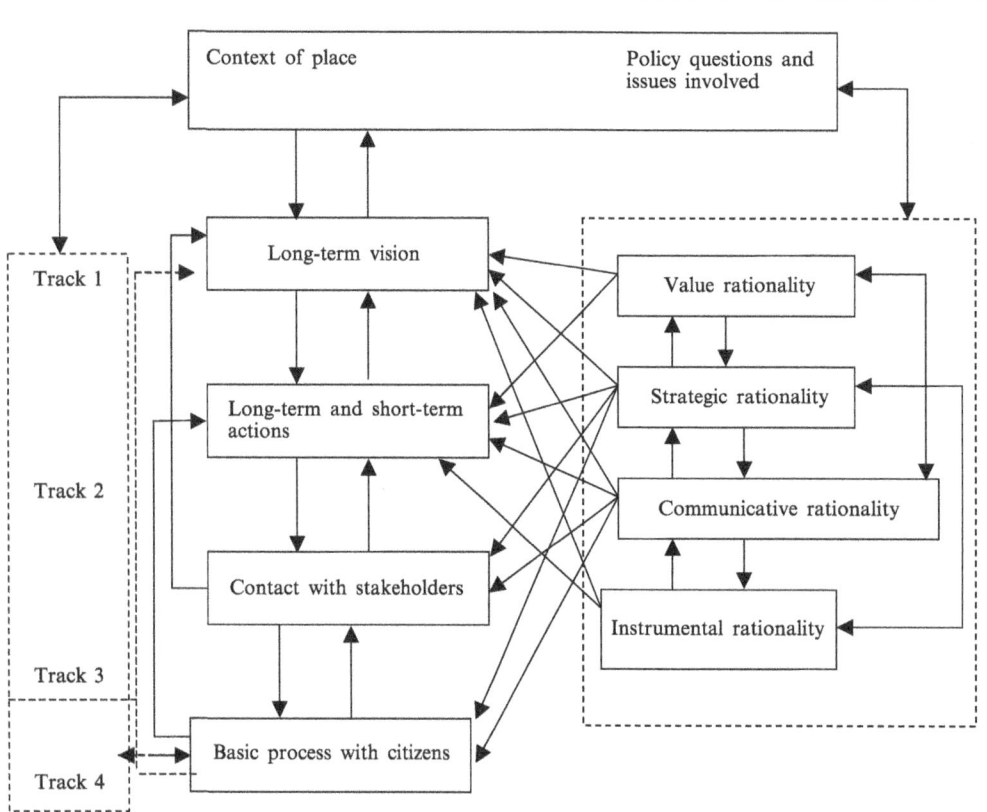

Figure 3. Four-track approach with tentative integration of different concepts of rationality.

As spatial planning has almost no potential for concretizing strategies, so track 3 involves relevant actors needed for their substantive contribution, their procedural competences, and the role they might play in acceptance, in getting basic support and in providing legitimacy. The technical skills and the power to allocate sufficient means to implement proposed actions are both usually spread over a number of diverse sectors, actors and departments. Integration in its three dimensions—substantive, organizational, and instrumental (legal, budget)—is at stake here.

The fourth track is about an inclusive and more permanent empowerment process (Forester, 1989; Friedmann, 1992) involving citizens in major decisions. In this process, citizens learn about one another and about different points of view, and they come to reflect on their own points of view. In this way a store of mutual understanding can be built up, a sort of 'social and intellectual capital' (see Innes, 1996; but see also the more critical view of Mayer, 2003).

The end product consists of an analysis of the main processes shaping our environment, a dynamic, integrated, and indicative long-term vision (frame), a plan for short-term and long-term actions, a budget, and a strategy for implementation. By introducing value rationality, this approach transcends mere contingency planning (for contingency planning see Alexander, 1988; Sager, 1994). The output of the process constitutes a consensus or (partial) disagreement between the key actors. For the implementation, credible commitments to action engagement (commitment package) are needed where planners, citizens, the private sector, and different levels of governance enter moral, administrative, and financial agreements to realize these actions (on the social contract/collective spatial agreement see also Albrechts, 1995; Granados Cabezas, 1995; Van den Broeck, 1996).

754   L Albrechts

To make formal decisionmaking and implementation more responsive to the context and to the agreements reached during the plan-making process, the four-track approach invites politicians, citizens, sector experts, and the arenas in which they meet to be active from start to finish in the entire process, including the agenda setting, the design of plans, the political ratification, and the practical implementation (see also Flyvbjerg, 2002). In this way, the arenas are used not as locations devoid of power, but rather as vehicles that acknowledge and account for the working of power and for the passionate commitment of planners and other actors who care deeply about the issues at hand (Flyvbjerg, 2002).

The proposed four-track approach cannot change the power relations, but I am confident (see also Forester, 1989; Healey, 1997b; Innes, 1994; Sager, 1994) that empowerment, as developed in track 4, supports wider, collective efforts to change such relations.

Preconditions
All of this requires a clear view of the mandates and responsibilities of all actors. Key actors who take part in the discussions bring the consensus or disagreement before the formal institutions (city council, parliament) and before sectoral departments and private institutions or agents in charge of implementation. This may stimulate a fair communication of the discussions that took place in the plan-making process, and a refined articulation of meaning, visions, and policy ideas constructed during the process (see also Bryson and Crosby, 1992).

At the start of the process, the relevant actors must become conscious of the unequal distribution of power in all stages of the planning process. This means that the strategic rationality of actors (individual, group, institutional, corporate) who calculate the feasibility of realizing their interests as opposed to the interests of their opponents must interrelate with the communicative rationality. A process solely dominated by experts and powerful actors must be avoided. This approach implies a minimal willingness to tackle the problems through interactive policy processes that allow others to have a say in their own policy domain. It also implies a willingness to accept decisions made through a network organization, including decisions that may depart from a generic policy. This is proving to be a very critical issue for traditional actors. This 'alternative' strategic planning demands a decisionmaking style in which the stakeholders become more actively involved in solving policy problems on the basis of a joint definition of the actual situation and of the sharing of interests, aims, and relevant knowledge. Active involvement, open dialogue, accountability, collaboration, and consensus building become key terms in most of the planning discourse. This implies collaboration on equal terms in all phases of the strategic planning process. 'Building trust' becomes a buzz phrase.

Epilogue
The 'alternative' strategic planning presented in this paper is conceived as a democratic, open, selective, and dynamic process. It produces a vision to frame problems, challenges, and short-term actions within a revised democratic tradition. A dissection of the process reveals the key elements that underlie this strategic planning: it involves content and process, statics and dynamics, constraints and aspiration, the cognitive and the collective, the planned and the learned, the socioeconomic and the political, the public and the private, the vision and the action, the local and the global, legitimacy and a revised democratic tradition, values and facts, selectivity and integrativity, equality and power, long term and short term.

2

I have applied the 'lenses' of a reflective practitioner and of the (strategic) planning literature in an effort to broaden the concept and provide an alternative to address the challenges of our postmodern world in a constructive and progressive way. Strategic planning case studies (Albrechts, 1999; Albrechts et al, 2003; Healey et al, 1997; Pascual and Esteve, 1997; Pugliese and Spaziante, 2003) illustrate innovative practices. Schön (1984) sees a need to inquire into the epistemology of practice, to make sense of what has been learned in action in relation to a wider context, and to test the depth and comprehensiveness of these practices. This would help efforts to evaluate and make sense of these practices in relation to a wider (theoretical) context. Abstract conceptualization and generalization of the accumulated knowledge of learning in action may help heorists to see some of what can be learned from practice. Strategic planners, on the other hand, can be inspired and guided by new emerging theories. The critical question of the leverage that these strategic spatial planning exercises will achieve over time must be raised. Do they have the persuasive power to shift territorial development trajectories or, as some argue (Kunzmann, 2001a; 2001b; 2001c), are they little more than a cosmetic veil to hide the growing disparities evolving within Europe? The European experiences (Albrechts et al, 2001; 2003; CEC, 1999; Healey et al, 1997; Pasqual and Esteve, 1997; Pugliese and Spaziante, 2003) provide a fertile laboratory for advancing the understanding of the nature and potential of strategic spatial frameworks and strategies for 21st-century conditions.

Acknowledgements. I would like to acknowledge the comments of two anonymous referees.

References
Albrechts L, 1995, "Bâtir le visage d'une région" [Constructing the image of a region] *DISP* **122** 29 – 34
Albrechts L, 1999, "Planners as catalysts and initiators of change: the new Structure Plan for Flanders", in *European Planning Studies* **7** 587 – 603
Albrechts L, 2002, "The planning community reflects on enhancing public involvement: views from academics and reflective practitioners" *Planning Theory and Practice* **3** 331 – 347
Albrechts L, 2003a, "Planning and power: towards an emancipatory planning approach" *Environment and Planning C: Government and Policy* **21** 905 – 924
Albrechts L, 2003b, "Reconstructing decision-making: planning versus politics" *Planning Theory* **2** 249 – 268
Albrechts L, Swyngedouw E, 1989, "The challenges for a regional policy under a flexible regime of accumulation", in *Regional Policy at the Crossroads* Eds L Albrechts, F Moulaert, P Roberts, E Swyngedouw (Jessica Kingsley, London) pp 67 – 89
Albrechts L, Leroy P, van den Broeck J, van Tatenhove J, Verachtert K, 1999, "Opstellen van een methodiek voor geintegreerd gebiedsgericht beleid" [A methodology for integrated area development] KU Leuven, KU Nijmegen, Leuven; copy available from author
Albrechts L, Alden J, Da Rosa Pires A (Eds), 2001 *The Changing Institutional Landscape of Planning* (Ashgate, Aldershot, Hants)
Albrechts L, Healey P, Kunzmann K, 2003, "Strategic spatial planning and regional governance in Europe" *Journal of the American Planning Association* **69** 113 – 129
Alexander E, 1988, "After rationality: towards a contingency theory for planning" *Society* **26**(1) 15 – 19
Balducci A, Fareri P, 1999, "Consensus-building, urban planning policies, and the problem of scale: examples from Italy", in *Participation and the Quality of Environmental Decision Making* Eds F Coenen, D Huitema, L O'Toole (Kluwer Academic, Dordrecht) pp 163 – 178
Barrett S, Fudge C (Eds), 1981 *Planning and Action* (Methuen, London)
Batley R, Stoker G (Eds), 1991 *Local Government in Europe* (Macmillan, London)
Benveniste G, 1989 *Mastering the Politics of Planning* (Jossey-Bass, San Francisco, CA)
Breheny M, 1991, "The renaissance of strategic planning?" *Environment and Planning B: Planning and Design* **18** 233 – 249
Bryson J M, 1995 *Strategic Planning for Public and Nonprofit Organizations* (Jossey-Bass, San Francisco, CA)

756   L Albrechts

Bryson J, Crosby B, 1992 *Leadership for the Common Good: Tackling Public Problems in a Shared-power World* (Jossey-Bass, San Francisco, CA)

Bryson J M, Roering W D, 1988, "Initiation of strategic planning by governments" *Public Administration Review* **48** 995 – 1004

Bryson J M, Roering W D, 1996, "Strategic planning options for the public sector", in *Handbook of Public Administration* Ed. J L Perry, 2nd edition (Jossey-Bass, San Francisco, CA)

CEC, 1997 *The EU Compendium of Spatial Planning Systems and Policies* Commission of the European Communities (Office for the Official Publications of the European Communities, Luxembourg)

CEC, 1999 *European Spatial Development Perspective: Towards Balanced and Sustainable Development of the Territory of the EU* Commission for the European Communities (Office for the Official Publications of the European Communities, Luxembourg)

Chapin F S, 1965 *Urban Land Use Planning* (University of Illinois Press, Urbana, IL)

Cullingworth J B, 1972 *Town and Country Planning in Britian* (Allen and Unwin, London)

Cullingworth J B, 1993 *Planning Control in Comparative Context* (Routledge, London)

Davies H, Edwards D, Hooper A, Punter J, 1989 *Planning Control in Western Europe* (HMSO, London)

De Jouvenel B, 1964 *L'art de la conjecture* [The art of conjecture] (Du Rocher, Monaco)

Dyrberg T B, 1997 *The Circular Structure of Power* (Verso, London)

Eadie D C, 1983, "Putting a powerful tool to practical use: the application of strategic planning in the public sector" *Public Administration Review* **43** 447 – 452

Elchardus M, Hooghe M, Smits W, 2000, "De vormen van middenveld particpatie" [Different forms of intermediary participation], in *Het maatschappelijk middenveld in Vlaanderen* [Civil society in Flanders] (VUB Press, Brussels) pp 15 – 46

Esping-Anderson G, 1990 *The Three Worlds of Welfare Capitalism* (Polity Press, Cambridge)

Faludi A, Korthals Altes W, 1994, "Evaluating communicative planning: a revised design for performance research" *European Planning Studies* **2** 403 – 418

Faludi A, Van der Valk A, 1994 *Rule and Order: Dutch Planning Doctrine in the Twentieth Century* (Kluwer Academic, Dordrecht)

Flyvbjerg B, 2002, "Bringing power to planning research: one researcher's praxis story" *Journal of Planning Education and Research* **21** 357 – 366

Forester J, 1989 *Planning in the Face of Power* (University of California Press, Berkeley, CA)

Foucault M, 1980 *The History of Sexuality* (Vintage, New York)

Friedmann J, 1992 *Empowerment: The Politics of Alternative Development* (Blackwell, Oxford)

Friedmann J, 2004, "Strategic planning and the longer range" *Planning Theory and Practice* **5**(1) 49 – 67

Friedmann J, Douglas M, 1998, "Editor's introduction", in *Cities for Citizens* Eds M Douglas, J Friedmann (John Wiley, Chichester, Sussex) pp 1 – 6

Friend J, Hickling H, 1987 *Planning under Pressure* (Pergamon, Oxford)

Fürst D, 2001, "Regional governance: Ein neues Paradigma der Regionalwissenschaften" [Regional governance: a new paradigm of the regional sciences] *Raumforschung und Raumordnung* number 5 – 6, 370 – 380

Granados Cabezas V, 1995, "Another methodology for local development? Selling places with packaging techniques: a view from the Spanish experience of city strategic planning" *European Planning Studies* **3** 173 – 187

Hall P, Pfeiffer U, 2000 *Urban Future 21* (Spon, London)

Hart D A, 1976 *Strategic Planning in London* (Pergamon, Oxford)

Healey P, 1997a, "An institutionalist approach to spatial planning", in *Making Strategic Spatial Plans: Innovation in Europe* Eds P Healey, A Khakee, A Motte, B Needham (UCL Press, London) pp 21 – 36

Healey P, 1997b *Collaborative Planning: Shaping Places in Fragmented Societies* (Macmillan, London)

Healey P, 2004, "The treatment of space and place in the new strategic spatial planning in Europe" *International Journal of Urban and Regional Research* **28** 45 – 67

Healey P, Khakee A, Motte A, Needham B, 1997 *Making Strategic Spatial Plans. Innovation in Europe* (UCL Press, London)

Hillier J, 2002 *Shadows of Power* (Routledge, London)

Innes J, 1994 *Planning Through Consensus-Building: A New View of the Comprehensive Planning Ideal* Institute of Urban and Regional Development, University of California, Berkeley, CA

2

Innes J, 1996, "Planning through consensus-building, a new view of the comprehensive planning ideal" *Journal of the American Planning Association* **62** 460 – 472

Kaufman J L, Jacobs H M, 1987, "A public planning perspective on strategic planning" *Journal of the American Planning Association* **53** 21 – 31

Kreukels A, 2000, "An institutional analysis of strategic spatial planning: the case of federal urban policies in Germany", in *The Revival of Strategic Spatial Planning* Eds W Salet, A Faludi (Royal Netherlands Academy of Arts and Sciences, Amsterdam) pp 53 – 65

Krumholz N, 1982, "A retrospective view of equity planning, Cleveland, 1969 – 1979" *Journal of the American Planning Association* **48** 163 – 174

Kunzmann K, 2000, "Strategic spatial development through information and communication", in *The Revival of Strategic Spatial Planning* Eds W Salet, A Faludi (Royal Netherlands Academy of Arts and Sciences, Amsterdam) pp 259 – 265

Kunzmann K, 2001a, "State planning: a German success story?" *International Planning Studies* **6** 153 – 166

Kunzmann K, 2001b, "L'aménagement du territoire en Allemagne: une discipline sans utopias?" [Spatial planning in Germany: a discipline without utopia?] *Territoires 2020* number 4, 101 – 116

Kunzmann K, 2001c, "The Ruhr in Germany: a laboratory for regional governance", in *The Changing Institutional Landscape of Planning* Eds L Albrechts, J Alden, A da Rosa Pires (Ashgate, Aldershot, Hants) pp 133 – 158

Mastop H, 1998, "National planning: new institutions for integration", paper for the XII AESOP Congress, Aveiro; copy available from the author, Nijmegen School of Management, University of Nijmegen

Mastop H, Faludi A, 1997, "Evaluation of strategic plans: the performance principle" *Environment and Planning B: Planning and Design* **24** 815 – 822

Mayer M, 2003, "The onward sweep of social capital: causes and consequences for understanding cities, communities and urban movements" *International Journal of Urban and Regional Research* **27** 110 – 132

Mintzberg H, 1994 *The Rise and Fall of Strategic Planning* (The Free Press, New York)

Mintzberg H, 2002, "Five Ps for strategy", in *The Strategy Process: Concepts, Contexts, Cases* Eds H Mintzberg, J Lampel, J B Quinn, S Goshal (Prentice-Hall, Englewood Cliffs, NJ) pp 3 – 9

Mintzberg H, Ahlstrand B, Lampel J, 1998 *Strategy Safari: A Guided Tour through the Wilds of Strategic Management* (The Free Press, New York)

Motte A, 1994, "Innovation in development plan-making in France 1967 – 1993", in *Trends in Development Plan-making in European Planning Systems* Ed. P Healey, WP 42, Department of Town and Country Planning, University of Newcastle upon Tyne, Newcastle upon Tyne, pp 90 – 103

Newman P, Thornley A, 1996 *Urban Planning in Europe* (Routledge, London)

Ozbekhan H, 1969, "Towards a general theory of planning", in *Perspective of Planning* Ed. E Jantsch (OECD, Paris) pp 45 – 155

Pascual I, Esteve J, 1997 *La estrategia de las ciudades. Planes estratégicos como instrumento: métodos, téchnicias y buenas practices* [Strategies of cities. Strategic plans as instruments: methods, techniques and good practices] (Diputación de Barcelona, Barcelona)

Poister T H, Streib G, 1999, "Strategic management in the public sector: concepts, models and processes" *Public Management and Productivity Review* **22** 308 – 325

Pugliese T, Spaziante A (Eds), 2003 *Pianificazione strategica per le città: riflessioni dale pratiche* [Strategic planning by cities: reflections from practice] (F Angeli, Milano)

Quinn J B, 1980 *Strategies for Change: Logical Incrementalism* (Down Jones – Irwin, Homewood, IL)

Sager T, 1994 *Communicative Planning Theory* (Avebury, Aldershot, Hants)

Salet W, Faludi A (Eds), 2000, "Three approaches to strategic planning", in *The Revival of Strategic Spatial Planning* (Royal Netherlands Academy of Arts and Sciences, Amsterdam) pp 1 – 10

Sandercock L, 1998 *Towards Cosmopolis. Planning for Multicultural Cities* (John Wiley, Chichester, Sussex)

Scharpf F W, Schnabel F, 1978, "Durchsetzungsprobleme der Raumordnung im öffentlichen Sektor" [Performance problems of spatial planning in the public sector] *Informationen zur Raumentwicklung* **1** 19 – 28

Schön D, 1984 *The Reflective Practitioner: How Professionals Think in Action* (Avebury, Aldershot, Hants)

Secchi B, 1986, "Una nuova forma di piano" [A new form of play] *Urbanistica* **82** 6 – 13

758   L Albrechts

Van den Broeck J, 1987, "Structuurplanning in praktijk: werken op drie sporen" [Structure planning in practice: workings on three tracks] *Ruimtelijke Planning ll* **A2C** 53 – 119

Van den Broeck J, 1996, "Pursuit of a collective urban pact between partners", paper for the 31st IsoCaRP International Congress, Sydney, September; copy available from the author, Department of Architecture, Urban and Regional Planning, Catholic University of Leuven, Leuven

Van den Broeck J, 2001, "Informal arenas and policy agreements changing institutional capacity", paper for the First World Planning School Congress, Shanghai; copy available from the author, Department of Architecture, Urban and Regional Planning, Catholic University of Leuven, Leuven

Webster, 1970 *Webster's Seventh New Collegiate Dictionary* (Merriam-Webster, Springfield, MA)

Weeks J, 1993, "Rediscovering values", in *Principled Positions. Postmodernism and the Rediscovery of Value* Ed. J Squires (Lawrence and Wishart, London) pp 189 – 211

Wilkinson D, Appelbee E, 1999 *Implementing Holistic Government* (Policy Press, Bristol)

p

Originaltext Healey 2009

Planning Theory & Practice, Vol. 10, No. 4, 439–457, December 2009

In Search of the "Strategic" in Spatial Strategy Making[1]

PATSY HEALEY

Global Urbanism Research Unit, School of Architecture, Planning and Landscape, Newcastle University, Newcastle, UK

ABSTRACT *In recent years, city governments and other entities concerned with urban futures have been exhorted to produce spatial strategies, indicating how their areas might develop in the future. But many of the resultant strategies do little "strategic work" in the sense of shaping future development trajectories. This paper reviews the meaning of "strategic work" in terms of mobilising attention to an urban area as a whole, and influencing the way that multiple actors involved in urban development shape their interventions. It emphasises the key judgements which those involved in promoting such strategic work have to make and discusses how the capacity for making such judgements may be cultivated among individuals, groups, and the wider political community involved in urban governance.*

Keywords: Spatial planning; practical judgement; strategy making; public realm; planning theory

Spatial Strategy Making in Complex Urban Contexts

During the 1990s and 2000s, there has been a wave of energetic effort in producing spatial strategies for city regions and metropolitan areas. This effort has been particularly evident in Europe,[2] but is also to be found in the USA (Wheeler, 2002), and in other parts of the world. Making such strategies, often equated with the production of some kind of document and visualisation, has been an important area of planning activity throughout the twentieth century (Ward, 2002). Advocates of strategy making stress its ability to locate specific interventions aimed at improving urban environments in a wider context, and its capacity to enhance the identity of an urban area. Their concerns include achieving greater national and international visibility for their locales; co-ordinating project initiatives; connecting interventions in different parts of an urban area, often in different jurisdictions; linking development locations to major infrastructure investments, thereby reducing spatial injustices; promoting economic opportunities; limiting threats to environmental balances; and working out which aspects of the past to conserve.

All kinds of planning instruments have evolved to give regulatory and political strength to such spatial strategies. It is often assumed that these instruments, for example, general regulatory plans (as in southern Europe and parts of Latin America), structure plans (as in north-west Europe), spatial development strategies (as promoted in the *European Spatial Development Perspective* (CSD, 1999)), schema directeur (in France), and comprehensive plans (USA), are effective carriers of such strategies. Funding bodies, such as the EU, the

Correspondence Address: Patsy Healey, Global Urbanism Research Unit, School of Architecture, Planning and Landscape, Newcastle University, Newcastle, UK. Email: Patsy.healey@ncl.ac.uk

1464-9357 Print/1470-000X On-line/09/040439-19 © 2009 Taylor & Francis
DOI: 10.1080/14649350903417191

international aid agencies, and many higher tier government funding programmes, have also encouraged the production of strategies and visions. Ideas promoted in the field of public management have been used by funders partly to control how their funds are used, but also to encourage the setting of project proposals by applicants in the context of wider programmes of intervention, so funders can assess the "value-added" of their aid contribution (Ferlie *et al.* 1996). One consequence has been that a lot of "strategies" have been produced by agencies involved in urban governance and urban development. However, only some of these actually produce significant effects other than ensuring formal compliance in order to attract funds or meet regulatory requirements. Some strategy statements may serve political purposes through a rhetorical flourish which displays the promises of a mayor or local regime. Other so-called strategies may merely record already well-established directions.

Only some of these strategies are "strategic" in the sense understood by Albrechts:

> Strategic planning is selective and oriented to issues that really matter. As it is impossible to do everything that needs to be done, "strategic" implies that some decisions and actions are considered more important than others and that much of the process lies in making the tough decisions about what is most important for the purpose of producing fair, structural responses to problems, challenges, aspirations, and diversity. (Albrechts, 2004, pp. 751–752)

Here, Albrechts is referring both to a general quality of planning work and to planning activity focused on a broad phenomenon such as a city region or larger territory. As Kitchen (2007) argues, all planning work has a strategic dimension in the sense that a wide range of considerations are "integrated" into the judgements made about any specific intervention. What distinguishes the spatial strategy-making referred to here is the concern to shape the dynamics through which larger urban regions evolve. In such situations, the intellectual challenge is to imagine the "entity" in question (what and where is our "city", see Healey, 2004a; Vigar *et al.*, 2005), its connectivities and the relation between its parts (people and groups, places and neighbourhoods) and the "whole" (the city, or urban region understood as an entity), and the relations with wider systems which flow through such an area. Such strategic initiatives also have to face the political challenge of mobilising attention to, and creating a "public" around, such an entity (Gualini, 2004; Healey, 2007).

How then, as analysts, can we distinguish efforts in spatial strategy making focused on large urban areas which are "strategic" in their intentions and in their actual and potential consequences? As practitioners, how can we tell when and how to do such "strategic" work and how can we learn how to do it? In this paper, I define "strategic work" in the planning field as both integrative and as geared to efforts to change direction, to open up new possibilities and potentials, and to move away from previous positions. Much planning work involves maintaining, managing and conserving local environments, within relatively stable parameters. Spatial strategy making, as understood by Albrechts, is explicitly transformative governance work. It is aimed at responding to changing contextual parameters and at making a contribution to those parameters, as an enduring piece of public infrastructure, in the manner of famous plans such as Patrick Abercrombie's plan for Greater London (1944) and Daniel Burnham's Chicago Plan of 1909 (Ward, 2002). It is in this sense that spatial strategies have a "structuring" dimension.

In this paper, I focus in particular on the critical judgements that those involved in such transformative strategy making face, and on the practical "art" of making such judgements (Vickers, 1965). These judgements involve technical expertise and

2

political astuteness. They combine analytical knowledge and moral considerations. They demand a broad grasp, a sensitivity to a plurality of issues and concerns, and an awareness of the detail of experience and of what it takes for a strategic idea to shape the flow of ongoing activity. Making spatial strategies that get to matter is not an easy assignment. Such strategies, although sometimes associated with a key politician (as with Mayor Maragall in Barcelona in the late twentieth century (Marshall, 2004)) or a planner (such as Abercrombie and Burnham, cited above),[3] are rarely the product of the effort of a single individual. They arise from a collective governance effort and they "work", in terms of having effects, because they become embedded in the governance culture of a locality.

This paper, then, is a review essay. It is partly an analysis of what is involved in making the judgements involved in spatial strategy making that "gets to matter". It is also an exercise in "normative abstraction", culling from my own research, reading and experience what seem to me to be critical dimensions of the "art of judgement" relevant to the activity of making spatial strategies.[4] I attempt a kind of bridge between the theorists of socio-spatial change and the work of acting to generate change, to transform place qualities and connectivities, in ways which might have the potential to encourage future benefits to emerge. Before embarking on the review itself, I expand below on the activity of spatial strategy making in the planning field.

Spatial Strategy Making as an Area of Planning Activity

Along with Albrechts (2004), I understand spatial strategies for large urban complexes as social products which emerge as an important part of the governance "infrastructure" of an area. By this I mean that they do "work" by helping to frame and focus the way people involved in urban development processes and projects come to think and act. By "governance", I mean the array of activities and agencies involved in collective action in an urban area (Cars *et al.*, 2002; Le Galès, 2002).[5] Spatial strategies get to "work" by providing an orientation, or reference frame, which gets shared by many stakeholders in urban development processes.[6] This orientation may be expressed in general principles or in spatial images such as maps and diagrams. Spatial strategies which get to have transformative effects accumulate the power to frame discourses and shape action through their resonance with issues and problems which are causing concern within a political community, or "polity",[7] and through the persuasive power of their core arguments and metaphors. If sufficient power is accumulated to give momentum to these strategic orientations, then the framing ideas may travel across significant institutional sites of urban and regional governance, to enrol others with the power to invent, invest and regulate subsequent development. In this way, strategic orientations may come to endure through time and have consequences in shaping future qualities and potentialities.[8]

Yet spatial strategy making is no simple activity which can be managed by procedural formulae. It is "a messy, back-and-forth process, with multiple layers of contestation and struggle" (Healey, 2007, p. 182). This "back and forth" process provides a good illustration of what policy analyst John Kingdon (2003) means when he refers to streams of policy "problems" and "solutions" jostling together, with clear "policies" emerging when a solution and a problem become attached. In the diffused governance landscapes which have emerged in western Europe in recent years, and which are common in metropolitan areas where urban relations stretch across administrative jurisdictions in complex ways,[9] spatial strategy making may involve something more than selecting what are the "critical" problems and identifying good ways of addressing them. It may also involve searching out and creating a political community which identifies with the urban complex in question.[10]

442   *P. Healey*

Strategies that make a difference, then, inherently transform, though such changes may emerge slowly over time and in unpredictable ways.[11] Strategies encourage a momentum towards some directions rather than others. In the public sphere, they are thus political acts, challenging established power dynamics and mobilising energy to move in different directions, although there are all kinds of ways in which such strategies are grounded in formal political jurisdictions. Because such strategies are social products formed to do governance work, they raise difficult political questions about their legitimacy and accountability. For those with expertise in their field, the associated technical knowledge and values draw them into the political struggle over ideas and directions. These values are not merely about making sure that some consideration is given to critical connections and relations in and beyond an urban area, that is to a systematic consideration of connectivities and the relations between "parts" and "wholes" (Churchman, 1979). Rather, planning expertise is also associated with a concern for social and environmental well being, or "flourishing" (Friedmann, 2002, 2008). Where experts and others committed to this project become involved in spatial strategy making, they will be faced not only with the pressure to keep an open, "comprehensive" orientation to complex interrelations which co-exist in the urban area in question (Hoch, 2007). They will also have a specific moral position about why certain strategic directions should be encouraged and others closed off. They become advocates of particular kinds of opportunity for future generations. They may add to the "stream" of problems and solutions in search of one another, and to what I have earlier called the "argumentative jumble" which has to be sorted through by those wishing to affect the type of spatial strategy making that has the potential to make a difference (Healey, 2006). Spatial strategy-making efforts thus have the potential to make a difference beyond framing the way many parties think about the future development of an urban complex. Such efforts also contribute to enlarging public debate about urban futures.

What, then, does "doing" such transformative strategic work involve? Drawing on my earlier work on the nature of strategic spatial planning (Healey, 2007), I emphasise four interacting dimensions (see Figure 1). The first centres on whether, why and how to mobilise attention to an urban complex, understood as a whole in some way. The second emphasises what is at stake, for whom and where, in the socio-spatial complex of an urban area, and how the power to act in relation to these stakes is or could be distributed. The third concerns the mobilisation and enrichment of the knowledge resources available to strategy-making work, and the fourth centres on how strategic ideas are created, as key framing concepts and projects with the power to shape future directions and actions.

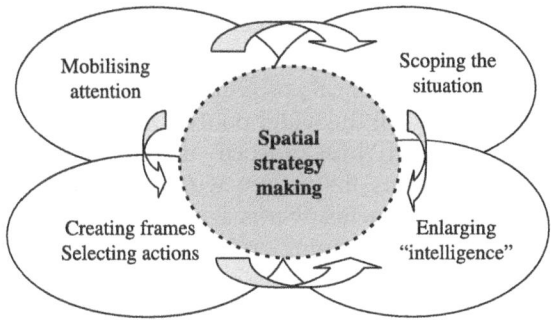

Figure 1. Dimensions of spatial strategy making.

2

The Momentum for Spatial Strategy-Making Initiatives

Despite the elision in popular perceptions between planning work and the widespread presence of city plans, transformative spatial strategies in the sense outlined here are not necessarily appropriate for every urban area. The making of a city plan may merely consolidate well-established ideas and practices as a basis for exercising formal legal powers, such as those enabling site assembly and land-use regulation. Any wider strategy may be left as implicit, or carried in the flow of decisions about small-scale changes, or emerge from the way in which larger place transformation projects are designed and developed. Explicit strategy-making work that is more than procedural compliance and political rhetoric emerges in place development and management work when some parties become frustrated by established ideas, regulatory practices and the design and decision making around major projects. For example, rising concerns about noise and air pollution around large airports may force changes in spatial strategies. In the Amsterdam area, locations designated for development near Schiphol international airport had to be re-designated as open areas when regulations governing noise pollution were tightened (Healey, 2007). Or some particular interests may become frustrated about the lack of attention to the significance of a major emerging challenge, such as addressing the multiple implications of climate change prognoses or re-configuring transport infrastructures. Salet and Thornley (2007) argue that the momentum for spatial strategy making in metropolitan areas has been driven in western European cities especially by urban and regional transport agencies. Alternatively, they may result from a sense of fragmented and overlapping initiatives, which, instead of producing creative synergies, undermine each other in a destructive manner.[12]

A key area of judgment relates, then, to an assessment of the institutional moment, or "opportunity structure" (Tarrow, 1994), for spatial strategy making. This leads spatial strategy makers to consider such questions as: what is the momentum for an explicit spatial strategy-making initiative? What forces and actors are driving this? What is the scope for the transformation of discourses and practices through such an initiative? How strong is the momentum? Can it be strengthened and what might weaken it? What kind of process is already underway, what might evolve and what could be created? What seems to be at stake and around which issues will critical judgements have to be made? How are the initiators situated in relation to this momentum, and how am "I" as an actor in such a process situated, in terms of role, skills, potential to exert influence and legitimacy?

On the face of it, the first judgemental step in undertaking strategy making would seem to be defining the "opportunity structure" for such an activity, and assessing whether it is likely to be "worth the effort". In the days when spatial strategy making was conceived as a linear process, with each step following consequentially on the one before, much of the planning literature assumed not just that spatial strategy-making initiatives started with such a step, but that this step was taken by those in formal political positions, charged with the responsibility for defining public values and concerns for the wider polity (Davidoff & Reiner, 1962). "Pattern books" for decision processes and "best practice" manuals also seem to offer ways of reducing the scope for judgements that actors within strategic planning processes have to make. But the experience of practice warns against this, and intellectual resources in theory and philosophy suggest that leaning on such guides may distort the capacity for judging what is going on and what is at stake in a specific context.[13]

Experientially, those involved in strategy making in the public domain know that the "opportunity structure" is not a static phenomenon, but a complex of relations that shifts and changes through time. Nor is it something that can be easily known; instead,

444   *P. Healey*

knowledge about it develops through experience. An opportunity structure is not a clearly delineated and contained "action space". It grows and expands, or contracts and fades away, in part through the momentum developed as the practice itself gets underway, as transformative energy enlarges the initial "cracks" for change to allow a more substantial momentum to develop and new trajectories to be cultivated. Doubts and dilemmas are in continual play, as strategy-making practices are challenged and critiqued by those immediately involved, those potentially affected, and the wider polity. As a spatial strategy evolves, key actors in such practices may therefore find themselves "called to judgement" about whether and how to proceed at all kinds of times and institutional places. They learn from coming to understand the specificities and potentialities of the context.

Theory too has caught up with this dynamic, continually evolving quality of governance practices (Gualini, 2004; Briggs, 2008). In traditional public administration it was assumed that governance activity occurred, or should occur, according to the laws and procedures laid down by legislatures. It was expected to take place in the formal arenas specified for government activity, and to be conducted by personnel given formal positions in these procedures and arenas. Judgements were anticipated via formal processes—the council chamber vote, the appeal to law and the arbitrating judge. But judicial "balancing" in relation to formal law has difficulty when it comes to doing the integrative intellectual work which is involved in spatial strategy making. In the 1960s an alternative, supposedly more "scientific" process was proposed. This encouraged strategy makers to follow a sequential, methodological process in such "analysis", mirroring what was seen to be a "scientific method" of arriving at a strategy (Hillier & Healey, 2008). These days, a more sociological view prevails, which stresses the power of "agency" in interaction with broader forces (see Hajer & Wagenaar, 2003; Healey, 2006). This examines the way informal relations and networks create the practices through which the formal procedures are enacted. And a more political view underlines the power struggles which are played out in these practices, as actors struggle with each other, and as dynamic struggles develop between the constraints of wider forces as experienced in the specifics of a situation, and the resisting and inventive power of specific constellations of actors to challenge and change these forces (Flyvbjerg, 1998; Healey, 2007). These developments in theorising planning and policy-making activity are thus positioned in a large field of recent social scientific discussion which has been searching for productive ways of understanding the complex interactions of structuring dynamics and agency effort.[14] These "in-between" contributions underline the importance of grasping the unique situatedness of particular instances of practice, rather than reading instances through the generalised lenses of procedural rules, best practice manuals, or of general structural forces. Methodologically, such a perspective underlines the value of the "thick" descriptions of "longitudinal" case studies.[15]

But none of this takes away the intensely political and ethical challenge of actually making judgements about spatial strategies for urban areas in the public domain. However broad the search for knowledge to ground judgements, and however shrewd and sensitive the understanding of institutional dynamics, judgements are always made in the context of limited knowledge and understanding, and of moral hazard. Though those involved may carefully aim to put prior conceptions and biases to one side, their judgements are nonetheless always shaped by implicit assumptions that somehow did not surface or were pushed aside. However hard those involved try to keep in play a rich and multi-dimensional conception of urban dynamics when shaping a strategy, their assumptions and their framing of what is at stake and what is strategic are inherently selective simplifications. If a strategy has enduring power, it cannot avoid "reductive effects" on future ways of understanding and acting. Spatial strategy making involves exercising the power to select and simplify (Albrechts, 2004; Healey, 2007). It involves taking risks, the consequences of which can be thought about, but cannot be known (Hillier, 2008). It is this frightening prospect which so often pulls actors back to easier short-term compromises, or encourages them to hand over the judgements to someone else—another tier of government perhaps, or to the courts, and sometimes to citizens in referenda and other forms of consultation. Those who argue in a specific situation for undertaking an effort in transformative spatial strategy making need to be prepared and able to justify such a step, in terms of both why "going on as usual" will not suffice, and where momentum already exists which could be harnessed towards a new strategic orientation. They need to be able to clarify how the urban area in general could benefit, and how the "parts"—people, places and connectivities—may be affected.

Situating and Scoping

Those who take a central role in deliberate initiatives in spatial strategy making are sometimes portrayed as "authors" creating a "script" for others to enact, or as "artists" creating an idea of the future to inspire others (Beauregard, 1998). This suggests that they are somehow different from their readers and viewers, that they are producers of a strategy, not consumers. This presents such strategy makers as located in a different place from those who are to be influenced, maybe as a small and highly centralised decision-making elite governing the rest of us. Or it assumes that strategy makers can be situated as outside experts, looking in on the challenges the rest of us face as we create our urban futures. Positioning strategy makers in this way makes little sense in situations where governance activity is complexly divided up into sectors and levels of attention, diffused among diverse arenas and relations between state, economy and civil society, and continually contested. These days, those involved in spatial strategy making are more usually located within the complexities they are seeking to shape, "inside" the politics they are seeking to influence (Healey, 2007). As a former planning director has noted, "planning is a collective art...it is an evolving set of dreams, not one person's dreams".[16] Viewed in this way, an initiative in spatial strategy making involves some kind of collective effort to mobilise attention to an urban area as a whole. Such initiatives are likely to be responding to matters which have created some public concern, and the work involved in them may itself reshape and generate the agendas of a polity.

Developing an idea of the "opportunity structure" for explicit spatial strategy making as discussed above is a necessary step in situating such an effort. But much more is needed to situate and scope practice adequately. One aspect of this relates to the identity of the "strategy makers". Formally, they could be the leaders and advisers of a new political regime, as was the case in the 1970s in Portland, Oregon (Abbott, 2001) or Barcelona (Marshall, 2004). They could be consultants, hired by municipalities or some other multi-agency partnership to prepare a strategy, as has become common in the UK in recent years. They could be the strategic sections of a municipal planning department or a local elite business group.[17] Whatever their starting position, the assumptions made about their institutional identity by spatial strategy makers may have a critical impact on how an initiative unfolds. Who are the "we", as initiators? What is "our" formal legitimacy and what other kinds of legitimacy may buttress our efforts? How are "we" related to others who are promoting ideas about urban futures, or deploying resources which result in place development? What opportunities are available to us to influence events and how

446   *P. Healey*

could we get nearer to other important levers in the process? It is often only during the process of strategy making that those engaged in spatial strategy making become aware of the nature and potentialities of their identity. In Amsterdam, for example, the city council planners involved in metropolitan initiatives realised that they had to change the domineering image they had acquired if they were to achieve a productive collaboration with other municipalities. In the Cambridge area of the UK, the local elite lobby group involved in promoting a spatial strategy initiative had to recognise that they needed to connect to the local planning machinery in order to have influence over the power to regulate land use and development (Healey, 2007). The impact of the assumptions that people make at the start of a process affects how subsequent activities are framed. But it also influences how the initiators present themselves to others, and how this may affect the responses their work generates. A key area, then, where judgement needs to be exercised is the way in which those involved at the centre of a strategy-making initiative present their identities, both to themselves and to others, and how they keep this under review.

A second aspect relates to the position of strategy makers in a governance context. Where are they situated? Within which landscapes of power dynamics? Within which debates and arguments? Answering these questions involves making some kind of institutional audit of the agency field. Which are the key arenas where strategic ideas about place development are already evolving? How are they related to each other, and to the levers that link to key powers, such as investment resources of various kinds and the power to regulate land use and development? Such assessments might lead to judgements about who should be involved in early-stage discussions, and who should be sought out later; about which networks to work with, and when; and about the ways of establishing new networks to create more mobilisation power around a strategic initiative. The judgements made here help to build connections between ideas about critical interventions, the impacts they may have, and the actions needed to realise them.

A third aspect relates to the power of any kind of governance effort to shape ongoing place dynamics. This emphasises the importance for spatial strategy making of a rich, multi-dimensional understanding of urban development dynamics and trajectories. This means much more than just developing an evidence base, to use a term currently made fashionable in public policy in the UK (Davoudi, 2006). It involves having some understanding of the material and cultural history of an urban area, of how this shapes conceptions of what trajectories are desirable and possible, and frames what is seen to be at stake or at issue. Such understanding feeds judgements about the extent to which a strategy aims to have resonance with, and realise the potential of, emergent trajectories which can be identified as having an endogenous dynamic (locally based political energy). It also allows planners to estimate how far a particular strategy may have an intention to roll out another trajectory, derived exogenously in, around and over pre-existing cultures and practices (externally based political energy). Both possibilities involve complex assessments of the relative power of a particular strategy-making initiative in relation to ongoing urban dynamics, of the power of collective action to shape these and of the institutional dynamics of that power.

Attention to these aspects feeds judgements about what is at stake behind a particular spatial strategy-making initiative. What is at issue? For whom is it an issue? What are current conflicts and struggles about? Who are the critical "stakeholders" active now? Who else could be persuaded to become an active stakeholder? What are the critical arenas and responsibilities in the institutional terrain in which the initiative exists? How far away is the initiative from these, and what could be done (and where and when) to reduce the distances which limit influence? What might need to be done to mobilise more attention to

2

an initiative and to build new networks and arenas to refocus governance attention more emphatically on urban area development agendas?

Initiatives in spatial strategy making are thus "inside politics". They also involve the exercise of the craft of "practical politics", shaping agendas and extending influence, enhancing attention to some issues and thereby pulling attention away from others (Albrechts, 2001). All those involved in spatial strategy making that aims to shape urban development futures are therefore placed in morally complex situations. Playing the game of practical politics according to rules which focus merely on extending power and influence is hardly a sufficient basis for the craft of such a practice (Albrechts, 2003). Instead, spatial strategy making demands a good awareness of how collective action operates as a dimension of urban development dynamics. It needs a holistic or comprehensive consciousness that can focus on the way relations need to be integrated into a strategy. It also demands an expansive moral sensibility, capable of relating specific questions about who and what might gain and lose in such a process to wider issues of social and environmental well being.

It is here that all kinds of resources can help to feed the imagination of strategy makers. One is, of course, the social science literature on urban and regional dynamics, with its interweaving of research about particular phenomena and its debates about how to conceptualise the dimensions of place dynamics and how to identify particular cause-effect relations. This body of theory and research offers not foundations which can be leaned on as an alternative to first-hand thinking about a specific situation, but inspirations for enlarging the imagination which strategy makers bring to their struggle to grasp what is going on and what is at stake in a particular situation. But there are other resources too. Behind seeking out good practice examples from elsewhere, something many practitioners do, is a deeper question about "how do we go about the difficult task we face here and now?" When considering an example, practitioners want to know "how did they do it there?" Experiential accounts (autobiographies) of planners (and others) about how they went about their work are particularly valuable as aids to learning the "craft" of spatial strategy making (Forester, 1999; Kitchen, 1997; Krumholz & Forester, 1990). These accounts convey the difficulties of making situated judgements in ways that highlight how much is unknown and "at risk" in spatial strategy-making work. Such accounts emphasise the learning process through which any strategy making initiative will develop, drawing knowledge from failures as well as successes. But this consolidated resource—"book knowledge" as we might term it—is not the only source which can or should feed strategic imagination.

Enlarging Perspectives and Knowledgeability

In the past, it was often thought that the work of spatial strategy making required planning expertise in its most scientific form. The "experts", in this conception, were those skilled in analysis and systematic research, well-informed by the scientific literature on urban and regional analysis. Yet, in the conception of spatial strategy making portrayed here, the "art" becomes a practice that does not rely too much on one particular field of systematised knowledge or area of expertise. Urban development dynamics are too complex to be grasped through work in any single discipline. Knowledge from urban and regional science, engineering, the physical and natural sciences, architecture and the humanities can provide inspirational insight. Yet many examples of the dilemmas of strategic planning in practice highlight difficulties which arise as these different fields of expertise, with their particular intellectual and practice cultures, encounter each other.

448   *P. Healey*

But beyond the difficulties of working across disciplinary and professional boundaries, it is important to recognise the limitations of such systematised knowledge in grasping what is going on in the particularities of a place. It is here that we find a critical resource in the experiential knowledge of those who live, work, enjoy themselves, suffer or invest in a place, those who manage it, argue about it, and get involved in collective action on its behalf; all activities that underline the particularities and specificities of a place, its qualities and meanings, and its dynamics. Critical judgements have to be made in spatial strategy-making initiatives about how to access such expertise, and how to weave this together with systematised knowledge, which itself is related in complex ways to experiential understandings (Healey, 2007). In doing so, it may not help to systematise experiences, for example through surveys about opinions and issues, as this may well smooth away the experiential depth out of which clues about what is going on and what is important may emerge. As in all planning work, spatial strategy makers need to get out and about in the streets and meeting places of their urban areas. There is no substitute for becoming "street-wise" as a way of accessing experiential knowledge, or for recognising that any place has many different streets and spaces of encounter. As Forester (1989) emphasises, the skills of noticing, listening and learning are critical to accessing this kind of experiential learning (Kitchen, 2007).

Once the value of recognising multiple sources of knowledge about places and their development is appreciated, the importance of maintaining an "open-minded" stance towards what is going on and what is at stake becomes obvious. Single-minded, one-issue agendas become difficult to justify. Instead, spatial strategy makers learn to anticipate a plurality of often overlapping and clashing agendas.[18] In recognition of the complexity and diversity of concerns about urban area development that may be asserted (or that could be held if people were aware of the trajectories that might evolve from the process), open-mindedness means actively stretching out to access multiple perspectives, challenging established assumptions and cultivating debate and argument among different viewpoints (Amin *et al.*, 2000).

There are all kinds of ways of encouraging this capacity (Albrechts, 2004, 2005). Above all, it demands an expansive open-mindedness, a capacious and flexible imagination, able to recognise multiple connections between people and places through time. As Hoch suggests, it is this quality of reaching for a holistic grasp of the situation that informs the old idea of a "comprehensive" planning approach. Reality, in this model, cannot be comprehensively trapped, mapped and managed in a single model of an urban "system" (Hoch, 2007). Instead, an open-minded approach involves searching out all kinds of "systemic" qualities, that is, locating the connections through which cause-effect relations may flow to create impacts. It helps to see such systems as open and meandering threads, with evolving patterns of flow shaped by various driving forces, creating diverse patterns of nodes and networks, with fuzzy and indistinct margins rather than clear boundaries containing tightly integrated and bounded sets of relations (Churchman, 1979). Exploring such dynamics to find out what is at stake and what action possibilities exist to affect the outcome involves explorative probing as much as precisely structured inquiry, and imaginative learning as much as the production of "sound evidence".[19] Approached in this way, spatial strategy making creates a "community of inquiry" around itself, which draws on and in turn helps to cultivate the overall "intelligence" of a polity (Dewey, 1991).

Many judgements are involved in developing such a "community of inquiry". What strands should be followed up first? How should the field of search be widened? How can areas of inquiry be related to each other? When and how can debate and contestation be encouraged? How can we prevent less visible strands being lost while others are

2

vigorously debated? An expansive, pluralistic, holistic imagination becomes a very valuable resource in making such judgements, especially if this develops within the community of inquiry. But perhaps the most critical judgement of all in relation to knowledgeability lies in making decisions about when to stop enlarging and expanding understanding, in order to get what is at stake back into focus. There can be no hard and fast rules about how to make such a judgement, no indisputable "reality checks" which can provide some external "stopping point" for probing inquiry. Instead, those centrally involved at this stage have to come to a judgement that their understanding of the context, of what is settled and what remains in doubt and contested, of what is at issue, is "enough for now". They must decide what, in the field of doubt and uncertainty, can now be put to rest, and determine in which direction the compass, which had seemed to be swinging wildly, now seems to point.[20] Having exercised an expansive, holistic imagination and keeping this continually in play, those making critical judgements in the public domain may find it useful to ask: in the circumstances we are in, with the resources and capacities we have, could we do more to expand our knowledge and, if we did, would this make a significant difference?[21] This centres the quality of such strategic judgement not so much in the amount of knowledge acquired as in the range and depth of the imaginative, expansive probing which occurs. This could lead to better judgements about whether it would make much difference to "know more", and about the effort that would be needed to acquire such knowledge.

Framing Selectively

As Albrechts argues, any strategy involves a selective focus. It offers a way through the morass of issues, ideas, claims and arguments to identify one or more concepts, images and/or principles which are both meaningful and give direction. Such strategies may be arrived at by systematic search procedures, such as the construction and probing of scenarios (Albrechts, 2004, 2005). Alternatively, they may be the result of active campaigning by groups promoting particular interpretative frames, or they can be reached via an imaginative leap, or through the creative processes of collaborative encounter. However they appear, strategies come into existence through an act of recognition in which they are "summoned up", "seen", "named" and "framed" (Healey, 2007, pp. 188-189). They are "willed" into being. They make sense because they show how specific problems, issues and concerns can be addressed more effectively when located in a wider frame.

A spatial strategy for an urban area, in this view, combines a holistic sensibility about urban development dynamics with a politically-shrewd awareness about what is critically at issue and what could be done. It involves judgements about how far it is possible to shape these dynamics in a transformative way, and highlights which specific actions (taken now, or in a future where the parameters of action possibilities can be foreseen with some confidence) have the potential to make a real difference to what is at issue. It involves judgements about what to position "frontstage" when it comes to action, and what to leave "backstage" for the time being (Giddens, 1984; Healey, 2007). The heart of a strategy, therefore, lies in the way that it frames ideas.

Creating the frame and identifying critical actions are both acts of intense simplification and selectivity. They are thus acutely political, since simplification inherently involves reductive discrimination (see above). However hard a strategy-making initiative attempts to keep multiple dimensions and perspectives "in play", some inevitably become the dominant focus as strategy making proceeds. This focus, in turn, evolves into the framework through which other understandings are attached and interpreted. Framing helps to reduce

450   *P. Healey*

uncertainties, and to allow people to see and justify action possibilities. Because strategies simplify how urban development dynamics are understood and how possible future directions are identified, those who initiate spatial strategy-making efforts carry a burden of moral responsibility.[22] They need to make complex and challenging judgements that go beyond whether it is possible to inject a strategic orientation into a particular governance landscape: they must also keep in view the potential consequences of consolidating governance activity around a strategic direction at a particular moment in time.

However, as stressed earlier, spatial strategy-making processes rarely proceed in a linear way. Creating a strategic framing concept is a process which evolves interactively with the work of situating and scoping, and of developing knowledgeability. Sometimes a clear strategic frame dominates a strategy-making process from the start. Sometimes it only appears after action possibilities have emerged from scoping and probing work, allowing it to be seen as a background to the identification of specific projects and proposals. Alternatively, it may come into being during an imaginative leap, as a new possibility comes into view. However it is discovered, framing, as Schon and Rein (1994) insist, is an integrative, creative process. Strategic frames cannot be arrived at by an analytic procedure, by aggregating up from "parts to wholes", as some imagined during the strategic planning movement of the 1960s (Mintzberg, 1994). Nevertheless, strategy making involves thinking systemically, by which I mean that it focuses on systems of relationships, such as those connecting where we live and work, or those surrounding the dynamic relations of housing markets, or those connecting how we live and our environmental impacts.

However, an over-reliance on previously developed models of urban systems leads to reductive simplification, which may bear little relationship to, and have little resonance with, the complex dynamics and specificities of what is happening in a particular place and time (Churchman, 1979; Verma, 1998). Planning literature has long recognised this, yet remains full of proposals for idealised urban forms or models of urban systems, all of which appear to offer a ready-made, integrated image of a whole system.[23] The alternative to this is to develop strategic frames which embody a sensitivity to the complexity, plurality and indeterminacy of particular urban development dynamics as they emerge, and which generate sufficient energy to inspire and direct transformative actions within those dynamics with the aim of shaping what happens in a place. This inspiring and directing quality lies as much in the frame as in the selective specification of actions. The frame allows those involved in shaping their own actions to position their activities in a wider context, and to develop a way of thinking and valuing which inspires and justifies what they do. The strategic frame, if widely referred to, then becomes a piece of public infrastructure, part of the capacity of a particular political community.

It is sometimes argued that strategic frames that have significant effects can only be created through a process that allows a single designer or design group to create a synthesis from their efforts in scoping and probing. This may take the form of an image, or a spatial sketch, or a verbal statement about direction and qualities. Behind this approach lies the idea that strategic design can only be an individual process. Yet studies of strategy making suggest that creating frames is as much a social as an individual process (Vickers, 1965). It involves making syntheses, and demands systemic thinking rather than analytic thinking. This can happen in collective contexts as well as in individual brains. Some fear that such collective processes can only arrive at lowest common denominator compromises. This fear assumes that participants engage in a zero-sum game, playing out their conflicts of interest. It ignores the creative social learning that can occur in processes where probing inquiry and exploratory framing occur.

Strategic frames do not appear fully formed. Frames arise through processes of collective "sense making" (Forester, 1989), as those involved seek out some kind of coherence among the issues and understandings as they become aware of them. Frames may already be lying around in an early state of conception, or in debates surrounding communities of inquiry, waiting to be summoned up and receive explicit attention.[24] In situations where understandings have been very diffuse and fragmented, such meaning making often gets explored not just through defining alternative designs and options, but by the use of metaphor and analogy ("What is our situation similar to?" "What if we think like this?", "Let us suppose this is the case?") (Verma, 1998; Weston, 1996). Framing work, like the process of probing the available knowledge about situations and issues, requires an expansive yet integrative, pluralistic yet synthetic, collective imagination. Pursued in this way, efforts at probing and developing strategic frames have the effect of making participants transform the way they think about their particular interests and trajectories (Innes & Booher, 1999, 2003).[25] Strategic frames which have transformative power achieve this potential in part by changing how key actors think.[26]

Because frames that transform are so powerful, making strategic judgements about them is perhaps the most ethically challenging task in spatial strategy making. Dominant framing ideas have a tendency to emerge as a process proceeds, though this can create problems: sometimes one assertive voice may impose a frame which seems superficially adequate to all involved in the process, yet lacks adequate probing, with the result that issues that later turn out to be highly significant get relegated to the "back stage" of attention. It is for this reason that deciding whether to employ a strategic frame becomes such a key judgement. In some situations it may be enough to focus on a few key actions, since it may be too difficult (intellectually or politically) to provide an explicit justification for them even though the compass seems to point quite clearly in their direction. It is only later on that the strategic frame which lies behind the actions can perhaps be articulated. Alternatively, a key strategic action may unsettle established frames, creating doubt, and thus allowing new ideas and actions to emerge. In these cases it may not be helpful to spend much time producing an inspiring vision. But in other cases, a clear strategic frame may be exactly what is needed to help people move on and to reduce the incapacity which often accompanies doubts and uncertainties. In these scenarios, it matters a great deal that a frame is produced. The precise moment of the strategy-making process when it is summoned up is also significant, as is the issue of its contents and dimensions.

Arriving at a strategic frame, which many see as the key product of a spatial strategy-making process, is thus an intellectually challenging, politically risky and morally demanding task. Yet participants and those who have set up the process are likely to expect some kind of vision statement or an outline of an overall strategy, especially where the strategy-making process was designed to resolve doubts and allow people to find a way forward into the future. Those charged with the task of articulating the "strategy" may find it helpful to go back to the scoping issues, and ask again what specific governance work a particular strategic frame will perform, in the context in which it now sits. Will it inspire, justify or mobilise people? Will it focus significant attention on the problems, and thus have an effect on both subsequent governance action and urban and regional dynamics? And are these effects all desirable? Further, what is the implication of our thinking about these questions for the way a strategy is expressed, both in terms of its framing conception and its specific actions? What are the relative merits of a visual sketch, a policy statement, a policy argument, a narrative, an artistic image, an imaginative story, a geographical map or a computer game in conveying the meaning and implications of the frame and the selected actions proposed?

452   *P. Healey*

Above all, articulating a strategic frame for the spatial development of an urban area demands a synthetic imagination, which recognises that wholes and parts do not relate to each other mechanically but are instead design ideas, with each inspiring the other through relational resonance rather than specified links. Strategy making demands both the ability to see a whole reflected in a part, and the capacity to recognise in the whole the possibility of parts (Churchman, 1979). Because such syntheses are contestable simplifications that are nonetheless intended to do actual governance work, strategy making in the public domain demands a moral sensibility as well as an imaginative capacity. It requires an ethical awareness not just of what is promoted but of what is left out, what is pushed to the background, as the selected frame and actions are forced into the limelight. In other words, spatial strategy making demands the courage to step forward, combined with the ability to recognise the provisionality and revisability of both framing conceptions and actions, and of the underlying assumptions on which these are based. Some individuals may have a cast of mind which tends towards this combination of synthetic imagination and ethical reflexivity, and such people often come to be acknowledged as great planners. But these qualities can also be encouraged in a polity. Spatial strategy-making work thus both benefits from, and helps to cultivate, a political culture which can relate specific issues and interventions to wider qualities of the collective experience of living in urban areas.[27]

Conclusions

All planning work involves some ability to set specific actions in wider contexts. This, in turn, involves making explicit connections between current action options and their potential impacts on people and places in the future, in this place and elsewhere (Kitchen, 2007; Hoch, 2007). Planners need to maintain a broad awareness of the multiple dimensions of such wider relations, as these impact on the specifics of the issues at hand. Following Hoch, I have termed this a holistic or comprehensive sensibility, a faculty capable of grasping the broader context of a problem whilst selecting specific aspects and actions to guide current action. If we accept the justification for planning as an activity that seeks to improve social and environmental wellbeing in a world of diverse peoples who interconnect in complex ways with place, then such a consciousness needs to be infused with the ability to sustain open-minded and pluralistic views, while recognising the moral imperatives to pursue actions that promote just and sustainable outcomes.

In this paper, I have tried to show that engaging in spatial strategy making for urban areas is a particularly demanding field of planning work. The challenge, though, is not one of analytical capability, as theorists of the mid twentieth century argued. Nor is it one of design imagination, as the town and landscape designers who dominated planning in the early twentieth century suggested. Though both of these capabilities are useful, at the heart of skilful strategy making of the kind that has the potential to enhance the future possibilities of an urban area, are four skills. Firstly, spatial strategy making requires a capacity to know a place in all its complexities, in its collective expressions and relations, and in the fine grain of its social, environmental, political and physical fabric. Secondly, it needs an imaginative capacity to see opportunities now which can provide the momentum for future development and safeguard against the many challenges to well-being that emerge in contemporary urban life. Thirdly, it necessitates the intellectual and political courage to engage in synthetic thinking, drawing together understandings and insights to imagine future trajectories and select specific pathways. Above all, though, spatial strategy making demands a capacity for judgement which is situated within, and sensitive to, the contingencies of particular times and places, rather than drawing on generalised theories of urban change or accepted methodological protocols.[28]

Spatial strategy-making efforts which have transformative and structuring effects on future directions achieve this by becoming a part of the institutional infrastructure of a community. They act through their role as a reference frame for those with the resources to deploy, and over time they come to shape urban futures. However, they are neither a blueprint for building the future, nor a generalised dream about a utopia to come. Rather, they reflect a process of trying to tie together elements of the building and dreaming that constitute part of the ongoing life of a polity. In Table 1, I attempt to distinguish such transformative strategy-making efforts from the kind of routine "strategy production", which merely responds to some external demand for an urban plan, without any connection to local momentum which could turn such an exercise into a transformative frame to shape future directions. However, there is no sharp divide between the two: routine strategy production may merely follow some procedural rule book or best practice exemplar to meet compliance requirements, but it can have the modest ambition of articulating what the policy actually does, reflecting the overall direction that emerges from the individual actions of those on the ground.[29] Such an exercise may help to decide whether or not there are issues that need more fundamental attention, and, if so, may act as a starting point for a more transformative effort in spatial strategy making.

However, embarking on the type of spatial strategy making that makes a difference is no easy enterprise. It is both an art and a craft. It has elements of both scientific analysis and of design science, but also involves a sensibility that is not cultivated in either. Spatial strategy making, situated in the public domain and concerned for urban futures, demands a more historical, anthropological and geographical imagination, capable of grasping the relations between people and place through time. Rather than procedural pattern books, strategy makers are perhaps better advised to read accounts, histories and biographies of those involved in spatial strategy making, to enhance the antennae with which they learn about the specific challenges they face.[30] Spatial strategy making can make a material contribution to the social and environmental well being of many in an urban area, if approached with a political, economic and cultural imagination that can grasp the relation between people and the governance of place, and understand the formation of publics and

Table 1. Transformative versus responsive strategy making

Dimension of strategy making	Transformative strategy making	Responsive strategy making
Mobilise attention	Re-orient attention to issues which lie behind immediate agendas, where this would highlight neglected opportunities and challenges	Express what the current aims, values and directions of our agency seem to be, with respect to shaping urban futures
Scope the situation	Identify where the energy for change may lie and build coalitions for change which expand this energy	Identify what our agency can achieve
Enlarge intelligence	Explore and recast agendas of problems, issues and potential actions and stakes, through accessing multiple sources of knowledge	Summarise what we seem to know
Create frames and select actions	Articulate strategic ideas within which specific issues and actions can be prioritised and given some justification and coherence	Find a way to give some kind of explicit expression and coherence to the above

454   *P. Healey*

the qualities of a polity. Further, such strategy making may create an infrastructure for discussion, generating a public realm asset within which the difficult material and moral challenges faced by those planning urban areas can be debated as the future unfolds. If developed with knowledgeability, imaginative skill, courage and ethical sensibility, spatial strategy-making efforts with transformative ambitions have significant potential to contribute by enhancing the intelligence and capacity of a polity, and thus improve the future well being of urban citizens and stakeholders (Dewey, 1991).

Notes

1. This paper is a development of a version produced for the *Liber Amicorum* prepared to celebrate the retirement of Professor Louis Albrechts at the Universiteit Leuven (Van den Broeck & Moulaert, 2008). I would like to thank the Editor, Heather Campbell and four anonymous referees for helpful and insightful comments on an earlier version of this paper.
2. See Albrechts *et al.* (2001, 2003), Fedeli and Gastaldi (2004), Marshall (2004), Motte (2007), Salet and Faludi (2000), Salet *et al.* (2003).
3. Ward (2002) notes that it was common in the interwar years for a strategy to be known by the name of the planner who led the planning team. These days, it is more common for a strategy to be known by its place name.
4. For the empirical basis of my reflections, see Healey *et al.* (1997), Vigar *et al.* (2000), Healey (2007). I have also benefited greatly from discussions with skilled strategic planners over the years, and learned from my own experience as a "strategic manager".
5. The term "governance" is used in several ways in planning literature. For some, it implies a shift from undertaking collective action primarily through the institutions of formal government and resort to all kinds of other coalitions, alliances and "partnerships". My usage is as in the text above.
6. Note that my use of "strategy" here is not related to the Habermasian distinction between strategic and communicative "rationality" (Habermas, 1984). A strategy in the sense that I use it here could arise through the exercise of both rationalities, or some hybrid between them.
7. By the term "polity", I mean a governance "culture" in the sense of the qualities of the political life of some collectivity with a political identity.
8. These "orientations" are given different terms in the planning literature. The most common term, taken from the work of Schon and Rein (1994), is a "frame", or framing set of ideas. This gets linked to the idea of a "discourse" as used in work on struggles between alternative orientations and in the capacity to transform them as developed in interpretive policy analysis, drawn from Foucauldian inspiration (see Hajer, 1995). Faludi uses the term "doctrine", although he developed this to describe a very strongly developed conception of the Dutch landscape (Faludi & van der Valk, 1994).
9. See Esping Anderson, 1990; Gualini, 2004, 2006; Jessop, 2002; Le Gales, 2002.
10. This has always been a challenge for those promoting a perspective on urban dynamics which pays attention to the interplay of programmes and projects across an urban area. See studies in planning history, notably Ward (2002), Nasr and Volait (2003).
11. I think, therefore, that the difference between my understanding of strategy and that argued recently by Newman (2008) is very small.
12. Cities which have attracted funding from various national and EU programmes are particularly prone to such experiences.
13. There is a long tradition in planning thought which emphasises the importance of creatively responding to specific situations in experimental and exploratory ways, rather than relying on prior models and templates, see Lindblom (1990), Schon (1983), Forester (1989, 1999).
14. See Giddensian structuration theory, actor-network concepts, the varieties of institutionalist understanding. In philosophy also, the rediscovery of the USA pragmatic tradition, and also the lessons from European phenomenology and post-structuralist philosophy, underline the importance of situated practical judgement within social processes.
15. See Flyvbjerg (2001) and Yanow and Schwartz-Shea (2006) on case study method.
16. Talk by Larry Beasley, UBC, Vancouver, 9 June 2007.
17. See, for example, the cases of Amsterdam and the Cambridge sub-region in Healey (2007), and Salt Lake City in Briggs (2008). For the role of consultants see Healey (2008).

18. Here I draw on arguments which promote the positive benefits of discordant, conflicting viewpoints in resisting tendencies to univocal, monist policy discourses (see, for example, Connolly, 1987, 2005). The pluralism implied here is about a plurality of values and perspectives, not just a plurality of interests.

19. See Lindblom (1990), Healey (2007), see also Hillier (2007) for a similar conclusion arrived at through engaging with the poststructuralist theorists of complexity.

20. This image of a "compass" derives from the USA pragmatist writers (see Healey, 2009).

21. This draws on central arguments in the pragmatist tradition, see Thayer (1982).

22. See Campbell (2006) for a discussion of how general values, such as a concern for justice, are enacted in the making of practical judgements in the planning field.

23. See, for example, the new urbanism ideas (Grant, 2006), and the various ecologically inspired utopias that have appeared in recent years.

24. See Dewey (1982) for a very interesting discussion of these processes in an essay on the pattern of inquiry.

25. See James's idea of how we "heave" our will towards a new direction, cutting off another as we do so (in Thayer, 1982, pp. 184–185).

26. See Hajer (1995) on the re-framing that happens in the transformation of policy discourses.

27. See Healey (2004b), Briggs (2008).

28. Skilled practitioners have long known the reality of such contingencies. Social scientists are now giving these much more attention (Jessop, 2002; Sørensen & Torfing, 2007; Westbrook, 2005).

29. See Barrett and Fudge (1981) for a discussion of the interactive relation between policy and action, which argues that policies are often made as a consolidation of what is reflected in the flow of action.

30. See, for example, the classic case histories (Altshuler 1965; Meyerson & Banfield, 1955), and more recently also Flyvbjerg (1998). Other accounts by strategic planners include Albrechts (2001), Goodstadt and Buchan (2002), Krumholz and Forester (1990), Wannop (1985), while there are an increasing number of "longitudinal" accounts of urban development and planning experiences in individual cities (for example, Fainstein, 2001; Madanipour, 1998; Abbott, 2001; Punter, 2003; Hooper & Punter, 2006; Healey, 2007; Angotti, 2008; Briggs, 2008).

References

Abbott, C. (2001) *Greater Portland: Urban Life and Landscape in the Pacific Northwest* (Philadelphia, PA, University of Pennsylvania Press).

Albrechts, L. (2001) How to proceed from image to discourse to action: as applied to the Flemish, Diamond, *Urban Studies*, 38, pp. 733–745.

Albrechts, L. (2003) Reconstructing decision-making: planning versus politics, *Planning Theory*, 2, pp. 249–268.

Albrechts, L. (2004) Strategic (Spatial) Planning Reexamined, *Environment and Planning B: Planning and Design*, 31, pp. 743–758.

Albrechts, L. (2005) Creativity as a drive for change, *Planning Theory*, 4, pp. 247–269.

Albrechts, L., Alden, J. & Rosa Pires, A.D. (Eds) (2001) *The Changing Institutional Landscape of Planning* (Ashgate, Aldershot).

Albrechts, L., Healey, P. & Kunzmann, K. (2003) Strategic Spatial Planning and Regional Governance in Europe, *Journal of the American Planning Association*, 69, pp. 113–129.

Altshuler, A. (1965) *The City Planning Process: a political analysis* (Ithaca, New York, Cornell University Press).

Amin, A., Massey, D. & Thrift, N. (2000) *Cities for the Many not the Few* (Bristol, The Policy Press).

Angotti, T. (2008) *New York for Sale: community planning confronts global real estate* (Boston, Mass, MIT Press).

Barrett, S. & Fudge, C. (1981) *Policy and Action* (London, Methuen).

Beauregard, R.A. (1998) Writing the Planner, *Journal of Planning Education and Research*, 18, pp. 93–102.

Briggs, X.D.S. (2008) *Democracy as problem-solving* (Boston, Mass, MIT Press).

Campbell, H. (2006) Just planning: the art of situated ethical judgement, *Journal of Planning Education and Research*, 26, pp. 92–106.

Cars, G., Healey, P., Madanipour, A. & Magalhaes, C. (Eds) (2002) *Urban Governance, Institutional Capacity and Social Milieux* (Aldershot, Hants, Ashgate).

Churchman, C.W. (1979) *The Systems Approach (2nd edition)* (New York, Dell Publishing).

Connolly, W.E. (1987) *Politics and Ambiguity* (Madison, Wisconsin, University of Wisconsin Press).

Connolly, W.E. (2005) *Pluralism* (Durham, Duke University Press).

Committee for Spatial Development (CSD) (1999), The European Spatial Development Perspective. Strasbourg, European Commission.

456   *P. Healey*

Davidoff, P. & Reiner, T.A. (1962) A choice theory of planning, *Journal of the American Institute of Planners*, 28, pp. 103–115.

Davoudi, S. (2006) Evidence-based policy: rhetoric and reality, *DISP*, 165, pp. 14–24.

Dewey, J. (1982) The pattern of inquiry, in: H.S. Thayer (Ed.) *Pragmatism: the classic writings*, pp. 316–334 (Indianopolis, Hackett Publishing Co).

Dewey, J. (1991) *The public and its problems* (Athens, Ohio, Swallow Press/Ohio University Press).

Esping-Andersen, G. (1990) *The Three Worlds of Welfare Capitalism* (Cambridge, Polity Press).

Fainstein, S. (2001) *The City Builders: property development in New York and London 1980–2000* (Kansas, University of Kansas Press).

Faludi, A. & van der Valk, A. (1994) *Rule and Order in Dutch Planning Doctrine in the Twentieth Century* (Dordrecht, Kluwer Academic Publishers).

Fedeli, V. & Gastaldi, F. (Eds) (2004) *Pratiche strategiche di pianificazione: riflessione a partire da nuovi spazi urbani in costruzione* (Milan, Franco Angeli).

Ferlie, E., Ashburner, L., Fitzgerald, L. & Pettigrew, A. (1996) *The New Public Management in Action* (Oxford, Oxford University Press).

Flyvbjerg, B. (1998) *Rationality and Power* (Chicago, University of Chicago Press).

Flyvbjerg, B. (2001) *Making Social Science Matter: Why Social Inquiry Fails and How It Can Succeed Again* (Cambridge, Cambridge University Press).

Forester, J. (1989) *Planning in the face of power* (Berkeley, University of California Press).

Forester, J. (1999) *The Deliberative Practitioner: encouraging participatory planning processes* (London, MIT Press).

Friedmann, J. (2002) *The prospect of cities* (Minneapolis, University of Minnesota Press).

Friedmann, J. (2008) The Uses of Planning Theory: a bibliographic essay, *Journal of Planning Education and Research*, 28, pp. 247–257.

Giddens, A. (1984) *The Constitution of Society* (Cambridge, Polity Press).

Goodstadt, V. & Buchan, G. (2002) A statutory approach to community planning: repositioning the statutiry development plan, in: G. Cars, P. Healey, A. Madanipour & C. de Magalhaes (Eds) *Urban governance, institutional capacity and social milieux*, pp. 168–190 (Aldershot, Hants, Ashgate).

Grant, J. (2006) *Planning the good community: new urbanism in theory and practice* (London, Routledge).

Gualini, E. (2004) Integration, diversity and plurality: territorial governance and the reconstruction of legitimacy, *Geopolitics*, 9, pp. 542–563.

Gualini, E. (2006) The Rescaling of governance in Europe: new spatial and institutional rationales, *European Planning Studies*, 14(7), pp. 881–904.

Habermas, J. (1984) *The theory of Communicative Action: Vol 1: Reason and the Rationalisation of Society* (Cambridge, Polity Press).

Hajer, M. (1995) *The politics of environmental discourse* (Oxford, Oxford University Press).

Hajer, M. & Wagenaar, H. (Eds) (2003) *Deliberative Policy Analysis: Understanding Governance in the Network Society* (Cambridge, Cambridge University Press).

Healey, P. (2004a) The treatment of space and place in the new strategic spatial planning in Europe, *International Journal of Urban and Regional Research*, 28(1), pp. 45–67.

Healey, P. (2004b) Creativity and Urban Governance, *Policy Studies*, 25(2), pp. 87–102.

Healey, P. (2006) *Collaborative planning: shaping places in fragmented societies* (London, Macmillan).

Healey, P. (2007) *Urban Complexity and Spatial Strategies: towards a relational planning for our times* (London, Routledge).

Healey, P. (2008) Knowledge flows, spatial strategy-making and the roles of academics, *Environment and Planning C: Government and Policy*, 26, pp. 861–881.

Healey, P. (2009) The pragmatist tradition in planning thought, *Journal of Planning Education and Research*, 28(3), pp. 277–292.

Healey, P., Khakee, A., Motte, A. & Needham, B. (Eds) (1997) *Making strategic spatial plans: innovation in Europe* (London, UCL Press).

Hillier, J. (2007) *Stretching beyond the horizon: a multiplanar theory of spatial planning and governance* (Aldershot, Ashgate).

Hillier, J. & Healey, P. (Eds) (2008) *Foundations of the planning enterprise: Critical Readings in Planning Theory, Volume 1* (Aldershot, Hampshire, Ashgate).

Hillier, J (2008) Interplanary practice: towards a Deleuzian-inspired methodology for creative experimentation in strategic spatial planning, in: J. Van den Broeck, F. Moulaert & S. Oosterlynck (Eds) *Empowering the Planning Fields: Ethics, Creativity and Action*, pp. 43–77 (Leuven, Uitgeverij Acco).

Hoch, C. (2007) Making plans: representation and intention, *Planning Theory*, 6, pp. 16–35.

Hooper, A.J. & Punter, J. (Eds) (2006) *Capital Cardiff 1975–2020: regeneration, competitiveness and the urban environment* (Cardiff, University of Wales Press).

Innes, J. & Booher, D. (1999) Consensus-building as role-playing and bricolage, *Journal of the American Planning Association*, 65, pp. 9–26.

Innes, J. & Booher, D. (2003) Collaborative policy-making: governance through dialogue, in: M. Hajer & H. Wagenaar (Eds) *Deliberative Policy Analysis: Understanding governance in the network society*, pp. 33–59 (Cambridge, Cambridge University Press).

Jessop, B. (2002) *The Future of the Capitalist State* (Cambridge, Polity Press).

Kingdon, J.W. (2003) *Agendas, alternatives, and public policies* (New York, Longman).

Kitchen, J.E. (1997) *People, politics, policies and plans* (London, Paul Chapman).

Kitchen, T. (2007) *Skills for Planning Practice* (Basingstoke, Palgrave).

Krumholz, N. & Forester, J. (1990) *Making Equity Planning Work* (Philadelphia, Temple University Press).

Le Galès, P. (2002) *European Cities: Social Conflicts and Governance* (Oxford, Oxford University Press).

Lindblom, C.E. (1990) *Inquiry and Change: The troubled attempt to understand and shape society* (New Haven, Yale University Press).

Madanipour, A. (1998) *Tehran* (Chichester, John Wiley and Sons).

Marshall, T. (Ed.) (2004) *Transforming Barcelona* (London, Routledge).

Meyerson, M. & Banfield, E. (1955) *Politics, planning and the public interest* (New York, Free Press).

Mintzberg, H. (1994) *The Rise and Fall of Strategic Planning* (Edinburgh, Pearson Education Limited).

Motte, A. (Ed.) (2007) *Les agglomerations francaises face aux defis metropolitaines* (Paris, Economica/Anthropos).

Nasr, J. & Volait, M. (Eds) (2003) *Urbanism: imported or exported? Native aspirations and foreign plans* (Chichester, Sussex, Wiley-Academy).

Newman, P. (2008) Strategic Spatial Planning: Collective action and moments of opportunity, *European Planning Studies*, 16(10), pp. 1371–1383.

Punter, J. (2003) *The Vancouver Achievement* (Vancouver, B.C, UBC Press).

Salet, W. & Faludi, A. (2000) *The revival of strategic spatial planning*, Koninklijke Nederlandse Akademie van Wetenschappen (Royal Netherlands Academy of Arts and Sciences), Amsterdam

Salet, W., Thornley, A. & Kreukels, A. (Eds) (2003) (London, E&FN Spon), Metropolitan Governance and Spatial Planning: Comparative studies of European city-regions.

Salet, W. & Thornley, A. (2007) Institutional influences on the integration of multilevel governance and spatial policy in European City-Regions, *Journal of Planning Education and Research*, 27, pp. 188–198.

Schon, D. (1983) *The Reflective Practitioner* (New York, Basic Books).

Schon, D. & Rein, M. (1994) *Frame reflection: towards the resolution of intractable policy controversies* (New York, Basic Books).

Sørensen, E. & Torfing, J. (Eds) (2007) *Theories of Democratic Network Governance* (Basingstoke, Hants, Palgrave Macmillan).

Tarrow, S. (1994) *Power in movement* (Cambridge, Cambridge University Press).

Thayer, H.S. (Ed.) (1982) *Pragmatism: the classic writings* (Indianapolis, Hackett Publishing Co).

Van den Broeck, J., Moulaert, F. & Oosterlynck, S. (Eds) (2008) *Empowering the planning fields* (Leuven, Acco).

Verma, N. (1998) *Similarities, connections, systems: the search for a new rationality for planning and management* (Lanham, Maryland, Lexington Books).

Vickers, G. (1965) *The Art of Judgement: a study of policy-making* (London, Chapman Hall).

Vigar, G., Healey, P., Hull, A. & Davoudi, S. (2000) *Planning, governance and spatial strategy in Britain* (London, Macmillan).

Vigar, G., Graham, S. & Healey, P. (2005) In search of the city in Spatial Strategies: past legacies and future imaginings, *Urban Studies*, 42(8), pp. 1391–1410.

Wannop, U. (1985) The practice of rationality: the case of the Coventry-Solihull-Warwickshire Subregional Planning Study, in: M. Breheny & A.J. Hooper (Eds) *Rationality in Planning*, pp. 196–208 (London, Pion).

Ward, S. (2002) *Planning in the Twentieth Century: The Advanced Capitalist World* (London, Wiley).

Westbrook, R.B. (2005) *Democratic hope: pragmatism and the politics of truth* (Ithaca and London, Cornell University Press).

Weston, A. (1996) Before environmental ethics, in: A. Light & E. Katz (Eds) *Environmental Pragmatism*, pp. 139–160 (Oxford, Routledge).

Wheeler, S.M. (2002) The New Regionalism: characteristics of an emerging movement, *Journal of the American Planning Association*, 68(3), pp. 267–278.

Yanow, D. & Schwartz-Shea, P. (Eds) (2006) *Interpretation and Method: Empirical research methods and the interpretive turn* (New York, M.E.Sharpe).

Originaltext Mastop/Faludi 1997

Environment and Planning B: Planning and Design 1997, volume 24, pages 815–832

Evaluation of strategic plans: the performance principle

H Mastop
Sshool for Environment and Planning, Faculty of Policy Sciences, University of Nijmegen,
P O Box 9108, 6500 HK Nijmegen, The Netherlands; e-mail: H.Mastop@bw.kun.nl
A Faludi
Department of Physical Planning and Demography, Faculty of Spatial Sciences, University of
Amsterdam, Nieuwe Prinsengracht 130, 1018 VZ Amsterdam, The Netherlands;
e-mail: Faludi@frw.uva.nl
Received 18 March 1996; in revised form 16 May 1997

Abstract. Evaluation and implementation studies have a well-established tradition. For evaluation in general, this tradition seems to offer a clear-cut research design. This is not true for evaluation of strategic plans. First, most evaluation and implementation research deals with specific and well-defined operational policy and not with broad and sometimes vague indicative strategic planning. Second, the means–ends scheme underlying mainstream evaluation, in which conformance between a plan and final outcomes is the ultimate test of effectiveness, does not apply. In trying to establish conformance, we not only ask the wrong question but also use the wrong unit of analysis. Building on ideas from the planning and evaluation literature, we develop an alternative approach based on the notion that strategic plans serve the function of signposts for those involved in subsequent decisions. Our approach entails a test of the effectiveness of strategic plans which reflects their character; we suggest testing their performance. Empirical research on the role and purpose of strategic plans shows that 'performance' offers a promising way of understanding how strategic plans relate to intervention and of judging their usefulness.

> "... regulation [is] the maintenance of relationships over time ... . The goals we seek are changes in our relations or in our opportunities for relating; but the bulk of our activity consists in the 'relating' itself"
>
> (Vickers, 1965, page 33).

> "Yes, it is time for thoughts to become words. I have a plan. I have thought on it for many seasons. I have altered it many times to fit conditions as they have changed. Now there is little more changing that can be done and as the thoughts become words, the words must then become actions"
>
> (Eckert, 1992, page 429).

1 Introduction

The proof of the pudding is in the eating. And what is true of puddings is commonly held of plans. If implementation falls short of expectations, the plan (and the planner!) is said to have failed, whatever the reason, be it because others refuse to cooperate, because of lack of finance or simply because a forecast has been inaccurate. Common though it may be, this is a narrow view. Plans cannot be judged solely in terms of conformance between a plan and final outcomes. This then poses two related questions. How do plans relate to intervention? How should we assess them, if not in the light of outcomes?

As regards the first question, we reject the common, but misconceived, idea of a plan as a blueprint of which the specifications need to be followed. The relation between a plan and subsequent action is not prescriptive but conditional. In principle, plans are strategic instruments and their application is discretionary. Assessing plans requires another type of test of overall performance. Thus we propose a different perspective on

2

evaluating plans, including a research design. Taken together, the perspective and the research design constitute a research programme based on the performance principle. This performance research programme has been developed during ex-post evaluation studies in the Netherlands.

In section 2 we show how our approach is rooted in practical concerns, in section 3 we describe how we look at plans as both instrumental and conjectural, and in section 4 it is shown that conventional evaluation fails to take adequate account of this fact. In section 5 we distinguish between project and strategic plans. For project plans, conventional evaluation will do; however this is not the case for strategic plans. In section 6 we discuss the theoretical roots of our research programme. In section 7 we develop our main argument based on the view that plans are frames of reference for operational decisions and as such need not have any direct impact upon physical development. This implies that the conformance test is insufficient as an indicator of the effectiveness of plans; this is why we introduce the performance principle. The research design, presented in section 8, is robust. This is illustrated by recent Dutch research discussed in section 9. In the conclusions we boldly claim to have solved the problem of evaluating plans, even on a strategic level. The message is counterintuitive: lack of conformance between a plan and final outcomes does not mean poor performance.

2 Performance: roots in practical concerns

The introduction to this issue noted changes in Dutch planning in the mid-1980s. In this respect Dutch planning is no different from planning elsewhere; the systematic approaches, popular in the 1960s and 1970s, have given way to pragmatic market-oriented approaches. In the Dutch situation though, change was somewhat slower in coming and the survival of strategic planning was never in doubt. The Dutch still have faith in planning but want it in a form tailor-made to the present.

The change started with critiques of planning (see den Hoed et al, 1983; Verduijn and Puylaert, 1983; WBS, 1981; for similar critiques see Healey et al, 1982; Masser, 1983). Much of the criticism focused on the statutory system. Notwithstanding painstaking efforts to refine it, that system had failed to deliver the goods. Indeed the steering capacity seemed inversely related to the elaborateness of statutory planning (for similar observations, see Chadwick, 1978).

In any case, town planners have always been uneasy about the effectiveness of their plans. When compared with the position of experts dealing with housing, economic development or environmental protection, that of planners is weak. In these policy fields, the ultimate effects might very well differ from intentions but at least one has recourse to standard procedures for measuring them. This is precisely the problem with urban and regional planning. There have of course been evaluation studies, both in the Netherlands (see Bos, 1988; Glasbergen and Simonis, 1979; Nozeman, 1986) and abroad, to assess the impact of policies as laid down in plans. Some of them even measure goal attainment. However, to measure the effectiveness of plans as such is much more difficult. What if some policies are followed whilst others are disregarded? What must the conclusion be if policies are modified during implementation? What must the verdict be if they are departed from, but the departures are nothing but reasonable adaptations to an evolving situation?

Contemplating such issues, Dutch national planners have invented the rather vague notion of *doorwerking* (see Galle et al, 1987). This term has no direct English equivalent. The meaning is that of policies 'working through' by diffusion into the deliberations which follow their adoption. For reasons which will become clear when discussing the theory underlying it, we translate doorwerking as performance, a term from Barrett and Fudge (1981).

On the basis of this concept, planners claimed that success was indicated by acceptance of, and commitment to, intentions underlying, solutions propagated by, and principles enunciated in strategic plans (as evidenced by verbal correspondence between policies at the national levels and policies at other levels of government). How success could be measured, and whether such correspondence could be ascribed to plans, was not discussed. 'Performance' represented an idea, not an operational standard; and, albeit implicitly, that idea contradicted conventional evaluation.

3 Background

Although loosely defined, the notion of performance raised pressing issues for evaluation research. As indicated, plans are not like puddings. They are more like a hammer or a saw or a recipe. In other words, plans are instruments, things to use, something to work with, and not products for consumption. As with a hammer, a saw or a recipe for that matter (as indeed with any other instrument) the test is whether it is appropriate for the job. As in the case of other instruments, if a plan is not appropriate it should be discarded.

Plans are not only instrumental, but also conjectural in nature. In both respects, plans share some basic characteristics with scientific theories. Both are formulated to solve problematic situations, on the basis of uncertain information and on an incomplete body of knowledge. Both are intellectual constructs with an outcome that is contingent and both generate new problems (Majone, 1980; for similar views of plans as policy-theories-in-use, see Schön, 1982).

This mixed instrument-conjectural nature makes the assessment of plans troublesome. If a plan cannot get the job done, the instrumental line of thought indicates a negative assessment. However, the conjectural nature of plans should caution us not to dismiss it outright. Even from plans that have failed, we can learn (as we can from theories that do not stand up to testing) how we can (and cannot) influence outcomes. Invoking the plan might have enabled us to understand better the job at hand, to appreciate that the world is different from what we thought, and that we should reconsider policies.

We are by no means the first to come up with such conclusions. Majone and Quade (1980) compare planning or policy analysis with the work of a physician. They warn of the many 'pitfalls of analysis', which are thus also pitfalls in assessing the outcomes of planning. In a similar vein, in questioning the 'predict and prepare' paradigm of operational research (then prevalent in planning and policy analysis), Ackoff (1979a; 1979b) criticised a purely instrumental view. There need not be a one-to-one relationship between plans and final outcomes.

This need not be a problem. To control outcomes is not the only reason for making plans. Plans can be structuring devices which help us to know where we are heading, to guide our present and future actions. We also need them for coordinating our actions. The simple act of stating intentions, thus allowing others to react accordingly, is the first step towards such coordination.

This is why we try and imagine all sorts of situations and think about solutions to problems which we face, now or in the future. This is why we surmise how we might have to respond to actions by others. It is also the root cause why, as indicated, the purely instrumental view of plan evaluation, even though common and intuitively appealing, is wrong. This view fails to take account of the fundamentally conjectural nature of plans. In his book *The Quest for Control*, van Gunsteren (1976) argues that plans, and for that matter the law, are at best enabling structures. They must be reenacted every time they are used. Thus plan departures are a distinct possibility. On a more general level, Giddens (1979; 1984) has taken a similar line, based on a social interaction model

2

of society. Closer to our argument, the same idea comes through in Barrett and Fudge (1981). Confronting the widespread lack of conformance of outcomes to stated intentions, they develop the concept of performance as a regulatory idea in evaluation research. When applied to specific situations, the established policy or plan need never be followed blindly but needs to be reenacted and perhaps adjusted. Even when the plan is implemented, it might need to be justified anew.

4 Evaluation studies
With respect to the Dutch 1989 National Environmental Policy Plan, Ringeling states that:

> "Regulations, plans, even subsidies have hardly any direct bearing at all on the problems ... to which policy measures are addressed. Sometimes there is an autonomous effect. However, most of the time this effect is very weak or non-existent. It is only through the taking of measures that the real effects of policy come about. The emphasis in environmental policy should be on taking measures and invoking standards" (1990, page 4, translation by the authors).

The implication seems to be that the measures taken and the technical standards applied should be evaluated. However, we must not conclude that prior deliberations are futile. After all, it is during these deliberatons that the likely outcomes of measures to be taken and standards to be invoked are discussed, analysed, and decided. It is wrong to deprecate planning simply because outcomes are uncertain.

The heyday of systematic planning in the 1960s and 1970s saw surprisingly little effort put into ex-post evaluation of planning (as distinct from ex-ante evaluation; for example, see Faludi and Voogd, 1985; Lichfield et al, 1975; Masser, 1983; Voogd, 1983). Ex-post policy evaluation deals with outcomes. There are three basic approaches focused on goal attainment, impact, and effectiveness (see Winter et al, 1988). When assessing goal attainment, we measure conformance between goals and outcomes. The relevant criteria derive from stated goals; side effects are usually ignored. Impact assessment encompasses all outcomes. The criteria are not limited to stated goals which is why this is also called 'goal-free' evaluation (Patton, 1978). When assessing effectiveness, it is the causal links between outcomes and measures taken which are investigated; even in this case the stated goals form the criteria. Methodologically, the analysis of effectiveness is the most demanding of the three. It uses a combination of a before – after and a with – without analysis.

Both these approaches are based on ends – means logic. Be it implicitly or explicitly, the assumption is that the plan, or rather the policies enunciated in it, produces change in the real world (figure 1). More sophisticated approaches take external influences into account.

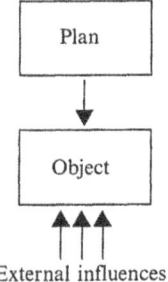

Figure 1. The conventional approach to plan evaluation.

The means–ends scheme has not gone unchallenged. Empirical research (including research on what is often called implementation, which is a misleading term because it implies that the plan has some innate right to be observed) shows that a social-interaction model offers a more powerful way of understanding practice (see Boelens, 1990; Bukkems, 1989; Edwards, 1991; Koningsveld and Mertens, 1986; van der Slaa, 1988; Zonneveld, 1991). For evaluating strategic plans in particular, the means–ends scheme is insufficient. In strategic planning, communication is the name of the game, and a social-interaction perspective is a more fruitful approach to understanding what is actually happening.

5 Two types of plan

In real life we encounter all sorts of plans differing with respect to subject matter, status, time horizon, comprehensiveness, and so forth. Notwithstanding this variety, the purpose of every plan is the same: to guide future decisions and measures. In short, to guide action. Where the problem is clearly defined and routine, a certain type of plan is appropriate; action is routine and outcomes certain. Measures can be seen as parts of the plan. Once the plan is adopted, it can be invoked without further ado. No further deliberations are necessary. At the other end of the scale, decisions and measures are deliberated in their own right. The plan merely forms a frame of reference for them. We describe these two ideal types as project and strategic plans (Faludi and van der Valk, 1994; see table 1).

Table 1. Project and strategic plans.

	Project plans	Strategic plans
Object	Material	Decisions
Interaction	Until adoption	Continuous
Future	Closed	Open
Time element	Limited to phasing	Central to problem
Form	Blueprint	Minutes of last meeting
Effect	Determinate	Frames of reference

Project plans provide blueprints of the intended end-state of the physical environment, including the measures necessary to achieve that state. The only important social interaction is when the plan is being adopted. Thereafter the plan forms an unambiguous guide to action precisely because the measures to be taken are routine so that we 'know' the outcomes. Adopting the plan thus implies closure of the image of the future. The time element in project planning is restricted to the phasing of works in line with production requirements. A project plan is expected to have a determinate effect. Thus evaluation can follow an ends–means logic, measuring conformance between what the plan states and the outcomes of intervention; for research purposes we can ignore the relation between the plan and any actions that follow. Technically, such evaluation can be complex; however, the logic is simple.

Strategic plans are different. These deal with the coordination of a multitude of actors. Such coordinating is a continuous concern. As all actors want to keep their options open, timing is of central importance. Rather than a finished product, a strategic plan is a fleeting record of agreements reached. It forms a frame of reference for negotiations and is indicative. The future remains open. For a strategic plan, subsequent action does not follow automatically. Each decision needs to be justified separately. It follows that the relation between the plan and subsequent action is crucial.

2

The logic underlying the evaluation of strategic plans is therefore more complex than is the case with project plans. Departures do not necessarily indicate ineffectiveness. A strategic plan may be interpreted freely, much as judges interpret (and thereby change!) the law in cases where strict adherence would create anomalies. These assertions will be discussed below.

6 Performance: theoretical considerations

The evaluation of strategic plans poses problems that conventional approaches cannot solve. To analyse these problems, we draw on earlier work (Barrett and Fudge, 1981; Faludi, 1987; Faludi and Mastop, 1982; Mastop, 1987). From Barrett and Fudge we borrow the distinction between performance and conformance. Conformance means concurrence between the original plan and changes in the outside world. Performance has to do with the way in which a strategic plan holds its own during the deliberations which follow its adoption. Critical questions are: Is it being used in the decision situations to which it relates? Does the plan shed light on those situations, that is, does it help in solving them?

In these situations, the prime concern should not be with whether or not the plan is followed, but with whether the plan plays a role in those decision situations in which it was meant to be used. Results are less important than the fact that the plan is invoked. In this, the use of strategic plans can be compared with the use of evaluation studies of which Pröpper (1987, page 117) says that we can: "... speak of policy evaluation studies being put to direct use wherever the research findings are shown to result in critical analyses of positions and in better argued decisions. Whether the research findings [or conclusions and recommendations] are followed or not, is neither here nor there. Even the reasoned rejection of other policy alternatives contributes to the testing and justification of the chosen alternative" (translation by the authors).

In earlier work we have identified the practice of day-to-day decisionmaking as the starting point for planning. In a logical sense, however, plans must precede these 'operational' decisions. Plans have thus been identified as 'prior investments' (Faludi, 1987) and as deriving from and contributing to operational decisionmaking (Mastop, 1987). To serve their purpose, plans have to be brought to bear on subsequent decisions. Plans as such never achieve anything. To put it differently, to have an effect, plans are dependent upon follow-on decisions (Bukkems, 1989), and 'in terms of intervention, plans as such mean nothing' (Geelhoed, 1983). Whatever a plan states, its effectiveness is always dependent upon it being used. Therefore whether it helps in interpreting and solving problems can only be judged in relation to the situations in which the plan is invoked. In terms of Toulmin's view of argumentation, plans are sources from which to derive grounds, warrants, backing, modal qualifiers, and possible rebuttals of claims (in this case, concerning the appropriateness of some measure) (Toulmin et al, 1979; for an example of the use of Toulmin's scheme to elucidate argumentation in planning, see Faludi, 1980; Faludi et al, 1981).

This position allows us to clarify the object of planning. This is another of those simple questions which on close consideration turn out to be complex. We contend that the object of planning is not society, physical development, social problems or any other formulation of the feature to be changed. The object of planning is the powers which the actor or actors, to whom a plan is addressed, have for intervening in society, physical development or social problems (see also Wedgewood Oppenheim in CES, 1970). We might also say that the object of planning is the set of operational decisions that its addressees can and must consider. This set of operational decisions we call the planning object, distinguishing it from the material object, the outside world which we seek to change.

Evaluation of strategic plans   821

We can now see the reason why conformance between outcomes of policy measures (in terms of changes in the material object) and the content of the plan is the wrong criterion of effectiveness of a strategic plan. It would amount to assessing strategic plans against the wrong object, that is, the material instead of the planning object.

In summary, mainstream evaluation is a long way from taking adequate account of the dynamics of day-to-day decisionmaking as it impinges upon the implementation of strategic plans. Operational decisions do not simply turn plans into reality. They involve deliberations in their own right about alternatives and their consequences. In doing so, decisionmakers use all available information. Plans are only parts of the information on which operational decisionmakers draw. A strategic plan serves as a guideline and source of information for subsequent decisionmaking. These follow-on decisions trigger changes in the outside world. If plans are to have any effect at all, they have to perform well during day-to-day decisionmaking. Figure 2 illustrates our point.

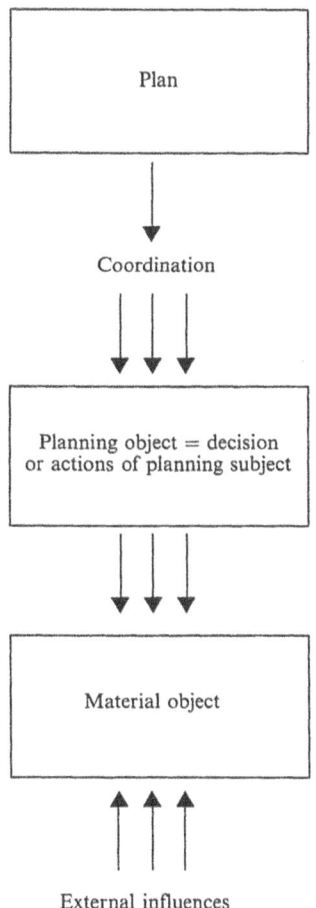

Figure 2. Plans according to the decision-centred view of planning.

7 Performance: beyond the means – ends scheme?

To show that the means – ends scheme is inadequate, we discuss the Dutch growth-centre policy of the early 1970s, which is the Dutch equivalent of overspill and new-town policies in other countries. In its final form in the *Urbanisation Report* (MHSP, 1976), the policy proposed fifteen growth centres. Since then, nearly 250 000 dwellings have been constructed in the growth centres and 500 000 people (3% of the Dutch population) have moved to where the policy said they should. The growth centres have developed more or less in accordance with the urbanisation report and thus the policy has been labelled a success (Faludi and van der Valk, 1990; Galle, 1990).

However, the critical question is whether a direct and causal link exists between the content of the urbanisation report and the tens of thousands of homes built in places such as Almere (a new town in the Lake IJssel Polders) or Zoetermeer (a former village east of The Hague, the population of which has recently passed the 100 000 mark)? To answer this question we need to establish that, the proposals in the urbanisation report did in fact apply to the situation at hand; the policymakers of each of the growth centres (perhaps in conjunction with developers, provincial authorities, and/or government departments other than the then Ministry of Housing and Physical Planning) deemed these proposals relevant; and, in making their choices, the policymakers in the growth centres did take into account the proposals.

Let us assume that things had gone differently. Suppose the behaviour of local actors had not been in accordance with the urbanisation report. Suppose therefore that the number of dwellings built had fallen short of the target. Would this give sufficient grounds for a negative assessment of the effectiveness of the urbanisation report? Not necessarily! We would first have to establish whether the departures had been reasonable. Remember that we have portrayed plans as conjectural in nature. They contain conditional statements referring to future decision situations surrounded by uncertainty. The situation could have changed as a consequence of revised population forecasts, new national building programmes, different urban growth strategies, and so on. Where this happens, there is every reason for reconsidering the initial idea. No one in their right mind would insist that policy be followed whatever the costs. The crucial question would be whether the initial plan had failed. The answer could not be an unequivocal yes. Suppose we found that local policymakers had used the plan (that is, that it had formed part of their considerations during decisionmaking) but that, after due consideration, they had concluded that the decision should be otherwise. In this case, it would be obvious that the plan had in effect helped to resolve some of the issues.

It is against this background that we can define performance as the regulative principle for evaluating strategic plans. A strategic plan is performing well, that is, serving its function, if and only if it plays a tangible role in the choices of the actors to whom it is addressed (including the subsequent choices of the plan-maker or plan-makers) and/or of other actors to whom the plan appeals, in either case irrespective of whether or not outcomes correspond with the plan.

A tangible role in subsequent choices of target groups is what plan performance is about. Essential in this view of performance is that it sees the target groups themselves as being responsible for their own policies. One could of course express this in terms of ends and means. Plans are then the means to an end, the end being to let ideas of the plan-maker play a role in subsequent decisions. Likewise, the actors involved in those decisions can be seen as means to the end of realising the goals of the plan-maker. However, the language seems inappropriate. The plans we are talking about are addressed to human beings whom we must not regard as means to an end. It would be better to talk about social interaction between the makers of a plan and the group or groups to whom the plan is addressed (figure 3).

Figure 3 can be read from the point of view of the plan-maker as well as of the recipients of the message. The plan-maker is responsible for the initial formulation of proposals. To make sure that the plan bears fruit, the plan-maker needs the cooperation of others whom he or she addresses through the medium of the plan. It is equally possible to look at the situation through the recipients' eyes. Evaluating the impact of

Evaluation of strategic plans   823

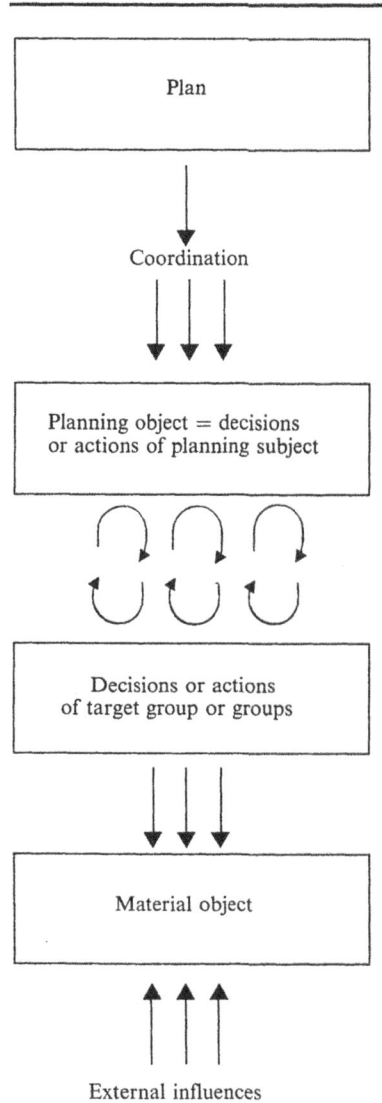

Figure 3. Plans according to the social-interaction perspective.

strategic plans from their point of view can enhance our understanding of how plans work or fail to work.

This implies, first, that recipients have a role to play in the assessment of whether a plan is reasonable, whether it is relevant to their situation; second, that strategic planning is characterised by interdependence between the maker of the plan and its recipients. Conventional evaluation neglects these interdependent relations and thus nonconforming decisions are deemed undesirable, which is a distinctly unhelpful attitude for understanding how strategic plans work. Interdependence means that plan-makers need to take account of the distinct possibility of subsequent decisions deviating from the plan. Likewise, recipients need to take account of the plan. After all, it has been made specifically for the purpose of helping their decisionmaking. Drawing on the classic works by Friend and Jessop (1977) and Friend et al (1974) we may say that, for the recipients, the plan must form part of their context of operations, or policy system. This view is basically the same as found in recent literature on implementation and on policy sciences in general, where network perspectives play a prominent role

2

(Alexander and Faludi, 1989; Simonis, 1983). Our perspective on performance does not necessarily imply a symmetrical relation, nor a direct interaction between plan-maker and recipients. However, we insist that both be viewed as human actors, each with its own perspective on its situation, and each competent to decide on what to do, albeit within constraints imposed by others [compare Giddens's (1979) notion of a competent actor as somebody who could have decided otherwise]. Perhaps strategic plans should be seen as an invitation by the plan-makers to the recipients to share the plan-maker's perception of problems and to contemplate the solutions and measures proposed. This invitation may be backed up by powers of the plan-maker to force, coax or tempt the recipient to comply, irrespective of whether the powers are legal, financial, moral, based on superior knowledge or whatever (for example, see Giddens, 1979; Goverde, 1987; Herweijer et al, 1990).

The social-interaction perspective underlines the point that planning involves communication, conveying intentions of the plan-maker to others. Those others (potential users, target groups, the general public) are competent human actors with views and aspirations of their own. They freely interpret the plan and can be quite creative in dealing, using, misusing or circumventing it. The key to plan performance is simply the way in which subsequent decisionmaking evolves.

8 Performance: research design

Our perspective on performance has two implications for research. First, the units of analysis are policy statements in a plan, and not the plan as such. Second, when measuring performance we need to look at decisions by those to whom planning statements are addressed. We begin with policy statements as the units of analysis.

Strategic plans are comprised of statements of various kinds. There are statements describing the current state of affairs or current policy (declaratory statements), statements describing possible future situations or policies (anticipatory statements), statements expressing intentions (intentional statements), and statements expressing commitments to concrete action (operational statements). More often than not, the statements in a strategic plan are addressed to various actors. Thus strategic plans are hybrid in nature. This means that the plan as such cannot form the object of evaluation. When assessing the performance of strategic plans, we need to focus on those statements that lend themselves to such a test (see Herweijer et al, 1990). More specifically, evaluation must focus on intentional and operational statements (evaluating declaratory and anticipatory statements is outside the scope of this paper).

With this established, we can now discuss the actual conduct of performance studies. Each study consists of two parts: analysing decisions subsequent to the plan; and establishing, measuring, and assessing final outcomes. A fully fledged performance study then implies a kind of effectiveness approach (see section 4), albeit one based on an unconventional view of the role of plans.

Mastop (1987) identified three conditions for strategic plans to be effective.
(1) The plan must specify the operational decisions for which it is intended to provide a framework. This condition helps to establish the quality of the plan as such. Is it specific enough to be evaluated?
(2) The recipient must judge the plan as of continuing relevance to the evolving situation. This condition will be interpreted differently depending on the specific perspective from which evaluation takes place.
(3) The plan must give significant assistance in defining operational decision situations which involve further measures. In other words, the subject must actually invoke the plan. This condition must be satisfied irrespective of outcome. Without this condition being satisfied, we can never consider a plan as being effective.

These conditions can be reduced to two, one necessary and one sufficient.
(a) Operational decisionmakers must be aware of planning statements relevant to them.
(b) Decisionmakers must accept these statements as part of their context of operations, or policy systems. In other words, planning statements must form part of the definition of subsequent operational decision situations (Faludi, 1986).

In view of these conditions, when analysing decisions by the recipients of a plan, one can distinguish between a *theoretical principle of performance* as outlined above, and a *methodological principle of conformance* (see Mastop et al, 1989; see also Barrett and Fudge, 1981). The theoretical principle of performance comes down to the statement that conformance between strategic planning statements and final outcomes is not the criterion for determining whether or not a plan has worked (this has already been discussed fully). The methodological principle of conformance relates to the research conducted; before we can discuss the level of performance, we have to determine whether there has been conformance or not.

Simple though it may seem, conformance is not easy to operationalise. As in the example of the growth towns, at least three questions need to be asked, each establishing a different sort of conformance.
(1) Do the declared intentions of the recipients of a policy statement conform to that statement (formal conformance)?
(2) Do the recipients behave in accordance with their declared intentions (behavioural conformance)?
(3) Do the outcomes of the measures taken by recipients conform to the initial intentions of the plan-maker (final conformance, which is the object of most conventional evaluation)?

Determining conformance can be more laborious than one might expect. Whether or not conformance has been the result of the planning statement being invoked is crucial. Here we follow the logic of with – without analysis as in the example of the growth-centres policy. We can speak of performance only if the outcomes of measures are demonstrably due to the planning statement invoked. Evidently, if all situations were like this, then the whole issue of the difference between performance and conformance would never have arisen. The theoretical principle of performance comes into its own only in situations of nonconformance. It then helps us to study the events which follow. Departures from a planning statement do not necessarily mean failure. Departures are perfectly normal phenomena which we must learn to live with (see Glastra-van Loon, 1989; Oosting, 1992). Thus when departed from, statements in strategic plans have to be judged in light of their function as frameworks for deliberating what to do. Research into these situations requires us to focus on why subsequent decisions, actions, and/or outcomes were different from what was intended by the plan-makers and, in particular, whether in the perception of the decisionmaker(s) concerned, the planning statement, even though not followed, was useful in bringing about the decision. This involves answering, once again from the point of view of the decisionmaker(s) concerned, the following questions:
(a) Does the policy statement under investigation have a bearing on the decision situation at hand?
(b) Is the statement in broad agreement with other current policies of the decision-makers concerned?
(c) What are the arguments for and against compliance with the plan?

Having answered these questions, we can form a considered opinion about whether the plan has worked as intended; whether the plan-maker has been able to inject some relevant considerations into subsequent deliberations. This assessment, more often than not involving qualitative judgments, is the final step in evaluating the performance of strategic plans.

2

9 Some research findings

As yet, no fully fledged performance study has been completed. However, the studies reported in this issue use several elements of our approach. A brief discussion of some of these, and of other salient findings of research in spatial planning, will illustrate the kind of questons raised by performance studies. One of the first performance studies concerned the *Structuurschema Verkeer en Vervoer* (structural outline scheme for traffic and transport), a strategic and indicative document issued by national government. de Lange et al (1997) discuss this study in this issue. The aim was to establish: whether the policies in the structural outline scheme had found their way into the plans and programmes of provinces and municipalities (a measure of conformance); whether the structural outline scheme was informative (a measure of its communicative quality); and who had used the structural outline scheme and in which way (an analysis of 'consumer behaviour').

The first question corresponds to what has been called formal conformance. The second and third questions relate to behavioural conformance. The analysis focused on specific aspects of the structural outline scheme, that is, whether it gave other authorities insight into national policies, in particular whether the other authorities understood the need to make reservations for land required for infrastructure. It transpired that the scheme served its function for the provinces but not the municipalities. Perhaps more importantly the structural outline scheme, although it had less impact than one might have hoped, proved to be a catalyst and a vehicle for regular consultations.

The starting point of the study by Postuma (1987) was theoretical rather than practical, namely performance as discussed in this paper. The study concerned the 1935 General Expansion Plan of Amsterdam, the first Dutch plan based on extensive surveys. The plan boldly reached out to the year 2000. Postuma established how the plan was used until 1955. He analysed housing, land acquisition, and compulsory purchase, in each instance establishing whether reference had been made to the plan. In the case of departures from the plan, Postuma assessed the role of the plan. Before World War 2, housing schemes conformed, but the port authority disregarded the plan. On this count it failed the performance test. After the war the general expansion plan proved useful, a depository of information for decisionmakers to draw upon, even though outcomes were different from what had been foreseen. The committee of Burgomaster and Aldermen used it, pointing out where departures occurred and why. Sometimes the council too referred to the plan in debating solutions. Thus the idea of a plan 'working' by assisting decisionmakers is applicable: "Based among other considerations on the information which (the plan) gives, alternative decisions concerning major as well as minor issues were being generated and evaluated. An additional merit of the General Expansion Plan is that it has contributed to keeping overall urban development on course" (Postuma, 1987, page 5; translated by the authors). Had the analysis been restricted to conformance, the conclusion would have been negative. By invoking the plan, those responsible were in fact able to make rapid assessments of the choices they faced.

A follow up to Postuma concerns the performance of the 1984 Amsterdam Structure Plan (Wallagh, 1988). The approach is as before, but Wallagh distinguishes four types of performance situation: conformance; departures based on arguments derived from the plan; departures, the consequences of which were analysed against the background of the plan; and plan revision in which the old plan was used as a frame of reference in making the new one. This is the ultimate consequence of the performance principle. A plan performs for as long as it is being used as a frame of reference, be it with conforming or nonconforming results.

In a similar vein, current structure plans of the province of North Holland (Borgman, 1989; Vernooij, 1990; Wortelboer, 1989) and Gelderland and Limburg (Eijck and Verhees, 1989) have been analysed for their performance. The analysis of the North Holland structure plans included all decisions about whether to approve municipal planning schemes submitted for approval to the provincial executive, as required under the planning act (see also Alexander and Faludi, 1989; van der Heiden, 1990; Postuma, 1991). All these studies involve analyses of decisions taken subsequent to a plan, assessing the way in which these plans act as frames of reference. The most important conclusion is that, if only for some years after their adoption, structure plans do perform the role of frames of reference. If the studies had been limited to an investigation of the formal decrees by the province, the conclusion would have been different. This is because many fail even to mention the structure plan. On the other hand, only a tiny fraction of decrees depart from it. Borgman and Vernooij find no departures whatsoever among the scores of decisions studied. Wortelboer identifies seven cases where the decision did depart from the structure plan. Of these, only one represented a clear departure. In terms of the three steps listed above for measuring conformance, we come to the surprising conclusion that there is much behavioural conformity but without formal conformity! The explanation lies in the procedure followed, which encourages early consultation, giving provincial planners ample opportunity to explain their policies. This clearly improves performance. At the same time, it explains why formal conformance, as evidenced by references in the decrees to the structure plan, is so low. All those concerned know what the plan states anyway, so why bother referring to it.[1]

The study of Eijck and Verhees (1989) draws similar conclusions. They analysed the use of the provincial structure plan in decisionmaking about local housing projects, depicting these as interagency negotiating processes. One of their conclusions is that there was a negative relationship between conformity and the complexity of the question at hand and a positive relationship between conformity and the intensity of communication between those involved. They also found that the use of the structure plan differs between the agencies involved, varying from nonuse at the one extreme (more often than not mainly by the municipalities) through departing, selective, and strictly conforming use at the other (mainly in the case of provincial authorities in simple clear-cut questions). Whatever its use, however, the structure plan was found to be first and foremost an instrument for the provincial executive and it is at their discretion to decide whether or not the plan should have a bearing on a specific decision.

It is important to note that all these studies concerned recent structure plans and may thus be expected to perform well. Follow-up studies of these same plans would shed light on whether they had ceased to perform. Some policy erosion, and thus less conforming use, is to be expected.

Bukkems (1989) points to another aspect of performance studies: the importance of breaking down strategic plans into various kinds of statements. He analysed the 1988

---

[1] The consequence is that failure to refer to the plan is no longer definite proof of nonconformance. Bac (1992) and van de Weijer (1992) distinguish four situations, depending on whether reference is made to the plan or not and on whether there is behavioural conformance or not. The four quadrants of the resulting 2×2 matrix describe situations in which (a) there is reference to the plan *and* the follow-up decision is taken accordingly; (b) there is no reference to the plan but the follow-up decision *conforms* to the plan *even so*; (c) there is explicit reference to the plan but the decision does *not conform*; (d) there is no reference to the plan *and* the decision does *not conform* to the plan. Situation (b) is the one found by Borgman, Vernooij, and Wortelboer. Situations (c) and (d) are the ones which require further analysis as to whether the follow-on decisions, even though nonconforming, were informed by the plan.

2

Fourth Report on Spatial Planning for statements lending themselves to performance analysis. He then established how their performance could be measured. His approach corresponds broadly with that developed more systematically by Herweijer et al (1990). Bukkems demonstrated that formal adoption is not necessary for strategic plans to have an impact. Policy 'casts its shadow' even before being adopted. Potential recipients anticipate it by applying (and creatively modifying) it to suit their circumstances. This applies particularly to parts of a plan offering new perspectives on existing problems, giving new (although not necessarily specific) directions in which to search for solutions. In the case of the fourth report, this applied to targeting a select few 'urban nodes' for investment and to identifying major urban regeneration projects in the cities of the Randstad. In fact, consultation concerning such proposals (something which is all pervasive in Dutch planning) is used for the purpose of letting policies take effect even before they have been formally adopted.

Likewise, research in progress concerning strategic planning in Amsterdam (Bolderheij et al, 1991; van der Heiden, 1990; Wallagh, 1988; 1990; 1992; 1994) shows strategic plans having an important influence even before being adopted. *Voorwerking*, the Dutch term coined for this phenomenon, has, like *doorwerking*, no adequate translation. The process of making a plan is important in itself and can cause other actors to adopt parts of its viewpoints or conform to some of the measures to be taken in the future even before the plan is adopted. This suggests that those concerned seem to view the statutory document as little more than a byproduct of ongoing deliberations. It is during plan-making that the all important consensus comes into being about the overall structure of the problem and how to approach it. In this way, an emergent plan casts its shadow long before its formal adoption. Obviously, this creates additional complications. Attention in research must extend beyond the decisions taken in the wake of the plan and must include plan-making as such. So far, performance research has been limited mostly to ex-post evaluation. Now, ex-ante evaluation of performance seems to be relevant and possible (see Faludi, 1989).

10 Conclusions

Plans performing even before their adoption opens a Pandora's box of issues. How can a plan work before it exists? The distinction by Friend and Hickling (1987) between 'visible' and 'invisible' products of planning comes to mind. The first refers to planning documents, the latter to insights shared between those participating in planning. These people do not wait until the plan has formally been adopted before acting upon such shared understanding. This poses a difficulty for evaluation. Invisible products are, by their very nature, difficult to perceive. An even more fundamental issue is whether shared understanding stems from those concerned having participated in plan-making. Maybe (as seems likely) there was some preexisting consensus based on tradition, professional ideology, shared culture, and so on, which only needed further specification. If so, then this means that plan-making is only the tip of the iceberg. There may be several layers of nested processes of exploration, communication, the assignment of meaning, and so on, which we need to look at before establishing plan performance (for a discussion on 'the deep structure' underlying planning, see Faludi and van der Valk, 1994).

This goes beyond the scope of this paper. Plan performance is difficult enough. The idea has been around for only a short while. However, it is an important addition to evaluation research. Insights gained correspond to those derived in other branches of policy research. Our discussion shows that evaluation is a complex affair. It has confronted us with the following issues.

(a) We cannot evaluate strategic plans. Instead the proper units of analysis are its component parts (individual policy statements).

(b) The first step is analysing the document concerned for the kinds of statements which it contains and the target groups to whom they are addressed.

(c) Final conformance is not the ultimate measure of whether a strategic plan has been effective, nor are formal, behavioural or final disconformity necessarily counterindications. Establishing conformance is just the beginning but never the end of the story.

(d) For achieving the goals of the plan, the plan-maker depends on the recipients. Performance analysis must therefore focus on the latter and ask: Have they received the message? Did the message form a relevant input into their deliberations?

(e) The answers to these questions, and they alone, form a basis for establishing whether, and to what extent, a strategic plan has performed its role as a frame of reference for subsequent decisions.

Analysing the effectiveness of strategic plans is much more difficult than analysing that of operational policies, and the latter is difficult enough. Simple solutions are not available. Our bold claim is that at least we have posed the right questions and given the right directions for seeking answers. This should bring an appropriate way for evaluating strategic plans within reach.

References

Ackoff R L, 1979a, "The future of operational research is past" *Journal of the Operational Research Society* **30**(2) 93–104

Ackoff R L, 1979b, "Resurrecting the future of operational research" *Journal of the Operational Research Society* **30** 189–199

Alexander E R, Faludi A, 1989, "Planning and plan implementation: notes on evaluation criteria" *Environment and Planning B: Planning and Design* **16** 127–140

Bac C, 1992, "Structuurschets Noord-Holland 1984: kader voor een keerpunt" (Structural outline scheme for North-Holland 1984; framework for change), Master's thesis, Department of Planning, University of Amsterdam, Amsterdam

Barrett S, Fudge C, 1981 *Policy and Action—Essays in the Implementation of Public Policy* (Methuen, London)

Boelens L, 1990 *Stedebouw en Planologie—Een Onvoltooid Project: Naar een Communicatief Handelen in de Ruimtelijke Planning en ont Werppraktijk* (Urban design and planning—the unfinished story: towards a communicative approach in spatial planning and the practice of design) (Delft University Press, Delft)

Bolderheij D et al, 1991, "Structuurplanning op de tweesprong in Amsterdam 1985–19toekomst" (Structure planning at the crossroads in Amsterdam 1985–19future) wp-139, Institute for Planning and Demography, University of Amsterdam, Amsterdam

Borgman F, 1989, "Planologisch toezicht in het Gooi en de Vechtstreek: doorwerking van het streekplanbeleid" (Planning supervision in the Gooi and Vecht areas: the performance of provincial structure plan policies) Master's thesis, University of Amsterdam, Amsterdam

Bos C A, 1988, "Successen en mislukkingen in de Nederlandse ruimtelijke ordening" (Successes and failures in Dutch spatial planning) *Planologische Studies* number 7, Institute for Planning and Demography, University of Amsterdam, Amsterdam

Bukkems G, 1989, "Doorwerking van het in nationale ruimtelijke plannen verwoorde beleid, in het bijzonder de Vierde nota over de ruimtelijke ordening" (The performance of national policies with special attention to the fourth report on spatial planning) *Nijmeegse Planologische Cahiers* number 30, University of Nijmegen, Nijmegen

CES, 1970, "The LOGIMP experiment", information paper 25, Centre for Environmental Studies, London

Chadwick G A, 1978 *A Systems View of Planning* second edition (Pergamon Press, Oxford)

Eckert A W, 1992 *A Sorrow in Our Heart: The Life of Tecumseh* (Bantam Books, New York)

Edwards A R, 1991, "Planning betwist: naar een rehabilitiatie van de communictieve planning-conceptie" (Planning in dispute: towards rehabilitating the concept of communicative planning), in *Planning: Praktijk en Onderzoek* (Planning: practice and research) Eds P de Jong, A F A Korsten, SISWO publication 358 (SISWO, Amsterdam) pp 59–63

2

Eijck Ch, Verhees F, 1989, "Planhantering: Het gebruik van het streekplan bij besluitvorming over gemeentelijke plannen" (Plan-use: the use of the provincial structure plan when deciding on municipal plans) *Nijmeegse Planologische Cahiers* number 31, University of Nijmegen, Nijmegen

Faludi A, 1980, "Strategic choice and the methodology of policy analysis", paper presented at the PTRC Summer School, Birmingham: copy available from A Faludi

Faludi A, 1986 *Critical Rationalism and Planning Methodology* (Pion, London)

Faludi A, 1987 *A Decision-centred View of Environmental Planning* (Pergamon Press, Oxford)

Faludi A, 1989, "Conformance vs performance: implications for evaluations" *Impact Assessment Bulletin* **7** 135 – 151

Faludi A, Mastop J M, 1982, "The 'IOR school': the development of a planning methodology" *Environment and Planning B: Planning and Design* **9** 214 – 256

Faludi A, Valk A J van der, 1990 *De Groeikernen als Hoekstenen van de Nederlandse Ruimtelijke Planningdoctrine* (New towns as key stones of Dutch planning doctrine) (Van Gorcum, Assen)

Faludi A, Valk A J van der, 1994 *Rule and Order: Dutch Planning Doctrine in the Twentieth Century* (Kluwer, Dordrecht)

Faludi A, Voogd H, 1985 *Evaluation of Complex Planning Problems* (Delftse Uitgevers Mij, Delft)

Faludi A, Mastop J M, Vermeulen A H, 1981, "Strategische keuze — uitgangspunten, concepten en methoden: een verantwoording van nieuwe inzichten" (Strategic choice — philosophy, concepts and methods: an account on new views) *Verkenningen in Planologie en Demografie* number 22A, University of Amsterdam, Amsterdam

Friend J K, Hickling A, 1987 *Planning Under Pressure: The Strategic Choice Approach* (Pergamon Press, Oxford)

Friend J K, Jessop W N, 1977 *Local Government and Strategic Choice* second edition (Pergamon Press, Oxford)

Friend J K, Power J, Yewlett C J L, 1974 *Public Planning — The Intercorporate Dimension* (Tavistock, London)

Galle M M A, 1990, "25 years of town and country planning achievements", in *Spatial Reconnaissances 1990* (English edition) National Spatial Planning Agency, Ministry of Housing, Spatial Planning and the Environment, PO Box 30940, 2500 GX The Hague, The Netherlands, chapter 1

Galle M, Mastop J M, Rigter N, 1987, "De bestuurlijke doorwerking van een rijksnota" (The political performance of a national report) *Bestuur* **6**(2) 44 – 49

Geelhoed L A, 1983 *De Interveniërende Staat* (The intervening state) (Staatsuitgeverij, The Hague)

Giddens A, 1979 *Central Problems in Social Theory* (Macmillan, London)

Giddens A, 1984 *The Constitution of Society* (Polity Press, Cambridge)

Glasbergen P, Simonis J B D, 1979 Ruimtelijk Beleid in de Vezorgingsstaat (Spatial policy in the welfare state) (Kobra, Amsterdam)

Glastra-van Loon J F, 1989, "Is besturen analoog aan sturen?" (Is governance analogous to steering?), in *Overheid op de (Terug) Tocht of naar een Nieuwe Profiel?* (Government in retreat or towards a new profile?) Eds A B Ringeling, I Th M Snellen (Vuga, The Hague) pp 163 – 171

Goverde H, 1987 *Macht over de Markerruimte* (Control over the Marker area) *Netherlands Geographical Studies* number 33, Royal Dutch Geographical Society, Amsterdam

Gunsteren H van, 1976 *The Quest for Control* (John Wiley, New York)

Healey P, McDougall G, Thomas M J (Eds), 1982 *Planning Theory — Prospects for the 1980s* (Pergamon Press, Oxford)

Heiden C N van der, 1990, "Bouwen met beleid: de doorwerking van het provinciaal ruimtelijk beleid in Noord-Holland 1926 – 1943" (Building policies: the performance of provincial spatial policy in North Holland 1926 – 1943), wp-116, Institute of Planning and Demography, University of Amsterdam, Amsterdam

Heiden N van der, et al, 1991, "Consensus-building as an essential element of the Dutch planning system", paper presented at the joint ACSP/AESOP conference, Oxford (mimeo); copy available from Institute of Planning and Demography, University of Amsterdam, Amsterdam

Herweijer M, Hummels G J A, Lohuizen C W W van, 1990, "Evaluatie van indicatieve planfiguren: handleiding en begrippen" (Evaluating indicative plans: a manual and concepts) research reports 50, National Spatial Planning Agency, Ministry of Housing, Spatial Planning and the Environment, PO Box 30940, 2500 GX The Hague, The Netherlands

Hoed P den, Salet W G M, Sluijs H van der, 1983 *Planning als Onderneming* (Entrepreneurial planning) (Staatsuitgeverij, The Hague)

Evaluation of strategic plans   831

Koningsveld H, Mertens J, 1986 *Communicatief en Strategisch Handelen: Inleiding tot de
Handelingstheorie van Habermas* (Communicative and strategic action: introducing Habermas'
action theory) (Coutinho, Muiderberg)

Lange M de, Mastop H, Spit T, 1997, "Performance of national policies" *Environment and
Planning B: Planning and Design* **24** 845–858

Lichfield N, Kettle P, Whitbread M, 1975 *Evaluation in the Planning Process* (Pergamon Press,
Oxford)

Majone G, 1980, "Policies as theories" *Omega* **8** 151–162

Magone G, Quade E S, 1980 *Pitfalls of Analysis* (John Wiley, New York)

Masser I, 1983 *Evaluating Urban Planning Efforts* (Gower, Aldershot, Hants)

Mastop J M, 1987, "Besluitvorming, handelen en normeren" (Decision making, action and the
setting of norms) *Planologische Studies* number 4, University of Amsterdam, Amsterdam

Mastop J M, Geest H J A M van, Goveride H J M, Leroy P, 1989, "Uitgangspunten voor een
evaluatie van de Wet en het Besluit op de ruimtelijke ordening" (Starting points for evaluating
the Act and Decree on spatial planning) *Nijmeegse Planologische Cahiers* number 34, University
of Nijmegen, Nijmegen

MHSP, 1976 *Urbanisation Report* Ministry of Housing and Spatial Planning (Staatsuitgeverij,
The Hague)

Nozeman Ed F, 1986, "Nieuwe bouwlokaties in het licht van enkele doelstellingen van ruimtelijke
ordening" (New housing estates in view of some of the objectives of spatial planning)
Planologische Studies number 1, Institute of Planning and Demography, University of
Amsterdam, Amsterdam

Oosting M, 1992, "Wat voert de uitvoerende macht uit als zij uitvoert?" (What does government
do when implementing policies?", in *Uitvoering van Beleid* (The implementation of policy)
Eds C Sas, M Herwijer (Platform Beleidsanalyse, Ministry of Finance, The Hague) pp 37–43

Patton M Q, 1978 *Utilization-focused Evaluation* (Sage, Beverly Hills, CA)

Postuma R, 1987, "Werken met het AUP: stadsuitbreiding van Amsterdam 1939–1955" (Working
with the AUP: urban expansion of Amsterdam 1939–1955), wp-97, Institute of Planning and
Demography, University of Amsterdam, Amsterdam

Postuma R, 1991, "The national plan: the taming of runaway ideas" *Built Environment* **17** 14–22

Pröpper I M A M, 1987, "Beleidsevaluatie als argumentatie" (Policy evaluation as argumentation)
Beleidwetenschap **1** 113–136

Ringeling A B, 1990, "Planning en organiseren: Het Nationaal Milieubeleidsplan" (Planning and
organisation: the National Environmental Policy Plan), in *Juridische en bestuurlijke
consequenties van het Nationaal Milieubeleidsplan* (Legal and administrative consequences of the
National Environmental Policy Plan) Eds A B Ringeling, Th G Drupsteen (Tjeenk Willink,
Zwolle) pp 3–11

Schön D A, 1982, "Some of what a planner knows: a case study of knowing-in-practice" *Journal
of the American Planning Association* **48** 351–364

SGBO, 1987, "Structuurschema Verkeer en Vervoer: evaluatie van een beleidsinstrument" (Structural
Outline Scheme on Traffic and Transport: evaluating a policy instrument), Sociaal Geografisch
en Bestuurskundig Onderzoek (Research department), Association of Dutch Municipalities,
P O Box 30435, 2500 GK The Hague, The Netherlands

Simonis J B D, 1983 *Uitvoering van Beleid als Probleem* (Policy implementation as problem)
(Kobra, Amsterdam)

Slaa P van der, 1988, "Naar een actorperspectief op technologieontwikkeling" (Towards an actor
perspective on the development of technology) *Kennis en Methode* **12** 22–42

Toulmin S, Rieke R, Janik A, 1979 *An Introduction to Reasoning* (Collier Macmillan, Basingstoke,
Hants)

Valk A J van der, 1989, "Amsterdam in aanleg" (Amsterdam under construction) *Planologische
Studies* number 8, Institute of Planning and Demography, University of Amsterdam,
Amsterdam

Verduijn A A, Puylaert H, 1983, "Het nationale ruimtelijke planningstelsel: pompen of verzuipen?"
(The national spatial planning system: sink or swim?) *Stedebouw en Volkshuisvesting* **64** 607–614

Vernooij Th, 1990, "Het streekplan als toetsingskader: onderzoek naar de doorwerking van het
streekplan Amsterdam-Noordzeekanaalgebied voor de region Groot-Amsterdam" (The
provincial structure plan as guideline: assessing the performance of the provincial structure
plan Amsterdam-Northsea-canal-region for the Greater Amsterdam area), Master's thesis,
Department of Planning, University of Amsterdam, Amsterdam

Vickers G, 1965 *The Art of Judgement* (Basic Books, New York)

2

Voogd H, 1983 *Multicriteria Evaluation for Urban and Regional Planning* (Pion, London)
Wallagh G, 1988, "Tussen wens en werking" (Between aspirations and reality), Master's thesis, Department of Planning, University of Amsterdam, Amsterdam
Wallagh G, 1990, "Naar een nieuwe tijd: strategische ruimtelijke planning in Amsterdam en omgeving, 1955 – 1969" (Towards a new era: strategic spatial planning in the Amsterdam area, 1955 – 1969) wp-126, Institute of Planning and Demography, University of Amsterdam, Amsterdam
Wallagh G, 1992, "Het structuurplan: toepassingen en mogelijkheden verkend" (The structure plan: an exploration of applications and possibilities) wp-149, Institute of Planning and Demography, University of Amsterdam, Amsterdam
Wallagh G, 1994 *Oog Voor het Onzichtbare: 50 Jaar Structuurplanning Amsterdam 1955 – 2005* (With an eye for the invisible: 50 years of structure planning in Amsterdam 1955 – 2005) (Van Gorcum, Assen)
WBS, 1981, "Politiek en planning" (Politics and planning) Volkshuisvesting en Ruimtelijke Ordening Notities 4 (Notes on Housing and Spatial Planning 4), Wiardi Beckman Foundation, Amsterdam
Weijer F van de, 1992, "Structuurvisie Noord-Holland 2015: lijnen naar nieuw streekplanbeleid" (Vision on the structure of North Holland 2015: perspectives on new structure plan policies), Master's thesis, Department of Planning, University of Amsterdam, Amsterdam
Winter H, Scheltema M, Oosting M, 1988, "Evaluatie van wetgeving", Tussenverslag (Evaluating legislation, interim report), Department of Administrative Law and Policy Science (University of Groningen Press, Groningen)
Wortelboer C, 1989, "Het streekplan als toetsingskader voor de goedkeuring van bestemmings-plannen" (The provincial structure plan as guideline for the approval of local land-use plans), Master's thesis, Department of Planning, University of Amsterdam, Amsterdam
Zonneveld W, 1991, "Conceptvorming in de ruimtelijke planning: patronen en processen" (Concepts in spatial planning: patterns and processes) *Planologische Studies* number 9A, Institute of Planning and Demography, University of Amsterdam, Amsterdam

Originaltext Kühn 2008

Manfred Kühn

Strategische Stadt- und Regionalplanung

Strategic urban and regional planning

Keywords: Stadt- und Regionalplanung, Strategische Leitbilder, Strategische Projekte, Planungsmodelle, Planungstheorie

Keywords: City and regional planning, Strategic visions, Strategic projects, Planning models, Planning theory

Kurzfassung

Strategische Planung ist in der planungswissenschaftlichen Debatte in Deutschland ein wieder häufiger verwendeter Begriff, der jedoch vieldeutig interpretiert und widersprüchlich bewertet wird. Für manche stellt strategische Planung einen neuen Hoffnungsträger dar, der als „dritter Weg" zwischen großen Plänen und kleinen Schritten eine Renaissance erlebt. Skeptiker betrachten strategische Planung dagegen eher als leeres Schlagwort, das für viele Formen der Planung relativ beliebig verwendet wird. Ziel des Beitrags ist es, ein Modell strategischer Stadt- und Regionalplanung zu entwickeln und auf dieser Grundlage Ansätze und Probleme der Anwendung empirisch zu analysieren. In Kapitel 1 wird strategische Planung zunächst in die theoretische Diskussion eingeordnet und von den Grundmodellen der „Integrierten Entwicklungsplanung" und des „Inkrementalismus" abgegrenzt. In Kapitel 2 wird ein Modell der strategischen Stadt- und Regionalplanung konstruiert, dessen Elemente und Zusammenhänge aus den normativen Ansprüchen der Literatur abgeleitet sind. Anhand dieses Modells werden im dritten Kapitel Rahmenbedingungen, Ansätze und Probleme der Anwendung beschrieben. Dabei wird besonders auf die Aufgabe eines Managements des Strukturwandels von schrumpfenden Städten und Regionen eingegangen. Abschließend erfolgt in Kapitel 4 eine kritische Diskussion des Modells.

Abstract

The term „strategic planning" is increasingly used in German planning theory but in a very ambiguous way. For ones, the revival of strategic planning offers hopes for a third way between comprehensive and incremental planning. For others strategic planning is actually a fuzzy concept and rather "old wine in new bottles". The main aim of the paper is to construct a new model of strategic urban and regional planning and to prove this by empirical analysis of approaches and problems in application. Therefore the first part of the paper tries a positioning of strategic planning within theoretical discourses: some definitions from international planning theories are presented, and differences between incremental and comprehensive planning models are pointed out. Part 2 provides a normative model of strategic planning with main elements: how should strategic planning be to fulfil different demands? Part 3 describes contexts, approaches and some experiences of the application of strategic visions and projects in the management of spatial changes of cities and regions in decline. Finally in part 4, the model of strategic urban and regional planning will be reflected in a critical way.

Manfred Kühn: Strategische Stadt- und Regionalplanung

1 Einordnung in die Planungstheorie

1.1 Begriffsverständnisse

Strategische Planung ist in den internationalen Planungswissenschaften ein vieldiskutierter Begriff, der sowohl für die Raumplanung („strategic spatial planning") als auch die Stadtplanung („strategic urban planning") verwendet wird. Eine anerkannte Definition und ein einheitliches Verständnis von strategischer Planung liegen bisher nicht vor. In der angelsächsisch geprägten Debatte wird strategische Planung relativ offen als ein sozialer Prozess verstanden, um den Strukturwandel von Räumen zu managen. Folgende ausgewählte Definitionen sollen dies verdeutlichen:

„(Strategic planning is) a social process through which a range of people in diverse institutional relations and positions come together to design a planmaking process and develop contents and strategies for the management of spatial change. This process generates not merely formal outputs in terms of policy and project proposals, but a decision framework that may influence relevant parties in their future investment and regulatory activities." (Healey 1997: 5)

"Strategic frameworks and visions for territorial development, with an emphasis on place qualities and the spatial impacts and integration of investments, complement and provide a context for specific development projects." (Albrechts/Healey/Kunzmann 2003: 113)

„Strategic spatial planning is a public-sector-led sociospatial process through which a vision, actions, and means for implementation are produced that shape and frame what a place is and may become." (Albrechts 2004: 747)

Auch in den deutschsprachigen Planungswissenschaften liegen bisher keine genaue Definition und kein einheitliches Verständnis von strategischer Planung vor. In den Wirtschaftswissenschaften wurde eine Übertragung der strategischen Planung aus privaten Unternehmen in die öffentliche Verwaltung bereits in den 1980er Jahren thematisiert (Seidel-Kwem 1983). In den Planungswissenschaften wurde der Begriff in der ersten Hälfte der 1990er Jahre eingeführt. Darin wird strategische Planung noch relativ vage als Nebeneinander unterschiedlicher Planungsebenen beschrieben (Fassbinder 1993). In den 1990er Jahren stand in den Planungswissenschaften weniger der Begriff der „strategischen Planung" im Vordergrund als die Begriffe „Planungsstrategie" (Häußermann/Siebel 1993) und „projektorientierte Planung" (Siebel/Ibert/Mayer 1999), die im Kontext der Internationalen Bauausstellung (IBA) Emscher Park geprägt wurden. In den letzten Jahren wird der Begriff der strategischen Planung in der deutschsprachigen Planungstheorie wieder häu-

figer verwendet und seine „Renaissance" (Altrock 2004; Ritter 2006; Wiechmann/Hutter 2008) thematisiert. Anstelle einer genaueren Begriffsbestimmung wird dabei jedoch die Offenheit und Vieldeutigkeit des Begriffs betont (Scholl 2005). Eine aktuelle Publikation verwendet den Begriff "strategieorientierte Planung" und nimmt damit den Anspruch an eine strategische Ausrichtung etwas zurück (Hamedinger/Frey/Dangschat/Breitfuss 2008).

1.2 Ein „dritter Weg" zwischen großen Plänen und kleinen Schritten?

Strategische Planung wird von vielen Autoren als Mix zwischen „großen Plänen" und „kleinen Schritten" bzw. den Modellen der integrierten Entwicklungsplanung und des Planungs-Inkrementalismus interpretiert (Frey/Hamedinger/Dangschat 2008: 26; Kuder 2008: 182, Wiechmann/Hutter 2008: 116). Brake interpretiert strategische Planung im Sinne eines dialektischen Geschichtsverständnisses als Synthese von Integrierter Entwicklungsplanung (These) und Inkrementalismus (Antithese): *„Wesentliches Merkmal der strategischen Entwicklungskonzepte ist die unbedingte Einheit von Orientierung und Umsetzung. Im historischen Vergleich heißt dies, dass das frühere Kaprizieren auf Leitbilder (etwa in Stadtentwicklungsplänen) nun darum ergänzt wird, wie die darin skizzierte zukünftige Realität herbeigeführt werden kann, bzw. dass der zwischenzeitlich favorisierte und autistische Inkrementalismus überwunden wird durch die Einbettung in orientierende Ziele und deren Verflechtungszusammenhänge."* (Brake 2000: 285)

Bevor ein Modell strategischer Planung als Synthese beider Grundmodelle entwickelt wird, sollen zunächst die wesentlichen Merkmale der integrierten Entwicklungsplanung und des Inkrementalismus dargestellt werden.[1]

Das Modell der *integrierten Entwicklungsplanung* („comprehensive planning") beruhte in der Phase der Planungseuphorie seit den 1960er Jahren auf dem Versuch, die wichtigsten öffentlichen Ressorts (einschließlich Finanzen) in einem „großen Plan" zu integrieren, die Stadt- und Regionalentwicklung durch Planwerke flächendeckend zu steuern und die verschiedenen Akteure auf einheitliche und langfristige Ziele zu orientieren (Ganser/Siebel/Sieverts 1993). In den 1970er und 1980er Jahren wurden die technokratische Rationalität und die Grenzen dieses „geschlossenen Modells" (Häußermann/Siebel 1993) deutlich. Angesichts unzutreffender Zukunftsprognosen, rückläufiger öffentlicher Finanzhaushalte sowie der Erfahrung, dass viele integrierte Entwicklungspläne mit ihren langen

Manfred Kühn: Strategische Stadt- und Regionalplanung

Maßnahmenkatalogen wieder in den Schubläden landeten, setzte sich eine planungsskeptische Sicht auf dieses Modell durch. Die Durchsetzungsprobleme langfristiger Planungsziele in politischen Prozessen wurden von der Politikwissenschaft schon früh aufgezeigt (Scharpf 1973). Die westeuropäischen Planungswissenschaften konstatierten angesichts der allgemeinen Trends zur Globalisierung, Liberalisierung und Deregulierung eine abnehmende Steuerungsfähigkeit der öffentlichen Stadt- und Regionalplanung. Neue paradigmatische Ansätze des „communicative turn" bzw. der „collaborative planning" (Healey 1997) relativierten die Planungshoheit der öffentlichen Verwaltung und thematisierten Kooperationsformen und Aushandlungsprozesse zwischen öffentlichen und privaten Akteuren.

Das Modell des *Inkrementalismus* („disjointed incrementalism") prägte nach Auffassung vieler Autoren das Planungsverständnis in der darauffolgenden Phase der 1970er und 1980er Jahre und hat sich als Gegensatz zur integrierten Entwicklungsplanung entwickelt. Das „offene Modell" (Häußermann/Siebel 1993) des Inkrementalismus ist durch die Merkmale der kleinen Schritte, teilräumige und punktuelle Lösungen, Lernfähigkeit und Flexibilität sowie Kurzfristigkeit des Handelns gekennzeichnet. Dieses Modell hat den Glauben an eine koordinierte Gesamtsteuerung aufgegeben, strebt hingegen Verbesserungen durch überschaubare Problemlösungen an. Die „Strategie der unkoordinierten, kleinen Schritte ist durch die typische Steuerungsorganisation pluralistischer Gesellschaften geprägt, im Gegensatz zur Masterplanung in totalitären Gesellschaften." (Braybrooke/Lindblom 1972: 164). Wesentliche Instrumente des Inkrementalismus sind nicht Planwerke, sondern Projekte. Doch auch die offensichtlichen Schwächen des Inkrementalismus wurden bald kritisiert. Die Kritik an der projektorientierten Planung bezog sich vor allem auf den Steuerungsverzicht zur Lösung komplexer und struktureller Probleme, die fehlende Einbettung und Gefahr der Verselbständigung von Projekten, die Orientierung an kurzfristigen Einzelinteressen der Akteure. Der Inkrementalismus wurde insgesamt als resignierte Anpassung an den Status quo charakterisiert (Häußermann/Siebel 1993: 143).

1.3 Eine Neuauflage des perspektivischen Inkrementalismus?

Wenn strategische Planung eine Synthese zwischen den beiden Modellen der integrierten Entwicklungsplanung und des Inkrementalismus darstellt, dann erfordert dies eine Kombination von gegensätzlichen Merkmalen in mehreren Dimensionen:

- Zeit: Langfristigkeit und Kurzfristigkeit
- Raum: Gesamtraum und Teilräume
- Akteure: Öffentliche und private Akteure
- Steuerungsform: Hierarchie und Netzwerke.

Hoffnungen auf einen sog. „dritten Weg", der die Schwächen beider Modelle meidet und ihre Stärken kombiniert, waren bereits mit dem Konstrukt des „perspektivischen Inkrementalismus" verbunden (Siebel/ Ibert/Mayer 1999: 163). Unter diesem Konstrukt haben die Gründer der IBA Emscher Park einige methodische Prinzipien zusammengefasst, ohne damit bereits eine konsistente Theorie zu entwickeln. Zu diesen Prinzipien zählen u.a.: allgemeine Zielvorgaben, Projekte statt Programme, Verzicht auf flächendeckende Realisierung, Integration der Instrumente statt Integration der Programme (Ganser/Siebel/Sieverts 1993). Wie lässt sich strategische Planung nun vom Ansatz des perspektivischen Inkrementalismus abgrenzen? Der perspektivische Inkrementalismus steht konzeptionell dem Modell des Inkrementalismus näher als dem Modell der integrierten Entwicklungsplanung (Wiechmann/Hutter 2008: 111). Der praktische Schwerpunkt der IBA Emscher Park lag auf der inkrementellen Durchführung von über einhundert Projekten. Das Perspektivische begrenzte sich auf allgemeine Zielvorgaben und Leitthemen. Bezeichnenderweise wurde das Modell der projektorientierten Planung im Kontext der IBA entwickelt. Strategische Planung beansprucht im Unterschied dazu, die Anforderungen beider Modelle gleichwertig zu kombinieren. Bevor in Kapitel 2 der Versuch unternommen wird, eine solche widersprüchliche Einheit zu konzeptualisieren, soll zuvor geklärt werden, wie sich strategische Planung vom Modell der integrierten Entwicklungsplanung abgrenzen lässt oder ob es sich dabei lediglich um eine „Renaissance" alter Ansätze handelt.

1.4 „Renaissance" der Entwicklungsplanung – alter Wein in neuen Schläuchen?

„The Revival of Strategic Spatial Planning" (Salet/ Faludi 2000) ist ein in der internationalen Debatte vielzitierter Buchtitel – ohne dass im Buch deutlich wird, auf welche früheren Planungsformen sich die Vorsilbe „Re" bezieht. Auch in Deutschland stellen heute viele Autoren die Frage, ob es eine „Renaissance" der strategischen Planung gibt (Altrock 2004; Meyer zum Alten Borgloh 2005; Hutter 2006). Diese „Renaissance" bezieht sich auf das Modell der integrierten Entwicklungsplanung der 1970er Jahre in Westdeutschland. In Fachpublikationen ist bereits provokativ von einer „Rückkehr der großen Pläne" die Rede (Frey/ Keller/Klotz/Koch/Selle 2003). Auch in der praktischen

Manfred Kühn: Strategische Stadt- und Regionalplanung

Anwendung lassen sich dafür Anzeichen finden. Im Rahmen des Bund-Länder-Programms Stadtumbau Ost werden seit 2001 wieder „integrierte Stadtentwicklungskonzepte" von den beteiligten Städten erarbeitet. Ein aktuelles Positionspapier des Deutschen Städtetags fordert wieder eine „langfristig orientierte, ressortübergreifende und umsetzungsorientierte integrierte Stadtentwicklungsplanung" (Deutscher Städtetag 2004). Die 2007 auf dem informellen Ministertreffen der für Raum- und Stadtentwicklung zuständigen Minister verabschiedete „Leipzig Charta zur nachhaltigen europäischen Stadt" empfiehlt den Städten die Erarbeitung „integrierter Stadtentwicklungsprogramme auf gesamtstädtischer Ebene". Ein Déjà-vu, das den Eindruck erweckt, als habe es die ganze Debatte um die Grenzen der Steuerungsfähigkeit der öffentlichen Planung nicht gegeben. Damit stellt sich die Frage, ob es sich beim Modell der strategischen Planung lediglich um eine Renaissance der integrierten Entwicklungsplanung handelt bzw. worin sich beide Modelle unterscheiden?

Ritter (2006: 129) setzt zunächst strategische Planung weitgehend mit der Stadtentwicklungsplanung gleich:

„Stadtentwicklungsplanung war und ist strategische Planung: Sie geht von einer Gesamtbetrachtung aus und sucht Teilaufgaben zu integrieren. Sie versteht sich als Richtungsangabe und steckt Entwicklungsrahmen ab. Sie ist grundsätzlich auf lange Sicht angelegt (...) Sie erhebt einen Steuerungsanspruch gegenüber nachgeordneten operativen Plänen."

Im Weiteren stellt er jedoch folgende Unterschiede heraus (Ritter 2006: 139–141):

• Das neue Steuerungsverständnis akzeptiert die Grenzen der Rationalität, indem es Ungewissheit, unvollständige Informationen und unkalkulierbare Verhaltensmöglichkeiten der Akteure einbezieht.

• Der neue Steuerungsanspruch kehrt sich vom „hoheitlichen Modell imperativer Planung" und der klassischen Hierarchie der Akteure ab.

• Die neue Steuerungsintensität ist nicht auf detaillierte Lenkung, sondern auf Rahmensetzung ausgerichtet.

• Der neue Steuerungsumfang reduziert den umfassenden Integrationsanspruch auf Schwerpunkte und Leitprojekte.

• Die neuen Steuerungsverfahren ergänzen hoheitliche durch kooperative Instrumente (ohne diese zu ersetzen).

• Die neue Reichweite der Steuerung begrenzt sich nicht auf administrative Grenzen, sondern kann stadtregional sein.

Zusammenfassend liegen Unterschiede der strategischen Planung gegenüber der integrierten Entwicklungsplanung vor allem in folgenden Punkten:

• Der Anspruch früherer Stadtentwicklungspläne, alle wichtigen Bereiche und Ressorts vollständig zu integrieren, wird zugunsten einer *Selektivität von Schlüsselthemen* aufgegeben. Strategische Planung reduziert die Komplexität und „Ganzheitlichkeit" der Stadt- und Regionalentwicklung, indem sie zwar in der Ausgangsanalyse von Stärken und Schwächen ressortübergreifend agiert, jedoch selektive Prioritäten in Orientierung und Umsetzung setzt.

• Der deduktive Charakter der technokratischen Planungsrationalität in der klassischen Abfolge von Bestandsaufnahme – Problemanalyse – Vision – Leitbild – Ziele – Maßnahmen – Umsetzung – Erfolgskontrolle wird aufgegeben zugunsten eines iterativen *Wechselspiels von Orientierung und Umsetzung*.[2] Projekte können aus Leitbildern abgeleitet werden, Leitbilder können auch aus Projekten entstehen.

• Die Orientierung der Akteure erfolgt nicht über detaillierte Zielkataloge und Handlungsprogramme aus den verschiedenen Ressorts, sondern über *Visionen und Leitbilder*, die lediglich einen Orientierungsrahmen für die zukünftige Entwicklung setzen wollen.

• Die Umsetzung erfolgt nicht über lange Maßnahmenlisten flächendeckend im Gesamtraum, sondern über eine überschaubare Anzahl von *Projekten* in ausgewählten Teilräumen der Städte und Regionen („area based strategy").

• Die gängige Arbeitsteilung zwischen der Zielbestimmung durch öffentliche Akteure aus Verwaltung und Politik und der Umsetzung durch private Akteure aus Wirtschaft und Zivilgesellschaft wird aufgegeben, indem neue *Governance-Formen* entwickelt und private Akteure aktivierend in Leitbildprozesse einbezogen werden.

Wichtige Gemeinsamkeiten zwischen strategischer Planung und integrierter Entwicklungsplanung liegen im ressortübergreifenden Anspruch und informellen Status. Wesentliche Unterschiede bestehen darin, dass strategische Planung einerseits die Integrationsansprüche reduziert, indem sie auf umfassende Ziel- und Maßnahmenkataloge verzichtet und fachliche wie räumliche Prioritäten setzt. Andererseits erhöht strategische Planung die Ansprüche an eine effektive Umsetzung durch eine stringente Ziel-Mittel-Kopplung, projektorientierte Vorgehensweise und Einbeziehung privater Akteure aus Wirtschaft und Bürgerschaft.

Manfred Kühn: Strategische Stadt- und Regionalplanung

2 Ein Modell strategischer Planung

„Strategie" meint allgemein ein planvolles Handeln von Akteuren und Institutionen, das die Bestimmung langfristiger Ziele aus der Analyse externer und interner Rahmenbedingungen ableitet und mit der Auswahl kurzfristiger und flexibler Schritte zur Realisierung dieser Ziele kombiniert. Strategien stellen damit eine Einheit aus den beiden Steuerungsfunktionen Orientierung und Umsetzung dar. Visionen, Leitbilder und Ziele auf der einen Seite wie auch Projekte und Maßnahmen auf der anderen Seite stellen allein noch keine Strategie dar.

Als konstitutives Merkmal der strategischen Planung wird von vielen Autoren die Gleichzeitigkeit von Orientierung und Umsetzung beschrieben:

„Wesentliches Merkmal der strategischen Entwicklungskonzepte ist die unbedingte Einheit von Orientierung und Umsetzung." (Brake 2000: 285)

„Strategisch stehen Gleichzeitigkeit und Wechselwirkung von Konzept- und Projektentwicklung im Vordergrund; das heißt, Leitkonzepte und Projekte bedingen sich wechselseitig, sind jeweils Folge oder Vorlauf." (Becker 1999: 464)

„Wer konkretes Handeln vor Ort in übergreifende Bezüge einbinden will, muss in der Lage sein, parallel Strategie, Prozess und Projekt auf verschiedenen räumlichen Ebenen zu gestalten. Eben dies ist (...) die Herausforderung für heutige Stadtentwicklungsplanung. Dabei kann also auch nicht mehr wie früher deduktiv vorgegangen werden, sondern programmatische Ziele, Pläne und Projekte werden iterativ und im ständigen Wechselspiel entwickelt." (Frey u. a. 2003: 16, 17)

„Dieser (...) ‚Dreiklang' von Leitbild, Gesamtkonzept und Einzelprojekt entspricht nicht mehr der klassischen Unterscheidung zwischen ‚Bottom-up' und ‚Top-down' in den Planungsstrategien. Er ist eher beschreibbar mit dem Bild eines laufenden Maßstabswechsels in der Planung, bei der die jeweils gerade als notwendig erachtete Planungsebene auf die anderen Ebenen zurückwirkt." (IBA 2005: 132)

Die Begriffe „Gleichzeitigkeit", „Parallelität" und „ständiges Wechselspiel" heben die zeitliche Dimension hervor, so dass sich strategische Planung auch als permanenter Lernprozess verstehen lässt. Das „Wechselspiel" zwischen Orientierung und Umsetzung soll dazu beitragen, einerseits die häufige Folgenlosigkeit von Leitbildprozessen, andererseits den bloßen Aktionismus von Projekten zu vermeiden. Strategische Planung lässt sich demzufolge in einem Modell[3] als Wechselspiel verschiedener Elemente konzipieren (siehe auch Grafik, S. 236):

- *Kooperationen zwischen strategischen Akteuren*

Sie bilden in zweifacher Hinsicht das institutionelle Dach, um Leitbilder und Projekte miteinander zu koppeln, und zwar *erstens* durch die Kooperation verschiedener Ressorts der öffentlichen Verwaltung und *zweitens* durch die Kooperation öffentlicher und privater Akteure.

Der erste Anspruch, ressortübergreifend zu arbeiten, wird vielfach an die strategische Planung formuliert (Brake 2000: 287; Altrock 2004: 235). Zugleich geht strategische Planung mit dem paradigmatischen Wandel vom „Government" zur „Governance" einher (Albrechts 2004: 751; Albrechts/Healey/Kunzmann 2003: 114). Die Steuerungsgrenzen der staatlichen Planung führen in stark marktwirtschaftlich ausgerichteten Ländern wie Großbritannien und den USA zu einer Institutionalisierung von öffentlich-privaten Kooperationen (Public-Private-Partnerships) zwischen Staat, Wirtschaft und Zivilgesellschaft. Damit verbunden ist vielfach die Hoffnung, dass öffentliche Fördermittel private Folgeinvestitionen nach sich ziehen. Bereits in Bezug auf das Modell der projektorientierten Planung wurde darauf hingewiesen, dass damit strategische Kooperationen zwischen der öffentlichen Verwaltung und privaten Unternehmen entstehen, die die klassische Arbeitsteilung zwischen der Zielbestimmung durch öffentliche Akteure und der Realisierung durch private Akteure aufgeben (Siebel/Ibert/Mayer 1999: 164).

- *Stärken-Schwächen-Analyse*

Eine Stärken-Schwächen-Analyse ist im strategischen Management privater Unternehmen Ausgangspunkt und Grundlage für die Strategiebestimmung (Mintzberg 2000: 36). Für eine Positionsbestimmung von Städten und Regionen wird das Instrument der SWOT-Analyse[4] seit den 1990er Jahren auch von der öffentlichen Verwaltung übernommen. Dazu werden in einem ersten Schritt die internen Stärken und Schwächen analysiert. Basis dafür ist in der Regel ein moderierter Diskussionsprozess mit dem Ziel, einen Gruppenkonsens zwischen den beteiligten Akteuren herzustellen (der besonders zwischen öffentlichen und privaten Akteuren nicht ohne weiteres vorausgesetzt werden kann). In einem zweiten Schritt werden die externen Chancen und Risiken der Umwelt ebenfalls in einem gemeinsamen Konsens analysiert. Schließlich werden in einer Matrix Stärken und Chancen, Schwächen und Gefahren gegenübergestellt. Ziel der SWOT-Analyse ist es, Stärken und Chancen zu maximieren und Schwächen und Gefahren zu minimieren. Die SWOT-Analyse beruht als ein Element der strategischen Planung am ehesten auf der rationalistischen

2

Prämisse, wonach eine lineare Ziel-Mittel-Ableitung auf der Grundlage umfassender Informationen explizit möglich ist (Wiechmann/Hutter 2008). Aus der Stärken-Schwächen-Analyse können sowohl Leitbilder als auch Projekte abgeleitet werden.

- *Strategische Leitbilder*

Strategische Leitbilder bestimmen langfristige Ziele und gemeinsame Zukunftsvisionen auf der gesamträumlichen Ebene der Stadt und Region. Die Stadtplanung in Deutschland arbeitet in der Regel mit städtebaulichen Leitbildern (Becker 1999), die Landes- und Regionalplanung mir raumordnerischen Leitbildern (Knieling 2006). Strategische Leitbilder unterscheiden sich von städtebaulichen und raumordnerischen Leitbildern, indem diese nicht nur flächenbezogene Aussagen zur baulich-räumlichen Struktur, sondern auch zu prägenden Standortprofilen oder tragenden Wirtschaftsbereichen einer Stadt bzw. Region enthalten (z. B. als „Kulturstadt", „Universitätsstadt" oder „Energieregion"). Strategische Leitbilder verknüpfen also sozioökonomische und baulich-räumliche Ziele der Stadt- und Regionalentwicklung. Internationale Studien zeigen die Bedeutung langfristiger Leitbilder für die Bewältigung des Strukturwandels besonders für deindustrialisierte Städte und Regionen (Kunzmann 1993; BBR 2005).

Im Modell strategischer Planung wird zwischen Leitbildprozessen und Leitbildergebnissen unterschieden. Denn strategische Leitbilder haben eine doppelte Funktion:

(1) Strategischen Leitbild*prozessen* wird nach innen eine Funktion zur Abstimmung und Konsensfindung der beteiligten Akteure zugeschrieben („shared future"). Sie basieren – im Unterschied zu reinen Imagekampagnen – auf sozialen Verständigungs- und Aushandlungsprozessen zwischen den Akteuren und sollen durch die eigene Mitwirkung das Handeln der Akteure binden und aktivieren (Knieling 2006: 480; Altrock 2004: 225).

(2) Strategische Leitbild*ergebnisse* haben nach außen die Funktion, einen Imagewandel von Räumen und eine Standortprofilierung im Wettbewerb um die Ansiedlung von Unternehmen und Bewohnern zu bewirken. Negative Images von schrumpfenden, belasteten und „verbrauchten" Standorten stellen häufig ein Hemmnis für die Ansiedlung neuer Unternehmen und qualifizierter Arbeitskräfte in altindustriellen Städten und Regionen dar. Leitbildergebnisse sind deshalb nicht zuletzt auch ein Marketing- und Werbeinstrument, um die Aufmerksamkeit von Medien, Investoren und Bürgern im Standortwettbewerb zu wecken.

- *Strategische Projekte*

Strategische Projekte sind größere Bauvorhaben (z. B. Sportstadien, Museen, Brücken) oder temporäre Events (z. B. Festivals, Olympische Spiele, Gartenschauen, Kulturhauptstadt), denen für die zukünftige Stadt- und Regionalentwicklung eine Initialfunktion zugeschrieben wird. In vielen Fällen handelt es sich um eine Wiedernutzung brachliegender Flächen oder Gebäudekomplexe in alten Industrie-, Gewerbe- oder Hafengebieten. Projekte sind kurz- oder mittelfristig angelegt und in der Regel zeitlich befristet. Der begrenzte Realisierungszeitraum schafft einen Handlungsdruck für die Akteure und beschleunigt damit die Umsetzung. Projekte intervenieren nicht flächendeckend, sondern räumlich punktuell („area based"). Dadurch stellen sie eine Abkehr von der umfassenden und flächendeckenden Steuerung dar und setzen klar definierte Trägerschaften, Eigentumsverhältnisse und Finanzierungskonzepte voraus. Träger der Projekte sind in vielen Fällen Sonderorganisationen, die aus der bestehenden Verwaltungsstruktur ausgegliedert werden. Dazu zählen u.a. ressortübergreifende Arbeitsgruppen, privatrechtlich organisierte Entwicklungsgesellschaften oder öffentlich-private Agenturen.

Einer projektorientierten Planung werden folgende Stärken zugeschrieben (vgl. Mayer 2004):

- Strategische Projekte konzentrieren sich auf privilegierte Themenfelder und Teilräume und reduzieren damit die Komplexität der Stadt- und Regionalentwicklung. Im Rahmen von Projekten lassen sich baulich-räumliche und sozioökonomische Ziele und Ressorts einfacher integrieren als im Rahmen von Programmen oder Planwerken.

- Eine überschaubare Anzahl definierter strategischer Projekte hat bessere Umsetzungschancen als lange Maßnahmenkataloge in traditionellen Stadtentwicklungskonzepten, die häufig addierte „Wunschlisten" darstellen und wieder in den Schubläden verschwinden.

- Durch den konkreten Umsetzungsanspruch von Projekten lassen sich Akteure leichter mobilisieren und motivieren als durch allgemeine Leitbilder. Mit Projekten kann deshalb im stärkeren Maße politische Handlungsfähigkeit demonstriert werden.

- Das Management von Projekten erfolgt oftmals außerhalb von Verwaltungsroutinen und ermöglicht dadurch in vielen Fällen flexiblere, kreativere und qualitätsvollere Lösungsansätze sowie offenere Lernprozesse.

- Strategische Projekte symbolisieren den Strukturwandel von Städten und Regionen. Sog. „Leuchtturm"-Projekte („flagships") können einen Image-

Manfred Kühn: Strategische Stadt- und Regionalplanung

wandel bewirken als Basis für eine neue Attraktivität der Räume. Ein bekanntes Beispiel dafür ist das Guggenheim-Museum in Bilbao.

- Strategische Projekte können durch die Zusammenarbeit öffentlicher und privater Akteure sichtbare Erfolge in der Stadt- und Regionalentwicklung erzeugen und damit die Trennung von Planung und Umsetzung im Modell der integrierten Entwicklungsplanung überwinden.

- Strategische Projekte werden als „Schlüsselprojekte" oder „Impulsprojekte" (BBR 2005: 14) bezeichnet, wenn von ihnen Ausstrahlungswirkungen auf das Umfeld erwartet werden (Siebel/Ibert/ Mayer 1999: 169). Diese Wirkungen können entweder räumlich streuend (metastasierend) oder zeitlich beschleunigend (katalysatorisch) sein.

- *Erfolgskontrolle*

Die Erfolgskontrolle bzw. Evaluation ist im Rahmen der strategischen Planung ein begleitender Prozess („ongoing evaluation"), um das iterative Wechselspiel zwischen Leitbildern und Projekten regelmäßig rückzukoppeln. Methodisches Ziel ist dabei keine Wirkungsanalyse, da diese ein zeitliches Nacheinander voraussetzt. Im „Wechselspiel" von Leitbildern und Projekten lassen sich Ursachen und Wirkungen schwer unterscheiden. Eine begleitende Evaluation erfordert – im Unterschied zu einer Ex-post-Evaluation durch Wirkungsanalysen – ein laufendes Monitoring zur Kontrolle, um eine Anpassung der Leitbilder und Projekte an veränderte Rahmenbedingungen zu ermöglichen. Die Kunst dieses „lernenden Systems" besteht darin, die richtige Balance zwischen Kontinuität und Flexibilität zu finden. Während Leitbilder einerseits eine langfristige Gültigkeit beanspruchen und damit eine gewisse Kontinuität der Akteure und ihrer gemeinsamen Handlungsorientierungen voraussetzen, soll das Projektmanagement andererseits flexibel genug sein, um auf veränderte Umsetzungsbedingungen durch Anpassungen rasch zu reagieren. Insgesamt gewinnt strategische Planung damit den Charakter eines permanenten Lernprozesses.

- *Orientierung und Umsetzung*

Orientierung und Umsetzung bilden schließlich die beiden gegensätzlichen Steuerungsfunktionen der strategischen Planung ab. Während die Orientierungsfunktion der Akteure über Visionen, Leitbilder oder Ziele einen langfristigen Zeithorizont und eine flächendeckende Raumbezug beansprucht, ist die Umsetzungsfunktion durch die gegenteilige Merkmale kurzfristiger Projekte bzw. Maßnahmen und teilräumiger Interventionen gekennzeichnet.[5]

Die Pfeile in der Grafik stellen den dynamischen Prozesscharakter strategischer Planung dar. Insgesamt bildet das Modell eine widersprüchliche Einheit aus gegensätzlichen Elementen. Die Vermittlung zwischen diesen beiden erfolgt hauptsächlich durch Akteurs-Kooperationen und „integrierte Entwicklungskonzepte".

Modell der strategischen Stadt- und Regionalplanung

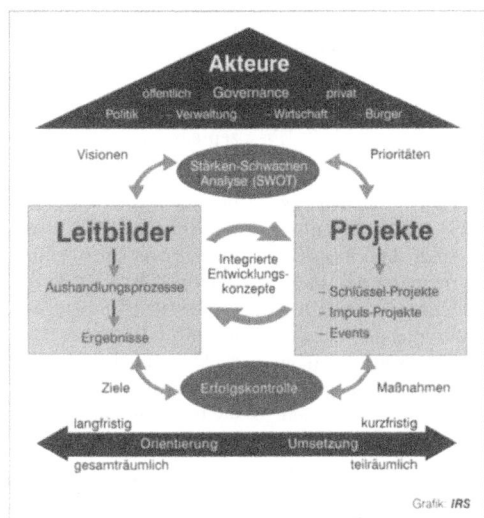

Grafik: *IRS*

3 Zur Anwendung strategischer Planung

3.1 Wachstum und Schrumpfung als Rahmenbedingung

Strategische Planung wurde ursprünglich als ein Ansatz im Management privater Unternehmen entwickelt, um neue Geschäftsfelder zu erschließen. Es war damit ein Instrument unternehmerischer Wachstumspolitik. Mit der Übertragung auf die öffentliche Planung wurde der Ansatz jedoch unter ganz unterschiedlichen Rahmenbedingungen in Städten und Regionen angewendet. In einigen europäischen Ländern diente strategische Planung insbesondere zur Bewältigung des Niedergangs von Städten und Regionen (Albrechts 2004: 743). Einzelne europäische Großstädte wie Barcelona haben das Instrument in den 1990er Jahren zur gezielten Forcierung des postindustriellen Strukturwandels angewandt (Meyer zum Alten Borgloh 2005). Aber auch in einigen prosperierenden europäischen Metropolen wie Wien und London wurden in dieser Zeit Strategiepläne neuen Typs entwickelt (Brake 2000). In den internationalen Planungswissenschaften wird „managing of spatial change" seit den 1990er Jahren als eine zentrale Aufgabe der strategischen Planung definiert

Manfred Kühn: Strategische Stadt- und Regionalplanung

2

(Healey 1997). In Großbritannien ist sie heute eine konzeptionelle Grundlage für die Politik der Urban Regeneration (Roberts/Sykes 2000). In Westdeutschland ging die Wachstumsphase der Nachkriegszeit zunächst mit einer Konjunktur integrierter Entwicklungsplanungen einher. „Große Pläne" fanden vor allem dort Anwendung, wo es demographische und ökonomische Wachstumsimpulse zu verteilen gab. Als Reflex auf die Deindustrialisierung an der Ruhr und die allgemeine Wachstumskrise seit den 1970er Jahren wurde erstmals durch die IBA Emscher Park die neue Aufgabe an Planung und Politik explizit formuliert, „Innovationen in altindustrialisierten Regionen und schrumpfenden Städten" (Häußermann/Siebel 1993: 143) zu organisieren.[6] Die Strategie der Erneuerung eines alten Industriegebiets stärkte inkrementelle Ansätze der „Planung durch Projekte". Mit der zunehmenden Thematisierung schrumpfender Städte und Regionen seit der Jahrtausendwende wird den Stadt- und Regionalplanern vermehrt die Rolle eines „manager of change" (Doehler-Behzadi u. a. 2005: 76) zugeschrieben. Der starke Anpassungsdruck in schrumpfenden Städten und Regionen, der verschärfte Standortwettbewerb um die Ansiedlung von Unternehmen und Haushalten und die knappen öffentlichen Kassen scheinen als Rahmenbedingungen eine Anwendung strategischer Planung zu begünstigen (DASL 2002). Eine europaweite Untersuchung von Bewältigungsstrategien in Städten mit Strukturkrise belegt, dass in allen untersuchten Fällen in Großbritannien, Dänemark, Schweden, Italien, Spanien und Frankreich projektorientierte Strategien verfolgt werden, die Gesamtkonzepte und Impulsprojekte eng verknüpfen (BBR 2005: 10).

3.2 Ansätze und Beispiele

Durch die zunehmende Aufgabe der Bewältigung von Schrumpfungsprozessen in Städten und Regionen und Anreize verschiedener Förderprogramme der EU, des Bundes und der Länder bildet sich eine Vielzahl von Ansätzen strategischer Planung in Deutschland heraus. Dies belegen folgende Beispiele:

• Im Rahmen des Bund-Länder-Programms „Stadtumbau Ost" werden seit 2001 integrierte Stadtentwicklungskonzepte seitens der Fördergeber gefordert und inzwischen in den meisten der 295 beteiligten Städte bereits fortgeschrieben. Die Integration bezieht sich auf städtebauliche und wohnungswirtschaftliche Ziele. Die Akteurs-Kooperationen begrenzen sich weitgehend auf die Stadtverwaltungen und (halb-)öffentlichen Wohnungsgesellschaften. Die Stadtumbaukonzepte greifen in den meisten Fällen auf bereits vorhandene Leit-

bilder zurück. In der Umsetzung werden in der Regel keine Projekte, sondern flächenbezogene Maßnahmen des Rückbaus und der Aufwertung definiert (BMVBS 2001).

• Im Rahmen des Bund-Länder-Programms „Stadtteile mit besonderem Entwicklungsbedarf – die soziale Stadt" werden integrierte Handlungskonzepte für insgesamt 360 benachteiligte Stadtteile in 252 Städten und Gemeinden gefordert, die im strategischen Teil ein Leitbild, im operativen Teil Maßnahmen und Projekte definieren sollen. Die Zwischenevaluierung des Programms hat bereits deutliche Defizite im integrativen Ansatz und der strategischen Kohärenz dieser Konzepte benannt (IfS 2004).

• Die IBA Stadtumbau Sachsen-Anhalt 2010 ist eine bis zum Jahr 2010 befristete Strategie des Landes Sachsen-Anhalt, an der insgesamt 17 Städte beteiligt sind und die von einem gesondert eingerichteten IBA-Büro durchgeführt wird. Ziel ist es, eine stärkere Profilierung der schrumpfenden Städte zu erreichen, um begrenzte Kräfte und knappe Mittel auf zukunftsträchtige Entwicklungspfade zu lenken. Dazu werden in jeder Stadt einige wenige Impuls- bzw. Schlüsselprojekte definiert, die bis 2010 realisiert werden sollen (IBA Stadtumbau 2006).

• Das Ministerium für Infrastruktur und Raumordnung des Landes Brandenburg fordert im Zuge einer stärkeren Konzentration der Förderpolitik von den Städten des Landes Integrierte Stadtentwicklungskonzepte, die aus der Definition von Leitbildern und Entwicklungszielen sog. „Schlüsselmaßnahmen" ableiten. Diese Schlüsselmaßnahmen sollen ausgewählte Maßnahmenbereiche bündeln. Die Umsetzung erfolgt im Rahmen von Projekten (MIR 2006).

In der Regionalentwicklung bilden sich strategische Planungsansätze durch Internationale Bauausstellungen, neue informelle Instrumente der Regionalplanung und die Einrichtung von Agenturen des Regionalmanagements im Rahmen der europäischen und nationalen Strukturpolitik heraus. Beispiele dafür sind:

• Die IBA Emscher Park (1989–1999) hatte als Ansatz einer „modellhaften baulichen, sozialen, ökologischen und ökonomischen Erneuerung eines alten Industriegebietes" (Ganser/Siebel/Sieverts 1993, S. 12) in Nordrhein-Westfalen die planungstheoretische Debatte der 1990er Jahre in Deutschland wesentlich bestimmt. In der Praxis arbeitete die IBA mit sechs Leitthemen und 120 Projekten (Kilper 1999). Wegen ihrer privilegierten Finanzierung durch Landesmittel blieb sie letztlich jedoch eine solitäre Initiative.

Manfred Kühn: Strategische Stadt- und Regionalplanung

• Die IBA Fürst-Pückler-Land (2000–2010) ist ein nach-folgender Ansatz im Land Brandenburg, um den Struk-turwandel in der Lausitz von einem Braunkohlerevier zu einer touristischen Seenlandschaft zu bewältigen und zu qualifizieren. Die Planungsstrategie dieser IBA ist bestimmt durch die Verknüpfung einer langfristigen Vision mit der kurzfristigen Verwirklichung von in-zwischen 25 Projekten (IBA 2006). Die IBA basiert auf dem Planungskonzept eines Wechselspiels zwischen langfristig orientiertem regionalem Leitbild, regionalen Entwicklungskonzepten und Projekten als Inseln der Innovation (IBA 2005: 122).

• Die REGIONALEN sind als Nachfolger der IBA Em-scher Park ein strategischer Ansatz in Nordrhein-West-falen, um durch die Förderung innovativer Projekte, Er-eignisse und Initiativen zur Profilierung der Regionen beizutragen. In einem Wettbewerbsverfahren können sich die einzelnen Regionen um die Austragung im zweijährigen Turnus bewerben. Träger sind gesondert gegründete und zeitlich befristete Agenturen. Alle RE-GIONALEN arbeiten dabei mit der Kombination von Leitthemen und Projekten (ILS 2006).

• Regionale Entwicklungskonzepte werden in vielen Bundesländern seit den 1990er Jahren als informelle Instrumente der Regionalplanung eingesetzt, um zur regionalen Strukturverbesserung und interkommu-nalen Kooperation beizutragen. Regionale Entwick-lungskonzepte ergänzen die formellen Planwerke und sollen laut Raumordnungsgesetz zur „Verwirklichung der Raumordnungspläne" beizutragen. Sie werden häufig im Rahmen der regionalen Strukturpolitik (u.a. LEADER-Programme, Gemeinschaftsaufgabe zur För-derung der regionalen Wirtschaftsstruktur) gefördert. Als umsetzungsorientierte Instrumente umfassen Re-gionale Entwicklungskonzepte die methodischen Be-standteile SWOT-Analysen, Leitbilder, Maßnahmen und Projekte (Knieling/Weick 2005).

Wie diese Beispiele zeigen, werden viele Ansätze ei-ner strategischen Stadt- und Regionalplanung durch Förderprogramme der Ebenen EU, Bund und Länder quasi „von oben" bzw. von außen initiiert. Förderan-reize stellen wesentliche Motive zur Erarbeitung strate-gischer Pläne dar (Altrock 2008: 70). Scheinbar seltener noch beruht strategische Planung auf der eigenstän-digen Initiative von Städten und Regionen. Dies deutet darauf hin, dass strategische Planung bisher eher eine Sonderform der Mehrebenenpolitik darstellt, die im Planungsalltag von Städten und Regionen noch nicht entsprechend etabliert ist.

3.3 Probleme der Anwendung

Im Folgenden werden Probleme der Anwendung stra-tegischer Leitbilder und Projekte in der Praxis der Stadt- und Regionalplanung beschrieben, ohne einen Anspruch auf Vollständigkeit zu erheben. Da bisher nur wenige kritische Analysen und Evaluationen vorliegen, wird auch auf eigene Erfahrungen zurückgegriffen.[7]

Strategische Leitbilder

In der Praxis haben sich Leitbilder in den letzten Jahren als fester Bestandteil vieler Entwicklungskonzepte, Ma-nagementansätze und Planwerke etabliert. Allerdings lassen sich große Diskrepanzen zwischen den hohen normativen Erwartungen an Leitbilder und ihrer häu-fig geringen Relevanz in der Praxis feststellen (Becker 1999). In der praktischen Anwendung von Leitbildern können u.a. folgende Probleme auftreten:

• Wenn Leitbilder für Städte und Regionen aus-schließlich durch Stadt- oder Regionalplaner erarbei-tet werden, stellen diese oft die städtebaulich oder raumordnerisch erwünschte Siedlungsstruktur in den Mittelpunkt. Strategische Leitbilder umfassen jedoch über Aussagen zur Fläche hinaus die Darstellung spe-zifischer Standortprofile und sozioökonomischer Stär-ken und setzen daher die Mitwirkung der Wirtschaft voraus.

• Um einen Konsens zwischen unterschiedlichen Akteuren aus Stadt und Region über gemeinsame Zu-kunftsvisionen zu erzielen und Konflikte bei unter-schiedlichen Positionen der Akteure zu vermeiden, tragen Leitbilder im Ergebnis häufig einen Kompro-misscharakter. In diesen Fällen bleiben sie unspezifisch und abstrakt und bieten damit für die Akteure entspre-chend große Interpretationsspielräume.

• Weisen Städte und Regionen eine stark diversifi-zierte Wirtschaftsstruktur – z.B. in Fällen des Nieder-gangs einer früheren Monostruktur – und damit keine dominanten Stärken auf, ist es im Rahmen von Leit-bildprozessen oft schwierig, ein langfristiges Profil im Gruppenkonsens zu bestimmen. Leitbildergebnisse sind dann sehr breit, um einzelne Bereiche nicht zu bevorzugen und mögliche andere Entwicklungspoten-ziale zu vernachlässigen (Kuder 2008).

• Wird die Erarbeitung von Entwicklungsleitbildern zur Entlastung der öffentlichen Verwaltung an externe Büros vergeben, besteht die Gefahr, dass sich die Mit-wirkung der Akteure häufig auf die punktuelle Teilnah-me an Veranstaltungen beschränkt. In manchen Fäl-len werden Leitbilder als politische Beschlussvorlagen sogar extern vorformuliert. In solchen Fällen beruhen

Manfred Kühn: Strategische Stadt- und Regionalplanung

die Leitbildergebnisse nur im geringen Maß auf sozialen Aushandlungs- und Verständigungsprozessen und können deshalb nur bedingt zur Identifikation und Motivation der Akteure beitragen.

• In anderen Fällen versuchen politische Eliten oder Führungspersönlichkeiten, Leitbilder für Städte oder Regionen von oben durchzusetzen. Diese haben dann ebenso wenig eine soziale Verankerung und finden eine entsprechend geringe Akzeptanz und Resonanz. Dies verdeutlicht das Dilemma zwischen hierarchischer („leadership") und netzwerksförmiger Steuerung (Partizipation).

Eine der wenigen empirischen Untersuchungen strategischer Stadtplanungsansätze am Beispiel von Duisburg und Dortmund zeigt, dass die befragten Akteure retrospektiv den sozialen Diskussionsprozess zur Erarbeitung der Leitbilder als wichtiger bewerten wie die Inhalte der Leitbilder (Ziesemer 2004: 223). Dies unterstreicht die Bedeutung sozialer Kommunikations- und Aushandlungsprozesse.

Strategische Projekte

Auch an strategische Projekte werden hohe Erwartungen geknüpft. Sie sollen ein „Schlüssel" für die weitere Stadt- und Regionalentwicklung sein und „Impulse" auslösen. Welche Wirkungen von solchen Projekten tatsächlich ausgehen, ist bisher wenig erforscht. Die Stärken der projektorientierten Planung werden in der empirischen Forschung zugleich als ihre Schwächen beschrieben. Zu den Problemen zählen

– die Ausblendung schwieriger Probleme und Konflikte der Stadt- und Regionalentwicklung durch die Bevorzugung von Projekten, die besonders öffentlichkeitswirksam, konsensfähig und kurzfristig realisierbar sind (Mayer 2004);

– die Einrichtung oder Ausgliederung von Sonderorganisationen des Projektmanagements, die Reibungen mit der bestehenden Verwaltung schafft und häufig nicht mehr ausreichend über demokratische Verfahren politisch legitimiert und kontrolliert ist (Mayer 2004);

– die mögliche Verselbständigung von Zielen aufgrund vorhandener Vermarktungs- und Erfolgszwänge besonders in Großprojekten, die ihre Einbettung in übergeordnete und längerfristige Planungen in Frage stellen kann (Liebmann 2003), sowie

– die Möglichkeit, dass neben den intendierten „Ausstrahlungseffekten" besonders im Falle von Großprojekten und -ereignissen auch nichtintendierte „Oaseneffekte" entstehen können, die „andere Zei-

ten und andere Räume austrocknen" (Häußermann/Siebel 1993: 145). Großprojekte wie z.B. Einkaufs- oder Entertainment-Center können autarke, von ihrer Umwelt abgekoppelte „Inseln" bleiben (Mayer 2004: 141).

4 Kritische Diskussion des Modells der strategischen Stadt- und Regionalplanung

Aus der empirischen Analyse der Rahmenbedingungen, Ansätze und Probleme in der Anwendung wird in diesem Schlusskapitel das Modell der strategischen Stadt- und Regionalplanung kritisch diskutiert und reflektiert. Dabei kann es sich bei dem derzeitigen Forschungsstand nur um die Formulierung vorläufiger Hypothesen sowie die Benennung von Widersprüchen und offenen Fragen handeln. Eine eigene Positionierung zwischen den Polen der Planungsgläubigkeit und Planungsskepsis kann nur ansatzweise erfolgen.

Strategische Planung ist im Kern eine widersprüchliche Einheit von Gegensätzen: langfristige Leitbilder und kurzfristige Projekte, gesamträumliche Steuerung und teilräumliche Interventionen, öffentliche und private Akteure, hierarchische und netzwerksartige Steuerung. Das Modell der strategischen Planung sucht als Steuerungs-Mix erneut einen „dritten Weg" zwischen integrierter Entwicklungsplanung und Inkrementalismus. Im Folgenden soll deshalb kritisch diskutiert werden, welche konkreten Schnittfelder zwischen diesen Gegensätzen in der praktischen Anwendung bestehen. Nach dem Modell strategischer Stadt- und Regionalplanung beziehen sich diese Schnittfelder vor allem auf die Träger, Instrumente und Kompetenzen strategischer Planung.

4.1 Träger: Governance zwischen öffentlichen und privaten Akteuren

Eine kritische Frage lautet: Wer ist der personelle und institutionelle Träger strategischer Planung? Die eingangs dargestellten angelsächsischen Definitionen bleiben hier zum Teil sehr vage.[8] Mit dem modischen Begriff Governance werden zwar Kooperationsformen zwischen politisch-administrativen, wirtschaftlichen und zivilgesellschaftlichen Akteuren erfasst, weitgehend offen bleibt jedoch die Frage konkreter Trägerschaften. Wer das Subjekt der Steuerung ist, verschwimmt in der Governance-Forschung (Mayntz 2005). Selle wendet als Skeptiker gegen die strategische Planung ein: „Es gibt nicht die eine Organisation und es ist völlig unklar, wer denn für die Konzipierung einer strategischen Orientierung zuständig sein und wer über die Mittel zu deren Umsetzung verfügen könnte." (Selle 2007: 4). Dagegen gehen Kunzmann und Albrechts davon aus, dass strategische Planung vom öffentlichen Sektor gesteuert wird (vgl. Kunzmann 2000, S. 259; Albrechts 2004: 747). Einzelne empirische Studien über Stadtentwicklungsstrategien in deutschen Großstädten bestätigen, dass diese von den öffentlichen Verwaltungsspitzen initiiert werden (Schneider 1997; Glock 2006).

Die Erarbeitung langfristiger Visionen, Leitbilder und Ziele für den Gesamtraum von Städten oder Regionen ist im Rahmen eines kontinentaleuropäischen Staatsverständnisses eine originäre Aufgabe der mit einer Planungshoheit ausgestatteten und dem Gemeinwohl verpflichteten öffentlichen Politik und Verwaltung. Von privaten Akteuren aus Wirtschaft und Bürgerschaft kann diese komplexe Aufgabe nicht erwartet werden, da diese in der Regel an partikularen und kurzfristigen Interessen orientiert sind. Eine Trägerschaft strategischer Planung in öffentlicher Hand kann entweder durch übergeordnete Stabsstellen der Verwaltung (ressortübergreifende Arbeitsgruppen, wenn strategische Planung zur „Chefsache" erhoben wird) oder durch die Stadtentwicklungsämter bzw. Regionalplanungsstellen erfolgen. Dies setzt jedoch voraus, dass diese über Kompetenzen der Koordination anderer Ressorts und des Projektmanagements verfügen. Wenn öffentliche Verwaltungsstellen Träger strategischer Planung sind, wird der private Sektor an den sozialen Aushandlungsprozessen von Visionen, Leitbildern und Zielen lediglich „beteiligt". Umgekehrt wird jedoch im Projektmanagement immer wieder privaten Investoren, Eigentümern und Unternehmern eine tragende Rolle zugeschrieben (wobei in vielen Fällen dennoch die öffentlichen Investitionen überwiegen).

Neue Governance-Formen in institutionalisierter Form entstehen durch intermediäre Sonderorganisationen, die zwischen öffentlichem und privatem Sektor angesiedelt sind. Dazu zählen insbesondere strategische Partnerschaften, Entwicklungsgesellschaften, Entwicklungsagenturen oder Geschäftsstellen. Diese werden in der Regel aufgabenbezogen und befristet eingerichtet. Sind diese öffentlich-privaten Sonderorganisationen die Träger strategischer Planungsansätze, dann stellen sich wiederum schwierige Fragen politischer Legitimation: Wessen Interessen verfolgen diese Agenturen und inwieweit wird ihr Handeln noch demokratisch kontrolliert? Eine offene Forschungsfrage ist, welche Stärken und Schwächen neue Governance-Formen in Gestalt öffentlich-privater Partnerschaften im Vergleich zu der klassischen Arbeitsteilung zwischen der öffentlichen Ziel- und Rahmensetzung und der privaten Realisierung haben.

4.2 Instrumente: integrierte Entwicklungskonzepte

Das Wechselspiel zwischen Leitbildern und Projekten als ein zentrales Merkmal strategischer Planung soll durch das Instrument der integrierten Entwicklungskonzepte hergestellt werden. Solche Konzepte liegen auf der Stadt- und Regionalebene in verschiedenen Varianten vor, ohne dass es dafür bisher anerkannte Standards gäbe. Regionale Entwicklungskonzepte umfassen meist die Elemente SWOT-Analysen, Leitbilder, Maßnahmen und Projekte (Knieling/Weick 2005). Auch viele integrierte Stadtentwicklungskonzepte arbeiten mit diesen Elementen. Inwieweit diese Konzepte die Orientierungs- und Umsetzungsfunktion tatsächlich gleichwertig „integrieren", ist bisher kaum evaluiert. Das „ständige Wechselspiel" zwischen Leitbildprozessen und Projekten ist sicher ein hoher Anspruch, der sich in der alltäglichen Planungspraxis von Städten und Regionen nur schwer realisieren lässt. In der Praxis bleibt die Anwendung des Modells erfahrungsgemäß oft fragmentarisch: Manche Städte und Regionen initiieren Leitbildprozesse, ohne auf die Umsetzung der Leitbildergebnisse in Projekten und Maßnahmen die gleiche Energie zu verwenden. Andere Städte und Regionen setzen auf einzelne Schlüsselprojekte, ohne dass diese in übergeordnete Leitbilder eingebettet sind. Viele integrierte Stadtentwicklungskonzepte wurden in der Vergangenheit zwar ressortübergreifend erarbeitet, blieben jedoch Papiere mit geringer Wirkung. Hierzu wären genauere empirische Analysen über die Ursachen ihres Scheiterns bzw. lautlosen Verschwindens in den Schubläden erforderlich.

4.3 Kompetenzen: Hierarchie und Netzwerke

Strategische Planung steht quer zu den etablierten Ressortstrukturen öffentlicher Verwaltungen. Sie hat jedoch lediglich – wie schon die integrierte Entwicklungsplanung – einen informellen Status. Damit stellt sich die Frage, welche Durchsetzungskompetenzen strategische Planung zwischen den Steuerungsmodi Hierarchie und Netzwerke überhaupt hat? In der Privatwirtschaft ist die strategische Unternehmensplanung eine Aufgabe des Top-Managements. Auch innerhalb der öffentlichen Verwaltung erhebt strategische Planung einen Führungsanspruch gegenüber nachgeordneten Verwaltungen (Ritter 2006). Nimmt man die neue Rollenzuschreibung eines „managers of change" ernst, so setzt dies Führungskompetenzen und den „Schatten der Hierarchie" voraus. Strategische Planung ist deshalb auf Führungspersönlichkeiten der politischen Eliten (Oberbürgermeister, Landräte oder Dezernenten) angewiesen. Mit dieser Form von „lea-

2

dership" verbunden ist jedoch immer die Gefahr einer personellen Diskontinuität im Fall politischer Machtwechsel. Eine personelle Kontinuität wird eher durch die Mitwirkung leitender Mitarbeiter der Verwaltung gewährleistet.

Sind Stadtentwicklungsämter oder Regionalplanungsstellen Träger strategischer Planung, dann verfügen diese in vielen Fällen nicht über Kompetenzen der Koordination anderer Ressorts, die für einen ressortübergreifenden Ansatz erforderlich sind. Besonders gegenüber der Wirtschaftsförderung und anderen starken Ressorts (wie z. B. Verkehr) ist die Stadt- und Regionalplanung in der Praxis oft in einer relativ schwachen Position.

Wird die Trägerschaft der strategischen Planung von der Verwaltung an eine Sonderorganisation delegiert (z. B. eine Entwicklungsgesellschaft oder Agentur), dann besteht in der Praxis oft ein großes Missverhältnis zwischen den hohen Managementkompetenzen und der faktisch geringen Durchsetzungsmacht von „Stadt- und Regionalmanagern", die häufig als befristet Beschäftigte gegenüber der öffentlichen Verwaltung am viel kürzeren Hebel sitzen. In der Praxis der projektorientierten Planung kommt es bezeichnenderweise häufig zu Reibungen zwischen dem Projektmanagement und der Verwaltung (Mayer 2004: 141).

4.4 Ausblick

Das konzipierte Modell der strategischen Stadt- und Regionalplanung unterliegt der rationalistischen Prämisse, dass die komplexe Welt von Städten und Regionen durch Allianzen zwischen öffentlichen und privaten Akteuren, umfassende Analysen, soziale Verständigungs- und Aushandlungsprozesse sowie eine stringente Ziel-Mittel-Kopplung als Ganzes steuerbar ist. Das Modell rückt zwar vom früheren technokratischen Glauben an „große Pläne" ab, versteht Planung als sozialen und politischen Aushandlungs- bzw. Lernprozess und relativiert die Rolle politisch-administrativer Akteure gegenüber Wirtschaft und Zivilgesellschaft. Dennoch erscheint eine vollständige Anwendung der Elemente des Modells sehr voraussetzungsvoll. Einige Probleme, Hemmnisse und Widersprüche in der Praxis wurden bereits beschrieben. Erst genauere empirische Studien können zeigen, unter welchen Rahmenbedingungen eine erfolgreiche Anwendung des Modells überhaupt möglich ist bzw. welche Gründe zum Scheitern führen.

Anmerkungen

(1)
Da diese beiden Modelle in der Literatur bereits häufig beschrieben wurden, handelt es sich lediglich um eine kurze Skizzierung.

(2)
Einige Autoren verwenden dennoch deduktive Modelle, um den Ablauf strategischer Planungsverfahren darzustellen, vgl. Ziesemer 2004, S. 15; Meyer zum Alten Borgloh 2005, S. 49.

(3)
Modelle dienen allgemein der vereinfachten Darstellung von Elementen der Wirklichkeit und ihren Zusammenhängen. Durch Idealisierung und Abstraktion sind sie ein Mittel zur Erkenntnis der Wirklichkeit. Das vorliegende Modell ist aus verschiedenen planungstheoretisch formulierten Ansprüchen abgeleitet und konstruiert. Es beschreibt normativ, wie strategische Planung sein soll. Als Strukturmodell stellt es wesentliche Elemente und ihre Beziehungen in den Mittelpunkt. Durch seine Vereinfachung lässt sich das Modell trotz unterschiedlicher institutioneller und organisatorischer Bedingungen auf die Stadt- und Regionalplanung übertragen. Es beansprucht jedoch nicht, mögliche Varianten oder einen bestimmten Prozessablauf abzubilden. Das Modell dient im Weiteren als heuristischer Rahmen, um empirische Analysen zu strukturieren.

(4)
SWOT ist ein Akronym für *Strengths* (Stärken), *Weaknesses* (Schwächen), *Opportunities* (Chancen) und *Threats* (Gefahren).

(5)
Dies sind auch die wesentlichen Elemente der Definition „strategieorientierter Planung", vgl.: Frey/Hamedinger/Dangschat 2008, S. 27.

(6)
Dieses Zitat verweist auf die frühzeitigere Problemwahrnehmung von Schrumpfungsprozessen durch Vertreter der Stadt- und Regionalsoziologie, während dieses Thema erst seit der Jahrtausendwende – ausgelöst durch das wohnungswirtschaftliche Problem der Wohnungsleerstände – breiteren Eingang in die Diskurse der Stadt- und Regionalplanung gefunden hat.

(7)
Evaluierung und Erfolgskontrolle von Plänen, Programmen und Projekten werden zwar in der Planungstheorie regelmäßig als integraler Bestandteil der Planungsprozesse dargestellt, jedoch selten praktiziert und publiziert. Die eigenen Erfahrungen beziehen sich auf das Leitprojekt „Strategien der Regenerierung schrumpfender Städte" (Laufzeit 2005–2008) am IRS Erkner sowie das Studienprojekt „Strategische Planung am Beispiel der IBA Fürst-Pückler-Land" an der BTU Cottbus (Wintersemester 2006/07);

(8)
Vgl. bspw. die Definition von strategischer Raumplanung als „a social process through which a range of people in diverse institutional relations and positions (...)" (Healey 1997, S. 5). Hierbei wird nicht klar, welche Akteure oder Institutionen diesen anspruchsvollen Prozess tragen und steuern.

Manfred Kühn: Strategische Stadt- und Regionalplanung

Literatur

Albrechts, L.; Healey, P.; Kunzmann, K.R. (2003): Strategic spatial planning and regional governance in Europe. In: Journal of the American Planning Association, 69, 2, S. 113–129.

Albrechts, L. (2004): Strategic (spatial) planning reexamined. In: Environment and Planning B: Planning and Design, 31, S. 743–758.

Altrock, U. (2004): Anzeichen für eine „Renaissance" der strategischen Planung? In: Altrock, U.; Günther, S.; Huning, S.; Peters, D. (Hrsg.): Perspektiven der Planungstheorie. Berlin, S. 221–238.

Altrock, U. (2008): Strategieorientierte Planung in Zeiten des Attraktivitätsparadigmas. In: Hamedinger, A.; Frey, O.; Dangschat, J.S.; Breitfuss, A. (Hrsg.): Strategieorientierte Planung im kooperativen Staat. Wiesbaden, S. 61–86.

Becker, H. (1999): Städtebau zur Sprache bringen – Leitbildentwicklung und -umsetzung in Deutschland. In: Becker, H.; Jessen, J.; Sander, R.: Ohne Leitbild? Städtebau in Deutschland und Europa. Stuttgart, Zürich, S. 454–474.

BBR – Bundesamt für Bauwesen und Raumordnung (Hrsg.) (2005): Stadtumbau in europäischen Städten mit Strukturkrise. Anregungen aus 10 Städten. Bonn. = Werkstatt. Praxis, Heft 37.

BMVBS – Bundesministerium für Verkehr, Bau- und Wohnungswesen (Hrsg.) (2001): Stadtumbau in den neuen Ländern. Integrierte wohnungswirtschaftliche und städtebauliche Konzepte zur Gestaltung des Strukturwandels auf dem Wohnungsmarkt der neuen Länder. Berlin.

Brake, K. (2000): Strategische Entwicklungskonzepte für Großstädte – mehr als eine Renaissance der „Stadtentwicklungspläne"? In: Deutsche Zeitschrift für Kommunalwissenschaften, 2, S. 269–288.

Braybrooke, D.; Lindblom, C.E. (1972): Zur Strategie der unkoordinierten kleinen Schritte (Disjointed Incrementalism). In: Fehl, G.; Fester, M.; Kuhnert, N. (Hrsg.): Materialien zur Planungsforschung. Gütersloh, S. 139–166.

DASL – Deutsche Akademie für Städtebau und Landesplanung (Hrsg.) (2002): Schrumpfende Städte fordern neue Strategien für die Stadtentwicklung. Berlin.

Deutscher Städtetag (2004): Zukunftssicherung durch integrierter Stadtentwicklungsplanung und kooperatives Stadtentwicklungsmanagement. Positionspapier des DST zum II. Nationalen Städtebaukongress. o. O.

Doehler-Behzadi, M.; Keller, D.A.; Klemme, M.; Koch, M.; Lütke-Daltrup, E.; Reuther, I.; Selle, K. (2005): Planloses Schrumpfen? Steuerungskonzepte für widersprüchliche Stadtentwicklungen. In: DISP, 2, S. 71–78.

Fassbinder, H. (1993): Zum Begriff der strategischen Planung – Planungsmethodischer Durchbruch oder Legitimation notgedrungener Praxis? In: Strategien der Stadtentwicklung in europäischen Metropolen. Hamburg. = Harburger Berichte zur Stadtplanung, Bd. 1, S. 9–16.

Frey, O.; Hamedinger, A.; Dangschat, J.S. (2008): Strategieorientierte Planung im kooperativen Staat - eine Einführung. In: Hamedinger, A.; Frey, O.; Dangschat, J.S.; Breitfuss, A. (Hrsg.): Strategieorientierte Planung im kooperativen Staat. Wiesbaden, S. 14–33.

Frey, O.; Keller, D.A.; Klotz, A.; Koch, M.; Selle, K. (2003): Rückkehr der grossen Pläne? Ergebnisse eines internationalen Workshops in Wien. In: DISP 153, S. 13–18.

Ganser, K.; Siebel, W.; Sieverts, T. (1993): Die Planungsstrategie der IBA Emscher Park. In: RaumPlanung 61, S. 112–118.

Glock, B. (2006): Stadtpolitik in schrumpfenden Städten. Duisburg und Leipzig im Vergleich. Wiesbaden.

Häußermann, H.; Siebel, W. (1993): Wandel von Planungsaufgaben und Wandel der Planungsstrategie – Das Beispiel der Internationalen Bauausstellung Emscher-Park. In: AK Stadterneuerung (Hrsg.): Jahrbuch Stadterneuerung 1993. Berlin, S. 141–154.

Hamedinger, A.; Frey, O.; Dangschat, J.S.; Breitfuss, A. (Hrsg.) (2008): Strategieorientierte Planung im kooperativen Staat. Wiesbaden.

Healey, P. (1997): Collaborative Planning. Shaping Places in Fragmented Societies. London.

Hutter, G. (2006): Strategische Planung. Ein wiederentdeckter Planungsansatz zur Bestandsentwicklung von Städten. In: RaumPlanung 128, S. 210–214.

IBA Stadtumbau Sachsen-Anhalt 2010 (2006): Die anderen Städte. IBA Stadtumbau 2010. Berlin.

IBA – Internationale Bauausstellung Fürst-Pückler-Land (2005): Landschaften verwandeln. Empfehlungen am Beispiel dreier industriell gestörter Landschaften in Europa. Berlin.

IBA – Internationale Bauausstellung Fürst-Pückler-Land (2006): IBA-Halbzeitdokumentation 2000–2010. Großräschen.

IfS – Institut für Stadtforschung und Strukturpolitik GmbH (2004): Die soziale Stadt. Ergebnisse der Zwischenevaluierung. Hrsg.: Bundesministerium für Verkehr, Bau- und Wohnungswesen. Berlin.

ILS – Institut für Landes- und Stadtentwicklungsforschung und Bauwesen des Landes NRW (Hrsg.) (2006): Die REGIONALEN in NRW. Impulse für den Strukturwandel. Dortmund, Aachen.

Kilper, H. (1999): Die Internationale Bauausstellung Emscher Park. Eine Studie zur Steuerungsproblematik komplexer Erneuerungsprozesse in einer alten Industrieregion. Opladen.

Knieling, J.; Weick, T. (2005): Regionale Entwicklungskonzepte. In: Akademie für Raumforschung und Landesplanung (Hrsg.): Handwörterbuch der Raumordnung. Hannover, S. 928–933.

Knieling, J. (2006): Leitbilder und strategische Raumentwicklung. In: Raumforschung und Raumordnung 64, 6, S. 473–485.

Kuder, T. (2008): Leitbildprozesse in der strategischen Planung. In: Hamedinger, A.; Frey, O.; Dangschat, J.S.; Breitfuss, A. (Hrsg.): Strategieorientierte Planung im kooperativen Staat. Wiesbaden, S. 178–192.

Kunzmann, K.R. (1993): Pittsburgh – eine Erfolgsgeschichte? In: Kunzmann, K.; Lang, M.; Theisen, R. (Hrsg.): Pittsburgh – eine Erfolgsgeschichte? Dortmund. = Dortmunder Beiträge zur Raumplanung 65, S. 10–21.

Manfred Kühn: Strategische Stadt- und Regionalplanung

Kunzmann, K.R. (2000): Strategic Spatial Development through Information and Communication. In: Salet, W.; Faludi, A. (Hrsg.): The Revival of Strategic Spatial Planning. Amsterdam, S. 259–265.

Liebmann, H. (2003): Mut der Verzweiflung? Events und Großprojekte als Motor der Stadtentwicklung. In: Liebmann, H.; Robischon, T. (Hrsg.): Städtische Kreativität - Potential für den Stadtumbau. Erkner, Darmstadt, S. 133–145.

Mayer, H.-N. (2004): Projekte in der Stadtentwicklung - Chancen und Risiken einer projektorientierten Planung. In: Gestring, N. u.a. (Hrsg.): Jahrbuch StadtRegion 2003. Opladen, S. 133–143.

Mayntz, R. (2005): Governance-Theory als fortentwickelte Steuerungstheorie? In: Schuppert, G.F. (Hrsg.): Governance-Forschung. Baden-Baden, S. 11–20.

Meyer zum Alten Borgloh, C. (2005): Eine Renaissance der Stadtentwicklungsplanung? Die strategische Entwicklungsplanung - untersucht in den europäischen Dienstleistungsmetropolen Barcelona und Frankfurt/Main. Dortmund. = Dortmunder Beiträge zur Raumplanung 123.

Mintzberg, H. (2000): The Rise and Fall of Strategic Planning. Edinburgh.

MIR - Ministerium für Infrastruktur und Raumordnung des Landes Brandenburg (Hrsg.) (2006): Arbeitshilfe im Rahmen von Integrierten Stadtentwicklungskonzepten INSEK auf Grundlage des „Masterplan Starke Städte" des Landes Brandenburg. Potsdam.

Ritter, E.-H. (2006): Strategieentwicklung heute – zum integrativen Management konzeptioneller Politik. In: Selle, K. (Hrsg.): Planung neu denken. Bd. 1: Zur räumlichen Entwicklung beitragen. Dortmund. S. 129–145.

Roberts, P.; Sykes, H. (Hrsg.) (2000): Urban regeneration. A handbook. London.

Salet, W.; Faludi, A. (Hrsg.) (2000): The Revival of Strategic Spatial Planning. Amsterdam.

Sartorio, F.S. (2005): Strategic Spatial Planning. In: DISP 162, 3, S. 26–40.

Scharpf, F.W. (1973): Planung als politischer Prozeß. Aufsätze zur Theorie der planenden Demokratie. Frankfurt/Main.

Schneider, H. (1997): Stadtentwicklung als politischer Prozeß. Stadtentwicklungsstrategien in Heidelberg, Wuppertal, Dresden und Trier. Opladen.

Scholl, B. (2005): Strategische Planung. In: Akademie für Raumforschung und Landesplanung (Hrsg.): Handwörterbuch der Raumordnung. Hannover, S. 1122–1129.

Seidel-Kwem, B. (1983): Strategische Planung in öffentlichen Verwaltungen. Berlin. = Betriebswirtschaftliche Schriften, H. 114.

Selle, K. (2007): Wer? Was? Für wen? Wie? In der planungstheoretischen Fachdiskussion bleiben mehr einfache Fragen offen als diese für die Verständigung gut ist. In: PNDonline: www.planung-neu-denken.de.

Siebel, W.; Ibert, O.; Mayer, H.-N. (1999): Projektorientierte Planung – ein neues Paradigma? In: Informationen zur Raumentwicklung, H. 3/4, S. 163–172.

Wiechmann, T.; Hutter, G. (2008): Die Planung des Unplanbaren. Was kann die Raumplanung von der Strategieforschung lernen? In: Hamedinger, A.; Frey, O.; Dangschat, J.S.; Breitfuss, A. (Hrsg.) (2008): Strategieorientierte Planung im kooperativen Staat. Wiesbaden, S. 102–121.

Ziesemer, A. (2004): Strategische Stadtentwicklungsplanung im Ruhrgebiet. Eine Analyse am Beispiel der Städte Duisburg und Dortmund. Dortmund. = Duisburger Geographische Arbeiten, Bd. 25.

Der Aufsatz ist im Rahmen des von der Deutschen Forschungsgemeinschaft finanzierten Projekts „Strategische Stadtplanung – Ansätze zur Regenerierung schrumpfender Städte" (2007–2009) entstanden. Ich danke meinen Kolleginnen und Kollegen Susen Fischer, Roland Fröhlich, Thomas Kuder, Heike Liebmann und James Scott für Hinweise und Kritik.

Dr. Manfred Kühn
IRS Institut für Regionalentwicklung und Strukturplanung
Flakenstraße 28–31
15537 Erkner
E-Mail: kuehnm@irs-net.de

Originaltext Fürst 2012

Dietrich Fürst

Internationales Verständnis von „Strategischer Regionalplanung"

S. 18 bis 30

Aus:

Dirk Vallée (Hrsg.)

Strategische Regionalplanung

Forschungs- und Sitzungsberichte der ARL 237

Hannover 2012

AKADEMIE
FÜR RAUMFORSCHUNG
UND LANDESPLANUNG
LEIBNIZ-FORUM FÜR RAUMWISSENSCHAFTEN

Dietrich Fürst

2   Internationales Verständnis von „Strategischer Regionalplanung"

Gliederung

2.1   Begriff und Konzept

Strategische Planung ist formal ein Pleonasmus („weißer Schimmel"), denn *strategisch* heißt *zielgerichtet planend* und Planung ist zielgerichtet. Solche Pleonasmen sind in der angelsächsischen Literatur beliebt, um deutlich zu machen, dass sich qualitativ etwas geändert hat.[1] Insofern ist strategische Planung auch ein normatives Konzept: Ausdruck einer gewünschten Neuausrichtung.

Strategische Planung kann in erster Annäherung definiert werden als „a particular type of planning that seeks to 'join up' major goals, policies and actions into a cohesive entity that follows well-informed efforts of marshalling and allocating resources into a viable planned response to a set of challenges, undertaken following a critical appreciation of competencies and shortcomings, as well as anticipating changes in the context(s) in which the strategy is formulated" (Dimitriou, Thompson 2007: 5) oder kürzer: „a systematic, integrated approach to policy making which takes full account of context, resources and the long term" (ebenda: 3).

Nach dieser Definition wäre neu an dem Ansatz: die (gestaltende) Ausrichtung auf (wirtschaftliche) Regionalentwicklung über einen Prozess kollektiver Einigung auf Problemlage, Stärken und Schwächen einer Region (SWOT-Analyse) und Ziele der Entwicklungssteuerung unter expliziter Wahl der Handlungswege und Prioritätensetzung sowie Einbindung derer, die letztlich die Planung umsetzen sollen.

Blickt man aber genauer hin, so wird der Begriff unterschiedlich verwendet, und zwar offenbar in Abhängigkeit vom herrschenden Institutionenrahmen der Planung: Im britischen und niederländischen Kontext liegt dabei das Gewicht auf räumlicher (wirtschaftlicher) Entwicklungsplanung, im amerikanischen steht das Management im Vordergrund (handlungsorientierter Ansatz mit Ressourcenzuordnung und Zeitplanung: Poister, Streib 2005), im französischen Kontext liegt der Schwerpunkt auf wirtschaftlicher

---

[1] Ein anderes Beispiel für solche Pleonasmen ist das viel genutzte Wort *„pro-active"*. Damit will man signalisieren, dass die betreffende Aktivität eine starke Stoßrichtung hat.

Entwicklungssteuerung (mit eher nachrangiger Ausrichtung auf Raumordnung).[2] In der deutschsprachigen Diskussion verbindet man damit den Prozess kollektiver Selbststeuerung – und ist damit schnell bei der Diskussion zu „urban" resp. „regional governance".

Zudem wird manchmal etwas als „neuartig" dargestellt, was bei näherem Hinsehen nur eine Modifikation des Vorhandenen ist (Taylor 2010: 201 ff. für das britische Planungssystem). Gleichwohl gibt es einen wachsenden Bedarf, das traditionelle Planungsgeschäft stärker strategisch auszurichten und neue Ansätze zu entwickeln, die der Regionalentwicklung dienen können. Es gibt zahlreiche Auslöser für das gewachsene Interesse an strategischer Planung, u. a.

- die in den westlichen Staaten zu beobachtenden Regionalisierungstendenzen (Brenner 2004),

- die im Zuge von Globalisierung, Regionenkonkurrenz und regionalem Engagement von Großinvestoren veränderten Planungsaufgaben (Frey et al. 2008: 16 f.; Zibell 2008: Standort-Qualitätswettbewerb, Gestaltung von Strukturveränderungen[3], nachhaltige Raumentwicklung, Entwicklung von Wissensregionen, Umgang mit Großprojekten etc.),

- die zunehmende Verschränkung von Stadtentwicklungspolitik mit Regionalentwicklung,

- die wachsenden Koordinationsbedarfe über fachliche wie gebietskörperschaftliche Grenzen hinweg, insbesondere unter der Thematik „nachhaltige Raumentwicklung" und „Klimaschutz",

- aber auch die zunehmende Planungs- und Politikverdrossenheit mit Bemühungen, die Bevölkerung wieder aktiver in die Gemeinwesenarbeit einzubinden (Öffentlichkeitsbeteiligung: Healey 2006a),

- ferner neue rechtliche Regelungen wie § 13 ROG, der in einigen Landesplanungsgesetzen zu einem konkreten Handlungsauftrag wurde,[4]

- ebenso wie die EU-Impulse zu regional koordinierten Entwicklungsstrategien, einschließlich der Leipziger „Territoriale Agenda" in Verbindung mit der „Leipziger Charta zur nachhaltigen europäischen Stadt": Damit geht ein neues Planungsverständnis einher, das auf Entwicklung, Partnerschaft und strategisches kollektives Handeln auf Regionalebene ausgerichtet ist (Schön, Selke 2007).

---

[2] Die englische Raumplanung trifft wie die deutsche die Unterscheidung in „Regionalpolitik" und „Raumordnung" (physical planning) und verfolgt auch grob die gleichen Ziele: Abbau regionaler Disparitäten, Förderung der räumlichen Entwicklung, Ordnung des Raumes. Die französische Raumplanung wird stärker von der früheren nationalen „Planification" geprägt, die auf wirtschaftliche Entwicklung ausgerichtet ist und die Raumordnung eher als „Anpassungsplanung" betreibt. Sie hat aber wesentlich das europäische Denken beeinflusst – das EUREK ist dem französischen Ansatz deutlich ähnlicher als dem britischen.

[3] Strukturveränderungen erfordern ein synchronisiertes Vorgehen vieler unterschiedlicher Akteure, weil jeder Akteur für sich gelassen in einem „Gefangenendilemma" verharrt: Wenn er sich nicht auf komplementäre Leistungen anderer verlassen kann, muss er die für ihn risikoärmste Lösung wählen, die aber bezogen auf eine kollektive Lösung suboptimal bleibt.

[4] Beispiel: § 15 LPlG Ba-Wü.

2

Auch in der deutschen Planerpraxis hat sich die Regionalplanung vielerorts um Ansätze der strategischen Planung erweitert, etwa über projektbasierte Planungsansätze und fachliche Teilprogramme (z. B. regionales Einzelhandelskonzept, regionale Windenergiekonzepte, vorbeugender Hochwasserschutz) (vgl. Zusammenstellungen für Baden-Württemberg bei Schmitz 2008: 36 f.).

Die Diskussion in der Planungswissenschaft scheint aber eher auf andere Impulse zu reagieren:

- auf die Integrationsbedarfe sich zunehmend entgrenzender Fachdisziplinen, Institutionen und Handlungsfelder (Grenzen in den Handlungsfeldern verwischen oder überlappen sich),

- auf den mit der Verwaltungsreform (*new public management*, Strategisches Management in der Verwaltung) verbundenen Paradigmawechsel, der klarere Ziele, bedarfsorientiertes Handeln, mehr *leadership* und mehr Managementqualitäten fordert (Shaw, Lord 2007: 67),

- auf die neoliberale Skepsis gegen das Plänemachen – auch die Theoretiker der strategischen Planung wenden sich tendenziell gegen die traditionelle Planung, weil sie diese als überholtes Konzept in einem Umfeld mit weiten Interdependenzen der Variablen, daraus folgend: unbestimmten Problemen und sich schnell wandelnden Problemlösungen ansehen (Frey et al. 2008: 26 f.). Deshalb sprechen einige Autoren statt von *strategischer* lieber von *strategieorientierter* Planung (Hamedinger et al. 2008).

Konzeptionell geht deshalb der Ansatz – bei allen Unterschieden diverser Autoren – in der Grundstruktur in Richtung eines *handlungsorientierten Management*-Ansatzes (und wurde so auch aus der Betriebswirtschaft übernommen: Wiechmann 2008a: 266). Weiter ausdifferenziert hieße das: Es gibt keinen rechtsverbindlichen Plan, vielmehr stehen der Prozess und seine sorgfältige Gestaltung im Mittelpunkt der Planung. Der Prozess ist zudem strukturiert: *kollektive Zielformulierung* über *Leitbilder (mission)* auf der Basis einer mit *stakeholders* entwickelten Potenzialanalyse (*SWOT-Analysis*) unter Wahl geeigneter Lösungswege (*Prioritätensetzung über strategische Projekte*) mit *Kontrolle des Ergebnisses* (über Indikatoren) und (lernenden) *Korrekturmöglichkeiten (controlling)* in der folgenden Planungsrunde. Die Planung soll den Umsetzungsprozess einbeziehen, ihn zumindest mit den Umsetzungsakteuren vorklären (Ressourcenzuordnung). Aber dieser Umsetzungsbezug wird eher gefordert als methodisch aufgezeigt. Darin liegt noch eine Hürde für die Praxis. Denn dort findet häufig eine Trennung von Planung und Umsetzung statt – die Gruppe derer, die an der Umsetzung beteiligt ist, unterscheidet sich häufig von der, die an der übergeordneten Planung mitwirkt. Das birgt Konflikte, weil diejenigen, die Projekte planen, nicht unbedingt diejenigen sind, die schließlich die Kosten für die Projektumsetzung zu tragen haben.

Dieses lineare, an traditioneller Planungslogik ausgerichtete Konzept wird von Wiechmann (2008a: 270 f.) infrage gestellt, weil es den Wirklichkeitsbedingungen nicht gerecht werde: Im Gegensatz zu den Prämissen der traditionellen Planung hätten wir es i. d. R. nicht mit monokausalen Beziehungen, sondern mit Interdependenzen zu tun, ferner sei eine steuernde Kontrolle der Umwelt ausgeschlossen, der praktische

Planungsprozess sei iterativ angelegt, finde auch im Vollzug noch statt und impliziere Lernprozesse, die zu neuen Strategien führen können. Wiechmann setzt deshalb das Modell der *adaptiven strategischen Planung* dagegen, das erstens von den Handlungs-möglichkeiten und der Pfadabhängigkeit kollektiver Entscheidungen ausgeht, zweitens statt umfassender vorausschauender Strategie eher adaptives, graduelles Justieren auf Basis von Lernprozessen verwendet und drittens irrtumsfreundlich ist.

Generell geht die Diskussion zur Strategischen Planung davon aus, dass zwar zielbezo-gene Steuerung angestrebt wird, man sich aber der geringen Steuerungsmöglichkeiten in einem immer weniger von den Steuernden bestimmbaren Handlungsfeld bewusst ist und deshalb einen „dritten Weg" zwischen komprehensiver und inkrementalistischer Planung suchen muss (Frey et al. 2008: 15).

2.2 Bedeutung

Die Bemühungen um strategische Planung sind Ausdruck des wissenschaftlichen Unbehagens an der herkömmlichen Stadt- und Regionalplanung.

Aber auch planungspraktisch sucht man nach neuen Antworten

- auf den wirtschaftlichen Strukturwandel, der traditionelle Regionalpolitik obsolet macht (Rückbau der industriellen Basis zugunsten der wissens- und dienstleistungs-bezogenen Wirtschaft),

- auf die paradigmatische Veränderung von (wohlfahrtsstaatlicher) Ausgleichspolitik zu (neoliberaler) Befähigungspolitik,

- auf die neuen Aufgaben der „nachhaltigen Regionalentwicklung",

- auf die wachsende Bedeutung von Großinvestoren in der Regionalentwicklung,

- auf die immer engere Verflechtung von Arbeitsfeldern in einem arbeitsteilig frag-mentiertem Handlungsfeld,

- auf die neuen Herausforderungen der EU-Kommission u. Ä.

Planungstheoretisch drehen sich die Bemühungen um einen neuen Planungsansatz mit den typischen Planungsdilemmata:

- von Bedarf nach Planungssicherheit bei wachsender Unsicherheit in den Rahmen-bedingungen der Planung,

- von der Notwendigkeit integrativer Ansätze bei wachsender intersektoraler, inter-regionaler Verflechtung der Problemfelder,

- vom Bedarf nach „integrierter Planung", aber nicht im technokratischen Modell der „Monsterplanung" der früheren „integrierten Entwicklungspläne",

- von schnellen, einfachen Lösungen, ohne die sachlichen, sozialen, räumlichen und ökologischen Interdependenzen aus dem Blick zu verlieren,

- von langfristiger Planung und kurzfristigem Vollzug,

21

2

■ von wachsenden (politischen) Konsenskosten und beschleunigten (marktbeein-flussten) Entscheidungsbedarfen,

■ von Integration der *stake-holders* als Bündnispartner bei nicht mehr beherrschbarer Komplexität und Kompliziertheit des dahinter steckenden Pluralismus von Interessen und Belangen,

■ generell: von Spontaneität und Planungssicherheit.

Zur Lösung dieser Dilemmata gibt es praxistaugliche Ansätze eigentlich nur in der Form, dass bei der einen oder anderen Forderung Abstriche gemacht werden müssen, was sofort Kritik und Gegenkonzepte auslöst, die stattdessen bei anderen Forderungen Abstriche einklagen. Oder man flüchtet sich in utopische Vorstellungen, über „regulierte Selbststeuerung und Selbstorganisation" und „Planung des Nicht-Planens" die traditionelle Planung faktisch durch kommunitaristisch-diskursive Prozesse zu ersetzen (so Oliver Frey 2008).

Praxisrelevante Ansätze der strategischen Planung reichen von Etzionis „mixed scanning" über Ganser et al. „perspektivischen Inkrementalismus" zur „projektorientierten Planung" und zur Kombination von Regionalen Entwicklungskonzepten mit Regionalmanagement. Ihnen allen ist gemeinsam, dass es sich um iterative Planungsprozesse handelt (bottom-up und top-down in wechselnder Folge), die den breiteren regionalen Richtungskonsens über kollektiv entwickelte Leitbilder/Visionen herstellen, deren Umsetzung jedoch auf Projekte mit eingeschränkterer Beteiligung (Suche nach Bündnispartnern) konzentrieren und den Planungsprozess bis in den Vollzug verlängern. Was bei strategischer Planung jedoch verstärkt hinzukommt, sind Verfahren, die Wirkungskontrollen und kollektive Lernprozesse institutionalisieren (z. B. über Controlling-Verfahren).

Die Neuausrichtung der Planung (von der viele sagen, sie sei gar nicht so neu) und die in der Literatur erkennbare Vorliebe für diskursive Ansätze beziehen sich in der deutschsprachigen Diskussion implizit (seltener: explizit) primär auf die Stadtentwicklungsplanung, seltener auf Regionalplanung. Bei den Autoren wird dabei nicht immer klar, ob sie lediglich die (kooperative) Entwicklungsplanung oder auch die regulierende Ordnungsplanung meinen. Zumindest bleibt unklar, wie mit den Aufgaben der regulierenden Ordnungsplanung verfahren werden soll, wie also Regelsetzung im Umgang mit der knappen Ressource Raum (Verteilungsfragen) und Planungssicherheit (Bindungswirkung der Planung) in diskursiven Ansätzen vermittelt werden sollen.

Bei aller Unschärfe dessen, was strategische Planung sein kann, wird ihr positiv zugeschrieben (vgl. Poister, Streib[5] 2005: 51 f.),

■ dass Ziele, Handlungsrichtungen und Prioritäten klar definiert und kontrolliert werden,

---

[5] Die Untersuchung von Poister/Streib an amerikanischen Kommunen ist nur bedingt auf Regionalplanung zu übertragen: Die amerikanischen Städte nutzen „strategische Planung" als integrierten Managementansatz für die Steuerung der kommunalen Aktivitäten, analog zu dem, was die KGSt unter „strategischem Management" versteht (Heinz 2000) – das geht also weit über das hinaus, was strategische Regionalplanung in Deutschland je sein könnte.

- dass der Planungsprozess strukturiert ist und die Akteure enger zusammenschweißt, weil der gemeinsame Handlungsbezug jedem deutlicher wird,

- dass klarer an den Handlungspotenzialen angesetzt wird und auch über deren Engpässe und suboptimalen Einsatz in diesem Zusammenhang nachgedacht werden kann,

- dass strategische Planung für das Zusammenspiel mit der Öffentlichkeit wegen ihres Entwicklungsbezugs und ihrer Projektorientierung sehr viel lebendiger ist als die traditionelle regulative Planung.

2.3 Kontroversen in der Literatur

Wie eingangs gesagt, findet sich in der Literatur kein gemeinsames Konzept der strategischen Planung. Die dazu geführten Kontroversen spiegeln vielmehr wider, warum es keine ideale Konzeption der strategischen Planung geben kann, sondern diese stark kontextgebunden organisiert und inhaltlich gestaltet werden muss.

Lassen wir Diskussionen aus, die nur nationale Relevanz haben, weil sie sich an der nationalen Praxis der Planung reiben oder umgekehrt Folgerungen für die Planung aus nationalen Verwaltungsreformbemühungen diskutieren,[6] so geht es im Wesentlichen um Folgendes:

Erstens ist unklar, wie die typischen planerischen Dilemmata zu lösen sind (so Friedmann in einer Diskussion 2004,[7] ferner Altrock 2008: 73 f., Altrock 2004). Insbesondere geht es um institutionelle Friktionen der Planung:

- dass der „rationale Plan" gesucht wird, Planung sich aber mit „schlecht definierten Problemen" herumschlagen muss, die sich rationalen Algorithmen entziehen und deren „Lösungen" bei unvollständiger Information mit begrenzter Informationsverarbeitungskapazität nur pluralistisch ausgehandelt werden können,

- dass die Planung gesamträumlich und überfachlich ausgerichtet werden soll, aber in sektoralisierten sowie fragmentierten Politikstrukturen stattfindet,

---

[6] Interessant ist hier England: Die Verwaltungsreformen zur Umwandlung der Gemeinden in „Unternehmen Gemeinde" haben die Stadtentwicklungsplanung aufgewertet – sie wird immer mehr als integrative Klammer der „Unternehmensführung" betrachtet. Mit der Folge, dass den Gemeinden die Planer ausgehen – nach Berechnung der Audit Commission (2006) haben 66 % der befragten Gemeinden ernsthafte Probleme, geeignete Planer zu finden, 48 % sagen, dass sie Schwierigkeiten haben, ihre Planer zu halten (Shaw, Lord 2007: 72).

[7] *John Friedmann* erörterte auf Einladung von *Patsy Healey* in der Zeitschrift *Planning Theory & Practice (5(2004): 49-67)* über Planungen in HongKong, Vancouver u. a. das Verhältnis von strategischem Planen zu langfristigen Orientierungen, Planung und Governance, integrativen Ansätzen vs. sektoralen Politikstrukturen und vertrat einen prozessualen, konsensorientierten pluralistischen Ansatz. Partner waren *John Bryson* (der strategische Planung auf einen „Werkzeugkasten" reduzierte), *John Hyslop* (der mit Mintzberg das Wesen der strategischen Planung im neuen Denken sah, nämlich sich über Ziele und Handlungsmöglichkeiten Klarheit zu verschaffen), *Alessandro Balducci* (der dem prozessualen Ansatz Friedmanns die Notwendigkeit eines Planungsdokuments und der politischen Legitimation der Prozesse entgegenhielt – Planen sei politische Kunst), *Wim Wiewel* (der in Friedmanns Ansatz eigentlich nichts Neues gegenüber der holländischen Praxis sah) und *Louis Albrechts* (der das Strategische in der Planung im Einbezug der richtigen Akteure, im Prioritätensetzen und in der problemabhängigen Nutzung geeigneter Instrumente sah).

■ dass Regelsetzungen sich immer mehr zentralisieren und den dezentralen problemspezifischen Entscheidungsspielraum einengen.

Zweitens wird zunehmend auch das Problem der Entscheidungsspielräume in den für moderne Planungen typischen *Multilevel Governance*-Systemen aufgeworfen (Hutter 2007). Im weiteren Kontext geht es dabei um die Verbindung von strategischer Planung mit *regional governance* (Healey 2006, Dangschat 2008). Fragen, die dabei auftreten, sind:

■ Wie geht man mit *institutional congestion* um? (Roberts, Lloyd 1999)

■ Reichen netzwerkartige Strukturen aus oder braucht man verfasste Regionen, und wenn ja: wie müssten die Strukturen dann aussehen?

■ Ist die institutionelle Einbindung der strategischen Planung überhaupt losgelöst vom Grad der Verwaltungstradition, Regionalidentität etc. zu bestimmen? (Elcock 2008)

Dieser Diskussionsstrang öffnet weitere Perspektiven in Richtung auf eine Systematik aller Einflussfaktoren, die konkrete strategische Planungen bestimmen können (Hutter 2007: 13 f.).[8]

Drittens finden sich zahlreiche Arbeiten, die sich kritisch mit Verfahren der strategischen Planung auseinandersetzen (Hamedinger et al. 2008; Altrock 2008; Pirhofer 2005), weil Verfahren Inhalte restriktiv bestimmen können, Beteiligungen selektiv aussortieren lassen, den Innovationsgrad absenken können etc.

Viertens werden allgemeinere planerische Fragen neu diskutiert (Dimitriou 2007): Wie geht man mit Komplexität um, wie mit Unsicherheit und Risiken? Gefordert werden flexiblere Verfahren, die Optionen öffnen und längerfristig offenhalten können, sowie Strategien, die reversible Lösungen zulassen („Experimentieren", Lernen) (Mintzberg et al. 1998). Dabei sollte das Handlungsfeld danach differenziert werden, welche Variablen gut oder schlecht zu prognostizieren sind, welche Handlungsfolgen gut oder schlecht einzuschätzen sind, und es sollten entsprechend unterschiedliche strategische Ansätze damit verbunden werden (Dimitriou 2007: 50 f.: „shape the future", „adapt to the future", „reserve the right to play").

Fünftens werden vereinzelt (aber zunehmend) Erfahrungen und Methodenfragen diskutiert (vgl. Dimitriou 2007), wie strategische Planung in der Praxis betrieben wird resp. werden sollte. Die Themen sind vielfältig und beziehen sich auf Fragen,

■ wie strategische Planung angelegt werden sollte (Bryson 2000; Mastop 2000),

■ wer beteiligt werden sollte,

■ wie sich Planung durch veränderte Rahmenbedingungen wandelt (Thornley 2000; Altrock 2008: 76 f.),

---

[8] Hutter differenziert drei Dimensionen: die inhaltliche Dimension, die Prozessdimension (wie Strategien entwickelt werden, welcher Typus von strategischer Planung genutzt wird, wie Lernprozesse organisiert werden etc.) und die Kontextdimension (externer Kontext: politisch, rechtlich, sozial, ökonomisch, physisch; interner Kontext: Planungskultur, Initiatoren, Mikropolitik, Ressourcen, Fähigkeiten).

- wie strategische Planung institutionell eingebunden werden sollte (Planungsstab, Planungsverband, Regionale Entwicklungsgesellschaft),

- wie ihr ausreichend Handlungsressourcen zugeordnet werden können (hierarchische Einbindung, Entwicklungsfonds),

- wie das Verhältnis von Prozess zu Plan aussehen sollte[9] u. Ä.

Vor allem die Konzentration auf größere *strategische Projekte* (Leitprojekte) ist umstritten. Denn damit sind eine *charismatische Steuerung* (Ibert 2007), eine Einfluss- und Zielverschiebung zugunsten der Wirtschaft und der Immobilienbetreiber, eine Verlagerung von Entscheidungen aus den politischen Gremien in Vor-Entscheiderstrukturen/ Sonderorganisationen u. Ä. verbunden.

2.4 Eine Art gemeinsamer Nenner der strategischen Planung

Nach überwiegender Vorstellung derer, die sich im deutschsprachigen Raum zu strategischer Planung äußern, scheint sich ein Konzept herauszubilden, das grob die folgenden Charakteristika hat:

a) Klare Zielorientierung: Der Ansatz arbeitet mit operationalisierten Zielen im Kontext übergeordneter Leitvisionen, die auf der Basis einer Stärken-Schwäche-Analyse entwickelt werden. Dabei kommt es allerdings entscheidend auf den Prozess an: Alle relevanten *stakeholders* sollten kooperativ in den Prozess integriert werden, und dieser sollte iterativ organisiert werden – mit zunehmender Einengung auf operationale Ziele.

b) Diskursive Methodik: Situationsbewertung und Zielfindung werden über einen kollektiven Prozess unter breiter Beteiligung relevanter regionaler Akteure (*stakeholders*) durchgeführt. Szenariotechnik, Visualisierungstechniken und Internet-Interaktionen unterstützen den Prozess. Üblicherweise wird dazu eine professionelle Moderation eingesetzt.

c) Mehrstufiges Verfahren: Die Konkretisierung der Programmierung erfolgt über einen mehrstufigen Ansatz. Zunächst wird eine allgemeine Einordnung der Region im Wettbewerb mit anderen vorgenommen, häufig werden dabei auch *benchmarks* gesetzt (Beispiel: „Wir wollen zu den „top 20" der europäischen Regionen gehören"). Daraus werden Leitlinien für die Verbesserung der Region definiert. Diese werden im dritten Schritt über zentrale Herausforderungen konkretisiert, die von der Region zu meistern sein werden. Jeder dieser Herausforderung werden schließlich konkrete Projekte zugeordnet, die so weit konkretisiert werden, dass Kosten und Nutzen grob abgeschätzt werden können. Entweder wird der gesamte Prozess oder es werden einzelne Teilschritte über förmliche und feierliche Unterzeichnungen von Vereinbarungen (Verträge) abgeschlossen – diese (rechtlich unbestimmten)

---

[9] Das wird vor allem von Healey (1997), Friedmann (2004), Bryson (2003; 2004) immer wieder betont: Es kommt auf das Management von Interdependenzen und das permanente Austarieren von Spannungen an. Strategische Planung wird dann lediglich als "a set of concepts, procedures, and tools, that may be used selecltively for different purposes in different situations" angesehen (Bryson 2004: 57).

2

Festlegungen sind meist die einzigen Bindungen, denen sich die Akteure förmlich unterwerfen müssen.

d) Lernorientiertes Controlling: Die Umsetzung wird laufend kontrolliert und entsprechend wird die Planung fortgeschrieben. Auf diesen Lernprozess wird großer Wert gelegt – strategische Planung ist ein iteratives Vorgehen mit zunehmender Konsensdichte und Selbstbindung der Beteiligten.

e) Einsetzung der Öffentlichkeitsbeteiligung als Ressource: Der Öffentlichkeit wird große Bedeutung zugeordnet, nicht nur aus Partizipationsgründen, sondern um einen äußeren Druck aufzubauen, der die handelnden Akteure zusammenhalten soll.

2.5 Einschätzung

Bei der Vielfalt von sehr unterschiedlichen Ansätzen strategischer Planung in der Praxis (Überblick für den deutschsprachigen Raum: Zibell 2008) ist es vermessen, eine allgemeine Einschätzung *der* strategischen Planung zu geben. Jedoch drängt sich der Eindruck auf,

- dass dahinter kein Paradigmawechsel steht, sondern dass unter neo-liberalen Bedingungen dasjenige Paradigma weitergesponnen wird, das mit dem *communicative turn of planning* seit Beginn der 1990er Jahre (Healey 1992) bereits angelegt war: Strategische Planung wird als Entwicklungsplanung wahrgenommen, aber dazu gehört auch zwingend, sich intensiver mit Komplexitätsreduktion, Unsicherheitsreduktion, Risikobearbeitung und Konfliktmanagement zu befassen. Allerdings führen Fragen der Macht, der Ungleichheit in der Interessenwahrnehmung (Ungleichheiten in Organisations-, Artikulations- und Konfliktfähigkeit) und Gefahren der Deprofessionalisierung in solchen Prozessen, also der *dark side of planning theory* (Flyvbjerg, Richardson 2002), auch bei Anhängern der strategischen Planung zu Unbehagen: Sollte die Rolle des Planers nicht vielmehr diejenige des *deliberative practitioner* sein, der fachlich kompetent ist, Interessen abwägt und normative Aussagen macht? (Peters 2008),

- dass die geheime Messlatte die alte integrierte Entwicklungsplanung ist, jetzt allerdings anders aufgebaut: Es geht zwar um fachübergreifende Handlungskonzepte, die wirksamer Synergieeffekte erzeugen. Aber nicht der finale Plan ist entscheidend, sondern der diskursive kooperative Prozess mit dem Ziel, diejenigen Akteure, auf welche die Planung in der Umsetzung angewiesen ist, in den Planungsprozess als Bündnispartner einzubinden. Der Prozess soll Einstellungen, Werthaltungen und Denkmuster der Beteiligten – bezogen auf die anstehenden Aufgaben – stärker harmonisieren: Der Prozess soll dazu führen, dass die relevanten *stakeholders* gleichartige Vorstellungen („Paradigmen") von den Handlungsbedarfen und erforderlichen Lösungsrichtungen entwickeln,

- dass deshalb aus planerischen Diskussionskreisen der strategischen Raumplanung ein großes Potenzial zugeordnet wird, um kollektives und integriertes Handeln im Raum möglich zu machen (Newman 2008: 1374), wobei viele der Diskutanten der Literatur offenbar einem kommunitaristischen (auf Solidarität gründenden) Gesellschaftsbild, zumindest für Kommunen und Regionen, anhängen, ohne sich genügend mit dem Problem der Austragung von Interessenkonflikten und der korporatistischen Entscheidungsfindung „am gewählten Gremium vorbei" zu befassen.

Zudem ist zu konstatieren, dass erstens die wissenschaftliche Diskussion sich eher im Normativen (Forderungen zu Verfahren, Institutionalisierungen) und Kritischen zur strategischen Planung bewegt, weniger im Methodisch-Instrumentellen. Zweitens setzt sie sich explizit von der in den 1960er und 1970er Jahren populären *integrierten Planung* insofern ab, als strategische Planung heute (Albrechts 2006b: 1162 ff.):

- selektiver ist: Sie ist auf die Kooperation des Privatkapitals angewiesen und muss dieses über konkrete Projekte sowie bezogen auf konkrete Problemfelder gewinnen,

- weniger auf den Plan als auf den Prozess der Veränderung setzt – der Plan „is just one vehicel amongst others with the purpose of producing change" (ebenda: 1163),

- über den Prozess vor allem Änderungen der Denkmuster, Einstellungen und Werthaltungen, aber auch Vertrauen bewirken soll,

- weniger auf traditionelle Prognosen, eher auf Szenarien und Visionen setzt,

- umsetzungsorientiert *(action oriented)* agiert,

- ein erweitertes Demokratieverständnis zugrunde legt, das dem Bürger mehr Mitsprache einräumt, gleichzeitig privates Kapital stärker auf die Bereitstellung öffentlicher Güter lenkt und kollektivem Lernen einen hohen Stellenwert zumisst (ebenda: 1165).

Varianten der strategischen Planung gewinnen in der Planungspraxis wachsende Bedeutung:

- Erstens wurden vielerorts entsprechende Rechtsregelungen geschaffen und die EU-Kommission zwingt über ihre Strukturfonds und die *Strategische Umweltprüfung* faktisch zur strategischen Planung.

- Zweitens wächst die Bereitschaft, auf regionaler Ebene mit neuen Formen der *regional governance* zu experimentieren, zumal auch die größeren Unternehmen eine Hinwendung zur Regionalebene erkennen lassen, und zwar als Folge des wachsenden Drucks auf die *corporate social responsibility,* aber auch aus wohlverstandenem Eigennutz (*place* als Ressource, vgl. Florida 2002).

- Drittens wird die strategische Kooperation der regionalen Eliten faktisch erzwungen durch den wirtschaftlichen Regionenwettbewerb im Kontext der dominanten neoliberalen Denkmuster.

Allen Formen der strategischen Planung dürfte aber eigen sein, dass sie erhebliche Rückwirkungen auf das Akteursfeld und den Institutionenrahmen haben: Planer lösen sich vom Denken in traditionellen Raumnutzungsordnungen, Politiker müssen ihr kurzfristig orientiertes Politikmanagement in längerfristiges strategisches Handeln einbinden, enge institutionelle Eigeninteressen, die den Nutzen der Kooperation verkennen, nehmen die Beziehungen zu anderen Akteuren nicht mehr als Nullsummenspiele der Macht wahr, sondern als *strategische Partnerschaften* u. Ä.

Internationales Verständnis von „Strategischer Regionalplanung"

2

Dabei ist sorgfältig zu beobachten und kritisch zu werten, wie sich mit strategischer Planung die Planungslandschaft ändert. Denn es ist unverkennbar, dass

- sich die Gewichte zugunsten der wirtschaftlichen Regionalentwicklung verschieben – ökologische und soziale Belange werden eher untergeordnet,

- damit inhaltlich eine Hinwendung zur Gestaltung der Lebensqualität in der Region einhergeht (*place-making*),

- innovativere Ansätze im Planungsprozess gefordert werden, weil Routinehandeln infrage gestellt wird und die regionalen Eliten gezwungen werden, sich über die Stärken und Schwächen, Chancen und Risiken der Entwicklung ihrer Region Gedanken zu machen sowie eine mittelfristige Orientierung und Prioritäten des koordinierten Handelns festzulegen,

- Planer einen besseren Zugriff zur Umsetzungsebene gewinnen müssen als es der gegenwärtigen Regionalplanung in Deutschland möglich ist. Das hat Rückwirkungen auf Fragen der Institutionalisierung und Ressourcenausstattung der Planungsorganisation,

- die Planungslandschaft vielfältiger wird, weil strategische Planung noch intensiver als traditionelle Regionalplanung an die spezifischen regionalen Kontextbedingungen, Akteurskonstellationen und Handlungsbedarfe angepasst werden muss (Tewdwr-Jones, Allmendinger 2007: 34).

Aber auch strategische Planung kann die Grundprobleme integrierten, koordinierten Handelns nicht lösen, wenn die Beteiligten nicht ihre gemeinsamen Interessen und die Vorteile solidarischen Handelns erkennen. Es verwundert dann nicht, wenn die Protagonisten der *strategischen Planung* etwas enttäuscht feststellen, dass in der Praxis eigentlich nirgends die „reine" Form zu finden ist (Albrechts 2006).

Generell beobachtet man eine stärkere Verbindung von Planung und neuen Formen der *regional governance*, weil im Kontext der strategischen Planung immer mehr Planungshandeln außerhalb der formalen Institutionen stattfindet, aber eng damit verbunden ist, und weil eine Vielzahl von Netzwerken ineinandergreifen, die im strategischen Planen ihre Koordination finden (so Allmendinger, Haughton 2009 für das Projekt „Thames Gateway"). Das verwundert nicht, denn strategische Planung verlangt neue kooperative Handlungsformen. Das ist weniger ein organisatorisches Problem als ein mentales: Veränderungen in den paradigmatischen Orientierungen der Beteiligten auszulösen, um Gemeinsamkeiten in den Problemlagen und Lösungswegen klarer zu erkennen, eine längerfristige Perspektive des Handelns einzunehmen, weniger auf externe Hilfe zu setzen, sondern sich selbst stärker in den Problemlösungsprozess einzubringen etc.

Strategisches Handeln ist meist leichter über Projekte herzustellen, weil hier die Komplexität begrenzbar ist, Ziele leichter zu harmonisieren sind, der Planungshorizont überschaubar bleibt und generell die Vorteils-Kosten-Struktur der Interaktionen für jeden Beteiligten besser zu berechnen ist. Deshalb wird den sog. *strategischen Projekten* immer mehr Aufmerksamkeit gewidmet – sie sollen die Verbindung herstellen zwischen den anspruchsvolleren Formen der strategischen Planung und der strategischen Steuerung in der Praxis (Albrechts 2006a).

Internationales Verständnis von „Strategischer Regionalplanung" ▨

Literatur

Albrechts, L. (2006a): Shifts in strategic spatial planning? Some evidence from Europe and Australia. In: Environment and Planning A 38 (6), 1149-1170.

Albrechts, L. (2006b): Bridge the gap: From spatial planning to strategic projects, In: European Planning Studies 14 (10), 1487-1500.

Almendinger, P.; Haughton, G. (2009): Soft spaces, fuzzy boundaries, and metagovernance: The new spatial planning in the Thames Gateway. In: Environment and Planning A 41 (3), 617-633.

Altrock, U. (2004): Anzeichen für eine Renaissance der strategischen Planung? In: Altrock, U.; Güntner, S.; Huning, S.; Peters, D. (Hrsg.): Perspektiven der Planungstheorie. Planungsrundschau 10, 221-238.

Altrock, U. (2008): Strategieorientierte Planung in Zeiten des Attraktivitätsparadigmas. In: Hamedinger, A.; Frey, O.; Dangschat, J.S.; Breitfuss, A. (Hrsg.): Strategieorientierte Planung im kooperativen Staat. Wiesbaden, 61-86.

Brenner, N. (2004): New state spaces: Urban governance and the rescaling of statehood. Oxford.

Bryson, J. (1997): Strategic planning. In: Shafritz, J.M. (Hrsg.): The International Encyclopedia of Public Policy and Administration. Boulder/Co., 2160-2169.

Bryson, J. (2000): Strategic planning and management for public and non profit organizations and communities. In: Salet, W.G.M.; Faludi, A. (Hrsg.): The revival of strategic spatial planning. Amsterdam, 205-217.

Bryson, J. (2004): Comment. In: Planning Theory & Practice 5 (1), 57-58.

Dangschat, J.S. (2008): Autobahnen ins Glück. Der Münchhausen-Effekt der Strategischen Raumplanung. In: Hamedinger, A.; Frey, O.; Dangschat, J.S.; Breitfuss, A. (Hrsg.): Strategieorientierte Planung im kooperativen Staat. Wiesbaden, 38-60.

Dimitriou, H.T. (2007): Strategic planning thought: Lessons from elsewhere. In: Dimitriou, H.T.; Thompson, R. (Hrsg.): Strategic Planning for Regional Development in the UK. A Review of Principles and Practices. London u.a., 43-65.

Dimitriou, H.T.; Thompson, R. (2007): Introduction. In: Dimitriou, H.T.; Thompson, R. (Hrsg.): Strategic Planning for Regional Development in the UK. A Review of Principles and Practices. London u.a., 3-9.

Elcock, H. (2008): Regional futures and strategic planning. In: Regional and Federal Studies 18 (1), 77-92.

Florida, R. (2002): The rise of the creative class. New York.

Flyvbjerg, B.; Richardson, T. (2002): Planning and Foucault. In search of the dark side of planning theory. In: Allmendinger, Ph.; Twedwr-Jones. M. (Hrsg.): Planning Futures. New Directions for Planning Theory. London u.a., 44-62.

Frey, O. (2008): Regulierte Selbststeuerung und Selbstorganisation in der Raumplanung. In: Hamedinger, A.; Frey, O.; Dangschat, J.S.; Breitfuss, A. (Hrsg.): Strategieorientierte Planung im kooperativen Staat. Wiesbaden, 224-249.

Frey, O.; Hamedinger, A.; Dangschat, J.S. (2008): Strategieorientierte Planung im kooperativen Staat – eine Einführung. In: Hamedinger, A.; Frey, O.; Dangschat, J.S.; Breitfuss, A. (Hrsg.): Strategieorientierte Planung im kooperativen Staat. Wiesbaden, 14-35.

Friedmann, J. (2004): HongKong, Vancouver and beyond: Strategic spatial planning and the longer range. In: Planning Theory & Practice 5 (1), 50-56.

Hamedinger, A.; Frey, O.; Dangschat, J.S.; Breitfuss, A. (Hrsg.) (2008): Strategieorientierte Planung im kooperativen Staat. Wiesbaden.

Healey, P. (1992): Planning through debate: The communicative turn in planning theory. In: Town Planning Review 20 (1), 9-20.

Healey, P. (1997): The revival of strategic planning in Europe. In: Healey, P.; Khakee, A.; Motte, A.; Needham, B. (Hrsg.): Making Strategic Spatial Plans. London u.a., 3-19.

2

Healey, P. (2006a): Transforming governance. Challenges of institutional adaptation and a new politics of space. In: European Planning Studies 14 (3), 299-320.

Healey, P. (2006b): Collaborative planning. Shaping places in fragmented societies. 2. Aufl., Baingstoke u. a.

Heinz, R. (2000): Kommunales Management. Überlegungen zu einem KGSt-Ansatz. Stuttgart.

Hutter, G. (2007): Stadtumbau und Governance. Strategische Planung zwischen maximaler Wahlfreiheit und institutionellem Determinismus. In: Städte im Umbruch 4, 11-21. http://www.schrumpfende-stadt.de/magazin/downloads/2007_4.pdf (15.02.2008).

Ibert, O. (2007): Megaprojekte und Partizipation. Konflikte zwischen handlungsorientierter und diskursiver Rationalität in der Stadtentwicklungsplanung. In: disP 171 (4), 50-63.

Mastop, H. J. M. (2000): The performance principle in strategic planning. In: Salet, W.; Faludi, A. (Hrsg.): The Revival of Strategic Planning. Amsterdam, 143-155.

Mintzberg, H.; Ahlstrand, B.; Lampel, J. (1998): Strategy Safari: The Complete Guide Through the Wilds of Strategic Management. Harlow u. a.

Newman, P. (2008): Strategic spatial planning: Collective action and moments of opportunity. In: European Planning Studies 16 (10), 1371-1383.

Peters, D. (2008): PlanerInnen als "deliberative practitioners". Auf dem Weg zu einem neuen, diskursiven Pragmatismus in der Planung(stheorie)? In: Hamedinger, A.; Frey, O.; Dangschat, J. S.; Breitfuss, A. (Hrsg.): Strategieorientierte Planung im kooperativen Staat. Wiesbaden, 309-321.

Poister, T. H.; Streib, G. (2005): Elements of strategic planning and management in municipal government: Status after two decades. In: Public Administration Review 65 (1), 45-56.

Roberts, P.; Lloyd, M. G. (1999): Institutional aspects of regional planning, management and development: Models and lessons from the English experience. In: Environment and Planning B 26 (4), 517-531.

Schmitz, G. (2008): Aktuelle und künftige Anforderungen an die Regionalplanung. In: Köhler, St.; Schulze, U.; Wille, V. (Hrsg.): Landes- und Regionalplanung in Baden-Württemberg. = Arbeitsmaterialien der ARL 342. Hannover, 32-52.

Schön, K. P.; Selke, W. (2007): Territoriale Agenda der EU – ein Ansatz für ein neues Planungs- und Entwicklungsverständnis in Europa. In: Informationen zur Raumentwicklung 7/8, 435-440.

Shaw, D.; Lord, A. (2007): The cultural turn? Cultural change and what it means for spatial planning in England. In: Planning, Practice and Research 22 (1), 63-78.

Taylor, N. (2010): What is this thing called spatial planning? An analysis of the British government's view. In: Town Planning Review 81 (2), 193-208.

Tewdwr-Jones, M.; Allmendiger, P. (2007): Regional institutions, governance and the planning system. In: Dimitriou, H. T.; Thompson, R. (Hrsg.): Strategic Planning for Regional Development in the UK. A Review of Principles and Practices. London u. a., 28-40.

Thornley, A. (2000): Strategic planning in the face of urban competition. In: Salet, W. G. M.; Faludi, A. (Hrsg.): The Revival of Strategic Spatial Planning. Amsterdam, 39-52.

Wiechmann, T. (2008a): Planung und Adaption. Strategieentwicklung in Regionen, Organisationen und Netzwerken. Dortmund.

Wiechmann, T. (2008b): Strategische Planung. In: Fürst, D.; Scholles, F. (Hrsg.): Handbuch Theorien und Methoden der Raum- und Umweltplanung. Dortmund, 265-275.

Zibell, B. (2008): Strategieorientierung in der Planung – eine neue Idee? In: Hamedinger, A.; Frey, O.; Dangschat, J. S.; Breitfuss, A. (Hrsg.): Strategieorientierte Planung im kooperativen Staat. Wiesbaden, 322-351.

Originaltext Ritter 2007

PNDonline I|2007

 PNDonline - eine Plattform des Lehrstuhls für Planungstheorie und Stadtentwicklung mit Texten und Diskussionen zur Entwicklung von Stadt und Region

Strategieentwicklung heute

Zum integrativen Management konzeptioneller Politik (am Beispiel der Stadtentwicklungsplanung)

Dr. iur **Ernst-Hasso Ritter**, berufliche Tätigkeiten in der Hochschule (Uni. Bochum), in der Kommunalverwaltung, in der Landesverwaltung Nordrhein-Westfalen, Mitglied u.a. in der Akademie für Raumforschung und Landesplanung und bei der Deutschen Akademie für Städtebau und Landesplanung.

1. Aufstieg und Fall der integrierten Stadtentwicklungsplanung

Als nach dem zweiten Weltkrieg die gröbste Not behoben war, stellte sich die Frage, nach welchem Zukunftskonzept die zerstörten Städte wieder aufgebaut werden sollten. Dabei zeigte sich ziemlich bald, dass es kaum um die bloße Wiederherstellung des ehemals Gewesenen und dass es nicht allein um die reine Fachaufgabe »Städtebau« gehen konnte. Verlangt waren vielmehr konzeptionelle Vorstellungen darüber, wie das soziale, wirtschaftliche und kulturelle Leben der Menschen in den Städten künftig aussehen sollte, wie die verschiedenen Fachkompetenzen in den Städten daran mitwirken konnten und auf wie die Maßnahmen zu finanzieren war. Es gab das Bedürfnis nach einer konzeptionellen Stadtpolitik. Schon Ende der 1950er Jahre hatte Norbert J. Lenort die gemeindliche »Entwicklungsplanung als vorausschauende Gestaltung des gemeindlichen Lebens« zur »Aufgabe aller Zweige der Selbstverwaltung« erklärt, die dem »Leitbild von der wohlgestalteten Gemeinde« folgen müsse (Lenort 1958, 9 f.). So entstand allmählich die kommunale Aufgabe »Stadtentwicklungsplanung« (dazu auch Laux 1972) , zunächst in den Großstädten, später auch in vielen mittleren und kleineren Städten (zur Historie vgl. Keppel 2004,11 ff.).

Stadtentwicklungsplanung war und ist strategische Planung:

- Sie geht von einer Gesamtperspektive aus und sucht Teilaufgaben zu integrieren.

- Sie versteht sich als Richtungsangabe und steckt Entwicklungsrahmen ab.

- Sie ist grundsätzlich auf lange Sicht angelegt (ohne dass dies gegenüber den operativen Plänen ein Alleinstellungsmerkmal wäre – »ewige« Bebauungspläne).

- Sie erhebt einen Steuerungsanspruch gegenüber »nachgeordneten« operativen Plänen.

Sieht man von der Sonderentwicklung in Ostdeutschland ab, dann wurde integrierte Planung zur einem reflektierten und umfassenden politischen Ansatz allerdings erst, als in den 1960er Jahren der Steuerungsanspruch umfassend erhoben wurde. Die »politische Planung« entstand. Sie fand ihren Ausdruck z.B. in der Neustrukturierung der Arbeit der Bundesregierung, in den großen Plänen zur Landesentwicklung oder eben in den integrierten Stadtentwicklungskonzepten auf kommunale Ebene. Planung im Sinne von individueller gedanklicher Vorwegnahme künftigen Handelns hatte es zwar schon immer gegeben, neu war jedoch der das ganze

(Langfassung aus: Klaus Selle (Hrsg.), 2006: Planung neu denken|Bd. 1, Zur räumlichen Entwicklung beitragen, Dortmund: Verlag Dorothea Rohn, S. 129-146)

Staatswesen erfassende generelle Schwenk von der reaktiven Vergangenheitsorientierung zur aktiven, zukunftsorientierten Umgestaltung. Steuerung wurde definierte durch die Fähigkeit, neue Probleme wirksam voraus zu sehen und ihnen durch konzeptionelle Gestaltung zuvor zu kommen. Politische Planung verstand sich demgemäß als integrierte und integrierende Steuerung des politischen Handelns (zum Ganzen E.-H. Ritter, 1987).

Nach dem Ölpreisschock Mitte der siebziger Jahre trat indes bald ein spürbarer Stimmungsumschwung ein. Allenthalben begannen sich »Grenzen des Wachstums« zu zeigen und damit war auch das Ende jener großzügigen Entwicklungsplanungen angesagt, die auf ständigem Ressourcenwachstum gebaut hatten. Der gesellschaftliche Grundkonsens über die langfristigen Ziele zerbrach; das Vertrauen in die zukunftsgestaltenden Fähigkeiten der Politik war erschüttert. Zugleich wurde immer deutlicher, dass die politische Planung letztlich noch einem mechanistisch-deterministischen Weltbild verhaftet gewesen war. Sie hatte sich in der Kontinuität eines abendländischen Rationalisierungsprozesses in Staat und Gesellschaft gesehen, in dem schließlich Politik zur Planung geworden und Planung nichts weniger war als der »systematische Entwurf einer rationalen Ordnung auf der Grundlage allen verfügbaren Wissens« (Kaiser 1965, 7). Dies mechanistisch-deterministische Weltbild ist in den vergangenen Jahrzehnten durch neue Erkenntnisse und Sichtweisen in Physik, Chemie und Biologie aus den Angeln gehoben worden (Quantentheorie, fraktale Mathematik; Chaostheorie, Theorie der Selbststeuerung). Ein anderes Weltbild brach sich Bahn und dessen Stichworte lauten nun: Komplexität, Nichtlinearität, Unsicherheit, Instabilität und Selbstorganisation (vgl. auch Ritter 1998, 10 ff.).

Das neue naturwissenschaftliche Denken konnte zwar nicht direkt auf Gesellschaft und Politik übertragen werden. Es entzog den bisherigen Steuerungsvorstellungen jedoch ihren Boden und veränderte nachhaltig die Optionen für politische Lösungen. Vor diesem Hintergrund sind die Schwächen des alten Planungsmodells klar zu benennen:

- Es ging prinzipiell davon aus, alle relevanten Informationen erfassen und zu einem rationalen Gesamtbild verarbeiten zu können (Synoptische Planung).

- Bei aller Unsicherheit, die Prognosen generell anhaftet, wurde die Zukunft doch als erkennbar und als umfassend gestaltbar angesehen.

- Das alte Planungssystem war zentralistisch ausgerichtet; es dachte in den etablierten staatlichen Hierarchien und in geschlossenen Kreisläufen.

- Subjekt der Planung war der Staat (einschließlich der Gemeinden); Planungsobjekt war die Gesellschaft. Planung folgte dem Modell der Über- und Unterordnung.

Im Ergebnis fand integrierte Regierungsplanung fortan nicht mehr statt; die umfassenden Pläne zur Landesentwicklung verschwanden; auch die integrierte Stadtentwicklungsplanung löste sich allmählich auf (Göb 1989). Der Deutsche Städtetag teilte im Februar 1982 seinen Mitgliedern lapidar mit, die Erarbeitung geschlossener und umfangreicher gesamtstädtischer Entwicklungspläne brauche nicht mehr im Vordergrund der Aufgabe zu stehen. Die gesetzliche Erwähnung der »Entwicklungsplanung«, die mit der üblichen legislativen Verspätung 1976 im Bundesbaugesetz erschienen war, wurde mit dem Baugesetzbuch 1987 wieder gestrichen.

2. Der Pendelumschwung zur inkrementalistischen Stadtpolitik

Die Desillusionierung war komplett: Weg von der integrierenden Planung – hin zur schrittweisen Abarbeitung der aktuellen Tagesnotwendigkeiten. Ein pragmatischer Inkrementalismus breitete sich aus (Waterhouse/Engel 1979). An die Stelle der Stadtentwicklung durch große Pläne trat die Stadtentwicklung durch Projekte. Wo dazu Pläne notwendig waren, beschränkten sie sich auf punktuelle Absicherung. Gefragt war nicht länger der integrative Ansatz (allen) kommunalpolitischen Handelns, sondern der fachliche städtebauliche Planungsansatz. Aber auch der fachliche Ansatz war nicht mehr der einer extensive Stadtentwicklung (Flächensanierungen, Neugründung von Stadtteilen) sondern der einer Entwicklung im Bestand (behutsame Erneuerung, Stadtumbau). Als ideelle Erhöhung des pragmatischen Ansatzes wurde nun der »perspektivische« Inkrementalismus propagiert. Damit ist eine »Vielzahl von kleinen Schritten gemeint, die sich auf einen perspektivischen Weg machen« (Ganser 1991, 59); seine Merkmale sind:

3

- Bewusster Verzicht auf Operationalisierung der Zielperspektiven, statt dessen Prinzipientreue im Einzelfall;

- Projekte statt Programme.

- Dieses inkrementalistische Vorgehen bestimmte das Planungsverständnis bis weit in die 1990er Jahre (vgl. etwa Albers 1995, 883).

Der Wandel in der Stadtentwicklungspolitik wurde verursacht und geprägt von einem tiefreichenden Wandel von Politik und Staat. Die Vorstellung vom Staat als dem mächtigen Akteur der Steuerung gesellschaftlicher Prozesse hatte abgedankt; sie wurde ersetzt durch die Leitideen des kooperativen (Ritter 1979) und des informalen (Bohne 1981) Staates. Das ist der Staat, der nicht mehr seine einseitig-hoheitlichen Machtmittel in den Vordergrund stellt, sondern den Konsens und die Zusammenarbeit mit den Betroffenen sucht. Das ist der Staat, in dem folglich die strengen und förmlichen Verfahren zur Erzeugung der hoheitlichen Machtmittel überlagert werden durch offene und informale Verfahren der Kooperation. Selbstverständlich trifft dieser Politikwandel das Staatswesen insgesamt: Bund, Länder und Kommunen. Ja gerade die Kommunen waren es, die den Druck der Partizipationswelle, die aus den unterschiedlichsten Teilen der Gesellschaft kam und der die unterschiedlichsten Motive zugrunde lagen, am stärksten verspürten.

Das blieb für das kommunale Planungsverständnis nicht ohne Folgen. Gewiss kannte die städtebauliche Planung schon seit dem Bundesbaugesetz 1960 Anhörungs- und Beteiligungsregeln und diese wurden im Laufe der Jahre immer stärker ausgebaut. Doch die Auswirkungen des Wandels reichten sehr viel weiter. Vielen galt die öffentliche Hand von ihren Ressourcen, ihren Fähigkeiten und ihren Steuerungsinstrumenten her inzwischen schlechthin als überfordert. Privatisierung und Deregulierung waren die Antwort. Staatsverwaltung und Kommunen verlagerten einen Teil ihrer Aufgaben in die Gesellschaft zurück, überführten sie zumindest in privatrechtliche Formen (z.B. Wohnungsunternehmen, Entwicklungsgesellschaften, privatrechtliche Projektträger) oder gründeten mit Privaten zusammen gemischtwirtschaftliche Unternehmungen (public-private-partnership). Auch in der Bauleitplanung wurden private Interessen unmittelbar (z.B. Vorhaben- und Erschließungsplan nach § 12 BauGB 1997) oder mittelbar (z.B. über städtebauliche Ver-

träge nach § 11 BauGB) institutionalisiert. Die kommunale Planung verstand sich schon lange nicht mehr als rein imperative Planung mit einseitig-hoheitlichem Anspruch sondern als kooperative Planung, die sich in einer »kommunikativen Wende« (vgl. Beiträge bei Selle 1996) den Mitgestaltungs-anspüchen der Gesellschaft öffnet und für die nicht allein die materielle Rationalität des Planungsinhalts sondern ebenso die prozessuale Rationalität des Planungsvorgangs Bedeutung erlangt. In diesem Sinne ist Planung ein Steuerungs-, Kommunikations- und Konsensbildungsprozess zugleich (Ritter 1998, 18).

Insgesamt begreift sich der Staat nicht mehr als »Erfüllungsstaat«, der alle von der Gesellschaft geforderten Leistungen in eigener Regie erbringt, sondern als »Gewährleistungsstaat«, der seine Verantwortung mit der Gesellschaft teilt und nur dort selbst eintritt, wo Aufgaben von den gesellschaftlichen Trägern nicht oder nicht richtig wahrgenommen werden können (vgl. zusammenfassend Schuppert 2003, 289 ff.). Für diese Art des Regierens hat sich inzwischen der Begriff »Governance« eingebürgert; was meint, dass die politische Steuerung auf die öffentlichen wie privaten Akteure in Netzwerken und Verhandlungssystemen angewiesen ist – oder zugespitzt formuliert: ohne eindeutiges Zentrum und ohne klare Hierarchien steuert. Die zunehmende Vernetzung ist charakterisiert durch eine Vielzahl von Beteiligten mit durchaus unterschiedlichen Interessenkonstellationen. In diesen Verbünden auf Zeit herrschen keine einfachen, reziproken Austauschverhältnisse zwischen den Beteiligten; demgemäß bestehen zwischen den Beteiligten nicht immer direkte, lineare Beziehungen; vielmehr laufen die Wirkungen über komplexe Beteiligungsketten. Steuerung wird multilateral und indirekt.

3. Rückbesinnung auf das strategische Niveau

Angesichts der realen Veränderungen in Staat und Gesellschaft mussten sich Zweifel einstellen, ob das an Teilverbesserungen und Einzelfallentscheidungen orientierte inkrementalistische Planungsverständnis noch den heutigen Anforderungen genügt, zumal sich die unsichtbare Hand der »perspektivischen« Steuerung allenfalls über die charismatischen Fähigkeiten einzelner Personen verwirklichte. Es breitete sich erneut ein Bedürfnis nach konzeptionellem Vorgehen, nach Wiedergewinnung der strategischen Dimension aus. Die wichtigsten Gründe dafür sind:

2

- Die hohe und immer noch steigende Komplexität der Lebensverhältnisse und die zunehmende Unübersichtlichkeit ihrer Entwicklung – im Allgemeinen.

- Im Besonderen: Die zunehmende Fragmentierung der öffentlichen Aufgaben durch Verlagerung und Privatisierung, die die Koordinationsnotwendigkeiten sprunghaft anschwellen lässt.

- Die enorme Verschuldung der öffentlichen Hand und die knappen öffentlichen Kassen, die gerade in den Kommunen die Investitionsspielräume gegen Null sinken lassen und dadurch die direkten Einflussmöglichkeiten auf die Stadtentwicklung drastisch reduzieren, so dass eine durchdachte Schwerpunktsetzung unumgänglich wird.

- Inhaltlich ist es die Leitidee der Nachhaltigen Stadtentwicklung, die ein Bedürfnis nach gesamthaften Sichtweisen auslöst, weil anders die schwierige Balance zwischen ökologischen, ökonomischen und sozialen Aspekten nicht zu halten ist.

- Und inhaltlich sind es die sich ausbreitenden Notwendigkeiten einer Politik unter Schrumpfungsbedingungen, die derart tiefe Einschnitte erfordern, dass angesichts der wachsenden innerkommunalen Disparitäten eine auf Einzellösungen angelegte Stadtpolitik alsbald überfordert wäre.

Mit den 1990er Jahren setzt die Rückbesinnung auf die strategische Einbettung der Stadtpolitik ein. Das gilt im international Maßstab auch allgemein (vgl. etwa Salet/Faludi 2003; Healey 2004; Poister/Strieb 2005). Man spricht schon von der »Rückkehr der großen Pläne« (Frey u.a. 2003). Der Begriff der »integrierten Stadtentwicklungspolitik« ist in Deutschland spätestens seit der Städtetagsveranstaltung »Stadt der Zukunft« 1998 (vgl. Deutscher Städtetag 1999) wieder hoffähig geworden und seit dem 2001 ausgelobten Bundes-Wettbewerb im Vorfeld des Programms Stadtumbau Ost sogar als Fördervoraussetzung rehabilitiert. Im Verlaufe des letzten fünfzehn Jahre entstanden in vielen Städten neue Stadtentwicklungsplanungen (Nachweise u.a. bei Schneider 1997; Keppel etc.2004; Difu 2004; Ziesemer 2004; Janzer/Schwedler 2005; Klotz/Frey 2005; Wuschansky 2005). Sie wurden mit angestoßen durch den Forschungsverbund »Stadt 2030«, mit dem das Bundesforschungsministerium seit 2000 entsprechende kommunale Bemühungen unterstützte (BMBF 2003); Impulse gingen ebenso von europäischen Programmen aus (z.B. URBAN oder dem Erfahrungsaustausch über URBACT). Mit der BauGB-Novelle 2004 tauchte sogar der Hinweis auf »städtebauliche Entwicklungskonzepte« wieder im Gesetz auf.

Man kann es durchaus als Ironie der Geschichte verstehen, dass zur Renaissance der Stadtentwicklungsplanung das gleiche sozi-ökonomische Kräftefeld beigetragen hat, an dem seinerzeit die (alte) politische Planung gescheitert war. Denn diese konnte sich nie von dem Odium eines staatlichen Dirigismus befreien, der Wirtschaft und Gesellschaft gemäß seinen Rationalitätskriterien steuern wollte. Nach dem Hinscheiden des sozialistischen Planungssystems und dem daraus interpretierten Sieg des Systems der wettbewerbsorientierten Marktwirtschaft hat sich die Situation dagegen umgedreht. In Politik und Gesellschaft hat ein ökonomisches Denken um sich gegriffen, das auch die Planungserfahrungen aus privaten Unternehmen für verallgemeinerungsfähig erklärt. Mithin geht es nicht länger um die Übertragung von Kategorien staatlicher Rationalität auf Wirtschaft und Gesellschaft, sondern umgekehrt um die gesellschaftlich akzeptierte Übernahme von Rationalitätskriterien aus der Wirtschaft durch den Staat. Während der öffentliche Bereich lange Jahre in Planungsphobie verharrte, entwickelten sich nämlich im privatwirtschaftlichen Bereich Planungstheorie und Planungspraxis weiter. Das gilt nicht zuletzt für die strategische Planung (vgl. etwa Eschenbach u.a. 2003; Hungenberg 2004). Wenngleich das strategische Vorgehen in der Unternehmensrealität keineswegs immer den hohen theoretischen Ansprüchen genügte, so ist doch heute die Wiedergewinnung der strategischen Dimension im öffentlichen Bereich ohne Rückgriff auf Modelle und Verfahren aus dem privatwirtschaftlichen Bereich unvorstellbar. Das belegt die Geschichte des New Public Management bzw. des Neuen Steuerungsmodells auch für die Bundesrepublik. Deshalb ist die Strategieentwicklung in diesen Rahmen zu fügen (zum Ganzen Ritter 2003).

4. Elemente der Strategieentwicklung

Strategieentwicklung im öffentlichen Bereich kann definiert werden als die Auseinandersetzung um Sinn und Zweck des Gemeinwesens und um die Zielrichtung seines Handelns. Es geht also um die langfristigen und

PNDonline I|2007   5

grundsätzlichen Orientierungen, die Identität geben im Bewusstsein der Organisationen von sich selbst, die Sinn stiften in der Ausrichtung ihres Handelns und die Leitplanken sind für das operative Geschäft. Dabei schwingen fast immer drei Begriffsinhalte von »Strategie« mit:

- Strategiefähigkeit (mentale Voraussetzungen);

- Strategie als Konzept (im engeren Sinne als strategischer Plan);

- Strategie als Art des Vorgehens (z.B. Strategien der Stadtentwicklung).

Die strategische Ebene ist charakterisiert durch ihr Niveau: sie liegt über der operativen Ebene, deren Handeln sie steuern soll. Ähnlich den geläufigen Planungsphasen gibt es einen idealtypischen Zyklus der Strategieentwicklung. Beide folgen einem systemtheoretischen Ansatz, dessen wichtigstes Merkmal bekanntlich im Einbau von Rückkopplungsschleifen besteht, so dass auch der Zyklus der Strategieentwicklung keine schematische Abfolge von Verfahrensschritten darstellt, sondern die Strukturlogik eines ansonsten flexiblen Prozesses:

Standortbestimmung: In der ersten Phase geht es darum, ein fundiertes Bild der relevanten Welt, wie sie zur Zeit ist und vermutlich in Zukunft sein wird, zu gewinnen (strategische Beobachtung). Dazu gehören: Die Diagnose und Prognose der Entwicklungen auf den jeweils wichtigen Tätigkeitsfeldern, die Ermittlung der Einstellungen und Erwartungen der Bürger (Verwaltungskunden), die Ermittlung der gesetzlichen Vorgaben für des Handeln der Gemeinden. Bei der Beobachtung des relevanten Umfelds sind die Kommunen zunehmend auf die Zulieferung fremden Wissens angewiesen; immer mehr Informationen sind nur über die Kooperation mit den Informationsbesitzern zu erlangen. Damit wird die Grenze zwischen Informationshingabe und Interessenvertretung fließend. Ferner leidet die Informationsaufnahme häufig unter einer Datenflut, die es schwer macht, das wirklich Wichtige herauszufiltern. Deshalb muss sich gerade die strategische Beobachtung auf ausgewählte Schwellenwerte konzentrieren, die problematische Veränderungen anzeigen, und sie muss funktionsfähige Frühwarnsysteme aufbauen. In verschiedenen Städten wird in diese Richtung experimentiert (z.B. Berlin). Ein Verbundforschungsprojekt aus Wissenschaft, kommunaler Praxis und Wirtschaft, das vom Bundesforschungsministerium Anfang 2004 aufgelegt worden ist, soll zum Aufbau von Frühwarn- und Kontrollsystemen beitragen.

Optionenentwicklung: Die eigentliche Strategiebildung beginnt mit der Frage, welche Entscheidungsnotwendigkeiten oder -möglichkeiten sich aus der Standortbestimmung ergeben, welche strategischen Optionen sich anbieten. In der Optionsphase sind die einzuschlagenden Linien zu diskutieren und Szenarien für die einzelnen Varianten zu entwickeln. Die zweite Phase endet mit der Festlegung der obersten Ziele. In der Praxis mündet diese oft in die Formulierung eines Leitbildes, das eine Vision von Gestalt und Aufgaben der künftigen Stadt vermitteln soll (vgl. Spaltenberg/Schiller 2004). In manchen Städten wird die Erarbeitung des Leitbilds verselbständigt (z.B. Heidelberg, Köln). Es kann dann als Zukunftsvision im Stadtmarketingkonzept eingesetzt werden. In stadtregionalen Zusammenhängen erfüllen Leitbilder eine wichtige Kooperationsfunktion; sie können als erste gemeinsame Grundlage für das künftige strategische Vorgehen dienen (z.B. Leitbild für den Ballungsraum Frankfurt/Rhein-Main 2004).

Zielprogrammierung: Sind die denkbaren strategischen Pfade diskutiert und entschieden, kommt es darauf an, die obersten Ziele zu Maßnahmenbündeln zu schnüren. Dazu sind die Zielsetzungen in einen transparenten Sinn- und Aktionszusammenhang zu bringen, die Verantwortlichkeiten für die Umsetzung zu benennen. Für den Übergang von der Strategie zum operativen Geschäft spielen in der Theorie sog. Zielvereinbarungen eine große Rolle; während die Praxis hier noch Zurückhaltung übt (Bollhoff/Wewer 2005, 152). Notwendig ist allerdings, zwischen den strategischen und den operativen Entscheidungen Synapsen herzustellen, die flexible und wirkungsvolle Übergänge schaffen und so angelegt sind, dass die strategischen Ziele intentionsgerecht von der Führungs- zur Umsetzungsebene gelangen.

Optimale Umsetzungbedingungen zu schaffen: In der Theorie wird Planung oft nur als Entscheidungsvorbereitung verstanden (statt vieler Selle 2005, 31, 93), die Planverwirklichung jedenfalls ausgeklammert. Für die Planungspraxis ist jedoch völlig klar, dass der Planungserfolg sich mit der Umsetzung entscheidet. Deshalb gehört die Schaffung von optimalen

Rahmenbedingungen für die Umsetzung auch zum Auftrag der Strategieentwicklung. Die Umsetzung von Strategie in operatives Handeln erfolgt nicht im deduktiven Nachvollzug, sondern durch eigene Zielinterpretationen und Abwägungsentscheidungen der operativen Instanzen, von denen Kreativität, Eigeninitiative und Eigenverantwortung verlangt sind. Jede Zielumsetzung ist zugleich ein Stück (konkreterer) Zielfindung. Um die Intentionen der Zielsetzung bis in das operative Geschäft hinein zu wahren, müssen deshalb neben dem Wortlaut der Ziele auch der Geist und die normativen Absichten, die sich mit der Zielsetzung verbinden, der operativen Ebene vermittelt werden. Dies geschieht am besten dadurch, dass im Gegenstromverfahren die operativen Instanzen bei der Strategiefindung mit einbezogen werden. Strategieentwicklung ist somit auch ein Prozess von unten nach oben.

Strategische Kontrolle: Ihr Sinn ist die Rückmeldung der Ergebnisse und Wirkungen, damit Planung nachgesteuert werden kann. Die strategische Kontrolle hat wenig gemein mit der traditionellen Verwaltungskontrolle (Rechnungsprüfung, Disziplinarrecht); sie richtet sich an den Anforderungen zukunftsorientierter Steuerung aus und greift fehlerhafte Verläufe nicht auf, um diese zu ahnden, sondern um aus ihnen zu lernen. Strategisch steht sie unter drei Fragen:

- Sind Umsetzungslinien nicht verfolgt worden (Fortschrittskontrolle)?

- Was hat die Umsetzung gebracht (Ergebnis- und Wirkungskontrolle)?

- Stimmen die Grundannahmen noch (Prämissenkontrolle)?

Wegen der besonderen Aufgabe und um Verwechselungen mit dem traditionellen Kontrollbegriff zu vermeiden, hat sich der Begriff »Controlling« durchgesetzt; im Hinblick auf den Strategieentwicklungsprozess spricht man von strategischem Controlling (Ritter 2003; Weber 2005). Im modernen Steuerungskreislauf ist das strategische Controlling das Herzstück; denn von ihm hängen die Anpassungs- und Innovationsfähigkeit des Planungssystems ab. In der Praxis freilich ist strategisches Controlling bisher nur sporadisch ausgebaut (Bogumil 2005, 496; vgl. auch Janzer/Schwedler 2005, 95 f.). Das Planungscontrolling mündet ein in die periodische Fortschreibung der Pläne. Das gilt insbesondere

auch für die Stadtentwicklungsplanung; hier sind Fortschreibungsperioden von vier bis fünf Jahren angebracht.

5. Integration von Planung und Management in der Stadtentwicklung

Management ist die verantwortliche Steuerung komplexer Organisationen; auch die Stadtentwicklung muss modernen Managementanforderungen genügen:

- Das Management muss strategiefähig sein: langfristig und in größeren Zusammenhängen denkend, perspektivisch vorgehend, Richtung gebend.

- Es muss führungsfähig sein: Prioritäten und Posterioritäten setzend, entscheidungs- und verantwortungsbereit, verlässlich.

- Es muss integrativ sein: die verschiedenen Leitungsfunktionen zusammenbindend, koordinierend und ordnend, die Gesamtwirkung beachtend.

Solche Formen des integrativen Managements haben insbesondere mit der Verwirklichung des Neuen Steuerungsmodells seit den 1990er Jahren in der Kommunalverwaltung Einzug gehalten. Stand zunächst entsprechend dem betriebswirtschaftlichen Ansatz das operative Geschäft (Produktdefinition, Kostenkontrolle, Rechenhaftigkeit der Vorgänge) im Vordergrund, so wandte sich doch um die Jahrhundertwende die Aufmerksamkeit verstärkt den strategischen Anforderungen zu (KGSt-Berichte 2000; Heinz 2000). Dabei ist deutlich geworden, dass integrierte Stadtentwicklung nicht bloß eine strategische Planungsaufgabe sein darf, sondern dass alle Führungsfunktionen einzubringen sind: neben Planung eben auch Organisation, Personal und Kontrolle. Die Funktion Planung hat das Handlungsumfeld einschließlich aller Vorgaben, die an das kommunale Handeln gestellt werden, aufzubereiten, über die Handlungsziele zu entscheiden und die strategisch erforderlichen Zielvernetzungen zu knüpfen. Gegenstand der Organisationsfunktion sind nicht zuletzt die Prozesse der Planverwirklichung, in denen Leistungsziele und Verfahrensabläufe gekoppelt werden. Die Funktion Personal bezieht sich auf die Mitarbeiterinnen und Mitarbeiter, deren Arbeitskraft, deren Qualität und deren Engagement den Wert und Erfolg der Aufgabe »Stadtentwicklung« wesentlich bestimmen. Die Funktion Kontrolle erfasst als Wirkungskontrolle das Ende des

Managementprozesses und schlägt als Controlling zugleich den Kreis wieder zum Beginn (Anpassung der Ziele). Insofern sind im Prozess moderner Strategieentwicklung Planung, Management und Controlling nicht zu trennen. Weder ist die Anwendung von Managementmethoden ein zureichender Planungsersatz, noch kommt Planung ohne adäquates Management aus. Das wird in der Praxis noch nicht überall wahrgenommen. Deshalb wird es darauf ankommen, die Erkenntnisse aus dem Neuen Steuerungsmodell mit den Planungsprozessen in der Stadtentwicklung enger zu verbinden (vgl. Stadt Saarbrücken).

6. Was unterscheidet die neue Stadtentwicklungsplanung von der alten?

Nach der Wiederentdeckung der strategischen Planung wurde die Bezeichnung »Plan« zunächst vielfach vermieden, um keine falschen Assoziationen an Vergangenes zu wecken; oft wurde nur von Leitbildern, von Leitlinien, von Konzepten gesprochen oder es wurden blumige, am Stadtmarketing orientierte Titel gewählt. Heute ist der Sprachgebrauch wieder unbefangener geworden. Sogar vom »Strategieplan« ist die Rede (z.B. Wien 2000, 2004). Gleichwohl gibt es derzeit keine allgemein geltende und erkenntnisleitende Nomenklatur; die Begriffe sind so vielfältig wie das praktische Vorgehen in den Städten vielgestaltig ist. Zumindest gegenwärtig macht es keinen Sinn, krampfhaft in der Praxis nach kategorisierenden Schubladen zu suchen. Das ist nicht nur eine Folge des immer noch recht experimentellen Charakters der Stadtentwicklungsplanung, sondern entspricht auch deren informaler Natur. Immerhin lassen sich einige Merkmale benennen, in welchen die Unterschiede zur alten Entwicklungsplanung deutlich werden.

Steuerungsverständnis: Die neue Stadtentwicklungsplanung fußt nicht länger auf den Idealvorstellungen vollständiger Rationalität in einem geschlossenen System (allumfassende Informationen, eingehende Problemkenntnis mit allen Lösungsalternativen, geradliniger Einsatz der Instrumente, Beherrschung der Eingriffsfolgen). Sie hat vielmehr mit den Grenzen der Rationalität bei offenen Steuerungskreisläufen leben gelernt: mit genereller Ungewissheit, unvollständigen Informationen und den unkalkulierbaren Verhaltensmöglichkeiten anderer Akteure.

Steuerungsanspruch: Hier liegt der Unterschied nicht so sehr in den juristischen Bindungswir-

kungen, denn auch die alte Entwicklungsplanung war unter rechtlichen Aspekten informal. Neu ist vielmehr, dass die heutigen Stadtentwicklungsstrategien generell nicht mehr dem hoheitlichen Modell imperativer Planung folgen, sondern in erster Linie auf Information, Überzeugung und Selbstbindung der Beteiligten vertrauen. Zwar ist die strategische Ebene der operativen Ebene sachlogisch übergeordnet. Dies ist jedoch keine klassisch hierarchische Beziehung mit linearen Weisungssträngen, sondern wird eher durch zwei sich überlappende Regelkreise versinnbildlicht, in deren Schnittfläche die Beziehungen sowohl von oben nach unten wie von unten nach oben laufen.

Steuerungsintensität: Die Steuerung ist nicht auf detaillierte Lenkung ausgerichtet sondern auf Rahmengebung, auf das Aufzeigen von Entwicklungsszenarios und auf die Vermittlung von Intentionen. Steuerung ist hier im Wesentlichen dezentrale Kontextsteuerung. Und in diesem Kontext besitzen die operativen Instanzen relativ weite Räume zur Eigeninitiative und Selbstverantwortung; das gilt auch für den Einsatz der Finanzmittel (dezentrale Budgetverantwortung) bis hin zum Einwerben privater Ressourcen. Demgemäß arbeitet das Planungscontrolling überwiegend mit offenen Prüfverfahren zur »Aktivitätsfolgenabschätzung« (Fürst 2004a, 249) und kaum mit detaillierten Rechenschaftspflichten.

Steuerungsumfang: Die Stadtentwicklungsplanung hat Abschied genommen vom synoptischen Planungsmodell; mithin ist auch ihr Integrationsanspruch nicht mehr umfassend. Sie ist heute schwerpunktorientiert und stellt (nur) das in der gegebenen Situation Wichtige in ihre Gesamtsicht ein. In diesem Rahmen können Projekte durchaus bedeutsam sein. Auch strategisch eingebundene Leitprojekte wirken sinnvermittelnd.

Steuerungsverfahren: Die Steuerung erfolgt vorzugsweise auf kooperativer Basis. Bei der Zielfindung stehen die oft beschriebenen Verhandlungsstrategien (vgl. schon BENZ 1994) im Vordergrund. Bei der Umsetzung herrschen Verträge (wenn die Bindung fester sein soll) oder Absprachen vor; einseitig-hoheitliche Machtinstrumente (Satzungen, Verwaltungsakte) werden nur noch dort eingesetzt, wo kooperative Instrumente z.B. wegen der ungewissen Vielzahl der Adressaten nicht greifen (Bauleitplanung) oder Druck ausgeübt werden soll.

Reichweite der Steuerung: Stadtentwicklungsplanung kann immer weniger an den territorialen Grenzen der örtlichen Gemeinschaft Halt machen. Will sie aus einer Gesamtsicht Lösungen anstreben, muss sie sich an der räumlichen Ausdehnung der Probleme und nicht an administrativen Grenzlinien ausrichten. Das zeigt sich namentlich bei den Kernstadt-Umland-Beziehungen, aber ebenso auf vielen sektoralen Feldern interkommunaler Kooperation (gemeinsame Gewerbegebiete). Stadtentwicklungsplanung ist damit nicht nur erhöhten Abstimmungspflichten unterworfen, sondern im zunehmenden Maße auch in ihren Inhalten stadt-regional vernetzt. Ihr informaler Charakter erleichtert diese Grenzüberschreitung.

7. Versuch einer zusammenfassenden Bewertung

Strategische Planung war – selbst in den Zeiten der Planungsphobie – nie ganz tot, auch wenn sie optisch kaum in Erscheinung trat. Insofern gibt die »Planungsgeschichte« keine scharfen Zäsuren wieder, sondern lediglich die großen Linien bei ansonsten differenzierten Entwicklungen. Heute kann man jedenfalls feststellen, dass sich sowohl staatliche wie kommunale Politik wieder offen zur Notwendigkeit strategischen Planens bekennen. Stand und Herangehensweisen der Strategieentwicklung sind freilich gerade im kommunalen Bereich höchst unterschiedlich, was die Planungstheorie in ihrem Bestreben nach exakter Vermessung noch irritiert (vgl. Altrock 2004, 233 ff.). Des ungeachtet lassen sich drei fundamentale Bestandteile herauskristallisieren, die für eine erfolgreiche Strategieentwicklung unverzichtbar sind: Konzept, Struktur, Kultur.

Mit *Konzept* ist letztlich das dokumentierte Planwerk gemeint, das den Stand der Willensbildung nachprüfbar festhält und im Gegensatz zur geheimnisumwitterten Mechanik des »perspektivischen Inkrementalismus« das notwendige Mindestmaß an Transparenz aufweist. Strategieentwicklung geht entsprechend ihrem Prozesscharakter gewiss auch schrittweise und insofern »inkremental« vor (vgl. Hungenberg, 14 f.); sie muss dabei jedoch die Legitimationsgrundsätze, an die eine demokratische Verwaltung gebunden ist, beachten und ihre Prozesse nachvollziehbar gestalten.

Struktur bedeutet, dass eine Strategie nicht nebenbei »irgendwie« aus dem Alltagsgeschäft herauswächst, sondern bestimmter Institutionen und Zuständigkeiten bedarf. Strategieentwicklung muss initiiert, gemanagt und verantwortet werden. Deshalb ist die Abschaffung der Planungsämter (z.B. Duisburg) ein falsches Signal.

Strategieentwicklung ist eingebettet in eine politische und administrative *Kultur*, die Konsens, Vertrauen und Verlässlichkeit, die Selbstbestimmung, Eigeninitiative und Mitgestaltungsbereitschaft fordert. Gleichwohl ist sie auf eine übergeordnete Zielsetzung ausgerichtet und kann auf einen top-down-Ansatz nicht völlig verzichten. Sie verlangt also einen eigenen Planungsstil und muss eine eigene Autorität ausprägen.

Die (neue) ressortübergreifende Stadtentwicklungsplanung ist ein gutes Beispiel für eine moderne Strategieentwicklung. Allerdings sind noch einige wichtige Fragen ungeklärt. Wenn der Stadtentwicklungsplanung vorgehalten wird, ihr mangele es (noch?) »an stadtgesellschaftlicher Basis, Akzeptanz und Durchsetzungsfähigkeit« (Wékel 2004, 84), ist damit wieder auf die nach wie vor bestehenden Legitimationsprobleme verwiesen. Unter den Bedingungen von public private partnership mit ihren keineswegs immer durchsichtigen Beziehungen und ihrer selektiven Interessenwahrnehmung haben sich die Probleme noch verstärkt. Hinzu kommt, dass sich aus den in der Literatur dokumentierten Fallbeispielen eine gewisse Neigung erkennen lässt, die kommunalen Vertretungen (Räte) erst ziemlich spät oder manchmal auch gar nicht in die Strategieentwicklung einzubinden. Auch die Beauftragung von privaten Büros (in NRW 2004 bei ca. 75% der befragten Gemeinden; vgl. Wuschansky, 12. Ähnliche Zahlen für BaWü; vgl. Janzer/Schwedler, 59) ist kritisch zu sehen. Das wirft generell die Frage auf, wie in einem mit der Zivilgesellschaft eng verflochtenen Staat die »kooperative Gemeinwohlkonkretisierung« (Schuppert 2000, 810 ff.) denn ausgestaltet sein muss, um den verfassungsrechtlichen Anforderungen an eine demokratische Willensbildung zu genügen. Patentrezepte sind nicht in Sicht. Vermutlich muss ein neuer Mix verschiedenartiger Legitimationszugänge gefunden werden:

- Dazu gehört zuerst, die Organe der repräsentativen Demokratie, nämlich die plural zusammengesetzten Entscheidungsgremien (Gemeinderäte) im Prozess zu stärken. Denn die Idee der Repräsentation ist gerade in der heutigen Situation einer weiten Interessenausdifferenzierung besonders geeignet, die angemessene Vereinheitlichung im Prozess der Willensbildung herzustellen.

▪ Ergänzend scheint es nötig, besondere Foren zu schaffen, in denen die Interessen der Gruppen zur Sprache kommen können, die bekanntlich nicht zu den aktiven Trägern der Zivilgesellschaft gehören (Alte, Arme, Arbeitslose etc.). Die bisher propagierten Modelle etwa der Advocatenplanung, der Stadtforen oder der »Planungszelle« haben die in sie gesetzten Erwartungen noch nicht erfüllen können.

▪ An geeigneten Eckpunkten sollten die kommunalen Planungen durchaus mehr Gebrauch machen von den plebiszitären Möglichkeiten, die alle Kommunalverfassungen inzwischen anbieten. Geeignete Fragestellungen zu finden, ist auch im Rahmen von Strategieentwicklungsprozessen nicht unmöglich.

Die Legitimationsanforderungen, die an alles Handeln der öffentlichen Hand gestellt werden müssen, sind letztlich Schranken auch der Informalität und der Kooperation im Strategieentwicklungsprozess. Das ändert nichts daran, dass gerade auf der strategischen Ebene der kooperative und informale Steuerungsmodus Sinn macht. Denn hier geht es darum, einen breiten Grundkonsens über die zukünftige Stadtentwicklung zu erreichen, der längerfristig überzeugen muss, aber andererseits offen genug ist, bei der operativen Umsetzung genügend Raum für dann fällige Interessendifferenzierungen zu lassen. Strategische Pläne sind Rahmenkonsense und nicht auf detailgetreuen Vollzug angelegt.

Eine zweite Schwachstelle der Stadtentwicklungsplanung ist die nicht immer vorhandene Bereitschaft der Kommunalpolitik zu langfristiger Selbstbindung. Es besteht eine gewisse Abneigung, der jeweiligen Opposition »Abhaklisten« über Erfolge oder Nichterfolge der «Stadtregierung« zu liefern. Hier hilft auf Dauer nur eine größere Einsicht in die Notwendigkeit, grundlegende Gemeinwohlprozesse transparent und öffentlich zu halten. Der dazu erforderliche Bewusstseinswandel kann gerade durch den Druck der Öffentlichkeit auf breite Beteiligung verstärkt werden.

Die dritte Schwachstelle der Stadtentwicklungsplanung liegt in ihrer territorialen Begrenztheit; sie ist noch zu sehr »punktzent-

riert« auf Problemlösungen »innerhalb der Stadtmauern« (in NRW leiteten 2004 unter 10% der Gemeinden stadtübergreifende Beteiligungsverfahren ein; vgl. Wuschansky, 12). Problemlösungen für regionale Zusammenhänge (z.B. Wissens- und Innovationsverbünde, Arbeitsmärkte, Zuliefererpools, Kommunikations- und Transportinfrastruktur, Ver- und Entsorgung) geraten dann suboptimal. Insbesondere die großen Städte müssen sich stärker als Knoten regionaler Netzwerke verstehen und ihre zukünftigen Aufgaben funktional an den für die jeweiligen Problemlösungen angemessenen und je unterschiedlichen Gebietszuschnitten orientieren. Damit wird der traditionelle kommunale Hoheitsraum immer häufiger zugunsten stadtregionaler Verbünde überschritten. Administrativer Gebietszuschnitt und funktionale Gebietszuschnitte sind immer weniger deckungsgleich. Die überkommunale Zusammenarbeit in den bisherigen Formen kann dem nur teilweise gerecht werden. Neue stadtregionale Organisationsformen sind notwendig (z.B. ansatzweise etwa Region Hannover). Insofern ist das weitere Schicksal der strategischen Stadtentwicklungsplanung eng verknüpft auch mit der fortschreitenden Diskussion um die Neujustierung des föderativen Systems in der Bundesrepublik Deutschland.

Fazit

Die strategische Dimension der Stadtentwicklung hat heute und in überschaubarer Zukunft wieder ihren festen Platz in der Stadtplanung (ebenso Albers 2005, 1070). Mehr noch, strategische Stadtentwicklungsplanung ist angesichts des tief reichenden Wandels im Verhältnis von Staat und Gesellschaft als Clearingstelle für eine demokratischen Grundsätzen entsprechende Gemeinwohlkonkretisierung auf kommunaler Ebene unentbehrlich. Sie ist eine Art »Gesellschaftsvertrag« über die längerfristigen Gemeinwohlvorstellungen (vgl. Fürst 2004b, 242). Sie arbeitet kooperativ und informal. Das erlaubt längerfristig akzeptierbare Grundorientierungen bei genügend Flexibilität im Vollzugsfall. Damit insbesondere dürfte sich die strategische Planung künftig von der operativen Planung unterscheiden, die unter Schrumpfungsbedingungen wohl mehr zu einem hoheitlich-imperativen Modus (Sanierungsplanung) zurückkehren wird, weil negative Verteilungskonflikte, sobald sie konkret werden, sich nur schwer im Konsens lösen lassen.

2

Literatur

ALBERS, GERD (1995): Stadtentwicklungspla-nung, in: Akademie f. Raumforschung u. Lan-desplanung (Hrsg.), Handwörterbuch der Raumordnung, 3. Aufl. Hannover [Verlag der ARL], S. 881 ff.

ALBERS, GERD (2005): Stadtentwicklungs-planung, in: Akademie f. Raumforschung u. Landesplanung (Hrsg.), Handwörterbuch der Raumordnung, 4. Aufl. Hannover [Verlag der ARL], S. 1067 ff.

ALTROCK, UWE (2004): Anzeichen für eine Renaissance der strategischen Planung?, in: Altrock, Uwe/Güntner, Simon/Huning, Sand-ra/Peters, Deike (Hrsg.), Perspektiven der Pla-nungstheorie, Berlin [Leue Verlag], S. 221 ff.

BENZ, ARTHUR (1994): Kooperative Verwal-tung, Baden-Baden [Nomos]

BÖLLHOFF, DOMINIK/WEWER, GÖTTRICK (2005): Zieldefinitionen, in: Blanke, Bern-hard/Bandemer, Stephan von/Nullmeier, Frank/Wewer, Göttrick (Hrsg.), Handbuch zur Verwaltungsreform, 3. Aufl., Wiesbaden [VS Verlag für Sozialwissenschaften], S. 147 ff.

BOGUMIL, JÖRG (2005): Die Umgestaltung des Verhältnisses von Politik und Verwaltung, in: Blanke, Bernhard/Bandemer, Stephan von/Nullmeier, Frank/Wewer, Göttrick (Hrsg.), Handbuch zur Verwaltungsreform, 3.Aufl., Wiesbaden [VS Verlag für Sozialwis-senschaften], S. 494 ff.

BOHNE, EBERHARD (1981): Der informale Rechtsstaat, Berlin [Duncker & Humblot]

Bundesministerium für Bildung und For-schung (BMBF) (2003): Abschlusskongress „Auf dem Weg zur Stadt 2030", Infobrief 13, unter: www.newsletter.stadt2030.de/index13.shtml (aufgerufen 25.7.05).

DEUTSCHER STÄDTETAG (Hrsg.) (1999): Stadt der Zukunft – Verwaltung der Zukunft. DST-Beiträge zur Stadtentwicklung Reihe E, Heft 28, Köln [DST]

DIFU (Deutsches Institut für Urbanistik) (2004): Integrierte Konzepte der Stadtentwick-lungsplanung, Difu-Berichte 1/04, Berlin [Di-fu]

ESCHENBACH, ROLF/ESCHENBACH, SE-BASTIAN/KUNESCH, HERMANN (2003): Strategische Konzepte, 4. Aufl. Stuttgart [Schäffer u. Poeschel]

FREY, OTTO/KELLER, DONALD A./KLOTZ, ARNOLD/KOCH, MICHAEL/SELLE, KLAUS (2003): Rückkehr der großen Pläne?, DISP 153, S. 13 ff.

FÜRST, DIETRICH (2004a): Mentalitäts- und Paradigmawechsel in der Stadtentwicklungs-planung, in: Neues Archiv für Niedersachsen 1/2004, S. 67 ff.

FÜRST, DIETRICH (2004b): Planungstheorie – die offenen Stellen, in: Altrock, U-we/Güntner, Simon/Huning, Sandra/Peters, Deike (Hrsg.), Perspektiven der Planungstheo-rie, Berlin [Leue Verlag], S. 239 ff.

GANSER, KARL (1991): Instrumente von ges-tern für die Städte von morgen?, in: Ganser, Karl/Hesse, Jens Joachim/Zöpel, Christoph (Hrsg.), Die Zukunft der Städte, Baden-Baden [Nomos], S. 54 ff.

GÖB, RÜDIGER (1998): Abschied von der Stadtentwicklungsplanung?, Raumforschung und Raumordnung, S. 289 ff.

HEALEY, PATSY (2004): The Treatment of Space and Place in the new Strategic Spatial Planning in Europe, in: Müller, Bernhard/Löb, Stephan/Zimmermann, Karsten (Hrsg.), Steuerung und Planung im Wandel, Fest-schrift für Dietrich Fürst, Wiesbaden [VS Ver-lag], S. 297 ff.

HEINZ, Rainer (2000): Kommunales Mana-gement, Stuttgart [Schäffer u. Poeschel]

HUNGENBERG, HARALD (2004): Strategi-sches Management im Unternehmen, 3. Aufl. Wiesbaden [Gabler]

KAISER, JOSEPH (1965): Planung I, Vorwort, Baden-Baden [Nomos]

JANZER, HELENE/SCHWEDLER, ANDREA (2005): Renaissance der Stadtentwicklungs-planung? Ergebnisse einer Städteumfrage in Baden-Württemberg, Rottenburg [Marc Oliver Kersting]

KEPPEL, HOLGER (2004): Die Renaissance eines Traumes und seine praktische Wirklich-keit, in: Stadtentwicklungsplanung in der Pra-xis, Rottenburg [Marc Oliver Kersting], S. 11 ff.

PNDonline I|2007   11

KGSt (Kommunale Gemeinschaftsstelle für Verwaltungsvereinfachung) (2000): Bericht 8/2000 Strategisches Management I, Bericht 9/2000 Strategisches Management II, Köln

KLOTZ, ARNOLD/FREY, OTTO (Hrsg.) (2005): Verständigungsversuche zum Wandel der Planung, Wien/New York [Springer]

LAUX, EBERHARD (1972): Entwicklungsplanung in der Kommunalverwaltung, in: Akademie f. Raumforschung u. Landesplanung (Hrsg.), Entwicklungsplanung in der Kommunalverwaltung, Forschungs- und Sitzungsberichte Bd. 80, Hannover [Verlag der ARL], S. 83 ff.

LENORT, NORBERT J. (1958): Wirtschaftliche Überlegungen zur Kommunalen Entwicklungsplanung, Bd. 29 der Schriften des Deutschen Verbandes für Wohnungswesen, Städtebau und Raumplanung, Köln

POISTER, THEDORE H./STREIB, GREGORY (2005): Elements of Strategic Planning and Management in Municipal Government, Public Administration Review, Bd. 65, Nr. 1, Washington

RITTER, ERNST-HASSO (1979): Der kooperative Staat, Archiv des öffentlichen Rechts, 104. Bd., S. 389 ff.

RITTER, ERNST-HASSO (1987): Staatliche Steuerung bei vermindertem Rationalitätsanspruch? Zur Praxis der politischen Planung in der Bundesrepublik Deutschland, Jahrbuch zur Staats- und Verwaltungswissenschaft, Bd. 1, Baden-Baden [Nomos], S. 321 ff.

RITTER, ERNST-HASSO (1998): Stellenwert der Planung in Staat und Gesellschaft, in: Akademie für Raumforschung und Landesplanung (Hrsg.), Methoden und Instrumente räumlicher Planung, Hannover [Verlag der ARL], S. 6 ff.

RITTER, ERNST-HASSO (2003): Integratives Management und Strategieentwicklung in der staatlichen Verwaltung, Die Öffentliche Verwaltung, S. 93 ff.

SALET, WILLEM./FALUDI, ANDREAS (eds.) (2000): The revival of strategic spatial planning, Amsterdam [Royal Netherlands Academy of Sciences]

SCHNEIDER, HERBERT (1997): Stadtentwicklung als politischer Prozeß, Wiesbaden [VS Verlag für Sozialwissenschaften]

SCHUPPERT, GUNNAR FOLKE (2000): Verwaltungswissenschaft, Baden-Baden [Nomos]

SCHUPPERT, GUNNAR FOLKE (2003): Staatswissenschaft, Baden-Baden [Nomos]

SELLE, KLAUS (Hrsg.) (1996): Planung und Kommunikation, Wiesbaden u. Berlin [Bauverlag]

SELLE, KLAUS (2000): Perspektivenwechsel - Überlegungen zum Wandel im Planungsverständnis, in: Fürst, Dietrich/Müller, Bernhard (Hrsg.), Wandel der Planung im Wandel der Gesellschaft, IÖR-Schriften, Bd. 33, Dresden, S. 53 ff.

SELLE, KLAUS (2005): Planen – Steuern – Entwickeln. Über den Beitrag öffentlicher Akteure zur Entwicklung von Stadt und Land, Dortmund [Dortmunder Vertrieb für Bau- und Planungsliteratur]

SPALTENBERGER, TOBIAS/SCHILLER, TOBIAS (2004): Stadtleitbilder – Eine Bestandsaufnahme mit Vergleich sieben baden-württembergischer Städte, in:

Keppel, Holger (Hrsg.), Stadtentwicklungsplanung in der Praxis, Rottenburg [Marc Oliver Kersting], S. 189 ff.

Stadt Saarbrücken (o.J.): Saarbrücken – Stadtentwicklung und Verwaltungsmodernisierung, Nr. 2.2, unter: www.stadt2030.de/pdf/saarbruecken-skizze (aufgerufen 24.4.05)

WATERHOUSE, ALAN/ENGEL, GABRIELE (1979): Pragmatischer Inkrementalismus in der Planung komplexer Stadtprobleme, Stadtbauwelt 61, S. 51 ff.

WEBER, JÜRGEN (2005): Strategisches Controlling, Weinheim [WILEY-VCH]

WÉKEL, JULIAN (2004): Wachstum und Schrumpfung – Chancen und Risiken für die Stadt, in: Deutsche Akademie für Städtebau und Landesplanung (Hrsg.), Beschleunigung und Stillstand. Vom Umgang mit Diskontinuität in der Stadtplanung. Bericht 2003, Berlin [Selbstverlag der DASL]

2

WUSCHANSKY, BERND (2005): Stadtentwicklungskonzepte in Nordrhein-Westfalen;
in: ILS – Institut f. Landes- u. Stadtentwicklungsforschung und Bauwesen NRW (Hrsg.),
Werkstattreihe „Integrierte Stadtentwicklungskonzepte", Dokumentation des 2. Werk-
stattgesprächs am 14.4.2005 in Dortmund, unter: www.ils.nrw.de (aufgerufen am 9.9.05)

ZIESEMER, ALEXANDER (2004): Strategische Stadtentwicklungsplanung im Ruhrgebiet, Dortmund [Dortmunder Vertrieb für Bau-
und Planungsliteratur]

Originaltext Mintzberg 1987

11

The Strategy Concept I:
Five Ps For Strategy

Henry Mintzberg

uman nature insists on a definition for every concept. The field of strategic management cannot afford to rely on a single definition of strategy, indeed the word has long been used implicitly in different ways even if it has traditionally been defined formally in only one. Explicit recognition of multiple definitions can help practitioners and researchers alike to maneuver through this difficult field. Accordingly, this article presents five definitions of strategy—as plan, ploy, pattern, position, and perspective—and considers some of their interrelationships.

Strategy as Plan

To almost anyone you care to ask, *strategy is a plan*—some sort of *consciously intended* course of action, a guideline (or set of guidelines) to deal with a situation. A kid has a "strategy" to get over a fence, a corporation has one to capture a market. By this definition, strategies have two essential characteristics: they are made in advance of the actions to which they apply, and they are developed consciously and purposefully. (They may, in addition, be stated explicitly, sometimes in formal documents known as "plans," although it need not be taken here as a necessary condition for "strategy as plan.") To Drucker, strategy is "purposeful action"[1]; to Moore "design for action," in essence, "conception preceding action."[2] A host of definitions in a variety of fields reinforce this view. For example:

- in the military: Strategy is concerned with "draft[ing] the plan of war . . . shap[ing] the individual campaigns and within these, decid[ing] on the individual engagements."[3]

2

- in Game Theory: Strategy is "a complete plan: a plan which specifies what choices [the player] will make in every possible situation."[4]
- in management: "Strategy is a unified, comprehensive, and integrated plan . . . designed to ensure that the basic objectives of the enterprise are achieved."[5]
- and in the dictionary: strategy is (among other things) "a plan, method, or series of maneuvers or stratagems for obtaining a specific goal or result."[6]

As plans, strategies may be general or they can be specific. There is one use of the word in the specific sense that should be identified here. As plan, *a strategy can be a ploy*, too, really just a specific "maneuver" intended to outwit an opponent or competitor. The kid may use the fence as a ploy to draw a bully into his yard, where his Doberman Pincher awaits intruders. Likewise, a corporation may threaten to expand plant capacity to discourage a competitor from building a new plant. Here the real strategy (as plan, that is, the real intention) is the threat, not the expansion itself, and as such is a ploy.

In fact, there is a growing literature in the field of strategic management, as well as on the general process of bargaining, that views strategy in this way and so focusses attention on its most dynamic and competitive aspects. For example, in his popular book, *Competitive Strategy*, Porter devotes one chapter to "Market Signals" (including discussion of the effects of announcing moves, the use of "the fighting brand," and the use of threats of private antitrust suits) and another to "Competitive Moves" (including actions to preempt competitive response).[7] Likewise in his subsequent book, *Competitive Advantage*, there is a chapter on "Defensive Strategy" that discusses a variety of ploys for reducing the probability of competitor retaliation (or increasing his perception of your own).[8] And Schelling devotes much of his famous book, *The Strategy of Conflict*, to the topic of ploys to outwit rivals in a competitive or bargaining situation.[9]

Strategy as Pattern

But if strategies can be intended (whether as general plans or specific ploys), surely they can also be realized. In other words, defining strategy as a plan is not sufficient; we also need a definition that encompasses the resulting behavior. Thus a third definition is proposed: *strategy is a pattern*—specifically, a pattern in a stream of actions.[10] By this definition, when Picasso painted blue for a time, that was a strategy, just as was the behavior of the Ford Motor Company when Henry Ford offered his Model T only in black. In other words, by this definition, strategy is *consistency* in behavior, *whether or not* intended.

This may sound like a strange definition for a word that has been so bound up with free will ("strategos" in Greek, the art of the army general[11]). But the fact of the matter is that while hardly anyone defines strategy in this way,[12] many people seem at one time or another to so use it. Consider this quotation form a business executive:

> Gradually the successful approaches merge into a pattern of action that becomes our strategy. We certainly don't have an overall strategy on this.[13]

This comment is inconsistent only if we restrict ourselves to one definition of strategy: what this man seems to be saying is that his firm has strategy as pattern, but not as plan. Or consider this comment in *Business Week* on a joint venture between General Motors and Toyota:

> The tentative Toyota deal may be most significant because it is another example of how GM's strategy boils down to doing a little bit of everything until the market decides where it is going.[14]

A journalist has inferred a pattern in the behavior of a corporation, and labelled it strategy.

The point is that every time a journalist imputes a strategy to a corporation or to a government, and every time a manager does the same thing to a competitor or even to the senior management of his own firm, they are implicitly defining strategy as pattern in action—that is, inferring consistency in behavior and labelling it strategy. They may, of course, go further and impute intention to that consistency—that is, assume there is a plan behind the pattern. But that is an assumption, which may prove false.

Thus, the definitions of strategy as plan and pattern can be quite independent of each other: plans may go unrealized, while patterns may appear without preconception. To paraphrase Hume, strategies may result from human actions but not human designs.[15] If we label the first definition *intended* strategy and the second *realized* strategy, as shown in Figure 1, then we can distinguish *deliberate* strategies, where intentions that existed previously were realized, from *emergent* strategies, where patterns developed in the absence of intentions, or despite them (which went *unrealized*).

Strategies About What?—Labelling strategies as plans or patterns still begs one basic question: *strategies about what?* Many writers respond by discussing the deployment of resources (e.g., Chandler, in one of the best known definitions[16]), but the question remains: which resources and for what purposes? An army may plan to reduce the number of nails in its shoes, or a corporation may realize a pattern of marketing only products painted black, but these hardly meet the lofty label "strategy." Or do they?

As the word has been handed down from the military, "strategy" refers to the important things, "tactics" to the details (more formally, "tactics teaches the use of armed forces in the engagement, strategy the use of engagements for the object of the war"[17]). Nails in shoes, colors of cars: these are certainly details. The problem is that in retrospect details can sometimes prove "strategic." Even in the military: "For want of a Nail, the Shoe was lost; for want of a Shoe the Horse was lost . . . " and so on through

2

Figure 1. Deliberate and Emergent Strategies

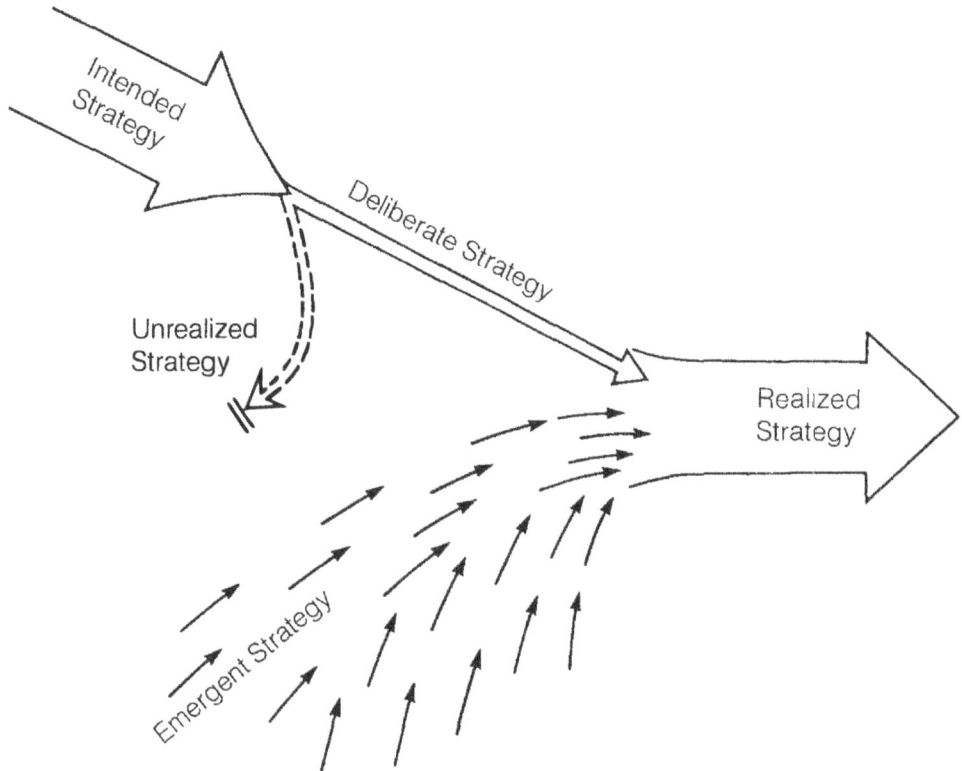

the rider and general to the battle, "all for want of Care about a Horseshoe Nail."[18] Indeed one of the reasons Henry Ford lost his war with General Motors was that he refused to paint his cars anything but black.

Rumelt notes that "one person's strategies are another's tactics—that what is strategic depends on where you sit."[19] It also depends on *when* you sit: what seems tactical today may prove strategic tomorrow. The point is that these sorts of distinctions can be arbitrary and misleading, that labels should not be used to imply that some issues are *inevitably* more important than others. There are times when it pays to manage the details and let the strategies emerge for themselves. Thus there is good reason to drop the word "tactics" altogether and simply refer to issues as more or less "strategic," in other words, more or less "important" in some context, whether as intended before acting or as realized after it.[20] Accordingly, the answer to the question, strategy about what, is: potentially about anything. About products and processes, customers and citizens, social responsibilities and self interests, control and color.

Two aspects of the content of strategies must, however, be singled out because they are of particular importance and, accordingly, play major roles in the literature.

Strategy as Position

The fourth definition is that *strategy is a position*—specifically, a means of locating an organization in what organization theorists like to call an "environment." By this definition, strategy becomes the mediating force—or "match," according to Hofer and Schendel[21]—between organization and environment, that is, between the internal and the external context. In ecological terms, strategy becomes a "niche"; in economic terms, a place that generates "rent" (that is "returns to [being] in a 'unique' place"[22]); in management terms, formally, a product-market "domain,"[23] the place in the environment where resources are concentrated (leading McNichols to call this "root strategy"[24]).

Note that this definition of strategy can be compatible with either (or all) of the preceding ones: a position can be preselected and aspired to through a plan (or ploy) and/or it can be reached, perhaps even found, through a pattern of behavior ("the concept of strategy need not be tied to rational planning or even conscious decision-making assumptions. Strategy is essentially a descriptive idea that includes an organization's choice of niche and its primary decision rules . . . for coping with that niche"[25]).

In military and game theory views of strategy, it is generally used in the context of what is called a "two-person game," better known in business as head-on competition (where ploys are especially common). The definition of strategy as position, however, implicitly allows us to open up the concept, to so-called n-person games (that is, many players), and beyond. In other words, while position can always be defined with respect to a single competitor (literally so in the military, where position becomes the site of battle), it can also be considered in the context of a number of competitors or simply with respect to markets or an environment at large.[26] Since head-on competition is not the usual case in business, management theorists have generally focussed on the n-person situation, although they have tended to retain the notion of economic competition.[27] But strategy as position can extend beyond competition too, economic and otherwise. Indeed, what is the meaning of the word "niche" but a position that is occupied to *avoid* competition.

Thus, we can move from the definition employed by General Ulysses Grant in the 1860s, "Strategy [is] the deployment of one's resources in a manner which is most likely to defeat the enemy," to that of Professor Rumelt in the 1980s, "Strategy is creating situations for economic rents and finding ways to sustain them,"[28] that is, any viable position, whether or not directly competitive.

Astley and Fombrun, in fact, take the next logical step by introducing the notion of "collective" strategy, that is, strategy pursued to promote cooperation between organizations, even would-be competitors (equivalent in biology to animals herding together for protection).[29] Such strategies can

2

range "from informal arrangements and discussions to formal devices such as interlocking directorates, joint ventures, and mergers."[30] In fact, considered from a slightly different angle, these can sometimes be described as *political* strategies, that is strategies to subvert the legitimate forces of competition.

Strategy as Perspective

While the fourth definition of strategy looks out, seeking to locate the organization in the external environment, the fifth looks inside the organization, indeed inside the heads of the collective strategist. Here, *strategy is a perspective*, its content consisting not just of a chosen position, but of an ingrained way of perceiving the world. Some organizations, for example, are aggressive pacesetters, creating new technologies and exploiting new markets; others perceive the world as set and stable, and so sit back in long established markets and build protective shells around themselves, relying more on political influence than economic efficiency. There are organizations that favor marketing and build a whole ideology around that (an IBM); others treat engineering in this way (a Hewlett-Packard); and then there are those that concentrate on sheer productive efficiency (a McDonald's).

Strategy in this respect is to the organization what personality is to the individual. Indeed, one of the earliest and most influential writers on strategy (at least as his ideas have been reflected in more popular writings) was Philip Selznick, who wrote about the "character" of an organization—distinct and integrated "commitments to ways of acting and responding" that are built right into it.[31] A variety of concepts from other fields also capture this notion: psychologists refer to an individual's mental frame, cognitive structure, and a variety of other expressions for "relatively fixed patterns for experiencing [the] world"[32]; anthropologists refer to the "culture" of a society and sociologists to its "ideology"; military theorists write of the "grand strategy" of armies; while management theorists have used terms such as the "theory of the business"[33] and its "driving force"[34]; behavioral scientists who have read Kuhn[35] on the philosophy of science refer to the "paradigm" of a community of scholars; and Germans perhaps capture it best with their word "Weltanschauung," literally "worldview," meaning collective intuition about how the world works.

This fifth definition suggests above all that strategy is a *concept*. This has one important implication, namely, that all strategies are abstractions which exist only in the minds of interested parties—those who pursue them, are influenced by that pursuit, or care to observe others doing so. It is important to remember that no-one has ever seen a strategy or touched one; every strategy is an invention, a figment of someone's imagination, whether conceived of as intentions to regulate behavior before it takes place or inferred as patterns to describe behavior that has already occurred.

What is of key importance about this fifth definition, however, is that the perspective is *shared*. As implied in the words Weltanschauung, culture, and ideology (with respect to a society) or paradigm (with respect to a community of scholars), but not the word personality, strategy is a perspective shared by the members of an organization, through their intentions and/or by their actions. In effect, when we are talking of strategy in this context, we are entering the realm of the *collective mind*—individuals united by common thinking and/or behavior. A major issue in the study of strategy formation becomes, therefore, how to read that collective mind— to understand how intentions diffuse through the system called organization to become shared and how actions come to be exercised on a collective yet consistent basis.

Interrelating the Ps

As suggested above, strategy as both position and perspective can be compatible with strategy as plan and/or pattern. But, in fact, the relationships between these different definitions can be more involved than that. For example, while some consider perspective to *be* a plan (Lapierre writes of strategies as "dreams in search of reality"[36]; Summer, more prosaically, as "a comprehensive, holistic, gestalt, logical vision of some future alignment"[37]), others describe it as *giving rise* to plans (for example, as positions and/or patterns in some kind of implicit hierarchy). This is shown in Figure 2a. Thus, Majone writes of "basic principles, commitments, and norms" that form the "policy core," while "plans, programs, and decisions" serve as the "protective belt."[38] Likewise, Hedberg and Jonsson claim that strategies, by which they mean "more or less well integrated sets of ideas and constructs" (in our terms, perspectives) are "the causes that mold streams of decisions into patterns."[39] This is similar to Tregoe and Zimmerman who define strategy as "vision directed"—"the framework which guides those choices that determine the nature and direction of an organization."[40] Note in the second and third of these quotations that, strictly speaking, the hierarchy can skip a step, with perspective dictating pattern, not necessarily through formally intended plans.

Consider the example of the Honda Company, which has been described in one highly publicized consulting report[41] as parlaying a particular perspective (being a low cost producer, seeking to attack new markets in aggressive ways) into a plan, in the form of an intended position (to capture the traditional motorcycle market in the United States and create a new one for small family motorcycles), which was in turn realized through an integrated set of patterns (lining up distributorships, developing the appropriate advertising campaign of "You meet the nicest people on a Honda," etc.). All of this matches the conventional prescriptive view of how strategies are supposed to get made.[42]

Figure 2. Some Possible Relationships Between Strategy as Plan, Pattern, Position, Perspective

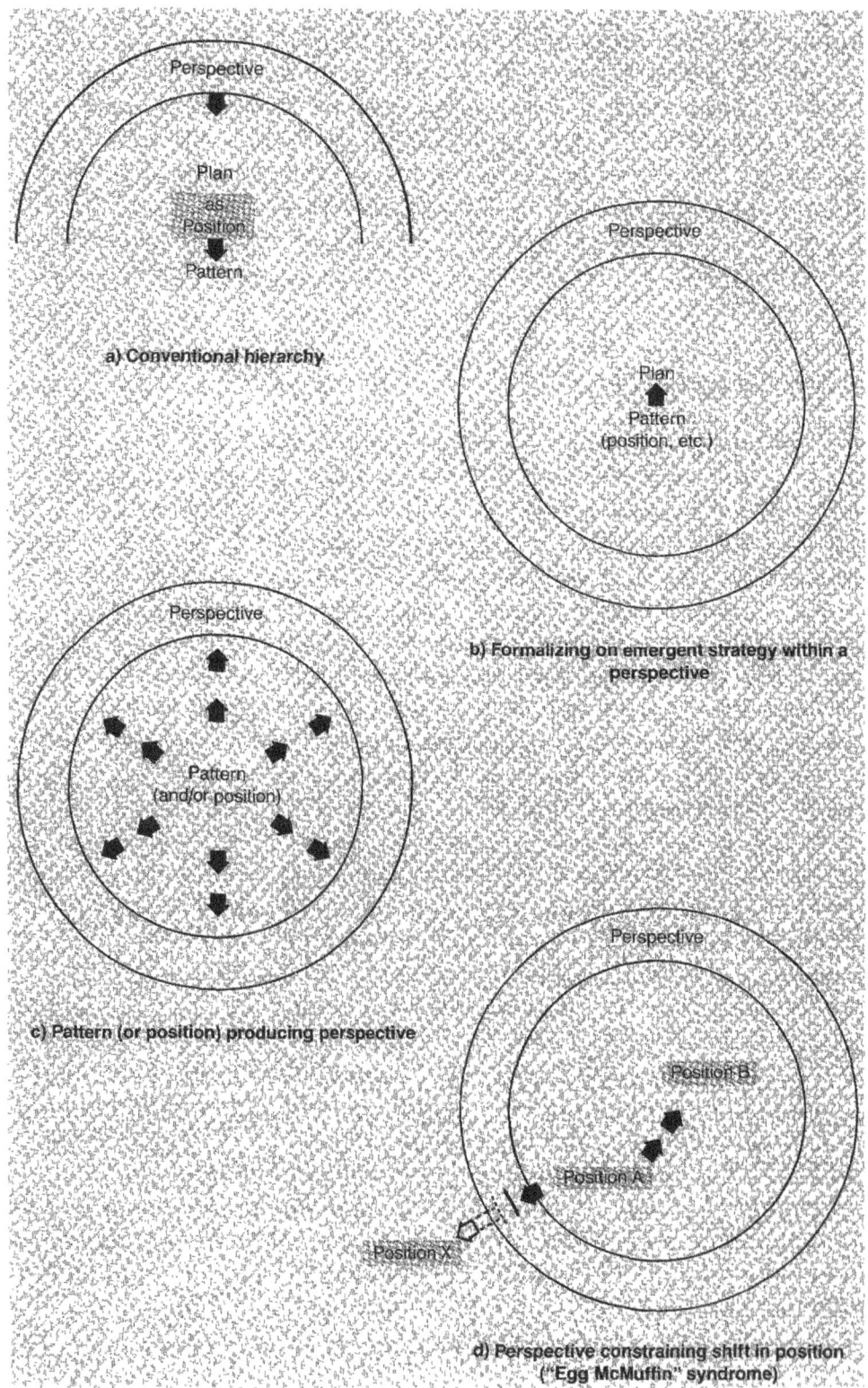

Five Ps for Strategy **19**

But a closer look at Honda's actual behavior suggests a very different story: it did not go to America with the main intention of selling small, family motorcycles at all; rather, the company seemed to fall into that market almost inadvertently.[43] But once it was clear to the Honda executives that they had wandered into such a lucrative strategic position, that presumably became their plan. In other words, their strategy emerged, step by step, but once recognized, was made deliberate. Honda, if you like, developed its intentions through its actions, another way of saying that pattern evoked plan. This is shown in Figure 2b.

Of course, an overall strategic perspective (Honda's way of doing things) seems to have underlaid all this, as shown in the figure as well. But we may still ask how that perspective arose in the first place. The answer seems to be that it did so in a similar way, through earlier experiences: the organization tried various things in its formative years and gradually consolidated a perspective around what worked.[44] In other words, organizations would appear to develop "character"—much as people develop personality—by interacting with the world as they find it through the use of their innate skills and natural propensities. Thus pattern can give rise to perspective too, as shown in Figure 2c. And so can position. Witness Perrow's discussion of the "wool men" and "silk men" of the textile trade, people who developed an almost religious dedication to the fibers they produced.[45]

No matter how they appear, however, there is reason to believe that while plans and positions may be dispensable, perspectives are immutable.[46] In other words, once they are established, perspectives become difficult to change. Indeed, a perspective may become so deeply ingrained in the behavior of an organization that the associated beliefs can become subconscious in the minds of its members. When that happens, perspective can come to look more like pattern than like plan—in other words, it can be found more in the consistency of behaviors than in the articulation of intentions.

Of course, if perspective is immutable, then change in plan and position is difficult unless compatible with the existing perspective. As shown in Figure 2d, the organization can shift easily from Position A to Position B but not to Position X. In this regard, it is interesting to take up the case of Egg McMuffin. Was this product when new—the American breakfast in a bun—a strategic change for the McDonald's fast food chain? Posed in MBA classes, this earth-shattering (or at least stomach-shattering) question inevitably evokes heated debate. Proponents (usually people sympathetic to fast food) argue that of course it was: it brought McDonald's into a new market, the breakfast one, extending the use of existing facilities. Opponents retort that this is nonsense, nothing changed but a few ingredients: this was the same old pap in a new package. Both sides are, of course, right—and wrong. It simply depends on how you define strategy. Position changed; perspective remained the same. Indeed—and this is the point— the position could be changed so easily because it was compatible with the

2

existing perspective. Egg McMuffin is pure McDonald's, not only in product and package, but also in production and propagation. But imagine a change of position at McDonald's that would require a change of perspective—say, to introduce candlelight dining with personal service (your McDuckling à l'Orange cooked to order) to capture the late evening market. We needn't say more, except perhaps to label this the "Egg McMuffin syndrome."

The Need for Eclecticism in Definition

While various relationships exist among the different definitions, no one relationship, nor any single definition for that matter, takes precedence over the others. In some ways, these definitions compete (in that they can substitute for each other), but in perhaps more important ways, they complement. Not all plans become patterns nor are all patterns that develop planned; some ploys are less than positions, while other strategies are more than positions yet less than perspectives. Each definition adds important elements to our understanding of strategy, indeed encourages us to address fundamental questions about organizations in general.

As plan, strategy deals with how leaders try to establish direction for organizations, to set them on predetermined courses of action. Strategy as plan also raises the fundamental issue of cognition—how intentions are conceived in the human brain in the first place, indeed, what intentions really mean. Are we, for example, to take statements of intentions at face value? Do people always say what they mean, or mean what they say? Ostensible strategies as ploys can be stated just to fool competitors; sometimes, however, those who state them fool themselves. Thus, the road to hell in this field can be paved with those who take all stated intentions at face value. In studying strategy as plan, we must somehow get into the mind of the strategist, to find out what is really intended.

As ploy, strategy takes us into the realm of direct competition, where threats and feints and various other maneuvers are employed to gain advantage. This places the process of strategy formation in its most dynamic setting, with moves provoking countermoves and so on. Yet ironically, strategy itself is a concept rooted not in change but in stability—in set plans and established patterns. How then to reconcile the dynamic notions of strategy as ploy with the static ones of strategy as pattern and other forms of plan?

As pattern, strategy focusses on action, reminding us that the concept is an empty one if it does not take behavior into account. Strategy as pattern also introduces another important phenomenon in organizations, that of convergence, the achievement of consistency in behavior. How does this consistency form, where does it come from? Realized strategy is an important means of conceiving and describing the direction actually pursued by

Five Ps for Strategy   **21**

organizations, and when considered alongside strategy as plan, encourages us to consider the notion that strategies can emerge as well as be deliberately imposed.

As position, strategy encourages us to look at organizations in context, specifically in their competitive environments—how they find their positions and protect them in order to meet competition, avoid it, or subvert it. This enables us to think of organizations in ecological terms, as organisms in niches that struggle for survival in a world of hostility and uncertainty as well as symbiosis. How much choice do organizations have, how much room for maneuver?

And finally as perspective, strategy raises intriguing questions about intention and behavior in a collective context. If we define organization as collective action in the pursuit of common mission (a fancy way of saying that a group of people under a common label—whether an IBM or a United Nations or a Luigi's Body Shop—somehow find the means to cooperate in the production of specific goods and services), then strategy as perspective focusses our attention on the reflections and actions of the collectivity—how intentions diffuse through a group of people to become shared as norms and values, and how patterns of behavior become deeply ingrained in the group. Ultimately, it is this view of strategy that offers us the best hope of coming to grips with the most fascinating issue of all, that of the "organizational mind."

Thus, strategy is not just a notion of how to deal with an enemy or a set of competitors or a market, as it is treated in so much of the literature and in its popular usage. It also draws us into some of the most fundamental issues about organizations as instruments for collective perception and action.

To conclude, a good deal of the confusion in this field stems from contradictory and ill-defined uses of the term strategy, as we saw in the Egg McMuffin syndrome. By explicating and using five definitions, we may be able to remove some of this confusion, and thereby enrich our ability to understand and manage the processes by which strategies form.

References

1. P.F. Drucker, *Management: Tasks, Responsibilities, Practices* (New York, NY: Harper & Row, 1974), p. 104.

2. Moore, in fact, prefers not to associate the word strategy with the word plan per se: "The term *plan* is much too static for our purposes unless qualified. There is not enough of the idea of scheming or calculation with an end in view in it to satisfy us. Plans are used to build ships. Strategies are used to achieve ends among people. You simply do not deal strategically with inanimate objects." But Moore certainly supports the characteristics of intentionality. D.G. Moore, "Managerial Strategies," in W.L. Warner and N.H. Martin, eds., *Industrial Man: Businessmen and Business Organizations* (New York, NY: Harper & Row, 1959), pp. 220,226.

3. C. Von Clausewitz, *On War*, translated by M. Howard and P. Paret (Princeton, NJ: Princeton University Press, 1976), p. 177.

2

4. J. Von Newmann and O. Morgenstern, *Theory of Games and Economic Behavior* (Princeton, NJ: Princeton University Press, 1944), p. 79.

5. W.F. Glueck, *Business Policy and Strategic Management*, 3rd Edition (New York, NY: McGraw-Hill, 1980), p. 9.

6. *Random House Dictionary*.

7. M.E. Porter, *Competitive Strategy: Techniques for Analyzing Industries and Competitors* (New York, NY: The Free Press, 1980).

8. M.E. Porter, *Competitive Advantage: Creating and Sustaining Superior Performance* (New York, NY: The Free Press, 1985).

9. T.C. Schelling, *The Strategy of Conflict*, 2nd Edition (Cambridge, MA: Harvard University Press, 1980).

10. H. Mintzberg, "Research on Strategy-Making," *Proceedings after the 32nd Annual Meeting of the Academy of Management*, Minneapolis, 1972, pp. 90–94; M. Mintzberg, "Patterns in Strategy Formation," *Management Science*, 24/9 (1978):934–948; H. Mintzberg and J.A. Waters, "Of Strategies, Deliberate and Emergent," *Strategic Management Journal*, 6/3 (1985):257–272.

11. Evered discusses the Greek origins of the word and traces its entry into contemporary Western vocabulary through the military. R. Evered, "So What Is Strategy," *Long Range Planning*, 16/3 (1983):57–72.

12. As suggested in the results of a questionnaire by Ragab and Paterson; M. Ragab and W.E. Paterson, "An Exploratory Study of the Strategy Construct," Proceedings of the Administrative Sciences Association of Canada Conference, 1981. Two notable exceptions are Herbert Simon and Jerome Bruner and his colleagues; H.A. Simon, *Administrative Behavior*, 2nd Edition (New York, NY: Macmillan, 1957); J.S. Bruner, J.J. Goodnow, and G.A. Austin, *A Study of Thinking* (New York, NY: Wiley, 1956), pp. 54–55.

13. Quoted in J.B. Quinn, *Strategies for Change: Logical Incrementalism* (Homewood, IL: Richard D. Irwin, 1980), p. 35.

14. *Business Week*, October 31, 1983.

15. Via G. Majone, "The Uses of Policy Analysis," in *The Future and the Past: Essays on Programs*, Russell Sage Foundation Annual Report, 1976–1977, pp. 201–220.

16. A.D. Chandler, *Strategy and Structure: Chapters in the History of the Industrial Enterprise* (Cambridge, MA: M.I.T. Press, 1962), p. 13.

17. Von Clausewitz, op. cit., p. 128.

18. B. Franklin, *Poor Richard's Almanac* (New York, NY: Ballantine Books, 1977), p. 280.

19. R. P. Rumelt, "Evaluation of Strategy: Theory and Models," in D. E. Schendel and C. W. Hofer, eds., *Strategic Management: A New View of Business Policy and Planning* (Boston, MA: Little Brown, 1979), pp. 196–212.

20. We might note a similar problem with "policy," a word whose usage is terrible confused. In the military, the word has traditionally served one notch in the hierarchy above strategy, in business one notch below, and in public administration in general as a substitute. In the military, policy deals with the purposes for which wars are fought, which is supposed to be the responsibility of the politicians. In other words, the politicians make policy, the generals, strategy. But modern warfare has confused this usage (see Summers), so that today strategy in the military context has somehow come to be associated with the acquisition of nuclear weapons and their use against non-military targets. In business, while "policy" has been the label for the entire field of study of general management (at least until "strategic management" gained currency in the 1970s), its technical use was as a general rule to dictate decisions in a specific case, usually a standard and recurring situation, as in "Our policy is to require long-range forecasts every four months." Accordingly, management planning theorists, such as George Steiner, describe policies as deriving from strategies although some textbook

Why Organizations Need Strategy   **23**

writers (such as Leontiades, Chang and Campo-Flores, and Peter Drucker) have used the two words in exactly the opposite way, as in the military. This reflects the fact that "policy" was the common word in the management literature before "strategy" replaced it in the 1960s (see, for example, Jamison, and Gross and Gross). But in the public sector today, the words "policy" and "policymaking" correspond roughly to "strategy" and "strategy making." H. G. Summers, *On Strategy: The Vietnam War in Context* (Carlisle Barracks, PA: Strategic Studies Institute, U.S. Army War College, 1981); G. A. Steiner, *Top Management Planning* (New York, NY: Macmillan, 1969), p. 264 ff; M. Leontiades, *Management Policy, Strategy and Plans* (Boston, MA: Little Brown, 1982), p. 4; Y.N.A. Chang and F. Compo-Flores, *Business Policy and Strategy* (Goodyear, 1980), p. 7; Drucker, op. cit., p. 104; C.L. Jamison, *Business Policy* (Englewood Cliffs, NJ: Prentice-Hall, 1953); A. Gross and W. Gross, eds., *Business Policy: Selected Readings and Editorial Commentaries* (New York, NY: Ronald Press, 1967).

21. C.W. Hofer and D. Schendel, *Strategy Formulation: Analytical Concepts* (St. Paul, MN: West Publishing, 1978), p. 4.

22. E.H. Bowman, "Epistomology, Corporate Strategy, and Academe," *Sloan Management Review*, 15/2 (1974):47.

23. J.D. Thompson, *Organizations in Action* (New York, NY: McGraw-Hill, 1967).

24. T.J. McNichols, *Policy-Making and Executive Action* (New York, NY: McGraw-Hill, 1983), p. 257.

25. Rumelt, op. cit., p. 4.

26. R. P. Rumelt, "The Evaluation of Business Strategy," in W.F. Glueck, *Business Policy and Strategic Management*, 3rd Edition (New York, NY: McGraw-Hill, 1980), p. 361.

27. E.g., Porter, op. cit., (1980, 1985), except for his chapters noted earlier, which tend to have a 2-person competitive focus.

28. Expressed at the Strategic Management Society Conference, Paris, October 1982.

29. W.G. Astley and C.J. Fombrun, "Collective Strategy: Social Ecology of Organizational Environments," *Academy of Management Review*, 8/4 (1983):576–587.

30. Ibid., p. 577.

31. P. Selznick, *Leadership in Administration: A Sociological Interpretation* (New York, NY: Harper & Row, 1957), p. 47. A subsequent paper by the author (in process) on the "design school" of strategy formation shows the link of Selznick's early work to the writings of Kenneth Andrews in the Harvard policy textbook. K.R. Andrews, *The Concept of Corporate Strategy*, Revised Edition (Homewood, IL: Dow Jones-Irwin, 1987).

32. J. Bieri, "Cognitive Structures in Personality," in H.M. Schroder and P. Suedfeld, eds., *Personality: Theory and Information Processing* (New York, NY: Ronald Press, 1971), p. 178. By the same token, Bieri (p. 179) uses the word "strategy" in the context of psychology.

33. Drucker, op. cit.

34. B.B. Tregoe and J.W. Zimmerman, *Top Management Strategy* (New York, NY: Simon & Schuster, 1980).

35. T.S. Kuhn, *The Structure of Scientific Revolution*, 2nd Edition (Chicago, IL: University of Chicago Press, 1970).

36. My own translation of "un reve ou un bouquet de reves en quete de realite." L. Lapierre, "Le changement strategique: Un reve en quete de reel," Ph.D. Management Policy course paper, McGill University, Canada, 1980.

37. Summer, op. cit., p. 18.

38. G. Majone, op. cit.

39. B. Hedberg and S.A. Jonsson, "Strategy Formulation as a Discontinuous Process," *International Studies of Management and Organization*, 7/2 (1977):90.

40. Tregoe and Zimmerman, op. cit., p. 17.

41. Boston consulting Group, *Strategy Alternatives for the British Motorcycle Industry* (London: Her Majesty's Stationery Office, 1975).

2

42. E.g., H.I. Ansoff, *Corporate Strategy* (New York, NY: McGraw-Hill, 1965); Andrews, op. cit.; Steiner, op. cit.; D.E. Schendel and C.H. Hofer, eds., *Strategic Management: A New View of Business Policy and Planning* (Boston, MA: Little Brown, 1979), p. 15.

43. R.T. Pascale, "Perspectives on Strategy: The Real Story Behind Honda's Success," *California Management Review*, 26/3 (Spring 1984):47–72.

44. J.B. Quinn, "Honda Motor Company Case," in J.B. Quinn, H. Mintzberg, and B.G. James, *The Strategy Process: Concepts, Contexts, Cases* (Englewood Cliffs, NJ: Prentice-Hall, 1988).

45. C. Perrow, *Organizational Analysis: A Sociological View* (Belmont, CA: Wadsworth, 1970), p. 161.

46. E.g., N. Brunsson, "The Irrationality of Action and Action Rationality: Decisions, Ideologies, and Organizational Actions," *Journal of Management Studies*, 19/1 (1982):29–44.

Originaltext Weick 1987

10 SUBSTITUTES FOR STRATEGY

Karl E. Weick

A little strategy goes a long way. Too much can paralyze or splinter an organization. That conclusion derives from the possibility that strategy-like outcomes originate from sources other than strategy. Adding explicit strategy to these other tacit sources of strategy can be self-defeating and reduce effectiveness (Bresser and Bishop 1983). Thus, the focus of this chapter is substitutes for strategy.

The model for this exercise is the concept in the leadership literature of substitutes for leadership (Kerr and Jermier 1978). Substitutes are conditions that either neutralize what leaders do or perform many of the same functions they would. Substitutes include characteristics of subordinates (ability, knowledge, experience, training, professional orientation, indifference toward organizational rewards), characteristics of the task (unambiguous, routine, provides its own feedback, intrinsically satisfying), and characteristics of the organization (high formalization, highly specified staff functions, closely knit cohesive groups, organizational rewards not controlled by leaders, spatial distance between subordinates and superiors). Leadership has less impact when one or more of these conditions obtains. It is not that the situation is devoid of leadership; rather, the leadership is done by something else.

It seems reasonable to work analogically and investigate the extent to which it is possible to create substitutes for strategies.

If pressed to define *strategy*, I am tempted to adopt DeBono's (1984: 143) statement that "strategy is good luck rationalized in hindsight," but I am also comfortable with a definition much like Robert Burgelmann's (1983)—namely, "strategy is a theory about the reasons for past and

222   THE THEORETICAL CONTEXT OF STRATEGIC MANAGEMENT

current success of the firm." Both of my definitional preferences differ sharply from Chandler's (1962) classic definition of *strategy*—"The determination of the basic long-term goals and objectives of an enterprise, and the adoption of courses of action and the allocation of resources necessary for carrying out these goals."

Definitions notwithstanding, I can best show what I think strategy is by describing an incident that happened during military maneuvers in Switzerland. The young lieutenant of a small Hungarian detachment in the Alps sent a reconnaissance unit into the icy wilderness. It began to snow immediately, snowed for two days, and the unit did not return. The lieutenant suffered, fearing that he had dispatched his own people to death. But the third day the unit came back. Where had they been? How had they made their way? Yes, they said, we considered ourselves lost and waited for the end. And then one of us found a map in his pocket. That calmed us down. We pitched camp, lasted out the snowstorm, and then with the map we discovered our bearings. And here we are. The lieutenant borrowed this remarkable map and had a good look at it. He discovered to his astonishment that it was not a map of the Alps, but a map of the Pyrenees.

This incident raises the intriguing possibility that when you are lost, any old map will do. Extended to the issue of strategy, maybe when you are confused, any old strategic plan will do.

Strategic plans are a lot like maps. They animate people and they orient people. Once people begin to act, they generate tangible outcomes in some context, and this helps them discover what is occurring, what needs to be explained, and what should be done next. Managers keep forgetting that it is what they do, not what they plan, that explains their success. They keep giving credit to the wrong thing—namely, the plan—and having made this error, they then spend more time planning and less time acting. They are astonished when more planning improves nothing.

Kirk Downey has suggested that the Alps example is a success story for two quite specific reasons. First, the troops found a specific map that was relevant to their problem. Had they found a map of Disneyland rather than a map of the Pyrenees their problem would have deepened materially. Second, the troops had a purpose—that is, they wanted to go back to their base camp—and it was in the context of this purpose that the map took on meaning as a means to get them back. These conditions, however, do not negate the basic theme that meaning lies in the path of the action. A map of Disneyland makes it harder to develop a shared understanding of what has happened and where we have been, but if it does not inhibit action and observation, some clearer sense of the situation may emerge as action proceeds.

SUBSTITUTES FOR CORPORATE STRATEGY   223

When I described the incident of using a map of the Pyrenees to find a way out of the Alps to Bob Engel, the executive vice president and treasurer of Morgan Guaranty, he said, "Now, that story would have been really neat if the leader out with the lost troops had known it was the wrong map and still been able to lead them back."

What is interesting about Engel's twist to the story is that he has described the basic situation that most leaders face. Followers are often lost and even the leader is not sure where to go. All the leader knows is that the plan or the map he has in front of him is not sufficient by itself to get them out. What he has to do, when faced with this situation, is instill some confidence in people, get them moving in some general direction, and be sure they look closely at what actually happens, so that they learn where they were and get some better idea of where they are and where they want to be.

If you get people moving, thinking clearly, and watching closely, events often become more meaningful. For one thing, a map of the Pyrenees can still be a plausible map of the Alps because in a very general sense, if you have seen one mountain range, you have seen them all (readers can test this assertion for themselves by examining "A Traveler's Map of the Alps" in the April 1985 issue of *National Geographic Magazine*). The Pyrenees share some features with the Alps, and if people pay attention to these common features, they may find their way out. For example, most mountain ranges are wet on one side and dry on the other. Water flows down rather than up. There is a prevailing wind. There are peaks and valleys. There is a highest point, and then the peaks get lower and lower until there are foothills.

Just as it is true that if you have seen one mountain range you have seen them all, it also is true that if you have seen one organization you have seen them all. Any old plan will work in an organization because people usually learn by trial and error, some people listen and some people talk, people want to get somewhere and have some general sense of where they now are, 20 percent of the people will do 80 percent of the work (and vice versa), and if you do something for somebody, they are more likely to do something for you. Given these general features of most organizations, any old plan is often sufficient to get this whole mechanism moving, which then makes it possible to learn what is going on and what needs to be done next.

The generic process involved is that meaning is produced because the leader treats a vague map or plan as if it had some meaning, even though he knows full well that the real meaning will come only when people respond to the map and do something. The secret of leading with a bad map is to create a self-fulfilling prophecy. Having predicted that the group will find its way out, the leader creates the combination of optimism

2

and action that allows people to turn their confusion into meaning and find their way home.

There are plenty of examples in industry where vague plan and projects provide an excuse for people to act, learn, and create meaning.

The founders of Banana Republic, the successful mail order clothier started their business by acting in an improbable manner. They bought uniforms from overthrown armies in South America and advertised these items in a catalog, using drawings rather than photographs. All of these actions were labeled poor strategy by other mail order firms. When these three actions were set in motion, however, they generated responses that no one expected (because no one had tested them) and created a belated strategy as well as a distinct niche for Banana Republic.

Tuesday Morning, an off-price retailing chain that sells household and gift items, opens its stores when they have enough merchandise to sell and then closes them until they get the next batch. As managers followed this pattern, they discovered that customers love grand openings and that anticipation would build between closings over when the store would open again and what it would contain. These anticipations were sufficiently energizing that stores that opened intermittently for four to eight weeks sold more than equivalent stores that were open year round.

The Microelectronics and Computer Technology Corporation (MCC) consortium in Austin, Texas, is a clear example of the sequence in which vague projects trigger sufficient action that vagueness gets removed. A key Texas state official described MCC as "an event, not a company." Bidding for MCC to locate in Texas became a vehicle to pull competing Texas cities together. It also became a vehicle to tell out-of-state people, "We are a national and an international force, not just a regional force, and not just a land of cowboys and rednecks." MCC became a tangible indication that Texas was growing, maturing, and on its way up. MCC's criteria for a good site became defining characteristics of what Austin was as a city, though Austinites did not realize they had this identity before. MCC said in its specifications that it did not want to locate where everyone thinks they know how high-tech R&D should be done. Texas thus "discovered" that its backwardness was in fact one of its biggest assets.

Acquiring MCC became a strategy to strengthen Texas, but only quite late, when more and more problems were seen to be solved if it landed in Austin. The action of bidding for MCC fanned out in ways that people had not anticipated. The point is, if action is decoupled from strategy, then people have a better chance to be opportunistic, to discover missions and resources they had no idea existed.

So far three themes have been introduced: (1) that action clarifies meaning; (2) that the pretext for the action is of secondary importance;

(3) and that strategic planning is only one of many pretexts for meaning-generation in organizations. To clarify some ways in which action can substitute for strategy, we will look more closely at the dynamics of confidence and improvisation.

Confidence as Strategy

In managerial work, thought precedes action, but the kind of thought that often occurs is not detailed analytical thought addressed to imagined scenarios in which actions are tried and options chosen. Instead, thought precedes action in the form of much more general expectations about the orderliness of what will occur.

Order is present, not because extended prior analysis revealed it but because the manager anticipates sufficient order that she wades into the situation, imposes order among events, and then "discovers" what she had imposed. The manager "knew" all along that the situation would make sense. This was treated as a given. Having presumed that it would be sensible, the manager than acts confidently and implants the order that was anticipated.

Most managerial situations contain gaps, discontinuities, loose ties among people and events, indeterminacies, and uncertainties. These are the gaps that managers have to bridge. It is the contention of this argument that managers first think their way across these gaps and then, having tied the elements together cognitively, actually tie them together when they act and impose covariation. This sequence is similar to sequences associated with self-fulfilling prophecies (see Snyder, Tanke, and Berscheid 1977).

Thus presumptions of logic are forms of thought that are crucial for their evocative qualities. The presumption leads people to act more forcefully, the more certain the presumption. Strong presumptions (such as, "I know that these are the Pyrenees") lead to strong actions that impose considerable order. Weaker presumptions lead to more hesitant actions, which means either that the person will be more influenced by the circumstances that are already present or that only weak order will be created.

Presumptions of logic are evident in the chronic optimism often associated with managerial activity. This optimism is conspicuous in the case of companies that are in trouble, but it is also evident in more run-of-the-mill managing. Optimism is one manifestation of the belief that situations will have made sense. William James (1956) described the faith that life is worth living that generates the action that then makes life worth living. Optimism is not necessarily a denial of reality. Instead it may be the belief that makes reality possible.

Presumptions of logic should be prominent among managers because of the climate of rationality in organizations (Staw 1980). Presumptions should be especially prominent when beliefs about cause and effect linkages are unclear (Thompson 1964: 336). Thompson labels the kind of managing that occurs when there are unclear preferences and unclear cause/effect beliefs *inspiration*. It is precisely in the face of massive uncertainty that beliefs of some sort are necessary to evoke some action, which can then begin to consolidate the situations. To inspire is to affirm realities, which then are more likely to materialize if they are sought vigorously. That sequence may be the essence of managing.

Examples of the effect of presumptions are plentiful. A male who believes he is telephoning an attractive female speaks more warmly, which evokes a warm response from her, which confirms the original stereotype that attractive women are sociable (Snyder, Tanke, and Berscheid 1977). A new administrator, suspecting that old-timers are traditional, seeks ideas from other sources, which increases the suspicion of old-timers and confirms the administrator's original presumption (Warwick, 1975). People who presume that no one likes them approach a new gathering in a stiff, distrustful manner, which evokes the unsympathetic behavior they presumed would be there (Watzlawick, Beavin, and Jackson 1967: 98–99). A musician who doubts the competence of a composer plays his music lethargically and produces the ugly sound that confirms the original suspicion (Weick, Gilfillan, and Keith 1973).

In each case, an initial presumption (she is sociable, they are uncreative, people are hostile, he is incompetent) leads people to act forcibly (talk warmly, seek ideas elsewhere, behave defensively, ignore written music), which causes a situation to become more orderly (warmth is exchanged, ideas emerge, hostility is focused, music becomes simplistic), which then makes the situation easier to interpret, thereby confirming the original presumption that it will have been logical.

This sequence is common among managers because managerial actions are almost ideally suited to sustain self-fulfilling prophecies (Eden 1984). Managerial actions are primarily oral, face-to-face, symbolic, presumptive, brief, and spontaneous (McCall and Kaplan 1985). These actions have a deterministic effect on many organizational situations because those situations are less tightly coupled than are the confident actions directed at them. The situations are loosely coupled, subject to multiple interpretations, monitored regularly by only a handful of people, and deficient in structure.

Thus a situation of basic disorder becomes more orderly when people overlook the disorder and presume orderliness, then act on this presumption, and finally rearrange its elements into a more meaningful arrangement that confirms the original presumption. It is suggested that typical managerial behavior is more likely to create rather than disrupt this sequence. Thus, a manager's preoccupation with rationality may be significant less for its power as an analytic problemsolving tool than for its power to induce action that eventually implants the rationality that was presumed when the sequence started.

The lesson of self-fulfilling prophecies for students of strategy is that strong beliefs that single out and intensify consistent action can bring events into existence (see Snyder 1984). Whether people are called fanatics, true believers, or the currently popular phrase *idea champions*, they all embody what looks like strategy in their persistent behavior. Their persistence carries the strategy; the persistence is the strategy. True believers impose their view on the world and fulfill their own prophecies. Note that this makes strategy more of a motivational problem than a cognitive forecasting problem.

An argument can be made that the so-called computer revolution is an ideal exhibit of confidence as strategy. The revolution is as much vendor-driven as it is need-driven. The revolution can be viewed as solutions in search of problems people never knew they had. Vendors had more forcefulness, confidence, and focus than did their customers, who had only a vague sense that things were not running right, although they could not say why. Vendors defined the unease as a clear problem in control and information distribution, a definition that was no worse than any other diagnosis that was available.

To say that it was IBM's strategy to be forceful is to miss the core of what actually happened. The key point is that IBM's strategy worked after it became self-confirming, when it put an environment in place. A common error is that the strategic plan is valued because it looks like it correctly forecast a pent-up demand for computers. Actually, it did no such thing. Instead, the plan served as a pretext for people to act forcefully and impose their view of the world. Once they imposed, enacted, and stabilized that view and once it was accepted, then more traditional procedures of strategic planning could be made to work because they were directed at more predictable problems in a more stable environment. What gets missed by strategy analysts is that proaction precedes reaction. Strategic planning works only after forceful action has hammered the environment into shape so that it is less variable and so that conventional planning tools can now be made to work. Because the constrained environment contains demands, opportunities, and problems that were imposed during proaction, proaction, not planning, predicts what the organization has to contend with.

To see how self-fulfilling prophecies can mimic strategy and affect the direction of behavior, consider the problem of regulation. Although companies groan about the weight of regulation, data (McCaffrey 1982)

2

suggest that regulators do not have their act together and are loosely coupled relative to the tightly coupled organizations and lawyers they try to regulate. Thus, many organizations ironically create the regulators who control them. The way they do this is a microcosm of the point being made here about confidence.

If a firm treats regulators as if they are unified and have their act together, then the firm gets its own act together to cope with the focused demands that are anticipated from the regulators. As the firm gets its act together it becomes a clearer target that is easier for the regulators to monitor and control. Concerted action undertaken by the firm to meet anticipated action from regulators now makes it possible for regulators to do something they could not have done when the firms were more diffuse targets.

A confident definition of regulatory power, confidently imposed, stabilizes the regulation problem for a firm. The irony is that the faulty prophecy brings the problem into existence more sharply than it ever was before confident behavior was initiated. The firm has become easier to regulate by virtue of its efforts to prevent regulation.

Environments are more malleable than planners realize. Environments often crystallize around prophecies, presumptions, and actions that unfold while planners deliberate. Guidance by strategy often is secondary to guidance by prophecies. These prophecies are more likely to fulfill themselves when they are in the heads of fanatics who work in environments where the definition of what is occurring can be influenced by confident assertions.

Thus, presumptions can substitute for strategy. We assume co-workers know where they are going, they assume the same for us, and both of us presume that the directions in which we both are going are roughly similar. A presumption does not necessarily mean that whatever is presumed actually exists. We often assume that people agree with us without ever testing that assumption. Vague strategic plans help because we never have to confront the reality of our disagreements. And the fact that those disagreements persist undetected is not necessarily a problem because those very differences provide a repertoire of beliefs and skills that allow us to cope with changing environments. When environmental change is rapid, diverse skills and beliefs are the solution, not the problem.

Improvisation as Strategy

Much of my thinking about organizations (such as Weick 1979) uses the imagery of social evolution, but there is a consistent bias in the way I use that idea. I consistently argue that the likelihood of survival goes up when variation increases, when possibilities multiply, when trial and error become more diverse and less stylized, when people become less repetitious, and when creativity becomes supported. Notice that variation, trial and error, and doing things differently all imply that what you already know, including your strategic plan, is not sufficient to deal with present circumstances.

When it is assumed that survival depends on variation, then a strategic plan becomes a threat because it restricts experimentation and the chance to learn that old assumptions no longer work. Furthermore, I assume that whatever direction strategy gives can be achieved just as easily by improvisation.

Improvisation is a form of strategy that is misunderstood. When people use jazz or improvisational theater to illustrate improvising, they usually forget that jazz consists of variations on a theme and improvisational theater starts with a situation. Neither jazz nor improvisational theater are anarchic. Both contain some order, but it is underspecified.

To understand improvisation as strategy is to understand the order within it. And what we usually miss is the fact that a little order can go a long way.

For example, we keep underestimating the power of corporate culture because it seems improbable that something as small as a logo, a slogan, a preference (Geneen's one unshakable fact), a meeting agenda, or a Christmas party could have such a large effect. The reason these symbols are so powerful is that they give a general direction and a frame of reference that are sufficient. In the hands of bright, ambitious, confident people who have strong needs to control their destinies, general guidelines are sufficient to sustain and shape improvisation without reducing perceived control.

If improvisation is treated as a natural form of organizational life, then we become interested in a different form of strategy than we have seen before. This newer form I will call a *just-in-time strategy*. Just-in-time strategies are distinguished by less investment in front-end loading (try to anticipate everything that will happen or that you will need) and more investment in general knowledge, a large skill repertoire, the ability to do a quick study, trust in intuitions, and sophistication in cutting losses.

Like improvisation, a just-in-time strategy glosses, interprets, and enlarges some current event, gives it meaning, treats it as if it were sensible, and brings it to a conclusion. This form of activity looks very much like creating a stable small win (Weick 1984). And once an assortment of small wins is available, then these can be gathered together retrospectively and packaged as any one of several different directions, strategies, or policies.

Strategies are less accurately portrayed as episodes where people convene at one time to make a decision and more accurately portrayed as

2

small steps (writing a memo, answering an inquiry) that gradually fore-close alternative courses of action and limit what is possible. The strategy is made without anyone realizing it. The crucial activities for strategy-making are not separate episodes of analysis. Instead they are actions, the controlled execution of which consolidate fragments of policy that are lying around, give them direction, and close off other possible arrange-ments. The strategy-making *is* the memo-writing, *is* the answering, *is* the editing of drafts. These actions are not precursors to strategy; they *are* the strategy.

Strategies that are tied more closely to action are more likely to contain improvisations (Weiss 1980: 401):

> Many moves are improvisations. Faced with an event that calls for response, officials use their experience, judgment, and intuition to fashion the response for the issue at hand. That response becomes a precedent, and when similar—or not so similar—questions come up, the response is uncritically repeated. Consider the federal agency that receives a call from a local program asking how to deal with requests for enrollment in excess of the available number of slots. A staff member responds with off-the-cuff advice. Within the next few weeks, programs in three more cities call with similar questions, and staff repeat the advice. Soon what began as improvisation has hardened into policy.

Managers are said to avoid uncertainty, but one of the ironies implicit in the preceding analysis is that managers often create the very uncertainty they abhor. When they cannot presume order they hesitate, and this very hesitancy often creates events that are disordered and unfocused. This disorder confirms the initial doubts concerning order. What often is missed is that the failure to act, rather than the nature of the external world itself, creates the lack of order. When people act, they absorb uncertainty, they rearrange things, and they impose contingencies that might not have been there before. The presence of these contingencies is what is treated as evidence that the situation is orderly and certain.

Conclusion

The thread that runs through this chapter is that execution *is* analysis and implementation *is* formulation. The argument is an attempt to com-bine elements from a linear and adaptive view of strategy, with a largely interpretive view (Chaffee 1985: 95). Any old explanation, map, or plan is often sufficient because it stimulates focused, intense action that both creates meaning and stabilizes an environment so that conventional analysis now becomes more relevant. Organizational culture becomes influential in this scenario because it affects what people expect will be orderly. These expectations, in turn, often become self-fulfilling. Thus the adequacy of any explanation is determined in part by the intensity and structure it adds to potentially self-validating actions. More forcefulness leads to more validation. Accuracy becomes secondary to intensity. Because situations can support a variety of meanings, their actual content and meaning is dependent on the degree to which they are arranged into sensible, coherent configurations. More forcefulness imposes more coherence. Thus those explanations that induce greater forcefulness often become more valid, not because they are more accurate but because they have a higher potential for self-validation.

Applied to managerial activity, substitutes for strategy are more likely among executives because their actions are capable of a considerable range of intensity, the situations they deal with are loosely connected and capable of considerable rearrangement, and the underlying explanations that managers invoke (such as, "This is a cola war") have great potential to intensify whatever action is underway. All of these factors combine to produce self-validating situations in which managers are sure their diagnoses are correct. What they underestimate is the extent to which their own actions have implanted the correctness they discover.

What managers fail to see is that solid facts are an ongoing accomplishment sustained as much by intense action as by accurate diagnosis. If managers reduce the intensity of their own action or if another actor directs a more intense action at the malleable elements, the meaning of the situation will change. What managers seldom realize is that their inaction is as much responsible for the disappearance of facts as their action was for the appearance of those facts.

Gene Webb often quotes Edwin Boring's epigram: "Enthusiasm is the friend of action, the enemy of wisdom." Given the preceding arguments we can see reasons to question that statement. Enthusiasm can produce wisdom because action creates experience and meaning. Furthermore, enthusiasm can actually create wisdom when prophecies become self-fulfilling and factual.

One final example of a vague plan that leads to success when people respond to it and pay close attention to their response involves a religious ritual used by the Naskapi Indians in Labrador. Every day they ask the question, "Where should we hunt today?" That question is no different from, "Where is the base camp?" or "What should we do with these uniforms?" or "Should we open today?" or "Could this conceivably be the Silicon prairie?"

The Naskapi use an unusual procedure to learn where they should hunt. They take the shoulder bone of a caribou, hold it over a fire until the bone begins to crack, and then they hunt wherever the cracks point. The surprising thing is that this procedure works. The Naskapi almost always find game, which is rare among hunting bands.

Although there are several reasons why this procedure works, one is of special interest to us: The Naskapi spend most of each day actually hunting. Once the cracks appear, they go where the cracks point. What they do not do is sit around the campfire debating where the game are today based on where they were yesterday. If the Naskapi fail to find any game, which is rare, they have no one in the group to blame for the outcome. Instead, they simply say that the gods must be testing their faith.

The cracks in the bone get the Naskapi moving, just as the mountain paths drawn on the map get the soldiers moving, and just as high-tech backwardness gets Texans moving. In each case, movement multiplies the data available from which meaning can be constructed.

Because strategy is often a retrospective summary that lags behind action, and because the apparent coherence and rationality of strategy are often inflated by hindsight bias, strategic conclusions can be misleading summaries of what we can do right now and what we need to do in the future.

I do not suggest doing away with strategic plans altogether, but people can take a scarce resource, time, and allocate it between the activities of planning and acting. The combination of staffs looking for work, high-powered analytic MBAs, unused computer capability, the myth of quantitative superiority, and public pressure to account for everything in rational terms tempts managers to spend a great deal of time at their terminals doing analysis and a great deal less time anyplace else (Weick in press). It seems astonishing that one of the hottest managerial precepts to come along in some time (MBWA, management by walking around) simply urges managers to pull the plug on the terminal, go for a walk, and act like champions. One reason those recommendations receive such a sympathetic reception is that they legitimize key aspects of sensemaking that got lost when we thought we could plan meanings into existence. As we lost sight of the importance of action in sensemaking, we saw situations become senseless because the wrong tools were directed at them.

Strategic planning is today's pretext under which people act and generate meanings and so is the idea of organizational culture. Each one is beneficial as long as it encourages action. It is the action that is responsible for meaning, even though planning and symbols mistakenly get the credit. The moment that either pretext begins to stifle action meaning will suffer, and these two concepts will be replaced by some newer management tool that will work, not for the reasons claimed but because it restores the fundamental sensemaking process of motion and meaning.

SUBSTITUTES FOR CORPORATE STRATEGY   233

REFERENCES

Bresser, R.K., and R.C. Bishop. 1983. "Dysfunctional Effects of Formal Planning: Two Theoretical Explanations." *Academy of Management Review* 8: 588–99.

Burgelman, R.A. 1983. "A Model of the Interaction of Strategic Behavior, Corporate Context, and the Concept of Strategy." *Academy of Management Review* 8: 61–70.

Chaffee, E.E. 1985. "Three Models of Strategy." *Academy of Management Review* 10: 89–98.

Chandler, A.D. 1962. *Strategy and Structure.* Cambridge, Mass.: MIT Press.

de Bono, E. 1984. *Tactics: The Art and Science of Success.* Boston: Little, Brown.

Eden, D. 1984. "Self-Fulfilling Prophecy as a Management Tool: Harnessing Pygmalion." *Academy of Management Review* 9: 64–73.

James, W. 1956. "Is Life Worth Living?" In *The Will to Believe,* edited by W. James, pp. 32–62. New York: Dover.

Kerr, S., and J.M. Jermier. 1978. "Substitutes for Leadership: Their Meaning and Measurement." *Organizational Behavior and Human Performance* 22: 375–403.

McCaffrey, D.P. 1982. "Corporate Resources and Regulatory Pressures: Toward Explaining a Discrepancy." *Administrative Science Quarterly* 27: 398–419.

McCall, M.W., Jr., and R.W. Kaplan. 1985. *Whatever It Takes: Decision Makers at Work.* Englewood Cliffs, N.J.: Prentice Hall.

Snyder, M. 1984. "When Belief Creates Reality." In *Advances in Experimental Social Psychology, Vol. 18,* edited by L. Berkowitz, pp. 247–305. New York: Academic Press.

Snyder, M., E.D. Tanke, and E. Berscheid. 1977. "Social Perception and Interpersonal Behavior: On the Self-Fulfilling Nature of Social Stereotypes." *Journal of Personality and Social Psychology* 35: 656–66.

Staw, B.M. 1980. "Rationality and Justification in Organizational Life." In *Research in Organizational Behavior, Vol. 2,* edited by B.M. Staw and L.L. Cummings, pp. 45–80. Greenwich, Conn.: JAI.

Thompson, J.D. 1964. "Decision-making, the Firm, and the Market." In *New Perspectives in Organization Research,* edited by W.W. Cooper, H.J. Leavitt, and M.W. Sheely II, pp. 334–48. New York: Wiley.

Warwick, Donald P. 1975. *A Theory of Public Bureaucracy: Politics, Personality, and Organization in the State Department.* Cambridge, Mass.: Harvard University Press.

Watzlawick, P., J.H. Beavin, and D.D. Jackson. 1967. *Pragmatics of Human Communication.* New York: Norton.

Weick, K.E. 1979. *The Social Psychology of Organizing,* 2d ed. Reading, Mass.: Addison-Wesley.

———. 1984. "Small Wins: Redefining the Scale of Social Problems." *American Psychologist* 39: 40–49.

———. 1985. "Cosmos vs. Chaos: Sense and Nonsense in Electronic Contexts." *Organizational Dynamics* 14 (Autumn): 50–64.

Weick, K.E., D.P. Gilfillan, and T. Keith. 1973. "The Effect of Composer Credibility on Orchestra Performance." *Sociometry* 36: 435–62.

Weiss, C.H. 1980. "Knowledge Creep and Decision Accretion." *Knowledge: Creation, Diffusion, Utilization* 1(3): 381–404.

Originaltext Bryson 1988

Long Range Planning, Vol. 21, No. 1, pp. 73 to 81, 1988
Printed in Great Britain

0024–6301/88 $3.00 + .00
Pergamon Journals Ltd.

73

2

A Strategic Planning Process for Public and Non-profit Organizations

John M. Bryson

A pragmatic approach to strategic planning is presented for use by public and non-profit organizations. Benefits of the process are outlined and two examples of its application are presented—one involving a city government and the other a public health nursing service. Requirements for strategic planning success are discussed. Several conclusions are drawn, namely that: (1) strategic planning is likely to become part of the repertoire of public and non-profit planners; (2) planners must be very careful how they apply strategic planning to specific situations; (3) it makes sense to think of decision makers as strategic planners and strategic planners as facilitators of decision making across levels and functions; and (4) there are a number of theoretical and practical issues that still need to be explored.

I skate to where I think the puck will be.

Wayne Gretzky

Men, I want you to stand and fight vigorously and then run. And as I am a little bit lame, I'm going to start running now

General George Stedman
U.S. Army in the Civil War

Not all of the readers of *Long Range Planning* may be familiar with either Wayne Gretzky or George Stedman, but their two quotes capture the essence of strategic planning (often called corporate planning in Britain). Wayne Gretzky is perhaps the world's greatest offensive player in professional ice hockey. He holds the single-season scoring record for players in the National Hockey League—by such a wide margin that many consider him the greatest offensive player of all time. His quote emphasizes that *strategic thinking and acting*, not strategic planning *per se*, are most important. He does not skate around with a thick strategic plan in his back pocket. What

John M. Bryson is Associate Professor of Planning and Public Affairs in the Hubert H. Humphrey Institute of Public Affairs and Associate Director of the Strategic Management Research Center at the University of Minnesota, MN 55455, U.S.A.

he does is to think and act strategically every minute of the game, in keeping with a simple game plan worked out with his coaches and key teammates in advance.

Let us explore Gretzky's statement further. What must one know and be able to do in order to make—and act on—a comment like Gretsky's? One obviously needs to know the purpose and rules of the game, the strengths and weaknesses of one's own team, the opportunities and threats posed by the other team, the game plan, the arena, the officials, and so on. One also needs to be a well-equipped, superbly conditioned, strong and able hockey player—and it does not hurt to play for a very good team. In other words, anyone who can assert confidently that he or she 'skates to where the puck will be' knows basically everything there is to know about strategic thinking and acting in hockey games.

Wayne Gretzky is respected primarily for his extraordinary offensive scoring ability. But defensive abilities obviously are important, too. Whereas Gretzky is a great offensive strategist, General George Stedman of the U.S. Army in the Civil War was an experienced defensive strategist. At one point he and his men were badly outnumbered by Confederate soldiers. A hasty retreat was in order, but it made sense to give the lame and wounded—and the General, too!—a chance to put some distance between themselves and the enemy before a full-scale retreat was called. The General and his men then would be in a position to fight another day.

Stedman had no thick strategic plan in his back pocket, either. At most he probably had a general battle plan worked out with his fellow officers and recorded in pencil on a map. Again, strategic thinking and acting were what mattered, not any particular planning process.

How does this relate to public and non-profit organizations today? The answer is that strategic thought and action are increasingly important to the continued viability and effectiveness of governments, public agencies and non-profit organizations of all sorts. Without strategic planning it is unlikely that these organizations will be able to meet successfully the numerous challenges that face them.

The environments of public and non-profit organizations have changed dramatically in the last 10 years—as a result of oil crises, demographic shifts, changing values, taxing limits, privatization, centralization or decentralization of responsibilities, moves toward information and service-based economies, volatile macroeconomic performance, and so on. As a result, traditional sources of revenue for most governments are stable at best or highly unpredictable or declining at worst. Further, while the public may be against higher taxes, and while transfers of money from central to local governments are typically stable or declining, the public continues to demand a high level of government services. Non-profit organizations often are called on to take up the slack in the system left by the departure of public organizations or services, but may be hard-pressed to do so.

To cope with these various pressures, public and non-profit organizations must do at least three things. First, these organizations need to exercise as much discretion as they can in the areas under their control to ensure responsiveness to their stakeholders. Second, these organizations need to develop good strategies to deal with their changed circumstances. And third, they need to develop a coherent and defensible basis for decision making.

What is Strategic Planning?

Strategic planning is designed to help public and non-profit organizations (and communities) respond effectively to their new situations. It is *a disciplined effort to produce fundamental decisions and actions shaping the nature and direction of an organization's (or other entity's) activities within legal bounds.*[1] These decisions typically concern the organization's mandates, mission and product or service level and mix, cost, financing, management or organizational design. (Strategic planning was designed originally for use by *organizations*. In this article we will concentrate on its applicability to public and non-profit organizations. Strategic planning of course can be, and has been, applied to projects, functions —such as transportation, health care or education —and communities.)

What does strategic planning look like? Its most basic formal requirement is a series of discussions and decisions among key decision makers and managers about what is *truly* important for the organization. And those discussions are the *big* innovation that strategic planning brings to most organizations, because in most organizations key decision makers and managers from different levels and functions almost *never* get together to talk about what is truly important. They may come together periodically at staff meetings, but usually to discuss nothing more important than, for example, alternatives to the organization's sick leave policy. Or they may attend the same social functions, but there, too, it is rare to have sustained discussions of organizationally relevant topics.

Usually key decision makers need a reasonably structured process to help them identify and resolve the most important issues their organizations face. One such process that has proved effective in practice is outlined in Figure 1. The process consists of the following eight steps:

1. *Development of an initial agreement concerning the strategic planning effort.* The agreement should cover: the purpose of the effort; preferred steps in the process; the form and timing of reports; the role, functions and membership of a strategic planning coordinating committee; the role, functions and membership of the strategic planning team; and commitment of necessary resource to proceed with the effort.

2. *Identification and clarification of mandates.* The purpose of this step is to identify and clarify the externally imposed formal and informal mandates placed on the organization. These are the *musts* confronting the organization. For most public and non-profit organizations these mandates will be contained legislation, articles of incorporation or charters, regulations, and so on. Unless mandates are identified and clarified two difficulties are likely to arise: the mandates are unlikely to be met, and the organization is unlikely to know what pursuits are allowed and not allowed.

3. *Development and clarification of mission and values.* The third step is the development and clarification of the organization's mission and values. An organization's mission—in tandem with its mandates—provides its *raison d'être*, the social justification for its existence.

Prior to development of a mission statement, an organization should complete a stakeholder analysis. A *stakeholder* is defined as any person, group or organization that can place a claim on an organization's attention, resources or output, or is affected by that output. Examples of a government's stakeholders are citizens, taxpayers, service recipients, the governing body, employees, unions, interest groups, political parties, the financial community and other governments.

2

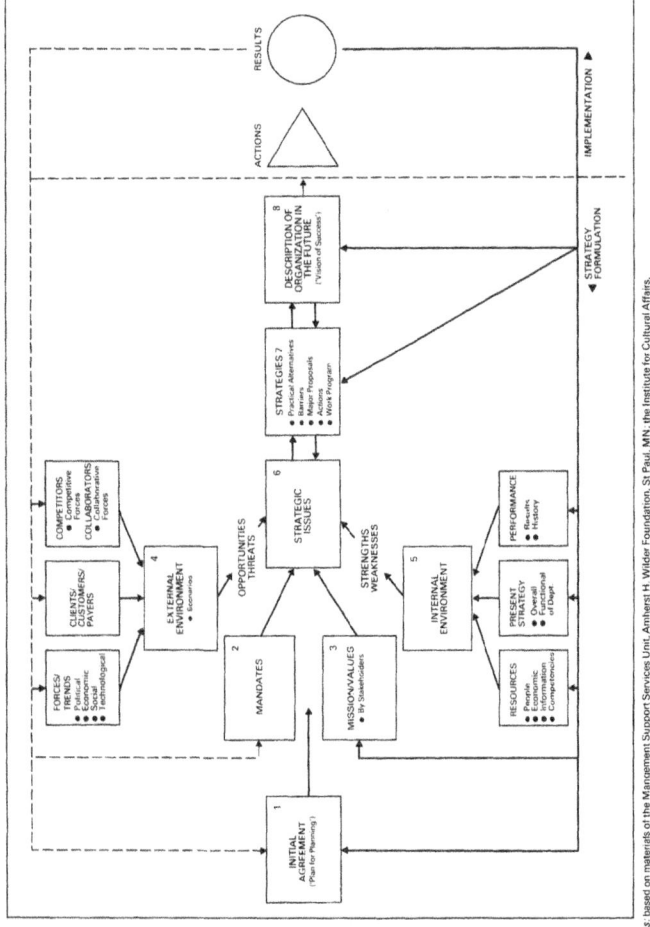

Sources: based on materials of the Mangement Support Services Unit, Amherst H. Wilder Foundation, St Paul, MN, the Institute for Cultural Affairs, Minneapolis, MN, and the Office of Planning and Development, Hennepin County, MN.

Figure 1. Strategic planning process

76   Long Range Planning Vol. 21   February 1988

In the simplest form of stakeholder analysis, the organization identifies its stakeholders and their 'stakes' in the organization, along with the stakeholders' criteria for judging the performance of the organization. The organization also explores how well it does against the stakeholders' criteria. Once a stakeholder analysis is completed, the organization can develop a mission statement that takes key stakeholder interests into account.

4. *External environmental assessment.* The fourth step is exploration of the environment outside the organization in order to identify the opportunities and threats the organization faces. Political, economic, social and technological trends and events might be assessed, along with the nature and status of various stakeholder groups, such as the organization's customers, clients or users, and actual or potential competitors or collaborators.

5. *Internal environmental assessment.* The next step is an assessment of the organization itself in order to identify its strengths and weaknesses. Three assessment categories include following a simple systems model—organizational resources (inputs), present strategy (process) and performance (outputs). Unfortunately, most organizations can tell you a great deal about the resources they have, much less about their current strategy, and even less about how well they perform. The nature of accountability is changing, however, in that public and nonprofit organizations are increasingly held accountable for their outputs as well as their inputs. A stakeholder analysis can help organizations adapt to this changed nature of accountability, because the analysis forces organizations to focus on the criteria stakeholders use to judge organizational performance. Those criteria are typically related to output. For example, stakeholders are increasingly concerned with whether or not state-financed schools are producing educated citizens. In many states in the United States, the ability of public schools to garner public financing is becoming contingent on the schools' ability to demonstrate that they do an effective job of educating their students.

The identification of strengths, weaknesses, opportunities and threats—or SWOT analysis—in Steps 4 and 5 is very important because every effective strategy will build on strengths and take advantage of opportunities, while it overcomes or minimizes weaknesses and threats.

6. *Strategic issue identification.* Together the first five elements of the process lead to the sixth, the identification of strategic issues. *Strategic issues* are fundamental policy questions affecting the organization's mandates; mission and values; product or service level and mix, clients, users or payers, cost, financing, management or organizational design. Usually, it is vital that strategic issues be dealt with expeditiously and effectively if the organization is to survive and prosper. An organization that does not address a strategic issue may be unable to head off a threat, unable to capitalize on an important opportunity, or both.

Strategic issues—virtually by definition—embody conflicts. The conflicts may be over ends (what); means (how), philosophy (why), location (where); timing (when); and who might be helped or hurt by different ways of resolving the issue (who). In order for the issues to be raised and resolved effectively, the organization must be prepared to deal with such conflicts.

A statement of a strategic issue should contain three elements. First, the issue should be described succinctly, preferably in a single paragraph. The issue itself should be framed as a question the organization can do something about. If the organization cannot do anything about it, it is not an issue—at least for the organization. An organization's attention is limited enough without wasting it on issues it cannot resolve.

Second, the factors that make the issue a fundamental policy question should be listed. In particular, what is it about mandates, mission, values or internal strengths and weaknesses and external opportunities and threats that make this a strategic issue? Listing these factors will become useful in the next step, strategy development.

Finally, the planning team should state the consequences of failure to address the issue. A review of the consequences will inform judgments of just how strategic, or important, various issues are. The strategic issue identification step therefore focuses organizational attention on what is truly important for the survival. prosperity and effectiveness of the organization—and provides useful advice on how to achieve these aims.

There are three basic approaches to the identification of strategic issues: the direct approach, the goals approach and the scenario approach.[2] The *direct approach*—in which strategic planners go straight from a view of mandates, mission and SWOTs to the identification of strategic issues—probably will work best for most governments and public agencies. The direct approach is best when one or more of the following conditions prevail: (1) there is no agreement on goals, or the goals on which there is agreement are too abstract to be useful; (2) there is no pre existing vision of success and developing a consensually based vision will be difficult; (3) there is no hierarchical authority that can impose goals on the other stakeholders; or (4) the environment is so turbulent that development of goals or visions seems unwise, and partial actions in response to immediate, important issues seem most prudent. The direct approach, in other words, can work in the pluralistic, partisan, politicized and relatively fragmented worlds of most public organizations—as long as there is a 'dominant coalition'³ strong enough and interested enough to make it work.

The *goals approach* is more in line with conventional planning theory which stipulates that an organization should establish goals and objectives for itself and then develop strategies to achieve those goals and objectives. The approach can work if there is fairly broad and deep agreement on the organization's goals and objectives—and if those goals and objectives themselves are detailed and specific enough to guide the identification of issues and development of strategies. This approach also is more likely to work in organizations with hierarchical authority structures where key decision makers can impose goals on others affected by the planning exercise. The approach, in other words, is more likely to work in public or non-profit organizations that are hierarchically organized, pursue narrowly defined missions and have few powerful stake-holders than it is in organizations with broad agendas and numerous powerful stakeholders.

Finally, there is the *scenario*—or 'vision of success'⁴—*approach,* whereby the organization develops a 'best' or 'ideal' picture of itself in the future as it successfully fulfills its mission and achieves success. The strategic issues then concern how the organization should move from the way it is now to how it would look and behave according to its vision. The vision of success approach is most useful if the organization will have difficulty identifying strategic issues directly; if no detailed and specific agreed-upon goals and objectives exist and will be difficult to develop; and if drastic change is likely to be necessary. As conception precedes perception⁵ development of a vision can provide the concepts that enable organizational members to see necessary changes. This approach is more likely to work in a non-profit organization than in a public-sector organization because public organizations are more likely to be tightly constrained by mandates.

7. *Strategy development.* In this step, strategies are developed to deal with the issues identified in the previous step. A *strategy* is a *pattern* of purposes, policies, programmes, actions, decisions and/or resource allocations that define what an organization is, what it does and why it does it. Strategies can vary by level, function and time frame.

This definition is purposely broad, in order to focus attention on the creation of consistency across *rhetoric* (what people say), *choices* (what people decide and are willing to pay for) and *actions* (what people do). Effective strategy formulation and implementation processes will link rhetoric, choices and actions into a coherent and consistent pattern across levels, functions and time.⁶

The author favours a five-part strategy development process (to which he was first introduced by the Institute for Cultural Affairs in Minneapolis).

Strategy development begins with identification of practical alternatives, dreams or visions for resolving the strategic issues. It is of course important to be practical, but if the organization is unwilling to entertain at least *some* 'dreams' or 'visions' for resolving its strategic issues, it probably should not be engaged in strategic planning.

Next, the planning team should enumerate the barriers to achieving those alternatives, dreams or visions, and not focus directly on their achievement. A focus on barriers at this point is not typical of most strategic planning processes. But doing so is one way of assuring that strategies deal with implementation difficulties directly rather than haphazardly.

Once alternatives, dreams and visions, along with barriers to their realization, are listed, the team should prepare or request major proposals for achieving the alternatives, dreams or visions directly, or else indirectly through overcoming the barriers. For example, a major city government did not begin to work on strategies to achieve its major ambitions until it had overhauled its archaic civil service system. That system clearly was a barrier that had to be confronted before the city government could have any hope of achieving its more important objectives.

After the strategic planning team prepares or receives major proposals, two final tasks must be completed. The team must identify the actions needed over the next one to two years to implement the major proposals. And finally, the team must spell out a detailed work programme, covering the next 6 months to a year, to implement the actions.

An effective strategy must meet several criteria. It must be technically workable, politically acceptable to key stakeholders, and must accord with the organization's philosophy and core values. It must also be ethical, moral and legal.

8. *Description of the organization in the future.* In the final (and not always necessary) step in the process the organization describes what it should look like as it successfully implements its strategies and achieves its full potential. This description is the organization's 'vision of success'. Few organizations have such a description or vision, yet the importance of such descriptions has long been recognized by well-managed companies and organizational psychologists.⁷ Typically included in such descriptions are the organization's mission, its basic strategies, its performance criteria, some important decision rules, and the ethical standards expected of all employees.

These eight steps complete the strategy formulation process. Next come actions and decisions to implement the strategies, and, finally, the evaluation of results. Although the steps are laid out in a linear, sequential manner, it must be emphasized that the process is iterative. Groups often have to repeat steps before satisfactory decisions can be reached and actions taken. Furthermore, implementation typically should not wait until the eight steps have been completed. As noted earlier, strategic thinking *and* acting are important, and all of the thinking does not have to occur before any actions are taken.

To return to Wayne Gretzky and George Stedman, one can easily imagine them zooming almost intuitively through the eight steps—while already on the move—in a rapid series of discussions, decisions and actions. The eight steps merely make the process of strategic thinking and acting more orderly and allow more people to participate in the process.

The process might be applied across levels and functions in an organization as outlined in Figure 2. The application is based on the system used by the 3M Corporation.[8] In the system's first cycle, there is 'bottom up' development of strategic plans within a framework established at the top, followed by reviews and reconciliations at each succeeding level. In the second cycle, operating plans are developed to implement the strategic plans. Depending on the situation, decisions at the top of the organizational hierarchy may or not require policy board approval, which explains why the line depicting the process flow diverges at the top.

The Benefits of Strategic Planning

What are the benefits of strategic planning? Government and non-profit organizations in the United States are finding that strategic planning can help them:

☆ think strategically;

☆ clarify future direction;

☆ make today's decisions in light of their future consequences;

☆ develop a coherent and defensible basis for decision making;

☆ exercise maximum discretion in the areas under organizational control;

☆ solve major organizational problems;

☆ improve performance;

☆ deal effectively with rapidly changing circumstances;

☆ build teamwork and expertise.

While there is no guarantee that strategic planning will produce these benefits, there are an increasing number of case example and studies that indicate it can help as long as key leaders and decision makers want it to work, and are willing to invest the time,

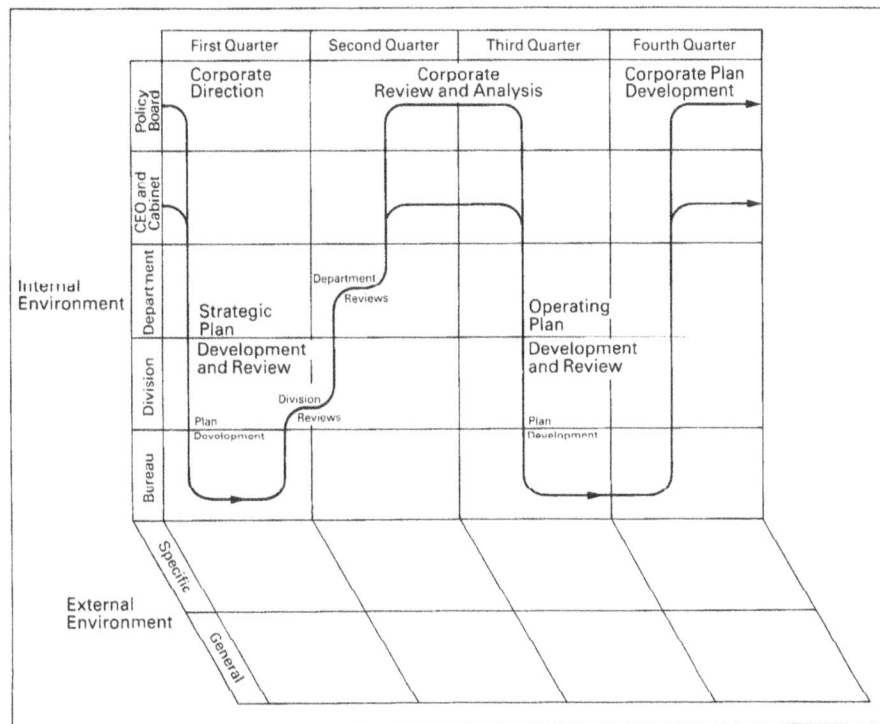

Figure 2. Annual strategic planning process

attention and resources necessary to make it work.[9] In the next two sections we will turn to two cases in which the strategic planning process outlined above produced desirable results. The author served as a strategic planning consultant in each case.

Case No. 1—Suburban City

Suburban City is an older, middle-class, 'first ring' suburb of a major metropolitan city in the American Midwest. Suburban City is regarded among city management professionals as one of the best-managed cities in the state. The city has 227 employees and an annual budget of $25·6m. The assistant city manager was the leader of the strategic planning team. The city manager was a strong supporter and member of the team. The team performed a stakeholder analysis, developed a mission statement, identified strategic issues, and developed strategies to deal with its most important issues. They are now implementing their strategies.

The following strategic issues were identified:

☆ What should the city do to enhance and improve its vehicular and pedestrian movements throughout its hierarchy of transportation facilities?

☆ What should the city do to improve its image as a place to live and work?

☆ What should the city do to attract high quality housing that meets the needs of a changing population and maintains the integrity of the existing housing stock?

☆ What should the city do to maintain its physical facilities while responding to changing demands for public services?

☆ What should the city do to restore confidence in its water quality and supply?

Strategies were developed to deal with all these issues, but we will consider the strategies stemming from the last two. The first step in responding to changing demands for public services was to undertake a major survey of households and businesses in the city to uncover preferences for services. Now that the survey is complete, city staff are rearranging and reorganizing services and delivery mechanisms to respond effectively.

Suburban City residents became worried, to the point of panic, when the city's water supply was found to be contaminated by uncontrolled seepage from a creosote plant. The city immediately closed down the affected wells and began a major cleanup effort. The water *quality* problem was cleared up, but the public *perception* that the city had a serious water quality problem persisted. City staff undertook a public education effort to deal with this misperception, and another effort was undertaken

to deal with the remaining—and real, not just perceived—problem of a water *quantity*.

The strategic planning team did not go on to draft a 'vision of success' for the city. One reason why this was not done was that the team had had real difficulty developing a mission statement that all could support. The difficulty was not over content, interestingly enough, but over style. The city manager felt that a mission statement should give a person 'goose bumps', and the team had trouble drafting a mission statement that did. Finally, the city manager relented and supported a mission statement that had less of a physiological effect.

An interesting result of the city's strategic planning effort has been the recognition by members of the city council that they have not been an effective policy-making board. As a result, they hired a nationally known consultant on effective governance to help them become better policymakers. The city manager and assistant city manager are convinced that as the council becomes more effective, strategic planning for the city also will become more effective.

Case No. 2—Public Health Nursing Service

Public Health Nursing Service (Nursing Service) is a unit of the government of a large, urban county in the same state as Suburban City. The county executive director decided to explore the utility of strategic planning for the county by asking several units of county government, including Nursing Service, to undertake strategic planning.

Nursing Service is required by statute to control communicable diseases, and it also provides a variety of public health services at its clinics throughout the county. In 1984 Nursing Service had over 80 staff members and a budget of approximately $3·5m.

The strategic planning team was led by the director of the service, who was a major supporter of the process. Other sponsors, though not strong supporters, included the county's executive director and the director of the department of public health, of which Nursing Service is a part. The department's health planner was an active and dedicated promoter of the process.

The director, deputy director and staff of Nursing Service saw strategic planning as an opportunity to rethink the service's mission and strategies in light of the rapidly changing health care environment. They were concerned, however, that they had been selected as 'guinea pigs' for the executive director's experiment in strategic planning. Nursing Service has always lived with the fear that it would be taken over, put out of business or otherwise circumvented by the county government's huge medical centre, a famous hospital that was considering entering the home health care field (Nursing Service's main 'business') at the same time that Nursing Service began its strategic planning process. Nursing Service was afraid that any information or arguments it created as part of its process might be used against it by the executive director and county board to benefit the medical centre. A number of reassurances from the executive director were necessary before Nursing Service would believe it was not being 'set up'.

As a result of the process, Nursing Service identified a number of strategic issues. The principal issue was what the mission of Nursing Service should be given the changing health care environment. After rethinking their mission, the Nursing Service team rethought their first set of strategic issues. The team identified a new set of strategic issues concerning how the new mission could be pursued. Those issues were:

☆ What is the role of Nursing Service in ensuring the health of the citizens of the county?

☆ How should Nursing Service deal with the growing health care needs for which there is inadequate or no reimbursement of services?

☆ What is the role of Nursing Service (and the county) in ensuring quality in community-based health care?

☆ What is the role of Nursing Service (and the county) in ensuring community health planning and health system development?

Nursing Service went on to develop a set of strategies designed to deal with these issues. The set includes:

☆ Differentiation and clarification of line and staff functions of Nursing Service's supervisors and administrators.

☆ Development of a process for programme development and change.

☆ Development of an organizational structure which will allow the agency to respond most effectively and efficiently to the needs of communities as well as individuals and families.

By the end of 1987 these strategies should be fully implemented. The strategies do not necessarily deal with the strategic issues directly. Instead, they focus primarily on overcoming the barriers to dealing with the issues. Once the agency is organized properly and has programme development and change procedures in place, it will be better able to address the health care needs of the citizens of the county.

Nursing Service also developed a 'vision of success'

for itself. The Service's idealized scenario of itself envisages an agency thoroughly responsive to community, family and individual health care needs.

Ironically, it was Nursing Service's strategic planning efforts that in part forced strategic planning on the county board. Nursing Service prepared its strategic issues and then was asked to make a presentation to the county board on the issues and desirable strategies to address them. The issues ultimately concerned the county government's role in the health care field and the board's willingness to pay for meeting the health care needs of the county's residents. County board members realized they were completely unprepared to deal with the issues raised by Nursing Service. The board also realized that they might soon be faced with similar vexing issues by other departments engaged in strategic planning. The board felt a need to think about the county government as a whole, and about how to establish priorities, before they were presented with any more policy questions for which they had no answers. The board decided to go on a retreat in order to clarify the county government's mission, to identify strategic issues and to agree on a process for resolving the issues. They identified eight key issues, including issues prompted by Nursing Service's questions concerning the county's role in health care.

Also ironically, partway through Nursing Service's planning efforts, the county board forced the county's executive director to resign. Nursing Service then saw the strategic planning process as a real opportunity to think through its position so that it could have the most impact on the thinking of the new executive director.

What it Takes to Initiate and Succeed with Strategic Planning

The two case histories and the growing body of literature on strategic planning for the public and non-profit sectors help us draw some conclusions about what appears to be necessary to initiate an effective strategic planning process. At a minimum, any organization that wishes to engage in strategic planning should have: (1) a process sponsor(s) in a position of power to legitimize the process; (2) a 'champion' to push the process along;[10] (3) a strategic planning team; (4) an expectation that there will be disruptions and delays; (5) a willingness to be flexible about what constitutes a strategic plan; (6) an ability to pull information and people together at key points for important discussions and decisions; and (7) a willingness to construct and consider arguments geared to very different evaluative criteria.

The criteria for judging the effectiveness of strategic planning for governments and public agencies probably should differ from those used to judge effectiveness in the private sector. The nature of the public sector prevents exact duplication of private sector practice.[11] The more numerous stakeholders, the conflicting criteria they often use to judge governmental performance, the presssures for public accountability and the idea that the public sector is meant to do what the private sector cannot, all militate against holding government strategic planning practice to private-sector standards. Until governments and public agencies (as well as non-profit organizations) gain more experience with strategic planning, it seems best to judge their strategic planning efforts according to the extent to which they: (1) focus the attention of key decision makers on what it important for their organizations, (2) help set priorities for action, and (3) generate those actions.

Conclusions

Strategic planning for public and non-profit organizations is important and probably will become part of the standard repertoire of public and non-profit planners. It is important, of course, for planners to be very careful about how they engage in strategic planning, since every situation is at least somewhat different and since planning can be effective only if it is tailored to the specific situation in which it is used.[12] The process outlined in this article, in other words, represents a generic guide to strategic thought and action, and must be adapted with care and understanding to be useful in any given situation.

To assert that strategic planning will increase in importance raises the question of who the strategic planners are. It is likely that within the organization they may not hold job titles that include the word 'planner'; instead, they may be in policy making or line management positions.[13] Since strategic planning tends to fuse planning and decision making, it makes sense to think of decision makers as strategic planners and to think of strategic planners as facilitators of decision making across levels and functions in organizations (and communities). The specific blend of technical knowledge and process expertise that the persons with the formal job title of planner should bring to strategic planning exercises, of course, will vary in different situations. The more the key decision makers already have the necessary technical knowledge, the more the planners will be relied upon to facilitate the process than to provide technical knowledge.

Finally, research must explore a number of theoretical and practical issues in order to advance the knowledge and practice of strategic planning for governments, public agencies and non-profit organizations. In particular, more detailed strategic planning models should specify key situational factors governing their use; provide specific advice

on how to formulate and implement strategies in different situations; be explicitly political; indicate how to deal with plural, ambiguous or conflicting goals or objectives; link content and process; indicate how collaboration as well as competition should be handled; and specify roles for the strategic planner. Progress has been made on all of those fronts[14] (to which, it is hoped, this article and the book from which it is drawn attest), but more is necessary if strategic planning is to help governments, public agencies and non-profit organizations, as well as communities and functions, fulfill their missions and serve their stakeholders effectively, efficiently and responsibly.

Acknowledgement—This article is based on a chapter in John M. Bryson, *Strategic Planning for Public and Nonprofit Organisations*, Jossey-Bass, San Francisco (1988).

References

(1) J. B. Olsen and D. C. Eadie, *The Game Plan: Governance With Foresight*, p. 4, Council of State Planning Agencies, Washington, D.C. (1982).

(2) B. Barry, *The Strategic Planning Workbook for Nonprofit Organizations*, The Amherst H. Wilder Foundation, St Paul, MN (1986).

(3) J. D. Thompson, *Organizations in Action*, McGraw-Hill, New York (1967).

(4) B. Taylor, Strategic planning—which style do you need?, *Long Range Planning* **17**, 51–62 (1984).

(5) R. May, *Love and Will*, Norton, New York (1969).

(6) P. Bromiley, Personal communication (1986).

(7) T. J. Peters and R. H. Waterman, Jr, *In Search of Excellence: Lessons from America's Best-Run Companies*, Harper & Row, New York (1982); E. A. Locke, K. W. Shaw, L. M. Saari and G. P. Latham, Goal setting and task performance: 1969–1980. *Psychological Bulletin* **90**, 125–152 (1981).

(8) M. A. Tita and R. J. Allio, 3M's strategy system—planning in an innovative organization, *Planning Review*, September, pp.10–15 (1984).

(9) J. M. Bryson and R. C. Einsweiler (Eds), *Strategic Planning for Public Purposes—Concepts, Tools and Cases*, The Planners' Press of the American Planning Association, Chicago, IL and Washington, D.C. (1988).

(10) P. Kotler, *Marketing Management*, p. 200, Prentice-Hall, Englewood Cliffs, N.J.; R. M. Kanter (1976), *The Changemasters*, p. 296, Simon & Schuster, New York (1983).

(11) P. S. Ring and J. L. Perry, Strategic management in public and private organizations: implications and distinctive contexts and constraints, *Academy of Management Review* **10**, 276–280 (1985).

(12) J. M. Bryson and A. L. Delbecq, A contingent approach to strategy and tactics in project planning, *Journal of the American Planning Association* **45**, 167–179 (1979); K. S. Christensen, Coping with uncertainty in planning, *Journal of the American Planning Association* **51**, 63–73 (1985).

(13) J. M. Bryson, A. H. Van de Ven and W. D. Roering, Strategic planning and the revitalization of the public service, in R. B. Denhardt and E. T. Jennings, Jr (Eds), *The Revitalization of the Public Service*, pp. 55–75, Extension Publications, University of Missouri, Columbia, MO (1987).

(14) B. Checkoway (Ed.), *Strategic Perspectives on Planning Practice*, Lexington Books, Lexington, MA (1986).

Planungskultur

Frank Othengrafen, Mario Reimer und Rainer Danielzyk

© Springer-Verlag GmbH Deutschland, ein Teil von Springer Nature 2019
T. Wiechmann (Hrsg.), *ARL Reader Planungstheorie Band 2*, https://doi.org/10.1007/978-3-662-57624-3_3

3.1 Einleitung

Indem räumliches Planen als kulturelle Praxis begriffen wird, wird Kultur zu einem hilfreichen Konzept auf dem Weg zu einem tieferen Verständnis von Planungspraxis. Planungskulturen können als institutionelle Muster verstanden werden, die die normativen Vorstellungen hinsichtlich der Aufgaben und Funktionen der räumlichen Planung in ihrer alltäglichen Praxis widerspiegeln. Die Art und Weise, wie die jeweiligen Akteure ihre Rollen und Aufgaben verstehen, wie sie Probleme wahrnehmen, damit umgehen und dabei bestimmte Regeln, Verfahren und Instrumente anwenden, sind somit Kennzeichen einer Planungskultur. Diese ist situationsspezifisch und kontextgebunden, sodass Planungskultur immer lokale und regionale Praktiken abbildet. Das bedeutet gleichzeitig, dass „Planungskulturen nicht per se stabil" sind, sondern immer wieder „neu produziert werden" müssen (Loepfe und Eisinger 2016, S. 45).

Für eine theoretisch fundierte Konzipierung von Planungskulturen und deren empirische Erforschung sprechen gewichtige Gründe. Planerische Alltagserfahrungen und planungswissenschaftliche Überlegungen führen rasch zu der Erkenntnis, dass sich das konkrete planerische Geschehen, z. B. die Breite des Spektrums der beteiligten Akteure, die Form ihrer Interaktionen, der Umgang mit kodifizierten Regularien oder auch die realisierten Ergebnisse im Sinne von Umsetzungen der Planung, nicht allein aus der Kenntnis der formal definierten Planungssysteme herleiten lassen. Vielmehr sind immer wieder räumlich, zeitlich und sachlich spezifische Planungspraktiken unter sonst gleichen Rahmenbedingungen zu beobachten. Damit handelt es sich bei Planungskultur um ein planungswissenschaftliches Konzept, das diese Differenzierungen besser erfassen und analysieren kann. Das kann als ein zentrales Anliegen der aktuellen planungswissenschaftlichen Diskussion zu Planungskulturen im deutschsprachigen Raum verstanden werden und beschreibt damit auch die Motivation zu diesem Beitrag.

Bishwapriya Sanyal: Hybrid Planning Cultures: The Search for the Global Cultural Commons (Sanyal 2005)

Die Debatte zu Planungskulturen hat im internationalen Kontext wesentliche Impulse durch den von Bishwapriya Sanyal (2005) herausgegebenen Sammelband *Comparative Planning Cultures* erfahren. Er geht zurück auf ein am MIT 2002 durchgeführtes Symposium zum Vergleich von Planungskulturen in zehn ausgewählten Ländern. Sanyal hat sein Interesse an planungskulturellen Fragestellungen insbesondere über den Austausch mit John Friedmann entwickelt, der sich bereits früh mit den Implikationen der Globalisierung für Planungspraktiken in unterschiedlichen räumlichen Kontexten beschäftigt hat. Basierend auf einem kursorischen Überblick über die großen Entwicklungszyklen räumlicher Planung – angefangen von den „goldenen Jahren" einer allumfassenden und rational orientierten Planung bis hin zu einer liberal motivierten Grundsatzkritik an Planung – stellt Sanyal in seinem einleitenden Beitrag die Frage, ob es einen kulturellen Nukleus gibt, der Planungspraktiken signifikant beeinflusst und gewissermaßen als „unabhängige Variable" begriffen werden kann. Darüber hinaus ist für ihn von Interesse, ob sich vor dem Hintergrund genereller gesellschaftlicher Entwicklungen auch die Praktiken der räumlichen Planung weltweit angleichen. Er verweist dabei auf die Rolle von gesellschaftlichen Kontexten und politischen Konstellationen, die wesentlich für die Ausprägung einer Planungskultur sind. Einen besonderen Wert erhält sein Beitrag aber vor allem dadurch, dass er auf die Dynamik von Planungskulturen verweist, die keinesfalls als starre kulturelle Formationen wirken, sondern in einem engen Wechselverhältnis mit sozialen, politischen und wirtschaftlichen Prozessen stehen und daher einem konstanten Wandel unterliegen. Er grenzt sich damit dezidiert von einem essenzialistischen Kulturverständnis ab.

Zur theoretisch-konzeptionell fundierten Herleitung des Begriffs von Planungskultur und seiner Diskussion ist das Kapitel in folgende Schritte gegliedert: Zunächst erfolgt der Versuch einer Begriffsbestimmung (▶ Abschn. 3.2). Daran anschließend wird ein kursorischer Überblick über die Diskussion des Begriffs in den Planungswissenschaften in den letzten Jahrzehnten gegeben (▶ Abschn. 3.3). In ▶ Abschn. 3.4 werden analytische Zugangsperspektiven zum Begriff der Planungskultur skizziert, wobei auch auf korrelierende Debatten und Konzepte in benachbarten Wissenschaften eingegangen wird. Abschließend wird erörtert, welche planungswissenschaftlichen Fragestellungen mit diesem Konzept besonders gut adressiert werden können (▶ Abschn. 3.5).

3.2 Begriffsbestimmung

Die Hinwendung zur Kultur erfolgt in verschiedenen Disziplinen oftmals als eine Art Gegenkonzept zum bestehenden Denk- und Ordnungssystem, d. h., mit der Einführung kultureller Aspekte und Dimensionen sollen Umbrüche in bestehenden wissenschaftlichen Erklärungsmustern eingeleitet werden (Landwehr und Stockhorst 2004; vgl. auch Casprig 2009). Dies ist auch im Bereich der räumlichen Planung zu beobachten – mit der Einführung des Planungskulturansatzes wird räumliches Planen als kulturelle Praxis begriffen, deren Handlungsrahmen sich in Abhängigkeit von kulturellen Strukturen und Diskursen herausbildet (vgl. Hölzl und Nuissl 2015; Levin-Keitel und Sondermann 2015; Othengrafen und Reimer 2013). Dabei wird räumliche Planung als Teil der Kultur einer Gesellschaft verstanden:

„Planungskultur ist also nicht ein irgendwie eingrenzbarer Bestandteil von (Raum-)Planung; vielmehr erfasst der Begriff der Planungskultur die (kulturelle) Praxis der Planung aus einer bestimmten Perspektive" (Nuissl 2008, S. 11).

Im Sinne von praxistheoretischen Ansätzen in den Kulturwissenschaften bzw. der Kultursoziologie (z. B. Nicolini 2013; Reckwitz 2003, 2006; Schatzki 1996; Schatzki et al. 2001; Schäfer 2016; Schmidt 2012; Shove et al. 2012) umfassen Praktiken u. a. Aktivitäten, Handlungsformen, Verhaltensmuster oder auch Interaktionsformen (Hirschauer 2016, S. 46), die in einem bestimmten Raum bzw. Kontext durch gemeinsame Verständnisse und Regeln organisiert sind (Schäfer 2016; Schatzki 2016). Diese sozialen Praktiken stellen den Bezugspunkt kultureller Analysen dar, in dem sie die öffentlich beobachtbaren, routinisierten Handlungsmuster von Akteuren als Ausgangspunkt nehmen (vgl. Reckwitz 2006, S. 558 ff.). Wichtig ist hierbei, dass es sich bei diesen Handlungsmustern um kollektive Muster handelt, die über die „persönlichen Besonderheiten von Individuen" hinweg existieren und damit öffentlich wahrnehmbar sind (Reckwitz 2006, S. 559). Gleichzeitig handelt es sich bei sozialen Praktiken immer auch um kulturelle Praktiken, die sich im Rahmen von „kulturell vorstrukturierten ways of doing" (Hirschauer 2016, S. 46) oder impliziten Wissensordnungen (Hörning und Reuter 2004, S. 11; Reckwitz 2016, S. 163; Schäfer 2016, S. 13) herausbilden. Dieser Haltung liegt ein bedeutungs- und wissensorientiertes Kulturverständnis zugrunde, d. h., Kultur bietet ein für eine Gesellschaft, Organisation oder Gruppe typisches Sinn- und Orientierungssystem. Kultur enthält mithin symbolische Ordnungen, „mit denen sich die Handelnden ihre Wirklichkeit als bedeutungsvoll erschaffen und die in Form von Wissensordnungen ihr Handeln ermöglichen und einschränken" (Reckwitz 2006, S. 84).

Philip Booth: The Cultural Dimension in Comparative Research: Making Sense of Development Control in France (Booth 1993)
Ausgebildet als Architekt und Planer, arbeitete Philip Booth zunächst auf der kommunalen Ebene in England. Hier machte er die Erfahrung, dass sich das Verhältnis von Politik und Planung – insbesondere die Festlegung räumlich bedeutsamer Ziele und Entwicklungsprioritäten sowie die Anwendung planerischer Steuerungsinstrumente – von Kommune zu Kommune unterscheidet. Diese Erfahrungen leiteten in der Folge seine Forschungsinteressen; so analysierte er vor allem planungsrechtliche Verfahren und Möglichkeiten der öffentlichen Hand, räumliche Entwicklungen in Städten und Regionen steuern zu können (insbesondere im Vergleich zwischen England und Frankreich). Der Artikel ist weitestgehend in diesem Kontext entstanden. Ausgangspunkt ist die Kritik an den bis dato vorherrschenden internationalen Studien und Vergleichen, die Planungsprozesse vor allem als zweckrationale und technische Verfahren verstehen die in allen Kommunen eines Staates theoretisch zu den gleichen Resultaten führen müssten. Diese Annahme widerlegt Philip Booth anhand verschiedener französischer Fallstudien, in denen er aufzeigt, dass kulturbedingte Wahrnehmungen und Vorstellungen zu unterschiedlichen Ergebnissen (Ziele, Konzepte, Handlungsempfehlungen etc.) führen. Vor diesem Hintergrund fordert er, Planung als kulturell definierten Prozess zu verstehen, in dem sich lokale Planungspraktiken herauszubilden, die sich durch das spezifische Zusammenspiel politischer Entscheidungsträger unterscheiden, ebenso wie durch die Bedeutung bzw. Erwartungen, die Politiker und Planer den verwendeten Instrumenten zukommen lassen. Die Beschreibungen von Verhaltensweisen und Reaktionen sind demnach zentral für (international) vergleichende Studien, da sie einen direkten Bezug zu Fragen der Landnutzungsallokation und der gebauten Umwelt aufweisen. Mit seiner dezidierten Aufarbeitung kulturbedingter Wahrnehmungen und Vorstellungen in der lokalen Planungspraxis legte Philip Booth nicht nur den Grundstein für die dann folgende Planungskulturforschung, sondern auch für die methodische Weiterentwicklung internationaler Vergleichsstudien.

Solche symbolischen Ordnungssysteme können auf Ebene mentaler Strukturen und Prozesse, auf Ebene von Diskursen und Texten sowie auf Ebene sozialer Praktiken bestehen (Reckwitz 2005, S. 96). Damit umfasst Kultur nicht nur „die kollektive Programmierung des Geistes, die Mitglieder einer Gruppe oder Kategorie von Menschen von einer anderen unterscheidet" (Hofstede 2001, S. 4), sondern auch eine vom interpretierenden Individuum abhängige Variable, die es erlaubt, dass die Verhaltensweisen und Reaktionen eines Individuums von den kollektiven Handlungsmustern abweichen können (Straub 2004, S. 573 ff.). Demnach hält eine kulturelle Ordnung zwar eine gewisse Struktur für das individuelle Handeln der jeweiligen Mitglieder bereit, wird aber von jedem Individuum interpretiert und durch individuelle bzw. institutionelle Kenntnisse und Fähigkeiten ausgestaltet (Hörning und Reuter 2004, S. 12). In diesem Sinn erfolgt eine „aktive interpretative Aneignung unterschiedlicher, einander ,überlagernder' Sinn- und Aktivitätselemente" (Reckwitz 2005, S. 100 f.). Kultur kann dann als soziale Praxis verstanden werden, in der Wissen, Kompetenzen, Alltagstechniken und Routinen inkorporiert und im Wechselspiel mit lokalen Kontexten, Artefakten und Sinnsystemen betrachtet werden (vgl. Schatzki 2001; Swidler 2001).

Dies gilt auch für den Bereich der räumlichen Planung: Die Festlegung räumlich bedeutsamer Ziele und Entwicklungsprioritäten, die Erstellung von Plänen, die Anwendung planerischer Steuerungsinstrumente oder die Einbeziehung Betroffener in Planungsprozesse können hier als Praktiken von Planern verstanden werden, die zu einer gegebenen Zeit und in einem bestimmten Raum/Kontext gelten. Planer wiederholen diese Praktiken häufig und regelmäßig und konstituieren damit bestimmte „räumliche Arrangements", die aus der Interaktion von Planern mit anderen Planern und Betroffenen sowie mit Artefakten (z. B. Plänen, bedeutsamen Gebäuden und Siedlungsstrukturen) entstehen. Gleichzeitig beeinflussen und prägen die spezifischen Arrangements auch die ausgeübten Praktiken (vgl. Schatzki 2016, S. 33). Zusammengefasst bedeutet dies, dass räumliches Planen als kulturelle Praxis begriffen werden kann, deren Handlungsrahmen sich in Abhängigkeit von individuellem bzw. institutionellem Handeln, kulturellen Strukturen und Diskursen sowie lokalen Kontexten ergibt (Levin-Keitel und Sondermann 2015; Othengrafen 2014; Othengrafen und Reimer 2013). Dadurch wird Kultur zu einem hilfreichen analytischen Konzept, um ein tieferes Verständnis der Planungspraxis und des Kontextes von planerischen Tätigkeiten zu ermöglichen (Loepfe und Eisinger 2016, S. 46).

Übergeordnetes Ziel einer so verstandenen Planungskulturforschung ist es, zu einem vertieften Verständnis und einer Reflexion der Planungspraxis beizutragen. Dies erfolgt vor allem durch den Versuch, die Denkmuster, Einstellungen und Handlungsroutinen von Planern und Planungsinstitutionen auf Basis von typischen (gesellschaftlichen) Orientierungen, Ideen und zugehörigen Werthaltungen zu identifizieren (Othengrafen 2014). Planungskulturen können somit explizit als Denk- und Handlungsmuster verstanden werden, die die normativen Vorstellungen hinsichtlich der Aufgaben und Funktionen der räumlichen Planung in der tatsächlichen Praxis von Planern widerspiegeln (Hölzl und Nuissl 2015; Othengrafen et al. 2015). Die Art und Weise, wie die jeweiligen Akteure ihre Rollen und Aufgaben verstehen, wie sie Probleme wahrnehmen, damit umgehen und dabei bestimmte kulturell geprägte Regeln, Verfahren und Instrumente anwenden, sind somit Kennzeichen einer Planungskultur (Neuman 2007; Othengrafen 2014).

Andreas Faludi: Patterns of Doctrinal Development (Faludi 1999)

Andreas Faludi hat als Planungsforscher über mehrere Jahrzehnte vor allem in den Niederlanden und Österreich gewirkt und sich insbesondere mit Raumplanung und Raumentwicklung auf europäischer Ebene, aber auch mit dem Vergleich nationaler Planungen befasst. Die planungswissenschaftliche Diskussion hat er vor allem mit seinem Konzept der „Planungsdoktrinen" beeinflusst. Demnach dienen Planungsdoktrinen der Rahmung des planerischen Handelns, indem sie möglichst überzeugend räumliche Strukturen und Entwicklungen analysieren und interpretieren sowie das darauf bezogene planerische Handeln sinnvoll anleiten. Nach Faludi benötigt erfolgreiche Raumplanung überzeugende Doktrinen, die sich möglichst in kraftvollen Metaphern verdichten sollten. Als überzeugende Beispiele gelten etwa der in der Mitte des 20. Jahrhunderts planerisch definierte Green Belt zur Eindämmung der Suburbanisierung Londons und das „Grüne Herz" im Kern der Randstad in den Niederlanden. An den beiden genannten Beispielen lässt sich zeigen, dass gute Planungsdoktrinen auch über die Raumplanung hinaus Wirkung entfalten.

In diesem Artikel diskutiert Faludi an vier Beispielen aus den 1990er Jahren (Strukturplan für Flandern, Raumstrukturkonzept für die Benelux-Staaten, Europäisches Raumentwicklungskonzept und Ansätze des Growth-Management im US-Bundesstaat Florida) insbesondere den Wandel von Planungsdoktrinen. Er kommt zu dem Ergebnis, dass sich dabei weniger ein „revolutionärer" paradigmatischer Wandel im Sinne von Thomas Kuhn, sondern eher evolutionäre Weiterentwicklungen beobachten lassen. Der Wert des vor allem von Faludi in den Vordergrund gerückten Ansatzes der „Planungsdoktrinen" für die Planungskulturforschung besteht darin, die Bedeutung der konzeptionellen Rahmung, auch durch Metaphern, zur Interpretation räumlicher Situationen und zur Anleitung planerischen Handelns überzeugend auf den Punkt gebracht zu haben. Dass dieser Ansatz nicht größere Verbreitung gefunden hat, könnte u. a. damit zusammenhängen, dass sich für viele raumplanerischen Aufgaben nicht derartig intern und extern überzeugende Metaphern wie das „Grüne Herz der Randstad" finden lassen.

3.3 Chronologischer Überblick

Die Auseinandersetzung mit dem Begriff bzw. Konzept „Planungskultur" – wenn auch nicht unter expliziter Bezugnahme auf praxistheoretische oder andere konzeptionelle Überlegungen – reicht zurück bis in die 1960er Jahre (vgl. Reimer 2016; Zimmermann et al. 2018). Interessanterweise wird der Begriff der Planungskultur erstmals bemüht, um die seinerzeit aufkommende Kritik am rationalen Planungsparadigma zu kommentieren. Es ist kein Zufall, dass gerade in einer Zeit, in der das zumindest in der theoretischen Debatte formulierte Scheitern des rationalen Modells und die daran anknüpfende Suche nach neuen Erklärungsansätzen an Dynamik gewinnen, kulturell orientierte Perspektiven und Ansätze Einzug in planungstheoretische Debatten erhalten. Der Kulturbegriff entfaltet seine Kraft und Attraktivität in dieser Zeit vor allem deshalb, weil er diejenigen Aspekte planerischen Handelns adressiert, die eine allein rationale Argumentationslogik konsequent ausblenden. Dazu gehören insbesondere die nicht direkt greifbaren Faktoren, die das Handeln bestimmen, also subjektive Werte und Wahrnehmungsmuster, ideologische

Grundhaltungen, Intuitionen, Traditionen und Wissens-ordnungen, die in ihrer Gesamtheit planerisches Handeln wesentlich beeinflussen (▶ Kap. 2, Bd. 1). Einer der Wegbereiter dieser Debatte ist John Friedmann, der sich bereits 1967 dezidiert mit der Rolle von rationalen und nichtrationalen Elementen (u. a. Werten, Traditionen, Einstellungen, Weisheiten) in planerischen Entscheidungsprozessen beschäftigt hat, ohne dafür den Begriff der Kultur explizit zu verwenden.

John Friedmann: A Conceptual Model for the Analysis of Planning Behavior (Friedmann 1967)

Der Artikel ist zu einer Zeit entstanden, in der John Friedmann viele Jahre als Regierungsberater und Hochschullehrer in Lateinamerika (u. a. in Brasilien, Chile und Mexiko) tätig war und die ihn hinsichtlich der kulturellen Unterschiede, des planerischen Selbstverständnisse und der institutionellen Einbindungen geprägt hat. Nach Friedmann werden alle politischen und planerischen Aktionen in unterschiedlichem Maße durch verschiedene Denk- und Handlungsweisen beeinflusst, die entweder rational („bounded" oder „non-bounded rationality") oder nichtrational („extra-rational thoughts") sind. Diesen kommt eine entscheidende Rolle bei der Ausprägung räumlicher Entwicklungsmuster, dem Verhalten von Planern und dem Verständnis von Planungsinstitutionen zu. Unter „bounded rationality" werden dabei der lokale Kontext und die spezifischen strukturellen Bedingungen zusammengefasst, die den Spielraum für die Planungsinstitutionen und den Einsatz von geeigneten Instrumenten und Verfahren bestimmen. Die „non-bounded rationality" umfasst utopische oder ideologische Leitvorstellungen, die Planung zu verwirklichen versucht (z. B. soziale Gerechtigkeit, partizipative Demokratie, nachhaltige Entwicklung). Neben den rationalen Elementen hebt Friedmann aber auch die Bedeutung der „extra-rational thoughts", d. h. der gesellschaftlichen Werte, Traditionen, Einstellungen, Weisheiten etc. hervor. Diese sind zwar nicht auf Basis zusammenhängender, logischer Strukturen abgeleitet, stellen in Entscheidungsfindungsprozessen aber die wichtigste (Legitimations-)Quelle für Planer dar. Damit bietet Friedmann ein Konzept an, Planungspraktiken und -verständnisse umfassend zu analysieren; allerdings wurde gerade der Bereich der „extra-rational thoughts" in den nachfolgenden Debatten weitestgehend ignoriert und erst später wieder stärker berücksichtigt.

Erstmals direkt adressiert wird der (politische) Kulturbegriff im Zusammenhang mit Planung 1969 von Richard Bolan, der mit ihm das direkte Umfeld kommunaler Entscheidungsfindungsprozesse umschreibt. Er konstatiert:

> » [...] rational planning procedures bear little relation to the governing of cities: the time horizon and issues that have preoccupied planners are largely irrelevant and local urban governments are so disorganized, fragmented, dispersed, and incompetent that no injection of rational planning (even when relevant) can survive such a political culture. (Bolan 1969, S. 301)

Der Kulturbegriff wird hier tendenziell eher negativ konnotiert, das Kulturelle in der Planung wird als Synonym für die Durchdringung von technisch-rationaler Informationsverarbeitung und politischer Aushandlung vereinnahmt.

Trotz der frühen Bezüge zum Kulturbegriff kann sich dieser in den nachfolgenden Jahren in der Planungsforschung kaum durchsetzen. Die in den 1960er Jahren einsetzende Institutionalisierung und Professionalisierung der Planung (Fürst 2005) basiert wesentlich auf einem naturwissenschaftlich-ingenieurtechnisch geprägten Weltbild abseits jeglicher kulturtheoretischer Perspektiven. Erst in den 1980er Jahren lässt sich im planungswissenschaftlichen Diskurs erneut eine intensive Auseinandersetzung mit grundlegenden Werthaltungen und Moralvorstellungen in der räumlichen Planung feststellen. Eine frühe Ausnahme stellt die bereits 1968 vorgelegte Studie von Heide Berndt zum Gesellschaftsbild von Stadtplanern dar (Berndt 1968). Darüber hinaus haben Elizabeth Howe und Jerome Kaufman in ihren Arbeiten die grundlegenden Werthaltungen und Moralvorstellungen von amerikanischen Planern erhoben und diese Studien später auch international vergleichend fortgeführt (Howe und Kaufman 1979; Kaufman 1985; Kaufman und Escuin 2000).

Elizabeth Howe und Jerome Kaufman: The Ethics of Contemporary American Planners (Howe und Kaufman 1979)

Jerome Kaufman hat als Professor am Department of Urban and Regional Planning der Universität Wisconsin-Madison gemeinsam mit seiner Kollegin Elizabeth Howe breit angelegte Studien zu ethischen Werthaltungen von Planern durchgeführt. Basierend auf einer 1978 durchgeführten Befragung von insgesamt 1178 Mitgliedern des American Institute of Planners (AIP) haben sie Szenarien konstruiert, die unterschiedliche ethische Grundhaltungen beinhalten und von den Planern bewertet wurden. Darüber hinaus haben sie sich intensiv mit der Frage auseinandergesetzt, welche Faktoren die ethische Grundhaltung von Planern bestimmen. Dazu zählen insbesondere unterschiedliche Rolleninterpretationen (Planung als technisch interpretierte Profession, Planer als politische Akteure oder „hybride" Planung, die zwischen beiden Extrempositionen je nach Situation wechselt). Die in ihrem Artikel im Detail dargelegte Studie stellt den Auftakt für eine Reihe weiterer

Studien dar, die überwiegend in den 1980er Jahren veröffentlicht wurden. Für die Planungskulturforschung stellen ihre Arbeiten einen reichen Fundus dar, auch wenn sie den Begriff an sich nicht benutzen. Allerdings adressieren sie jene Tiefendimensionen des planerischen Handelns, die auch für die Planungskulturforschung einen wesentlichen Erkenntnisgegenstand darstellen: ethische Grundhaltungen, Rolleninterpretationen, Werte und Normen von Planern etc. Von besonderem Wert sind ihre Arbeiten auch deshalb, weil sie der Planungs-kulturforschung wertvolle Hinweise hinsichtlich der empirischen Untersuchbarkeit liefern. Die Autoren operieren mit breit angelegten, quantitativen Befragungen, um professionelle Werthaltungen über statistische Auswertungen zu identifizieren. Darüber hinaus haben sie ihre Studien später sukzessive ausgeweitet und international vergleichend fortgeführt, um die (planungs-)kulturellen Unterschiede zwischen verschiedenen Ländern zu beleuchten (beispielsweise über den Vergleich von amerikanischen und israelischen Planern).

Der Planungskulturbegriff wird dann erst wieder Anfang der 1990er Jahre direkt adressiert, vor allem in einem von Donald Keller, Michael Koch und Klaus Selle herausgegebenes Themenheft in der Fachzeitschrift *disP* (Keller et al. 1993). Der Kulturbegriff wird hier verwendet, um die länderspezifischen Charakteristika der räumlichen Planung in vier europäischen Ländern (Deutschland, Frankreich, Italien, Schweiz) herauszuarbeiten. Diese sind letztlich Resultat des komplexen Wechselspiels zwischen übergeordneten Gesellschaftsmodellen und institutionalisierten Planungssystemen (Nadin und Stead 2008).

Vincent Nadin und Dominic Stead: European Spatial Planning Systems, Social Models and Learning (Nadin und Stead 2008)

Vincent Nadin und Dominic Stead haben über viele Jahre die international vergleichende Planungsforschung maßgeblich mitbestimmt. Vor allem der Vergleich von europäischen Planungssystemen und Planungspraktiken in unterschiedlichen räumlichen Kontexten nahm dabei ein besonderes Gewicht ein. In Ihrem 2008 gemeinsam publizierten Aufsatz „European Spatial Planning Systems, Social Models and Learning" setzen sie sich mit dem komplexen Zusammenspiel von Gesellschafts-modellen und Planungssystemen auseinander und zeigen eindrucksvoll auf, dass Planungssysteme eingebettet sind in gesellschaftskulturelle Traditionen und dass eben diese als erklärende Faktoren für die Transformation raumbezogener Planungsprozesse herangezogen werden müssen. Der Aufsatz ist für die Planungskulturforschung von Bedeutung, weil er nicht

nur eine systematische Übersicht über Typologien von Gesellschaftsmodellen und Planungssystemen bietet, sondern auch die Bedeutung gesellschaftlich verankerter Werthaltungen und kultureller Kontexte für den Wandel räumlicher Planung berücksichtigt. Die Autoren konstatieren, dass Konvergenztendenzen bei der Betrachtung europäischer Planungssysteme grundsätzlich feststellbar sind. Ursächlich dafür sind u. a. Prozesse der transnationalen Kooperation und des grenzüberschreitenden Lernens. Allerdings verweisen die Autoren auch auf divergente Bewegungen und eine damit verbundene Vielfalt von Planungspraktiken, die nicht über generalisierte Typologien abgebildet werden kann. Genau an diesem Punkt setzt die Planungskulturforschung an. Es gilt, auch die Aspekte abseits einer allein legal-administrativen Betrachtung von räumlicher Planung in den Blick zu nehmen. Darauf haben beide Autoren dezidiert hingewiesen, u. a. in dem 2013 veröffentlichten Aufsatz „Opening up the Compendium: An Evaluation of International Comparative Planning Research Methodologies":

> » The administrative and legal frameworks are important for the operation of planning, although planning systems can operate in similar ways under very different formal government and legal arrangements. Classifications based on these variables tend to follow the well-worn path of constitutional and legal families of nations but these tend to say very little about the way that planning performs in reality or about the culture and practice of planning. (Nadin und Stead 2008, S. 1558)

Die von Keller et al. (1993) zusammengestellten planerischen Erfahrungsberichte zeigen zum einen, dass es in den untersuchten Ländern verschiedene Phasen und Akzentuierungen von Planung (umfassende Planung, Projektplanung etc.) gegeben hat. Zum anderen wird dabei aber auch eine Verschiebung von Maßstäblichkeiten deutlich, da hier – aus Sicht der Planungskulturforschung – nicht mehr die lokal gebundenen Prozesse der institutionalisierten Entscheidungsfindung, sondern die auf einer nationalen Ebene aggregierten Prägungen des Planungshandelns in den Vordergrund rücken. Diese Sichtweise ist insofern problematisch, als dass sich mit ihr auch eine Ausblendung der vielfältigen Planungspraktiken und etablierten Routinen von Planern innerhalb einzelner Planungssysteme verbindet (vgl. Healey und Williams 1993; Stead und Cotella 2011; Nadin 2012; Reimer et al. 2014; Zimmermann et al. 2018). Damit entfällt auch eine stärkere Auseinandersetzung mit den systemimmanenten Praktiken innerhalb rechtlicher und administrativer Rahmenbedingungen, die das Planungshandeln zwar maßgeblich anleiten, aber nicht vollständig determinieren (Reimer und Blotevogel 2012).

Der Planungskulturbegriff hilft dabei, auf die kontextspezifischen und an konkrete Akteure gebundenen Praktiken innerhalb dieser Rahmenbedingungen zu verweisen und ist somit ein Hilfsmittel, um die lokalen Besonderheiten planerischer Denk- und Handlungsmuster zum Ausdruck zu bringen (Booth 1993).

Parallel zu dieser Diskussion, die den Kulturbegriff als analytisches Konzept für die Erklärung lokaler Besonderheiten der räumlichen Planung und ihrer tiefer liegenden Werte, Einstellungen, Traditionen und Wissensordnungen nutzt, entfaltet sich ein weiterer Diskursstrang, in dem der Begriff überwiegend normativ verwendet wird. Sein Ursprung liegt im „communicative turn" (Healey 1992; ▶ Kap. 2, Bd. 1), mit dem auf einen grundsätzlichen Wandel des Planungsverständnisses verwiesen wird. Planung wird demzufolge als offener, transparenter und inklusiver Prozess interpretiert, der Kernelement einer neuen Planungskultur (Ludwig 2005) ist. Der Begriff der Planungskultur wird hier – vor allem in der planerischen Praxis – normativ besetzt und im Sinne einer guten und besseren Planung verwendet (vgl. z. B. Loepfe und Eisinger 2016; Staatsministerium Baden-Württemberg 2014). Das ist aus wissenschaftlicher Sicht problematisch, weil so die analytische Kraft des Begriffs im Rahmen der Beschreibung und Interpretation der Tiefendimensionen des Planungshandelns verloren geht. In jüngerer Vergangenheit haben viele planungskulturell motivierte Forschungsarbeiten an diesem Punkt angesetzt und sich um die theoretisch-konzeptionelle Herleitung des Begriffs und seine empirische Nutzbarkeit bemüht (Levin-Keitel 2014; Othengrafen 2012; Reimer 2012). Gleichzeitig gibt es Versuche, diese beiden „Gegensätze" aufzuheben und Planungskultur praxisorientiert zu definieren, um analytische und normative Verständnisse zu verbinden:

» Genauer: Planungskultur in action äussert sich einerseits in einem Satz bestimmter explizit formulierter wie implizit angewendeter Normen, Wertvorstellungen, Ziele und Routinen, die im Planungsalltag zur Anwendung kommen. Gleichzeitig vollziehen sich in derartigen normativen Momenten gewissermaßen erste Analysen der räumlich-gesellschaftlichen Konstellationen eines Planungsvorhabens. Sie äußern sich in kollektiv geteilten Vorformungen der Wahrnehmung und Deutung von Problemlagen und Lösungsansätzen sowie bestimmten Verknüpfungen von Zielen mit Mitteln. (Loepfe und Eisinger 2016, S. 44)

3.4 Theoretisch-konzeptionelle Zugänge

Im Rahmen der Planungskulturforschung werden die komplexen institutionellen Muster, die das planerische Handeln anleiten, in den Blick genommen. Ihr Ziel ist es, unterschiedliche Denk- und Handlungsweisen von Planern sowie ihre materiellen Manifestationen zu identifizieren und in ein Verständnis räumlicher Planung(en) zu integrieren.

Planungskultur steht demnach für einen Erklärungsansatz in der Planungsforschung, der sich von anderen planungstheoretischen Ansätzen unterscheidet, indem er Prozesse der räumlichen Entwicklung als kulturell eingebettete Tätigkeiten versteht, sich damit von allein organisatorisch-strukturell orientierten Interpretationen räumlicher Planungspraktiken abgrenzt und dabei vor allem das Zusammenspiel manifestierter Elemente (z. B. rechtliche Grundlagen, administrative Organisationsstrukturen, Planwerke, Strategien und Konzepte) und nichtmanifestierter Elemente (z. B. individuelle und kollektive Wahrnehmungsweisen und internalisierte Handlungsmuster) in den Vordergrund stellt, um zu einem besseren Verständnis der Planungspraxis beizutragen.

Hinsichtlich ihrer theoretischen Verortung können bislang vor allem organisations- und kulturwissenschaftliche, praxistheoretische sowie governanceorientierte Ansätze zur Untersuchung von Planungskulturen unterschieden werden, wobei die im Folgenden vorgestellten Zugänge eher idealtypisch zu verstehen sind und teilweise auch Elemente aus verschiedenen Debatten miteinander kombinieren.

In Anlehnung an organisations- und kulturwissenschaftliche Ansätze (vgl. Hofstede 2001; Schein 2004) liegt ein Fokus der Planungskulturforschung auf der Analyse manifestierter und nichtmanifestierter kultureller Elemente der räumlichen Planung. Lokale Praktiken und Routinen von Planern entstehen hier im Zusammenspiel zwischen sichtbaren (räumlichen) Strukturen, z. B. der Architektur einer Stadt oder vorliegender Plandokumente („Planungsartefakte"), und den spezifischen planerischen Rahmenbedingungen und institutionellen Kontexten, in die diese eingebettet sind (▯ Abb. 3.1). Planungsartefakte werden hier als manifestierte Symbole oder Ausdruck der planerischen Praktiken sowie tiefer liegender gesellschaftlicher Werte, Einstellungen, Traditionen und Wissensordnungen gesehen (u. a. Levin-Keitel und Sondermann 2015; Knieling und Othengrafen 2009; Othengrafen 2012). Von besonderer Bedeutung ist dabei, dass die räumlichen bzw. planerischen Artefakte die Bedeutungen bzw. Erwartungen, die Politiker, Planer und andere Akteure den zu erreichenden

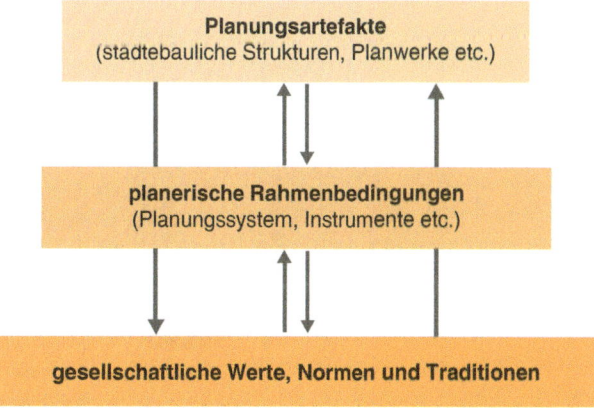

▯ **Abb. 3.1**   Die drei Dimensionen von Kultur. (Eigene Darstellung, in Anlehnung an Schein 2004)

Zielen oder den verwendeten Instrumenten beimessen, beeinflussen. Gleichzeitig setzen politische und planerische Entscheidungen den Rahmen für städtebauliche Strukturen, die Architektur einer Stadt oder andere sichtbare (physische) Artefakte, sodass hier eine gegenseitige Beeinflussung von Artefakten, planerischen Rahmenbedingungen und gesellschaftlichen Werten, Normen und Traditionen festzustellen ist. Über eine derart systematische Betrachtung von Planung zielt dieser Ansatz darauf ab, die verschiedenen Muster und Formen der planerischen Praxis in unterschiedlichen räumlichen und kulturellen Kontexten zu verstehen und zu analysieren (Kontextualisierung von Planungspraxis).

Neben Ansätzen, die eine Analyse manifestierter und nichtmanifestierter Elemente planerischer Praxis in den Vordergrund stellen, lassen sich solche identifizieren, die Planungskulturen im Sinne einer praxistheoretischen Lesart über die Beobachtung planerischer Praktiken betrachten (u. a. Forester 1993; Healey 1992; Hoch 1994; Zimmermann et al. 2018). Hier stehen insbesondere die Planer als handelnde Subjekte im Vordergrund, „denn es sind gerade die Aktionen im Sinne eingelebter Umgangsweisen und regelmäßiger Praktiken der Gesellschaftsmitglieder, die zu dem zentralen Bezugspunkt von Kulturanalysen avancieren" (Hörning und Reuter 2004, S. 10). Planungskulturen beziehen sich hier auf die Art und Weise, wie die jeweiligen Akteure ihre Rollen und Aufgaben verstehen, wie sie Probleme wahrnehmen, damit umgehen und dabei bestimmte Regeln, Verfahren und Instrumente anwenden (u. a. Birch 2001; Booth 1993). Dies erfolgt situationsspezifisch und kontextgebunden, sodass Planungskultur immer lokale und regionale Praktiken analysiert und abbildet. Planerische Praktiken – sowohl auf der individuellen als auch der kollektiven Ebene – entstehen hierbei vor allem durch das dynamische Zusammenspiel kognitiver Einstellungen und planerischer Leitvorstellungen, Interaktionen von Akteursgruppen sowie dem planungsrechtlichen Kontext (■ Abb. 3.2; vgl. auch Galler und Levin-Keitel 2016; Levin-Keitel und Sondermann 2015; Othengrafen 2014). Dadurch lassen sich Handlungsmaximen der beobachteten Planer, ihre spezifischen Wahrnehmungen

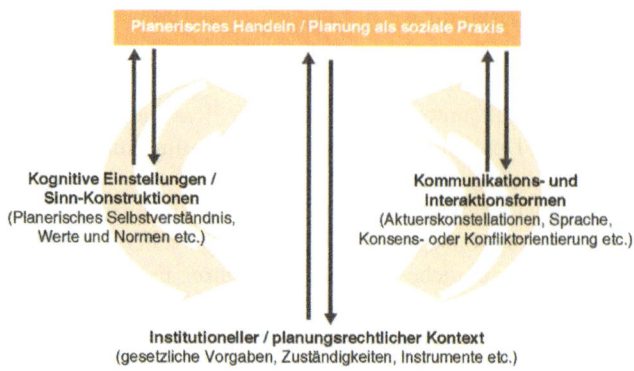

■ **Abb. 3.2**   Planungskultur als Analyse sozialer Praxis. (Eigene Darstellung)

und Bewertungen sowie handlungsbestimmende Selbstverständnisse erklären, die ein gemeinsames Prozessparadigma bilden (Levin-Keitel 2015, 2016) und damit eine Planungskultur beschreiben.

Mit Prozessparadigmen beschäftigt sich ein weiterer analytischer Ansatz (Wolff 2016), der davon ausgeht, dass sich soziale Praktiken vor allem durch die Interaktionen der beteiligten Individuen und „Kollektive" (vgl. auch Hansen 2011; Rathje 2009) ergeben. Diesem Verständnis folgend, gehören Individuen einer Vielzahl unterschiedlicher Gruppen bzw. Kollektive an, so dass aus der Interaktion von Individuen oder Kollektiven neue kollektive Strukturen und emergente Prozesskulturen entstehen (■ Abb. 3.3). Damit ist eine Planungskultur kein homogenes Gebilde, sondern vielmehr durch divergierende Interessen, Vorstellungen und vielfältige Aushandlungsprozesse gekennzeichnet. Das verbindende Element ist die Vertrautheit mit der Differenz (Hansen 2009; Rathje 2009). Bislang wurde dieser Ansatz für die Analyse kommunikativer Planungsprozesse eingesetzt (Wolff 2016). Planungskultur wird dann als ein „Prozessraum" (Hilpert 2002, S. 87 f.) gesehen, der von den Kulturen aller beteiligten Akteursgruppen geprägt ist. Es entsteht ein Spannungsfeld, in dem der räumlichen Planung von den jeweiligen Akteuren verschiedene Bedeutungen

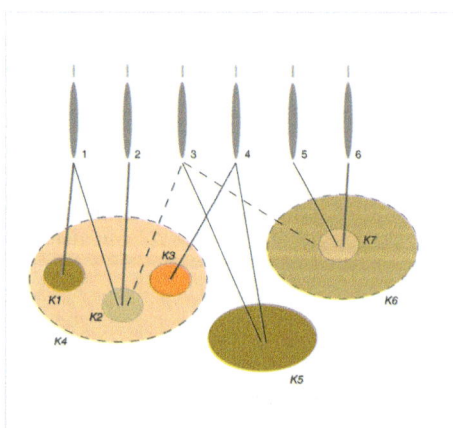
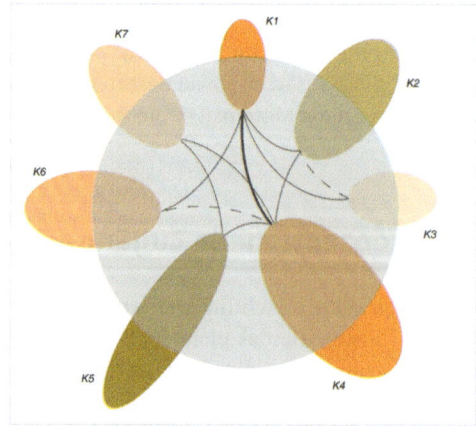

■ **Abb. 3.3**   Zugehörigkeit zu Kollektiven und Interaktionen von Kollektiven in projektspezifischen Planungskulturen (Wolff 2016)

zugeschrieben werden. Dies gilt aber nicht nur für das Planungsergebnis, sondern auch für den Prozess und die für geeignet erachteten Wege der Entscheidungsfindung. Die Unterschiedlichkeit dieser im Planungsprozess wirksam werdenden Bedeutungszuschreibungen, impliziten Annahmen und Handlungsmuster ist somit ein integraler Bestandteil von Planungskultur. Eine zentrale Annahme ist dabei, dass eine Planungskultur, die als Kultur aller involvierten Akteure verstanden wird, letztlich erst durch diese Interaktion erfahrbar wird (Wolff 2016).

Ein vierter Ansatz zur konzeptionellen Herleitung von Planungskulturen ist mit der Governance-Forschung verknüpft und bezieht die Schnittmengen zwischen Praxistheorien und neoinstitutionalistischen Perspektiven ein. Dabei geht der Ansatz von der Prämisse aus, dass Praktiken der Raumentwicklung nicht allein über die Analyse formeller Institutionen wie der rechtlichen und administrativen Rahmenbedingungen erklärt werden können, sondern auch „Wertsetzungen, Aufgabeninterpretationen und Rollenverständnisse, vor deren Hintergrund erst die Handhabung der Instrumente und Verfahren zu verstehen […] ist" (Keller et al. 2006, S. 279 f.), eine zentrale Rolle spielen (Hohn 2002, 2009; Reimer 2012; Hohn und Reimer 2014). Vor diesem Hintergrund lassen sich zwei „institutionelle Welten" oder „Schichten" von Planungskulturen unterscheiden, deren Elemente es zu analysieren und in ihrem Zusammenwirken zu betrachten gilt, um die Raumwirksamkeit von Planungskulturen, ihren Wandel und ihre Pfadabhängigkeiten besser zu verstehen (◘ Abb. 3.4).

Planungskulturen bilden hierbei höchst komplexe institutionelle Welten ab. Deren Charakteristikum besteht aus dem Wechselverhältnis zwischen dem jeweiligen Planungssystem als formell institutionalisiertem Handlungsrahmen (u. a. Gesetze und Verordnungen) und den kognitiven wie diskursiven Einflussfaktoren, welche den durch informelle Institutionen bestimmten Handlungskontext konstituieren (u. a. individuell und/oder kollektiv geteilte Problemwahrnehmungen und Sinndeutungen). Die Welt formeller Institutionen umfasst demnach die institutionellen Technologien, die den formellen Handlungskontext der Planungspraxis bestimmen (z. B. Planungsgesetze). Die institutionellen Technologien der Planung stehen in einem direkten Zusammenhang mit der Welt informeller Institutionen, d. h. den kognitiv verankerten Wertvorstellungen, die raumbezogene Planungsdiskurse und Planungspraktiken anleiten. Reimer (2012) greift diese Gedanken auf und unterscheidet zwischen Planungskulturen erster Ordnung (institutionelle Rahmenbedingungen) und zweiter Ordnung (Planungsdiskurse und ortsbezogene Praktiken), die erst in ihrem Zusammenspiel als planungskulturelle Konfigurationen greifbar werden.

3.5 Fazit

Mit dem Begriff der Planungskultur werden in den Planungswissenschaften verstärkt die Praktiken der räumlichen Planung adressiert, die in der Vergangenheit kaum

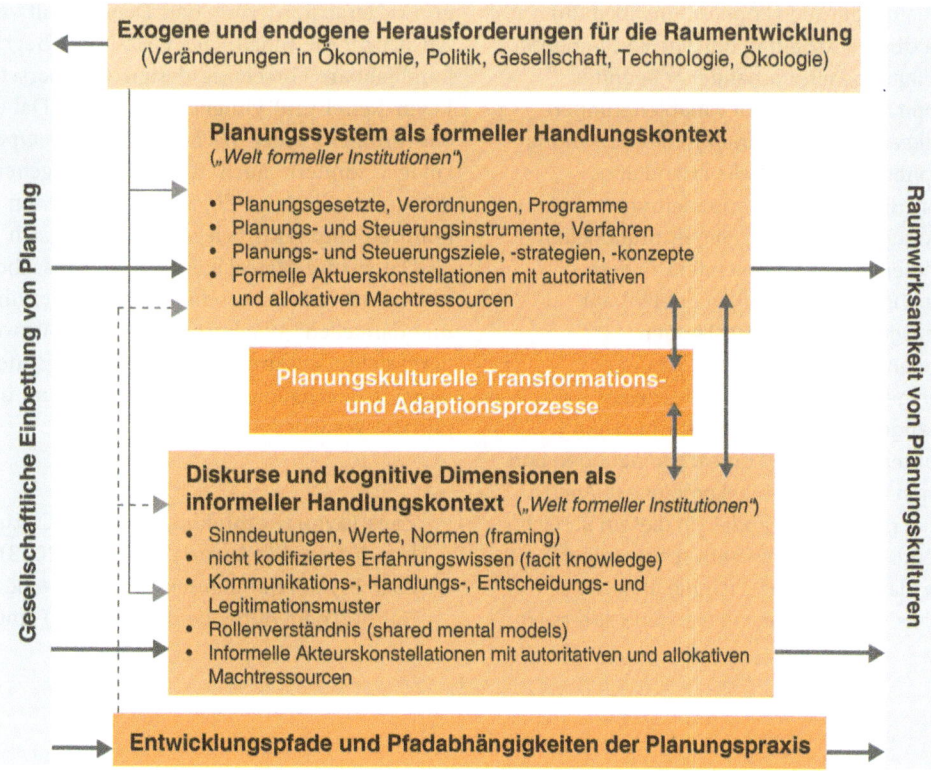

◘ Abb. 3.4 Zwei „institutionelle Welten" von Planungskulturen und ihre Interdependenzen (Hohn und Reimer 2014)

oder nur wenig systematisch untersucht wurden. Dabei stellen sie zweifelsfrei ein Forschungsdesiderat dar, das mit der Profilierung planungskultureller Forschungen in den vergangenen Jahren erkannt wurde. Ihr übergeordnetes Ziel ist dabei zunächst hermeneutischer Natur. Sie wollen dazu beitragen, die Praktiken der räumlichen Planung in ihren Tiefendimensionen besser zu verstehen. Dietrich Fürst (2007, 2016) stellt diese Zielsetzung ebenfalls in den Vordergrund, setzt sich dabei aber auch kritisch mit dem Begriff auseinander und skizziert dabei eindrucksvoll die Angriffsflächen der Planungskulturforschung.

Dietrich Fürst: Planungskultur: Auf dem Weg zu einem besseren Verständnis von Planungsprozessen? (Fürst 2007)

Im Rahmen einer über die Online-Plattform pnd 2007 initiierten Umfrage zum Begriff der Planungskultur hat Dietrich Fürst den damaligen Stand der Fachdebatte zusammengefasst. Er konstatiert durchaus skeptisch, dass „Planungskultur" an sich nicht den Rang eines wissenschaftlichen Begriffs für sich reklamieren kann. Dafür sei er zu unscharf definiert und auch kaum mit theoretischem Gehalt versehen. Dennoch lässt sich mit dem Begriff ein Untersuchungsfeld im Rahmen der Planungsforschung adressieren, dem bis dato kaum Aufmerksamkeit geschenkt wurde. So verweist der Begriff der Planungskultur auf die kulturellen Prägungen in Interaktionsprozessen, die als verdichtete Werthaltungen und Einschätzungen einen wesentlichen Einfluss auf das individuelle und kollektive Verhalten ausüben können. Fürst bemüht sich in seinem Überblicksbeitrag um eine konsequente Gegenüberstellung bereits vorhandener Begriffsdefinitionen und identifiziert fünf Variablengruppen, die offenkundig auch für die Planungskulturforschung Relevanz entfalten: die individuellen Handlungsorientierungen, die Interaktionsorientierungen, die spezifische Akteurskonstellation, den institutionellen Handlungsrahmen und das Problemfeld, in dem interagiert wird. Im weiteren Verlauf des Beitrags skizziert er – ausgehend von einschlägigen Vergleichsstudien zu institutionellen Planungssystemen – einige zentrale Problemfelder einer vergleichend angelegten Planungskulturforschung. Dazu zählen unterschiedliche Kontextbedingungen und eine nicht handhabbare Variablenkomplexität. Auch deswegen seien die meisten Studien in diesem Bereich eher hermeneutischer Natur. Im Kern schlussfolgert Fürst, dass die Planungskulturforschung ähnliche Sachverhalte wie die Governance-Forschung adressiert.

Als wissenschaftlicher Begriff kann Planungskultur erst in Ansätzen überzeugen, da er als „Passepartout-Begriff" (Casprig 2009; Selle 1993) so viele Aspekte adressiert, dass er überkomplex und – aus einer Forschungsperspektive heraus betrachtet – unhandlich wird. Auch wenn die oben skizzierten theoretisch-konzeptionellen Zugangsperspektiven ähnliche Beobachtungskategorien und Inhaltsdimensionen umschreiben, so mangelt es doch bisher an einem konsistenten Ansatz. Die Debatte zu Planungskulturen und insbesondere die jüngeren Bestrebungen einer theoretischen Fundierung sind schlicht zu jung, als dass sie dieses zum gegenwärtigen Zeitpunkt schon leisten könnten. Es muss sich erst noch zeigen, wie tragfähig die Begriffe und Konzepte sind und inwiefern sie tatsächlich andere Sachverhalte adressieren als bereits existierende Zugänge. Das ergibt sich vor allem aus dem Umstand, dass Planungskulturen einem stetigen Wandel unterliegen und neu verhandelt und etabliert werden müssen. Planungskulturen sind damit immer „vielfältig und individuell", was (als „fuzzy concept" im Sinne von Loepfe und Eisingern 2016, S. 53) eine Herausforderung für die Planungstheorie darstellt.

Dennoch, der ernst gemeinte Austausch und das kritische Reflektieren fachfremder Zugänge aus anderen Disziplinen stellen zweifelsfrei auch ein großes Potenzial für eine Weiterentwicklung in den Planungswissenschaften dar (Zimmermann et al. 2018). Dabei wird die Leistungsfähigkeit des Begriffs „Planungskultur" vermutlich dann deutlich, wenn es gelingt, ihn für empirische Forschungsarbeiten anschlussfähig zu gestalten. Gerade das Herunterbrechen seiner theoretischen (Über-)Komplexität auf die Wirklichkeiten der empirischen Planungsforschung stellt eine große Herausforderung dar. Denn damit verbindet sich der Anspruch, die inhaltliche Vielfalt des Begriffs in empirisch handhabbare Forschungsdesigns zu übersetzen (Fürst 2016; Levin-Keitel und Othengrafen 2016). Dabei tritt nicht nur das Problem einer nachvollziehbaren Operationalisierung zutage, sondern auch die noch weitgehend offene Frage der methodischen Zugänge. Wie lassen sich Planungskulturen empirisch belegen? Wie lässt sich die große Masse an Daten und Materialien, die sich aus überwiegend qualitativ motivierten Forschungsarbeiten ergibt, interpretieren und darstellen? Wie lassen sich möglicherweise interpretativ hergestellte Zusammenhänge kausal ordnen? Wie können letztlich überhaupt Kulturen studiert und insbesondere verglichen werden?

Zack Taylor: Rethinking Planning Culture: A New Institutionalist Approach (Taylor 2013)

Zack Taylor nimmt in seinem Artikel eine skeptische Haltung gegenüber dem Begriff der Planungskultur ein und konzentriert sich auf zwei wesentliche Kritikpunkte. Zum einen zeigt ein Blick in die Debatte zu Planungskulturen, dass der Begriff häufig unscharf bleibt und dabei sehr unterschiedliche Sachverhalte adressiert. Zum anderen ist er nicht geeignet, um Fragen nach der Stabilität und dem Wandel von planerischer Praxis ausreichend zu konzeptualisieren. Der Autor argumentiert, dass gerade die vergleichende Planungsforschung, deren Hauptinteresse das Verstehen der verschiedenen Muster und Formen der planerischen Praxis in unterschiedlichen zeitlichen, räumlichen und kulturellen Kontexten ist, nicht von der derzeitigen Debatte zu Planungskulturen profitiert. Zwar seien explizit kulturell motivierte Erklärungsansätze bereits in der vergleichenden Planungssystemforschung enthalten, jedoch bisher kaum systematisch präsentiert worden. Er stellt sodann die bereits seit Längerem etablierte Perspektive des historischen Institutionalismus als konzeptionelle Klammer für eine Erklärung dessen, was mit dem Begriff der Planungskultur eigentlich adressiert wird, in den Mittelpunkt der Diskussion. Dieser vermag deutlich besser die Genese der komplexen institutionellen Muster, die Planungskulturen prägen, zu erklären. Beispielhaft verdeutlicht er das anhand eines in der Planungskulturforschung wohl bekannten Beispiels, indem er den von Carl Abbott treffend dokumentierten „Oregon Planning Style" aus einer institutionentheoretischen Perspektive betrachtet und daran anschließend die Frage stellt, ob eine ähnlich tiefe Analyse auch auf der Basis der noch vorparadigmatischen Planungskulturforschung möglich gewesen wäre. Zumindest für die vergleichende Planungsforschung sei letztere nicht geeignet.

Trotz aller hier angedeuteten Schwierigkeiten handelt es sich bei der Planungskulturforschung um eine wichtige Erweiterung der bisherigen planungstheoretischen Debatte. Aus unserer Sicht stellt sich der Mehrwert der Planungskulturforschung auf drei Ebenen dar:

1. Der Begriff bietet das Potenzial, die scheinbar wenig greifbaren Einflussfaktoren der räumlichen Planungspraxis zu adressieren, die ansonsten unberücksichtigt blieben. Er hilft dabei, die „black box" der Planungspraktiken (Healey 2006) zu erhellen und eine kulturelle Position einzunehmen, die bisher in den Planungswissenschaften gefehlt hat.

2. Er trägt dazu bei, disziplinäre Grenzen zu überschreiten und auch andere Fachdebatten (z. B. in den Kultur-, Politik- und Sozialwissenschaften) in die Planungswissenschaften einzubeziehen. Allein die kursorische Darstellung der theoretisch-konzeptionellen Ansätze in diesem Kapitel spiegelt die große Bereitschaft, verschiedene Theorien und Konzepte miteinander zu kombinieren. Zwar adressieren auch andere in der Planungstheorie rezipierte Ansätze explizit die institutionellen Verankerungen des planerischen Handelns. Die Planungskulturforschung knüpft in ihrer theoretisch-konzeptionellen Verortung häufig an diese Debatten an und bearbeitet gewisse Schnittmengen, setzt jedoch andere Schwerpunkte, die ihr ein „Alleinstellungsmerkmal" verleihen. Sie erweitert beispielsweise die klassische Governance-Perspektive, die in ihrer raumwissenschaftlichen Verwendung häufig auf die formalisierten Regelsysteme in bestimmten Räumen und auf bestimmten Skalen (z. B. Formen der metropolitanen oder regionalen Governance) beschränkt wird, dabei aber die kulturelle Einbettung dieser Regelungsstrukturen aus dem Blick verliert. Der Neoinstitutionalismus verweist in seinen verschiedenen Lesarten auf die Bedeutung informeller Institutionen und ist daher ein wesentliches Korrektiv für solche eher auf formale Organisationsmuster gerichteten Perspektiven. Allerdings fehlt ihm oftmals ein konkreter Raumbezug, der wiederum im Zentrum der Planungskulturforschung steht.

3. Damit ist der besondere Mehrwert der Planungskulturforschung an sich angesprochen. Es gelingt ihr, die spezifischen Verräumlichungen der kulturellen Praxis – im Sinne der praxistheoretischen Perspektive – beispielsweise in Form von gebauten Strukturen als Manifestationen von Planungskulturen zu adressieren.

Literatur

Berndt, H. (1968). *Das Gesellschaftsbild bei Stadtplanern*. Stuttgart: Krämer.

Birch, E. L. (2001). Practitioners and the art of planning. *Journal of Planning Education and Research, 21*, 407–422.

Bolan, R. S. (1969). Community decision behavior: The culture of planning. *Journal of the American Planning Association, 35*(5), 301–310.

Booth, P. (1993). The cultural dimension in comparative research: Making sense of development control in France. *European Planning Studies, 1*(2), 217–229.

Casprig, A. (2009). Planungskultur: Plastikwort oder Passepartout? Planungskultur: Ursache oder Folge eines sprachlichen Missverständnisses. PNDonline II|2009. ▶ http://www.planung-neu-denken.de/images/stories/pnd/dokumente/Ausg_2_09/2_2009_casprig.pdf. Zugegriffen: 15. Apr. 2015.

Faludi, A. (1999). Patterns of doctrinal development. *Journal of Planning Education and Research, 18*(4), 333–344.

Forester, J. (1993). Learning from practice stories: The priority of practical judgement. In F. Fischer & J. Forester (Hrsg.), *The argumentative turn in policy analysis* (S. 186–209). Durham: Duke University Press.

Friedmann, J. (1967). A conceptual model for the analysis of planning behavior. *Administrative Science Quarterly, 12*(2), 225–252.

Fürst, D. (2005). Entwicklung und Stand des Steuerungsverständnisses in der Raumplanung. *disP, 163*, 16–27.

Fürst, D. (2007). Planungskultur. Auf dem Weg zu einem besseren Verständnis von Planungsprozessen? PNDonline III/2007. ▶ http://www.planung-neu-denken.de/images/stories/pnd/dokumente/pndonline3-2007-fuerst.pdf. Zugegriffen: 22. Apr. 2015.

Fürst, D. (2016). Planungskultur – Fruchtbare neue Konzeption? *disP – The Planning Review, 52*(4), 67–75.

Galler, C., & Levin-Keitel, M. (2016). Innerstädtische Flusslandschaften als integriertes Handlungsfeld – Planungspraktische Einflussfaktoren der Koordination und Kooperation. *Raumforschung und Raumordnung, 74*(1), 23–38.

Hansen, K. P. (2009). Kultur, Kollektiv, Nation. *Schriften der Forschungsstelle Grundlagen Kulturwissenschaft* (Bd. 1). Passau: Stutz.

Hansen, K. P. (2011). *Kultur und Kulturwissenschaft. Eine Einführung*. Tübingen: UTB & Francke.

Healey, P. (1992). Planning through debate: The communicative turn in planning theory. *Town Planning Review, 63*(2), 143–162.

Healey, P. (2006). Transforming governance: Challenges of institutional adaptation and a new politics of space. *European Planning Studies, 14*(3), 299–320.

Healey, P., & Williams, R. (1993). European urban planning systems: Diversity and convergence. *Urban Studies, 30*(4–5), 701–720.

Hilpert, M. (2002). Angewandte Sozialgeographie und Methode. Überlegungen zu Management und Umsetzung sozialräumlicher Gestaltungsprozesse. Habilitations-Schrift, Universität Augsburg. *Angewandte Sozialgeographie* (Bd. 47). Augsburg: Selbstverl & Lehrstuhl für Sozial- und Wirtschaftsgeographie University.

Hirschauer, S. (2016). Verhalten, Handeln, Interagieren. Zu den mikrosoziologischen Grundlagen der Praxistheorie. In H. Schäfer (Hrsg.), *Praxistheorie. Ein soziologisches Forschungsprogramm* (S. 45–70). Bielefeld: transcript.

Hoch, C. (1994). *What planners do: Power, politics and persuasion*. Chicago: Planners Press & American Planning Association.

Hofstede, G. (2001). *Culture's consequences: Comparing values, behaviors, institutions, and organizations across nations*. Thousand Oaks: Sage.

Hohn, U. (2002). Planungsstrukturen und Steuerungsformen in der japanischen Stadtplanung – Gegen-, Neben- und Miteinander von Top-down- und Bottom-up-Strategien. In A. Mayr, M. Meurer, & J. Vogt (Hrsg.), *Stadt und Region. Dynamik von Lebenswelten: Bd. 53. Tagungsbericht und wissenschaftliche Abhandlungen* (S. 70–713). Leipzig: Deutscher Geographentag.

Hohn, U. (2009). Zukunft wird gemacht: Urban Renaissance in der Global City Tokyo. In L. Basten (Hrsg.), *Metropolregionen – Restrukturierung und Governance. Deutsche und internationale Fallstudien: Bd. 3. Metropolis und Region* (S. 113–147). Dortmund: Verlag Dorothea Rohn.

Hohn, U., & Reimer, M. (2014). Formatorientierte Regionalentwicklung in der Zwischenstadt: Planungskulturelle Anpassungsfähigkeit an Rhein und Ruhr im Vergleich. In U. Altrock, S. Huning, T. Kuder, & H. Nuissl (Hrsg.), *Die Anpassungsfähigkeit von Städten. Zwischen Resilienz, Krisenreaktion und Zukunftsorientierung: Bd. 22. Planungsrundschau* (S. 315–342). Berlin: Altrock.

Hölzl, C., & Nuissl, H. (2015). Tragen Planungskonflikte zum Wandel der Planungskultur bei? Beobachtungen aus Santiago de Chile. In F. Othengrafen & M. Sondermann (Hrsg.), *Städtische Planungskulturen im Spiegel von Konflikten, Protesten und Initiativen: Bd. 23. Planungsrundschau*. Berlin: Altrock.

Hörning, K. H., & Reuter, J. (2004). Doing Culture: Kultur als Praxis. In K. H. Hörning & J. Reuter (Hrsg.), *Doing Culture. Neue Positionen zum Verhältnis von Kultur und sozialer Praxis* (S. 9–15). Bielefeld: transcript.

Howe, E., & Kaufman, J. (1979). The ethics of contemporary American planners. *Journal of the American Planning Association, 45*(3), 243–255.

Kaufman, J. (1985). American and Israeli planners: A cross-cultural comparison. *Journal of the American Planning Association, 51*(3), 352–363.

Kaufman, J., & Escuin, M. (2000). Thinking alike. Similarities in attitudes of Dutch, Spanish, and American planners. *Journal of the American Planning Association, 66*(1), 34–45.

Keller, D., Koch, M., & Selle, K. (Hrsg.). (1993). *Planungskulturen in Europa. Erkundungen in Deutschland, Frankreich, Italien und in der Schweiz: Bd. 115. Sonderdruck der disP Dokumente und Informationen zur Schweizerischen Orts-, Regional- und Landesplanung (ORL-Institut ETH Zürich)*. Verlag für Wissenschaftliche Publikationen: Darmstadt.

Keller, D., Koch, M., & Selle, K. (2006). Verständigungsversuche zum Wandel der Planungskulturen: Ein Lngzeit-Projekt. In K. Selle (Hrsg.), *Zur räumlichen Entwicklung beitragen. Konzepte. Theorien. Impulse* (S. 279–291). Dortmund: Verlag Dorothea Rohn.

Knieling, J., & Othengrafen, F. (Hrsg.). (2009). *Planning cultures in Europe. Decoding cultural phenomena in urban and regional planning*. Farnham: Ashgate.

Landwehr, A., & Stockhorst, S. (2004). *Einführung in die Europäische Kulturgeschichte*. Paderborn: Schöningh.

Levin-Keitel, M. (2014). Managing urban riverscapes: Towards a cultural perspective of land and water governance. *Water International, 39*(6), 842–857.

Levin-Keitel, M. (2015). Flusslandschaften in der Stadt. Einblicke in die empirische Erforschung von lokalen Planungskulturen. PND online, II/2015.

Levin-Keitel, M. (2016). Innerstädtische Flusslandschaften im Spiegel der lokalen Planungskultur. Planungskulturelle Perspektiven einer integrierten Stadtentwicklung im Umgang mit ihren Flusslandschaften. Dissertation, Leibniz Universität Hannover.

Levin-Keitel, M., & Othengrafen, F. (2016). Planungskultur – Auf der Suche nach einem kontemporären Verständnis räumlicher Planung? *disP – The Planning Review, 52*(4), 76–89.

Levin-Keitel, M., & Sondermann, M. (2015). Räumliches Planen als kulturelles Handeln: Planungskultur als analytischer Ansatz. In F. Othengrafen & M. Sondermann (Hrsg.), *Städtische Planungskulturen im Spiegel von Konflikten, Protesten und Initiativen: Bd. 23. Planungsrundschau* (S. 33–61). Berlin: Altrock.

Loepfe, M., & Eisinger, A. (2016). Planungskultur in action. Eine komplexitätstheoretische und praxisorientierte Annäherung an Planungskultur. *disP – The Planning Review, 52*(4), 43–54.

Ludwig, J. (2005). Die neue Planungskultur in der Regionalentwicklung – Eine Spurensuche. *Raumforschung und Raumordnung, 63*(5), 319–329.

Nadin, V. (2012). International comparative planning methodology: Introduction to the theme issue. *Planning Practice and Research, 27*(1), 1–5.

Nadin, V., & Stead, D. (2008). European spatial planning systems, social models and learning. *disP – The Planning Review, 172,* 35–47.

Nadin, V., & Stead, D. (2013). Opening up the compendium: An evaluation of international comparative planning research methodologies. *European Planning Studies, 21*(10), 1542–1561.

Neuman, M. (2007). How we use planning. Planning cultures and images of futures. In L. D. Hopkins & M. A. Zapata (Hrsg.), *Engaging the future. Forecasts, scenarios, plans, and projects* (S. 155–174). Hollis: Puritan Press Incorporated.

Nicolini, D. (2013). *Practice theory, work, and organization. An introduction.* Oxford: Oxford University Press.

Nuissl, H. (2008). Umfrage zur „Planungskultur". PNDonline I/2008. ▶ http://www.planung-neu-denken.de/images/stories/pnd/dokumente/pndonline1_2008_umfrage.pdf. Zugegriffen: 15. Apr. 2015.

Othengrafen, F. (2012). *Uncovering the unconscious dimensions of planning. Using culture as a tool to analyse spatial planning practices.* Farnham: Ashgate.

Othengrafen, F. (2014). The concept of planning culture: Analysing how planners construct practical judgements in a culturised context. *International Journal of E-Planning Research, 3*(2), 1–17.

Othengrafen, F., & Reimer, M. (2013). The embeddedness of planning in cultural contexts: Theoretical foundations for the analysis of dynamic planning cultures. *Environment and Planning A, 45,* 1269–1284.

Othengrafen, F., Reimer, M., & Sondermann, M. (2015). Städtische Planungskulturen im Wandel? Konflikte, Proteste, Initiativen und die demokratische Dimension räumlichen Planens. In F. Othengrafen & M. Sondermann (Hrsg.), *Städtische Planungskulturen im Spiegel von Konflikten, Protesten und Initiativen: Bd. 23. Planungsrundschau* (S. 371–391). Berlin: Altrock.

Rathje, S. (2009). Der Kulturbegriff – Ein anwendungsorientierter Vorschlag zur Generalüberholung. In A. Moosmüller (Hrsg.), *Konzepte kultureller Differenz: Bd. 22. Münchener Beiträge zur interkulturellen Kommunikation* (S. 83–106). Waxmann: Münster.

Reckwitz, A. (2003). Grundelemente einer Theorie sozialer Praktiken. Eine sozialtheoretische Perspektive. *Zeitschrift für Soziologie, 32*(4), 282–301.

Reckwitz, A. (2005). Kulturelle Differenzen aus praxeologischer Perspektive: Kulturelle Globalisierung jenseits von Modernisierungstheorie und Kulturessentialismus. In I. Srubar, J. Renn, & U. Wenzel (Hrsg.), *Kulturen Vergleichen. Sozial- und kulturwissenschaftliche Grundlagen und Kontroversen* (S. 9–111). Wiesbaden: Springer VS.

Reckwitz, A. (2006). *Das hybride Subjekt. Eine Theorie der Subjektkulturen von der bürgerlichen Moderne zur Postmoderne.* Weilerswist: Velbrück Wissenschaft.

Reckwitz, A. (2016). Praktiken und ihre Affekte. In H. Schäfer (Hrsg.), *Praxistheorie. Ein soziologisches Forschungsprogramm* (S. 163–180). Bielefeld: transcript.

Reimer, M. (2012). *Planungskultur im Wandel. Das Beispiel der REGIONALE 2010.* Detmold: Verlag Dorothea Rohn.

Reimer, M. (2016). Planungskultur – Eine Bestandsaufnahme. *disP – The Planning Review, 52*(4), 18–29.

Reimer, M., & Blotevogel, H. H. (2012). Comparing spatial planning practice in Europe: A plea for cultural sensitization. *Planning Practice and Research, 27*(1), 7–24.

Reimer, M., Getimis, P., & Blotevogel, H. H. (2014). *Spatial planning systems and practices in Europe. A comparative perspective on continuity and changes.* New York: Routledge.

Sanyal, B. (2005). Hybrid planning cultures: The search for the global cultural commons. In B. Sanyal (Hrsg.), *Comparative planning cultures* (S. 3–25). New York: Routledge.

Schäfer, H. (2016). Einleitung. Grundlagen, Rezeption und Forschungsperspektiven der Praxistheorie. In H. Schäfer (Hrsg.), *Praxistheorie. Ein soziologisches Forschungsprogramm* (S. 1–27). transcript: Bielefeld.

Schatzki, T. R. (1996). *Social practices. A Wittgensteinian approach to human activity and the social.* Cambridge: Cambridge University Press.

Schatzki, T. R. (2001). Introduction: Practice theory. In T. R. Schatzki, K. Knorr Cetina, & E. Savigny (Hrsg.), *The practice turn in contemporary theory* (S. 10–23). London: Routledge.

Schatzki, T. R. (2016). Praxistheorie als flache Ontologie. In H. Schäfer (Hrsg.), *Praxistheorie. Ein soziologisches Forschungsprogramm* (S. 29–44). Bielefeld: transcript.

Schatzki, T. R., Knorr Cetina, K., & Savigny, E. (2001). *The practice turn in contemporary theory.* London: Routledge.

Schein, E. H. (2004). *Organizational culture and leadership* (3. Aufl.). San Francisco: Jossey-Bass.

Schmidt, R. (2012). *Soziologie der Praktiken. Konzeptionelle Studien und empirische Analysen.* Berlin: Suhrkamp.

Selle, K. (1993). Versuch über Planungskultur – Zustandsbeschreibungen und Einordnungen. In J. Bärsch & J. Brech (Hrsg.), *Das Ende der Normalität im Wohnungs- und Städtebau? Theoretische Begegnungen mit Klaus Novy* (S. 195–219). Darmstadt: Verlag für wissenschaftliche Publikationen.

Shove, E., Pantzar, M., & Watson, M. (2012). *The dynamics of social practice. Everyday life and how it changes.* London: Sage.

Staatsministerium Baden-Württemberg. (2014). *Leitfaden für eine neue Planungskultur.* Stuttgart: Staatsministerium Baden-Württemberg.

Stead, D., & Cotella, G. (2011). Differential Europe: Domestic actors and their role in shaping spatial planning systems. *disP – The Planning Review, 186,* 13–21.

Straub, J. (2004). Kulturwissenschaftliche Psychologie. In F. Jäger & J. Straub (Hrsg.), *Handbuch der Kulturwissenschaften: Bd. 2. Paradigmen und Disziplinen* (S. 568–591). Stuttgart: Metzler.

Swidler, A. (2001). What anchors cultural practices. In T. R. Schatzki, K. Knorr Cetina, & E. Savigny (Hrsg.), *The practice turn in contemporary theory* (S. 83–101). London: Routledge.

Taylor, Z. (2013). Rethinking planning culture: A new institutionalist approach. *Town Planning Review, 84*(6), 683–702.

Wolff, A. (2016). Planung, Kollektive und Kulturen – Akteursperspektiven in der Planungskultur. *disP – The Planning Review, 52*(4), 55–66.

Zimmermann, K., Chang, R., & Putlitz, A. (2018). Planning culture: Research heuristics and explanatory value. In T. W. Sanchez (Hrsg.), *Planning knowledge and research.* London: Routledge.

CHAPTER I

3

HYBRID PLANNING CULTURES: THE SEARCH FOR THE GLOBAL CULTURAL COMMONS

BISHWAPRIYA SANYAL

INTRODUCTION

Are there significant variations in the ways planners in different nations have influenced urban, regional, and national development? Do such variations arise from differences in planning cultures, meaning the collective ethos and dominant attitude of professional planners in different nations toward the appropriate roles of the state, market forces, and civil society in urban, regional, and national development? How are such professional cultures formed? Are they indigenous and immutable, or do they evolve with social, political, and economic changes both within and outside the national territory? Particularly relevant for our times is the intensification of global interconnection in trade, capital flows, labor migration, and technological connectivity and its effect on national planning cultures. Are there signs that previously dominant planning cultures are being challenged as a result of such interconnection? And, if so, are such challenges leading to the formation of new, radically different planning cultures?

The contributors to this volume address these and related questions, drawing on planning experience in ten nations and at different territorial levels, ranging from the local to national level. The nations vary by degrees of urbanization and industrialization. The United States, the United Kingdom, the Netherlands, Japan, and Australia are relatively more industrialized and urbanized than China, India, Indonesia, Iran, and Mexico, which are industrializing

3

countries. The nations also vary in terms of their established political systems. On one end are the United Kingdom, the United States, and the Netherlands, with long political traditions of democracy; on the other end is China, ruled by a communist party, albeit with an administrative structure that has been decentralized recently. In between are India, democratic and with a federal structure of government; Australia, founded in the early part of the twentieth century, also with a federated governance structure; Mexico, democratic since the revolution in 1910 but led by one centralized political party until only recently; Iran, struggling with a unique blend of theocracy and democracy in a relatively centralized governance structure; and Indonesia, which until recently was ruled by an autocratic leader supported by the army. This complex political scenario makes the discussion of planning cultures difficult but also intriguing.

As a general background to the discussion of specific planning cultures in each nation, this volume contains two theoretical papers, from John Friedmann and from Manuel Castells, that attempt to capture broad global trends at the end of the twentieth century. Castells highlights the impact of technological changes–particularly in information and communication–and how such changes have radically altered the material basis for urbanism. Castells is arguing, implicitly, that contemporary planning practice in all nations must acknowledge and meet the challenges posed by the new technological dynamics influencing urbanism. Friedmann differentiates this global scenario into three different parts, highlighting the sharply varying quality of urban lives in industrialized nations, industrializing nations, and "transitional" nations attempting to transform their previously socialist economies to fully industrialized, market-driven economies anchored in private ownership of the productive forces. This differentiation suggests that global interconnections–of trade, investment, flows of labor, cultural symbols, and other ideas, which are grouped together all too often under the term *globalization*–are not leading toward a homogenization of planning cultures across the globe. The sharp differences in the levels of industrialization among the three groups of nations and the particularly different ways each group is linked to the global economy seem to be the crucial variables influencing different planning practices in the three sets of nations.

PLANNING CULTURE: THE GOLDEN YEARS

Why focus on the planning culture of a city, region, or nation if, indeed, its political economy is what ultimately shapes the particular characteristics of its planning endeavors? In this chapter we probe this question through a brief historical analysis of how and why the notion of planning cultures emerged from the discussion of planning practices in industrialized as well as industrializing countries. Such an analysis logically begins with the years immediately after World War II, when planning flourished in both industrialized and industrializing countries, so much so that Peter Hall described them as "the golden years of planning."[1] There was no discussion of planning cultures, however, during this period. What made it "golden" was the optimism among planners—urban, regional, as well as national—that planning efforts did not have to be based on the intuitive and aesthetic sensibilities of architects and urban designers of the past. In contrast, planning culture could be scientific and rational, based on accurate observations of statistically valid samples of reality, followed by dispassionate and value-neutral analysis of socioeconomic trends. Such analyses would lead to professionally crafted recommendations formulated through rigorous and objective assessment methodologies, such as cost-benefit analysis, planning-programming and budgeting systems, that had proven useful in conducting World War II.

The rational comprehensive model (RCM) of planning, about which much has already been written, reflected the aspirations of the postwar period.[2] It was backed intellectually by theories of location of firms, initially developed in Germany in the early part of the twentieth century and later introduced in the United States and elsewhere.[3] Earlier location theories took on a new intellectual power and persuasiveness when combined with analytical studies of transportation—in particular, the automobile and its impact on location of not only firms but also households. The result was a rapid growth in land use and transportation modeling that reinforced the role of planners as professionals with the necessary knowledge and expertise to shape the future in a scientific way.[4]

In industrializing countries emerging from colonial rule, the dominant planning culture was equally optimistic and technocratic and more centralized than in industrialized countries. Many industrializing countries drew their inspiration from the planning experience of the former Soviet Union.[5] Economists and statisticians dominated the planning process, which was conceived as a scientific and rational process requiring expert and technical knowledge. The topic of national culture was rarely, if ever, discussed. This was because, in part, the goal of planning was to change the national culture so as to rapidly modernize, both economically and politically. Though issues of national sovereignty, cultural autonomy, and economic self-sufficiency[6] were discussed regularly by political leaders in many newly decolonized nations, planners, on the whole, rarely incorporated particular cultural attributes in formulating plans. The only visible difference in planning cultures after World War II was between ex-British colonies and ex-French colonies–particularly in Africa. The French model of colonial governance had been more centralized than the British style of administration, and some differences lingered on even after the colonies were independent. Both types of ex-colonies, however, pursued the same technocratic and export-driven approach to planning, with one clearly defined objective–to estimate the need for bilateral and multinational aid to support the annual growth rate of their national economies.[7]

At the city level, planners pursued the Western style of comprehensive planning by creating new master plans that embodied the vision of modern cities with distinctly separated land uses connected by transportation arteries. Much has been already written about this effort.[8] One issue relevant for our purpose is that the actual culture of planning as practiced on a day-to-day basis was not as the planning documents described it.[9] Most city planning offices were poorly staffed, with limited resources. Usually there was not even the rudimentary infrastructure necessary for serious technocratic planning, which required large amounts of data, technological capabilities, and a cadre of well-qualified and well-paid staff. Nevertheless, the inspiration for modernization was so strong that some national governments invested large sums from export earnings and international aid to create new capital cities. Planning for many of these capital cities was led by foreign architects with little knowledge of local planning culture.[10] This lack of knowledge was not considered a drawback; on the contrary, since the goal was to interject a culture of modernization both in the physical form of the city and in its planning process, the lack of local knowledge was considered an asset, particularly because external experts who were to help modernize these cities were expected to be autonomous of traditional loyalties and local corruption.[11]

PARADIGM SHIFT IN PLANNING CULTURES

The golden years of planning lasted for almost two decades, if one acknowledges 1968 as the turning point when prevailing notions of planning came under attack in both industrialized and industrializing countries. Though this transition is well documented,[12] it is worth reminding ourselves that what came under attack were not only the results of planning but also the culture of planning practice. The criticism came from many quarters, including planners themselves–particularly those based in academia.[13] Attributes of planning that

3

had been viewed as strengths during the golden years were now seen as major drawbacks. Planning was now considered too technocratic, elitist, centralized, bureaucratic, pseudoscientific, hegemonic, and so on.[14] In industrializing countries the criticism of planning went even further. The critics argued that, rather than serving as a positive force for social change and modernization, planning had been the major hindrance to such change.[15] Drawing on criticisms of planning from both the right and left of the ideological spectrum, an eclectic argument was made that top-down, state-centered planning was inflexible, unresponsive to the needs of the people, and alien to local culture.[16]

There was much discussion in both industrialized and industrializing countries about the need for a paradigm shift in planning practice. According to the new paradigm, planning practice was to be "bottom-up" and "people-centered," relying no longer on economists, engineers, and statisticians, but on anthropologists, sociologists, scholars of cultural studies, and grassroots activists, who were closer to the people.[17] Institutionally, the focus was to shift from state agencies to nongovernmental organizations and private voluntary organizations, which were considered more efficient, equitable, flexible, and accountable.[18] In this new mode planning was to become more participatory, culturally sensitive, politically more explicit in advocating the needs of disadvantaged groups, and, overall, less technocratic and less reliant on modern technology, such as computers, for problem solving.[19] This paradigm shift in what was considered effective planning was more pronounced among academic planners than among practitioners, who could not change their style of practice as quickly as the academic discourse was changing. Nevertheless, with time, planning practice did change, producing a mixed outcome.[20]

On the positive side, planners became more concerned about environmental issues, sexism, and the impact of racism on urban form and planning practice. The civil rights movement had coincided with the paradigm shift in planning practice and raised the general awareness of planners regarding the multicultural composition of urban populations.[21] In general, the planning process became more open to public participation. In newly industrializing countries, the shift in planning practice was most noticeable in discussions of development. Until then, development had been equated with economic growth only. The new paradigm of planning from below stressed issues of income redistribution, poverty alleviation, and the critical roles of housing and the urban informal economy in meeting the basic needs of the urban poor.[22] This led to the recognition that the planning problems of industrializing countries were starkly different from those of industrialized countries. Hence, the old paradigm of modernization built on the experience of industrialized countries was not appropriate for the newly industrializing countries. Planning in industrializing countries required sensitivity to their cultural, economic, political, and institutional particularities.[23]

On the negative side, the shift in the dominant planning paradigm also created some problems. As traditional planning institutions came under attack, they lost not only legitimacy but also resources, weakening their power to intervene decisively in the socioeconomic and political processes influencing the urban built environment.[24] Though some alternative planning institutions did emerge in the process, they were not empowered to pursue a comprehensive approach to urban problems.[25] These new planning institutions focused on one or two problems of specific constituencies and were usually too small to address large-scale problems. Also, contrary to popular perception, they were not necessarily more efficient or accountable than traditional planning institutions.[26] True, the new paradigm opened up the planning process to public scrutiny. However, in some countries, this occurred to such an extent that the process of decision making became contentious. This forced planners to become negotiators, learning these skills on the job, through trial and error. In the process, planner-mediators often withheld their professional views to keep from "biasing" the deliberative process and, instead, searched for the common ground among contesting views, sometimes arriving at solutions that embraced the lowest common denominator.[27] This kind of planning process did not strengthen the claim that professional planners had valuable knowledge and training that others lacked.[28] Disagreements among planners themselves only deepened the ambivalence about what professional planners could contribute to decision making, which was reflected in growing disagreement among the planning theorists.[29] Lacking a professional consensus about how to plan well, professional planners reacted to planning problems with little certainty about their own effectiveness. This professional anxiety, combined with the threat of declining resources, led some to declare that the profession was in a state of crisis.[30]

PLANNING UNDER ATTACK

The 1980s interjected two new elements into the culture of planning practice. First, as globalization of industrial production became increasingly widespread, manufacturing industries were moving out of old industrial cities. The outflow of capital left behind cities with high unemployment, housing foreclosures,

3

and an underutilized infrastructure that could not be maintained on sharply declining revenues. Urban planners in the United States and other industrialized nations realized that the economic health of these deindustrialized cities could not be restored by traditional city planning.[31] Planners were at a loss for effective solutions, and some called for a national urban policy to tackle the effects of deindustrialization.[32]

Second, the ascendancy of neoliberal politics, led by President Reagan and Prime Minister Thatcher, radically changed the professional planning discourse.[33] For planners, what is important to remember about this major political turning point is how that historical moment tarnished the image of conventional planning by discrediting the role of government in general, and regulatory practices in particular, in influencing social outcomes. Politely marketed as "reinvention of government" or "new public management," neoliberal attacks on the state and planning were aimed at unraveling the social contracts among governments, market agents, and citizens that had been established earlier by the "welfare state" in industrialized countries. In industrializing countries, the attacks comprised three interconnected policy approaches, commonly known as stabilization, liberalization, and privatization.[34] The purpose of these policies was to counteract the lagging economies of industrializing countries, which were blamed on government intervention. Though the criticisms of state policies and planning practices in industrializing and industrialized nations varied, their objectives were similar—namely, to make all nations compete in the global economy by lowering the costs of production and accumulation. This required lean, flexible, and market-friendly states that were entrepreneurial as opposed to regulatory. The goal was to attract private investment by lowering the risk of such investments and decreasing taxes on profits. Thus, private–public partnerships became a key planning strategy for planners; and this strategy was pursued by bypassing traditional planning institutions, which had become an arena for contentious politics. New planning institutions emerged in the form of development corporations, rather than planning agencies, because what inspired the moment was entrepreneurship and development, not regulations and planning.[35]

Ironically, at a time when planning was under attack and losing its traditional power, there was a "communicative turn" among the planning theorists in industrializing countries.[36] At a time when the powers that be did not want to engage in serious planning, the planners were proposing that the legitimacy of planning could be restored via public deliberations organized by small-scale community groups and other nontraditional and grassroots organizations. The collapse of the Soviet Union in 1991 provided the last nail needed to seal the casket on the old planning paradigm. As mentioned earlier, the Soviet Union had inspired many industrializing countries to formulate national plans for rapid industrialization. For nearly seventy years, the Soviet Union, along with China, Cuba, and other communist countries, had also provided concrete examples of alternative institutional arrangements. These alternatives lost their initial appeal as the effects of authoritarianism came to be known, however, decreasing the resistance to a totally hegemonic discourse of the kind exemplified by Francis Fukuyama's (1989) declaration of "the end of history" with the collapse of the Soviet Union.[37]

POST–COLD WAR PLANNING

Fifteen years after Fukuyama's triumphant declaration, the world does not seem either more peaceful or more prosperous. The troika of neoliberalism—stabilization, liberalization, and privatization—along with the dismantling of traditional planning institutions did not generate a high rate of economic growth, except in China, which pursued a policy path of its own. The sluggishness of the economies of industrializing countries, even after many rounds of stabilization, liberalization, and privatization, is now being blamed on corruption.[38] To justify the failure of neoliberalism, some have reinvented the argument that certain cultural practices are the real barriers to economic growth.[39] In the industrialized countries, the rapid expansion of information and communication technologies did not really materialize into sustained economic growth. Moreover, the integrative power of the new technology has not brought the people of the world closer. Income inequalities within and among the nations of the world have increased since the Reagan–Thatcher effort to dismantle the welfare state in industrialized countries and the developmental state in industrializing countries.[40] The concurrent rise of religious fundamentalism in both types of countries has added a new anxiety about secular planning practices. Yet some of the benefits of social change achieved in the 1970s—such as environmental awareness, appreciation of racial and gender diversity, and recognition of global interconnectedness—continue to influence "planning conversations" in most countries.[41] This strange mix of social trends at the beginning of a new millennium in human history calls for serious reflection about the enterprise of planning and its validity, if any, under the new circumstances.

3

There are many ways to reflect on planning. One could study the effects of efforts to reinvent government and the concept of new public management, or one could focus on how neoliberal attacks on traditional planning institutions have altered planning styles.[42] One could highlight planning success stories, such as participatory budgeting in Porto Alegre, Brazil, or examples of successful infrastructure planning for the European Economic Union.[43] Conversely, one could focus on planning disasters and explore the reasons for such outcomes. Ironically, the number of case studies of "best planning practices" has increased significantly since the 1980s, when planning came under attack.[44] When read carefully, most such case studies demonstrate not so much the effectiveness of astute planning practice, but how either the market or, more commonly, the civil society contributed to the success of these projects. In other words, documentation of "best practices" did not strengthen the arguments for planning. On the contrary, it demonstrated that to achieve good results, traditional planning approaches relying on regulations must change to fit the demands of the market.[45]

The contributors to this volume are aware of the changing nature of planning practice, from its golden years immediately after World War II to its gradual loss of legitimacy over the last fifty years, as a unique professional service rooted in specialized knowledge and technical expertise for solving problems of spatial entities. The nature of change in planning practice has not been identical in all nations, however. Variations between industrialized, industrializing, and transitional nations certainly exist; even within each type of nation, one finds large variations in the ways traditional planning practices have changed, evolved, or declined over the last fifty years.[46] Traditional explanations for these variations point toward differences in political economies. But such explanations have come under scrutiny with the growing acknowledgment that the global interconnectedness of trade, finance, and managerial practices is inducing institutional isomorphism and beginning to erode distinctions among different territorial jurisdictions.[47]

The rapid expansion of information and communication technologies since the mid-1990s has strengthened the perception of a convergence in institutional forms and practices, even though, in reality, one can observe significant differences in the ways planners have coped with change. Country-specific evaluations of efforts to influence neoliberal policies clearly indicate that the way neoliberal rhetoric was translated into actual policies varied widely among nations.[48] Neither was the welfare state dismantled uniformly across all industrialized countries, nor was the developmental state disbanded in the same way in every industrializing country.[49] This large variation in out-comes raises the question whether neoclassical economists who predicted unifying and homogenizing effects of neoliberal policies overlooked the particularities of varying planning cultures.

FOCUS ON PLANNING CULTURES

The issues of culture in general and planning culture in particular have never been of interest to neoclassical economists, who dominate the current discourse on economic growth. During the golden years of planning, however, development economists and Keynesian economists dominated the discourse. But starting in the early 1970s, their theories came under attack, and the argument that some economies required specialized attention and state intervention began to wither away.[50] As planning came under attack for distorting the market, neoclassical economists argued that cultural differences among the peoples of the world were not relevant. They proposed that all individuals are "rational actors" continuously engaged in furthering their self-interest. According to their view, planners and policymakers should acknowledge this fundamental truth and create institutions that would facilitate, not hinder, the universal urge among people to maximize their self-interests.[51]

The purpose of this volume is to assess the validity of such universal proclamations about planning in light of concrete experiences in ten particular nations, which differ in their size, level of industrialization, resources, and political structures. Drawing on "thick descriptions" of planning practices, we have attempted to identify whether each case setting is characterized by unique institutional arrangements that have shaped its planning culture. We also probe the extent to which such cultural traits reflect what Paul Ricoeur once described as the "cultural nucleus" of a territory.[52] This is an important question because cultural identity is often viewed as comprising core cultural traits that are indigenous, inherited, and immutable. Much of the criticism of planning practice that emerged in the 1970s under the banner of multiculturalism argued that traditional planning had failed, in part, because it did not acknowledge this fundamental element in the way people formulate their own identities.[53]

Yet, as described earlier, planning culture in general seems to have changed over the last fifty years. In seeking to explain this change, we have focused, in particular, on whether and how the ongoing intensification of global interconnectedness in trade, capital flows, and technological connectivity

3

is affecting planning culture. Are there signs of a convergence of planning cultures since the golden years of the 1950s, when technical rationality, expert knowledge, comprehensiveness, and bureaucratic structures of administration were celebrated? How and why did this style change in different settings? Is a common planning style continuing to emerge as all nations compete for the benefits of globalization? Or is the planning style in each setting being shaped by its unique cultural practices?

The last question brings to the fore an old issue that planning theorists have grappled with since the early 1960s, when urban riots erupted, first in U.S. cities and later in Europe and elsewhere—namely, how politics influences planning style, and vice versa. The case studies in this volume confirm the changing nature of planning styles and cultures and raise the question whether planning culture should be regarded as a relatively autonomous and independent variable. And these case studies suggest that planning culture, much like the larger social culture in which it is embedded, changes and evolves with political-economic changes, sometimes becoming more democratic and participatory but at other times changing in the opposite direction. To be sure, planning culture is affected not only by political changes but also by other changes, such as technological innovations, demographic shifts, and the emergence of new problems or sudden deterioration of any one or more existing problems. International flow of planning ideas also affects planning styles, although not to the extent claimed by either its critics or its proponents.

How does one develop new insights about such a complex social process with multiple and interconnected causes and effects? The essays in this volume vary in the style with which they address the issue of planning cultures. The authors selected different time periods, problems, and intellectual approaches based on their experience and expertise. Such variations in methodology provide unique stories, which I have attempted to tie together under a set of broad themes.

VARIATIONS IN PLANNING CONTEXTS

The issue of contextual specificity seems obvious as one reads the descriptions of different planning practices in different nations: Indonesia is very different from India, which is very different from England, which, in turn, is different from France, and so on. Booth's comparative historical analysis of planning systems in Britain and France (Chapter 11: *The Nature of Difference: Traditions of Law and Government and Their Effects on Planning in Britain and*

France) demonstrates that even though both planning systems were inspired by German town planning in the nineteenth century, they evolved in very different ways, owing to differences in their legal systems (common law in Britain, in contrast to reliance on statutes in France), in state traditions (a relatively centralized state in France, which has a written constitution, in contrast to a relatively decentralized state in England, which does not), and in the ways private property rights are defined.

In other examples of institutional specifics, Sorensen (Chapter 10: *The Developmental State and the Extreme Narrowness of the Public Realm: The Twentieth Century Evolution of Japanese Planning Culture*) demonstrates the ways in which Japanese planning is shaped by a distinct state–society relationship characterized by a persistent notion of individual and collective sacrifice for the sake of national interests. Sorensen argues that although this uneven relationship between state and civil society was cultivated prior to World War II, it persisted during the postwar period of democratic governance. The distinctly centralized style of Japanese planning draws on this culture of sacrifice; and in this top-down approach, the Japanese planning bureaucracy is supported by both political parties and business elites, forming a mutually supportive triangular relationship.

All of the case studies in this volume reveal unique planning contexts. Faludi (Chapter 12: *The Netherlands: A Culture with a Soft Spot for Planning*), for example, describes how planning in the Netherlands is shaped by a set of circumstances created not only by its geography but also by its Protestant tradition, corporatist structure of decision making, and "a culture with a soft spot for planning." In sharp contrast to the Netherlands, planning in Australia, described by Sandercock (Chapter 13: *Picking the Paradoxes: A Historical Anatomy of Australian Planning Cultures*), is neither comprehensive nor anchored at the national level. This difference is explained by the unique history of Australia's emergence as a nation-state that consciously avoided reproducing both Britain's class antagonism and America's market-driven model.

CHANGING NATURE OF PLANNING CULTURES

It is widely known that planning contexts vary not only among different nations in the world, but also within nations, particularly those with federal governance structures. What is interesting, however, is to question the extent to which such contextual specifics can be attributed to indigenous cultural traits of planning. The studies in this volume demonstrate that the concept of *cultural essentialism,* in which culture is portrayed as static, homegrown, pure, and immutable, is inaccurate.[54] Rather, these planning cultures seem to have evolved with social, political, and economic influences, both internal and external, creating hybrid cultures whose complexity can only be understood through deep historical analyses.

Booth's study (Chapter 11), for example, documents well the German origin of French and British planning. Cowherd (Chapter 8: *Does Planning Culture Matter? Dutch and American Models in Indonesian Urban Transformations*) describes the influences of Dutch and American planning in Indonesia. Similarly, Davis (Chapter 9: *Contending Planning Cultures and the Urban Built Environment in Mexico City*) discusses the dual influence of French and American planning and design styles in Mexico City. Tajbakhsh (Chapter 4: *Planning Culture in Iran: Centralization, Decentralization, and Local Governance in the Twentieth Century*) mentions European (Belgian and French) influence on constitution writing in the early part of this century. Banerjee (Chapter 7: *Understanding Planning Cultures: The Kolkata Paradox*) illustrates the influence of Ford Foundation advisors familiar with American planning traditions in the planning culture of Kolkata, India. Similarly, Leaf (Chapter 5: *Modernity Confronts Tradition: The Professional Planner and Local Corporatism in the Rebuilding of China's Cities*) describes the impact of economic liberalization—a neoliberal initiative originally formulated in Washington, D.C., which, in conjunction with political pressure from inside, deeply affected planning practice in China. True, Leaf argues that all such changes were ultimately co-opted by China's established political hierarchy. But, as Ng (Chapter 6: *Planning Cultures in Two Chinese Transitional Cities: Hong Kong and Shenzhen*) describes, a new style of entrepreneurial planning did emerge in Shenzhen as a result of opening the Chinese economy. In other words, planning styles may evolve and change even without a radical change in the political system.

The most dramatic example of external influence on what Robert Fishman has called internal "planning conversations" is portrayed in Birch's case study (Chapter 14: *U.S. Planning Culture under Pressure: Major Elements Endure and Flourish in the Face of Crises*) of planning in lower Manhattan in the aftermath of the terrorist attacks on the World Trade Center. This case adds a new dimension to the growing conversation among planners regarding globalization—not only of capital and labor, but also of terror. Whether this unusual event will forever alter American planning culture is not clear from Birch's description of the many actors and institutions involved in the planning. But the way these numerous actors generated many planning responses, which at the end were also deeply influenced by a few wealthy individuals, is particularly North American. The reader may already be aware of the competition held to generate design options to rebuild on the site of the World Trade Center. To what extent do these design entries from private firms around the world reflect American planning culture? This question can only be answered by considering the contemporary cosmopolitan quality of New York City, whose planning culture has evolved over the last 125 years with waves of migration as well as investment from abroad. New York City is now rightly acknowledged as a world city.[55] Hence, it was appropriate to seek worldwide for design solutions for its destroyed landmarks. There is much support for this sentiment among New Yorkers, who recently applauded Santiago Calatrava's spectacular design for a new train terminal next to the site of the former World Trade Center. Although trained in Spain, Calatrava was able to capture the ethos of New York City, in part because New York is a global city with many cultural influences assimilated, over the years, in its built form.

GLOBALIZATION AND PLANNING CULTURE

Much has been written about the homogenizing impact of increasing global connectivity on culture.[56] The case studies presented here suggest, however, that both the promise and the threat of cultural homogenization through globalization may be exaggerated. Though these case studies provide many examples of global interactions, none of them demonstrate that such interactions are leading to a convergence in planning styles. True, decentralization of governance and planning is a trend described by Tajbakhsh (Chapter 4) in Iran as well as by Leaf (Chapter 5) in China. Similarly, what Sandercock (Chapter 13) describes as entrepreneurial planning (in contrast to regulatory planning), currently in vogue in Australia, also exists in many American cities as well as cities in other nations with very different planning histories. Nevertheless, those trends appear to have been adapted to local conditions, generating varied outcomes. For example, though local corruption seems to have increased with the reduced regulatory power of the local state in Indonesia (Chapter 8) and though Leaf (Chapter 5) notes a growing influence of Communist Party officials who might have benefited from new opportunities for corruption in China, similar outcomes were not reported for the other nations analyzed in this volume.

Although such variations in outcomes should be considered before we either criticize or praise the impact of globalization on planning culture, our studies indicate that the nations studied by the contributors are all making efforts to reap the benefits of globalization and that planning as a governmental activity is deeply engaged in such efforts. Planners are not resisting the growing interconnectedness of financial and information flows; instead, they are modifying planning practice to suit the needs of the moment.[57] Of course, planning is being transformed in different ways in different countries, but the intentions of planners worldwide are quite similar: to avoid parochial isolation and exclusion from the global movement of finance, trade, and technological advancement. Whether this trend is solely a result of the spread of communication and information technology, we do not know. But, as Castells argues in Chapter 3 (*Space of Flows, Space of Places: Materials for a Theory of Urbanism in the Information Age*), this new technology has definitely influenced the perceptions of planners around the world, who worry that if they are not part of what Castells calls "the Net," they will be left behind as the world moves forward.[58] Yet, as Tajbakhsh (Chapter 4), Cowherd (Chapter 8), Faludi (Chapter 12), and Leaf (Chapter 5) document, this trend has not homogenized planning cultures. Nations have been able to retain local planning characteristics that draw on their particular religious and political traditions.

Has globalization eroded the capacity of nation-states to plan and intervene to achieve particular social outcomes? Much has been written to suggest that nation-states have lost the ability to influence business cycles that had been part of the Keynesian approach since the 1930s Depression.[59] Some have argued that the taxing power of states, both national and local, has been decreased by the growing movement of capital across territories and the consequent increase in competition to attract external investment by lowering tax rates.[60] This, in turn, has reduced planning's resource base, making territorial entities more vulnerable to conditions set by global investment flows. In this volume, Castells' description (Chapter 3) of the growth of information technology and its adverse impact on the traditional planning capabilities of nation-states resonates with these predictions, although he is not as pessimistic as many others about the future of planning. As initially proposed by Peirce, Johnson, and Hall in 1993,[61] Castells suggests that an inadvertent but positive side effect of the decline of national capacity for planning may be the rise in the planning role of local states, particularly in large cities with diverse economic bases.

None of the authors writing for this volume attempt to verify the prevailing assertions about globalization's impact on planning capacity. Their discussions present some evidence, however, that the actual impact may be more complex and mixed than has been claimed by either the critics or proponents of globalization. For example, Sandercock (Chapter 13) describes how competition for global investment has influenced Australian planners to ignore some social justice issues and, instead, encourage the entrepreneurship of cities eager to offer joint investment ventures and reduced taxes to international corporations. Similarly, Cowherd (Chapter 8) provides some examples of how Indonesian planning has suffered from the de-emphasis on its regulatory functions, to provide opportunities for corporations as well as local politicians to profit from deregulation. In contrast, Ng's description of China's success (Chapter 6) suggests that cities and provinces are not totally at the mercy of private investors. In fact, Ng's comparison suggests that Hong Kong, once an entrepreneurial city-state, is now lagging behind Shenzhen, a municipality, mainly because of innovation and entrepreneurship by a new cadre of local young planners. Leaf's description of Chinese planning (Chapter 5) is not as optimistic. He is skeptical of the relative autonomy of local planners vis-à-vis the old political power elites. Nevertheless, the steady growth of the Chinese economy since around 1990 is an indicator of how local-level planning officials and locally based entrepreneurs can create the conditions for economic growth. Through administrative decentralization and other institutional mechanisms that have not yet been well analyzed, Chinese planners have managed to open the economy to external and internal corporate investors, thereby generating an unprecedented rate of economic growth, which has also increased the revenue base for planning. Part of the reason for China's success is that even though private investment can now roam the world at the press of a computer key, ultimately such investments need to settle in specific localities to generate further growth. As Krugman observed in 1995, localities with good physical and social infrastructures can be more attractive for private investment than cities that offer large tax concessions and cheap land but lack such infrastructures.[62] Local planners in China are aware of this comparative advantage. With relatively fewer dictates from central planners compared with the pre-reform period, these planners, along with local businesses, have ushered in a new entrepreneurial planning style that no one predicted in the mid-1990s.

The relationship between cities and prospective private investors, both internal and external, is also examined in Birch's analysis (Chapter 14) of New York City's effort to rebuild on the site of the World Trade Center. Although Birch describes a rather chaotic planning process, with many groups and institutions interacting simultaneously, her description does not suggest that private investors alone dominated the planning process. Planning, even if institutionally fractured in the classical North American way, still matters.

Banerjee's historical analysis (Chapter 7) of planning in Kolkata, the West Bengal state capital, just prior to a long Marxist rule provides yet another twist to the conventional understanding of the relationship between private investment and public planning culture. Banerjee describes vividly how the Ford Foundation devoted significant professional expertise and resources to broaden the scope and style of planning for the Kolkata metropolitan area. The chief minister of the state at that time was worried about rising unemployment in the region, which was being exploited by the Marxist opposition parties. One key objective of the Ford Foundation planning was to generate more employment by making the Kolkata metropolitan region more conducive to private investment. The strategy was not to reduce the state's capacity for planning in order to generate more employment. On the contrary, it expanded the city's planning capabilities by injecting elements of social and economic planning into the then-dominant physical planning efforts. The Ford Foundation's efforts did not immediately increase employment, and Ford eventually reduced its involvement as a Marxist-led coalition government came to power both in the state of West Bengal and in the city of Kolkata. Nevertheless, the dominant planning culture had been changed for the better because it was no longer confined to drawing traditional master plans. During a brief period after the collapse of the Marxist-led coalition government, the previous ruling political party returned to power and tried to resurrect the planning efforts initiated by the Ford Foundation. That effort led to the creation of a new planning institution, the Kolkata Metropolitan Development Authority, which remains active even after twenty-five years of continuous Marxist rule at the state level. The socioeconomic planning that the Ford Foundation had first brought to Kolkata to deter a Marxist takeover of the state is now being used by a Marxist government for the same purpose as was initially intended: to make Kolkata attractive for private investment through strategic public expenditures on physical and social infrastructure.

CULTURE MEETS POLITICS

The Kolkata planning story is only one example among many in this volume that suggest that to understand the planning culture of any place, one needs to understand the relationship between planning and the socioeconomic and political changes in that area. Leaf's exploration of Chinese planning culture (Chapter 5), Tajbakhsh's analysis of planning in Iran (Chapter 4), Cowherd's description of planning in Indonesia (Chapter 8), and Davis's portrayal of planning in Mexico City (Chapter 9) all provide the same lesson. Planning culture is not an independent variable, even though the word *culture* is often used to signify a domain separate from economy and politics. As was argued earlier, planning cultures, when subjected to historical analysis, reveal themselves to be in constant flux, sometimes resisting, while at other times facilitating social change in response to both internal and external pressures.

The impact of social, economic, and political changes on the planning culture of any one place is not predictable. As our case studies exemplify, in some countries, at certain historical moments, the impacts of such changes have been progressive. But there have been regressive outcomes as well, even within the same country. For example, as Sandercock (Chapter 13) describes, Australian planners' attitude towards native Australians has evolved significantly over the last 100 years. Also, they now have more awareness of gender inequalities and are more concerned about environmental degradation. Yet around the same time as environmental concerns first emerged in Australia, city planning began to move away from its traditional concerns with equity and social issues, toward "place marketing" to attract private investment. Cowherd (Chapter 8) describes similar mixed outcomes in Indonesia, where President Suharto's resignation led to administrative decentralization and increased the freedom of the press and participation by grassroots groups. At the same time, however, the planners implemented "American-style market liberalization," thereby decreasing regulatory controls and increasing stratification of the populace by race and class. Davis (Chapter 9) also describes mixed outcomes in Mexico City. On the one hand, the Mexican revolution terminated the dominance of the old oligarchy and ushered in a new era marked by new concern for the welfare of the poor. On the other hand, planning for Mexico City, in the aftermath of the revolution, was stifled by political and professional differences among competing groups of planners. Even in Japan, where the dominant ideology calls for sacrifice by the people in the interest of nation building, Sorensen (Chapter 10) mentions vigorous

3

opposition by civil society to the single-minded focus on industrial expansion, particularly after the collapse of Japan's bubble economy in the early 1990s.

One could point to more examples of varying outcomes within the same country. Leaf (Chapter 5), for example, describes radical changes in China during the Cultural Revolution, when Mao Zedong condemned professionalism as a form of elitism. Now there is a complete reversal. A new group of technocratic planners has emerged to manage the transition from socialist to entrepreneurial cities. In contrast, Hong Kong, as Ng (Chapter 6) notes, has lost its earlier entrepreneurial edge and is unable to restructure its planning culture to compete successfully with some emerging municipalities in Mainland China.

These examples highlight one issue particularly relevant for this volume—namely, to understand variations in social outcomes in any place, one needs to look beyond cultural attributes to political configurations and economic relations that constitute the specific political economy of that place. As Friedmann (Chapter 2) notes, the specific characteristics of planning institutions in each nation are shaped largely by their unique political-economic relationships. Using extensive historical analysis, Booth (Chapter 11) demonstrates that property relations, intergovernmental relations, and the legal framework of each nation are three areas with particular relevance for planning endeavors. Understanding the constitutional logic underlying these three elements and how they have evolved over time in each territorial jurisdiction can generate significant insights about the nature of planning cultures. Castells (Chapter 3) adds a fourth element specific to our times—namely, the role of information and communication technologies, which have created new economic as well as political linkages among territorial jurisdictions. Castells argues that such linkages have implications for planning from the top as well as from the bottom.

As the political and economic elites of nation-states are increasingly interconnected, there is a parallel connection among groups at the bottom who seek identity and recognition as they struggle to understand who is really affecting their quality of life. These movements from below, which Friedmann had described earlier as forms of radical planning,[63] did not receive adequate attention in our symposium. Although the authors in this volume do not ignore pressures from below and voices of dissent, such dissent is not the central focus of inquiry in any of the eleven case studies. One plausible reason for this unintended bias is that although planning is usually portrayed as a professional activity that engages all people, in practice it is still dominated by professional planners at the top, even though the nature of such domination has changed over the last 300 years. The cases presented here mostly describe the dominant planning practices in each country. Nevertheless, as all the case studies demonstrate, neither dominant planning practices nor the cultures underlying those practices are etched in stone. Both change, sometimes in a progressive direction, at other times regressively in response to political struggles. Understanding the origin and outcomes of such political struggles is essential if we are to go beyond the static conception of planning culture that only fuels social conservatism.

CONCLUSION

The eleven case studies of planning presented at the symposium (and in this volume) did not generate a precise formulation of how planning cultures affect planning practices. What emerges from them is a more complex understanding that planning culture should not be read as specifically demarcated and unchanging social attributes that clearly differentiate the planning practices of different countries. Instead, the focus of inquiry should be the continuous process of social, political, and technological change, which affects the way planners in different settings conceptualize problems and structure institutional responses to them. If planning culture is viewed in this dynamic way, in contrast to traditional notions of culture that are used to evoke a sense of immutability and inheritance, then we can go beyond "cultural essentialism," which, in essence, is exclusionary, parochial, and an inaccurate representation of history.

As the case studies in this volume document well, there is no cultural nucleus or core planning culture, no social gene that can be decoded to reveal the cultural DNA of planning practice. Planning culture, like the larger social culture in which it is embedded, is in constant flux. That is why it is so difficult to precisely demarcate the cultural elements in any process of social transformation. Cultural anthropologists now acknowledge this amorphous and changing nature of culture. As Shweder recently noted, "Cultural elements are too hard to define, too easily copied and too long detached from their points of original creation. Contact between cultures and processes such as borrowing, appropriation, migration, and diffusion have been ubiquitous for so long that little remains of the authentically indigenous."[64] Shweder's comment is valid for planning culture, which is deeply engaged in what Edward Said called "a complex traffic of ideas."[65] This is not to say that planning practice in

188   F. Othengrafen et al.

HYBRID PLANNING CULTURES   23

all nations is the same. The case studies here clearly demonstrate that each setting is distinct, but this distinct quality is the result of a complex process of social change, not the inevitable and predictable outcome of a static planning culture. Rather than searching for the cultural nucleus of planning practice in each nation, we need to understand how changes occur in planning practice in all nations, including our own. Lacking such a comparative and dynamic understanding of social change, which is a central objective of planning, we may inadvertently legitimize both the stereotypes we hold of others and those they hold of us.

To understand the impact of contemporary social change on planning culture, we must acknowledge the trend toward global connectivity through increasing movement of investment, trade, ideas, and people. Both the promise and the fear of this trend have been exaggerated, however. Our case studies demonstrate that even though global connectivity and the simultaneous ascendancy of neoliberal ideas have penetrated the planning discourse in all nations, their impacts have varied widely. Planning institutions have not been dismantled equally, nor have regulations been withdrawn to the same extent, in all nations. Similarly, the move away from comprehensive planning based on large data sets and technical analysis is not evident equally in all nations. On the contrary, the rapid advancement of information and communication technologies–in particular, the spread of geographic information systems–has resurrected the legitimacy of "scientific planning" at the local level.

To be sure, the dominant planning narrative in any setting is not free of opposition from below. The intensification of social and economic inequality with increased global connectivity has generated opposition to dominant planning narratives, in varying degrees, in many nations. These oppositional narratives are not articulated with equally strong voices in all nations, and they have not been integrated in a systematic way to create a global civil society. One plausible reason for this outcome is that planners worldwide are aware that external influences need to be tempered to fit local conditions. It is the changing politics of different settings–not of planning cultures–that have conditioned planners' responses to external forces. Nevertheless, planning as a professional activity has not lost legitimacy worldwide. On the contrary, the demand for planners' expertise is growing in many nations, although currently such expertise is sought after less to regulate and more to facilitate private investment, with minimal opposition from below.

This composite picture of planning practice, based on eleven case studies, merely suggests how planners in different nations are coping with multiple forces of social and spatial change. These examples do not lend themselves to rigorous comparisons among nations. There was never an intention on our part in launching this study to compare planning cultures by some well-calibrated criteria. In the past, efforts to make such comparisons have contributed not to better understanding but to cultural arrogance and parochialism. Our objective was to transcend such divisive outcomes by starting a global conversation about planning practice, using planning culture as a conceptual vocabulary for this open-ended discussion. This approach to the topic of culture–in particular, cultural differences among nations–is very different from the approach of those who fear an impending clash of civilizations. Our objective was not to confirm the stereotypes of planning cultures and thereby accentuate the differences among the peoples of the world, but to search for a common intellectual ground–a sort of "social commons"–that would provide a new context and meaning for planners at a time of significant social changes and increasing global connectivity.

Planners are not the only group searching for new meaning in their vocation at a time of rapid and uncertain changes. There are signs of such efforts in other domains of social action as well. The resurgence of religious identity emphasized by fundamentalist and orthodox groups is another indicator of how people distressed by the forces of social change are attempting to cultivate social meaning. Perhaps at the other extreme is the mobilization of social groups under the banner of multiculturalism. Unlike religious fundamentalists, the multiculturalists do not evaluate "others." Like the fundamentalists, however, they are not interested in seeking a common ground among different groups. In contrast, our effort to understand planning cultures in different nations was motivated by the intellectual need to seek such a common ground. This should not be misunderstood as an effort to create a universal culture for planners or a version of "Davos culture" for conversation among the planning elites of different nations.[66] Our goal, which became increasingly clear to me as I edited this volume, was to use intellectual encounters of people with very different planning experiences to create a global conversation about the role of planning in social change. The hope is that this kind of intellectual encounter will eventually lead to a more refined understanding of ourselves as well as others.

In the not-too-distant past, different cultures often encountered one another through armed confrontations and wars. We are still engaged in such encounters, and some are still trying to legitimize them by constructing theories based on cultural conflicts. Yet another way that different cultures

3

continue to encounter one another is through the exchange of goods and services in the ever-expanding market, now aided by new communication technologies. Our effort at understanding the planning cultures of ten nations was intended to encourage a different form of cultural encounter. We hope that, in the process, we have begun to mark the contours of the intellectual and social commons that form a common ground for the different peoples of the world. We may not have reached that destination as yet, but at least we have begun the journey.

Originaltext Booth 1993

European Planning Studies, Vol. 1, No. 2, 1993   217

The Cultural Dimension in Comparative Research: Making Sense of Development Control in France

PHILIP BOOTH

[Paper first received, September 1992;, in final form, December 1992]

ABSTRACT   _There is a tendency amongst planners to think of planning systems as phenomena with an independent existence that may be directly compared. Such a view lingers even where the concept of end-state planning has long been abandoned, and where it is understood that the context for plan-making and development control is an important factor in comparing systems of different countries. This paper will argue that systems of plans and development control procedures are essentially creatures of the cultures which give rise to them. They are to be understood as expressions of underlying beliefs about the way that decisions ought to be taken for the administration of a country. This paper presents three cases drawn from the author's own experience of research in France as indicative of the approach. The first, a development control case of the extension to a heavy goods vehicle depot, looks at the cultural factors that affected the way the application was determined. The second, concerning the formal development control agreements between state and commune in rural France, hinges on the understanding of the concept of a plan. The third, a consultation procedure in Lyon, is dependent for its explanation on the status of the participants in the procedure. Such an approach implies a mature understanding of the way in which a country works, not merely of its planning system. It implies an ability to communicate in the language of the country being studied. The paper concludes by suggesting that comparative research has important benefits for the understanding of planning in the home country, because of the way in which, if properly done, it questions the assumptions that are made about the nature and purpose of planning._

1. Introduction

The attractions of doing research that compares planning systems in different countries are obvious. Intellectual curiosity, the desire to test hypotheses about the nature of planning and the search for policy solutions to intractable problems of land use allocation are all spurs. The integration of the European Community and the developing links with eastern Europe makes studies across national boundaries an imperative. Indeed, the past 15 years have seen a rapid growth of comparative research and a wide debate on the problems of comparative planning methodology. The fruits of such work have included the Oxford–Leiden study (Thomas _et al_, 1983) and the Tri-National Inner Cities project (Davies, 1980) and the more limited but useful study of public inquiries in England and France (Macrory & Lafontaine, 1982). Conceptually, the approach has been described by Masser (1986) and the methodological problems have focused on the equivalence of the subject to be compared, the establishment of research teams,

P. Booth, Department of Town and Regional Planning, University of Sheffield, Sheffield S10 2TN, UK. An earlier version of this paper was given at the ACSP/AESOP Congress held at Oxford Polytechnic, 6–10 July 1991, under the title 'Undertaking Comparative Research: The Experience of Working in French Development Control'.

218   *Philip Booth*

the research framework, and the use of language. This author believes this approach to be flawed, and the intractability of the methodological problems to be evidence of a fundamental conceptual weakness.

This author's personal view is that the source of this conceptual weakness stems from an attitude towards planning, that holds it to be essentially a technical exercise, normative in intent. The attitude invites us to believe that the relationship between policy, its articulation in planning documents and its application through a control mechanism is essentially constant and understood by anyone, regardless of nationality, involved in landuse planning. That elevates plans and procedures (like planning inquiries) into objective phenomena whose purpose is unchanging. It further allows us to take the objects to which the plans and procedures are addressed as also constant phenomena. Such a concept then suggests that it is possible to isolate the impact of essentially external factors, like administration, history or a political system, and see how they modify the basic relationship between the plans, procedures and eventual development. Or expressed in another way, to take a policy area (such as it might be, the regeneration of inner cities) and compare different approaches with a view to transferring experience.

More recent comparative studies, it is true, have begun to confront this problem. Wakeford's study of development control in the US (1990) presents a personal exploration which, for all that it attempts to make a comparison between mechanisms, recognizes their rootedness in a particular way of life. The same is true of the major study of development control in five European countries by Davies and his colleagues (1989). While apparently presenting a mechanistic comparison of planning systems, the authors insist, notably in the concluding chapter, on the extent to which planning is a creative of administrative and political culture. And in his critique of the rational-central-rule approach to public policy-making, van Gunsteren (1976) had already argued that any system of government planning was doomed if it failed to recognize the role of administrative law in implementation.

Nevertheless, any study which starts by taking systems of plans or the instruments for controlling development runs the risk of reinforcing a traditional, technocratic view of planning, that ultimately leads to a blind alley. The purpose of this article is to explore the philosophical difficulties that must be overcome in doing research in other countries, in order that proper comparisons may be made.

1.1 The Problem of Language

Some inkling of the difficulties of the traditional conceptual framework for comparative study is gleaned both from the studies themselves and from those who comment upon them. One of the issues that confront comparative researchers is that of language. Here, the problem is clearly not a superficial one of translation of texts or of researchers being able to make themselves understood. Hantrais *et al.* (1985), writing from the perspective of social science research, warn:

> People not using the same language are known to differ in their subjective cultures. Methodologically this means that measurements must be made under different research conditions which take into account the effects of language, social structure and culture.

Both setting up a research project—in Hantrais *et al.*'s case the implication is for the collection of statistical data—and the analysis of the results are largely influenced by culture. At another level, the existence in English of the word 'policy' and its absence in German and French is a direct reflection of political culture which results in considerable conceptual difficulties in European social research (Hildenheimer, 1986). The problem does not go away, as Williams suggests, even when one language is used; the Tri-National Inner Cities project which was conducted in English nevertheless still faced difficulties in coping with the deeper meanings attached to the use of words (Williams, 1986). And, finally, the beguiling equivalence of the term '*enquête publique*' and 'public inquiry' in Macrory and Lafontaine's work (1982) appears to obscure the very real differences between the purpose of these two procedures in Britain and France. Language, deriving its force from the particularity of culture, is thus central both to setting up a comparative research project and to analyzing the project results.

1.2 Planning and the Cultural Dimension

Language may not be just a trivial problem, therefore, but providing there is a bedrock of a shared perception or of a shared problem, comparative cross-national studies of the kind referred to are still theoretically possible. The problem is that the things we choose to compare—plans, forms of development, procedures for control—in no sense derive from common perceptions. The conclusions of the Oxford–Leiden study present the problem clearly: (Thomas *et al.*, 1973, p. 261):

> ... despite the obvious differences in the legal and administrative characteristics of the Dutch and English planning system there was a tendency to assume initially that there was an overriding similarity in the types of plans produced and the relationship between plans and operational decisions. However, as the research progressed it became evident that the planning systems had a different emphasis ... it became clear that the links between plan-making and control can be wholly different ...

Setting up an *a priori* framework for the comparison of planning systems may therefore prove well nigh impossible if the starting point is procedures and documents because the assumptions made about them will not be shared. They signify different things about the nature of control and the power to take decisions that are culturally determined. The way forward must be to consider planning as a culturally determined process in which the interaction of decision-makers and the meaning they assign to the instruments they use will effect the outcome. A deep understanding of culture therefore becomes an essential prerequisite of cross-national research. Lisle (1985 p. 26) makes the point in respect of the social sciences in general:

> We are aware of such cultural differences in every day life. We are, strangely, less aware of the problem when, as social scientists, we are addressing an identical problem in different countries ... we overlook the fact that we ourselves, the scientists, are part of the cultural environments we are seeking to control.... To correct for this difference in perspective, there is no alternative but to attempt to adopt the alternative cultural perspective, to slip into the mindset and thought pattern of one's opposite number, to learn to look at the other country as one of its natives would, and to reconsider one's own country from the viewpoint of the skilled observer from the other.

Lisle is thus emphasizing the importance of the cultural dimension in research, but also suggesting an important interaction between the researcher and the subject being researched. The observer is not a neutral eye in the process.

To consider planning as culturally determined does, however, beg the question of what we actually mean by the word culture. Here, in an introduction to their book which stresses the importance of culture in the studies of cities, Agnew *et al.* (1985) offer helpful insights. One view of culture is that of an overarching structure which governs the way in which people

220   *Philip Booth*

behave. Apart from the difficult philosophical questions about choice and free will that such a view poses, Agnew and his colleagues suggest that it creates an insoluble dilemma in the relationship between structure and individual action. Rather, for them (pp. 1–2):

> Culture is created by thought and actions of both historical and living populations. Culture can change because it refers to material and symbolic contexts or limiting conditions for individual behaviour; it does not comprise an *entity* that governs what every human being thinks and does.

Culture for them is rooted in the action of individuals and is therefore not static, but is in a constant state of flux as people exploit the potential of known patterns of interaction. The anthropologist Geertz (1973) argues that culture is an effect in the signification we give to actions as much as in the actions themselves. Citing Ryle, he compares the contraction of the eyelids with a wink: to the casual observer both actions appear identical, but the wink is fraught with conspiratorial significance. Understanding the potential significance of winking leads to the possibility of parody, and a 'fake wink' to imply ridicule rather than conspiracy is possible. The physical gesture remains the same in all three cases; it is the meaning ascribed to it that changes. Geertz amplifies that with a story from his own case work of a Jewish trader, Berber tribesmen, and French colonialists in the highlands of Morocco. The Jewish trader, having been robbed by the Berbers, exacts and receives tribal retribution in the form of a flock of sheep, a gesture which is misunderstood by the French as implying that the Jew had acted as a spy for the Berbers, and thus had him jailed and the sheep confiscated. In this story, the outcome is a direct result of all the participants failing to understand the significance of the gestures and actions (Geertz, 1973).

Planning, however, is not about winking, sheep trading or robbery. What then is the basis on which this kind of cultural interpretation can proceed in cross-national planning research? It is argued here that the way forward comes from looking at the nature of the planning process. The decision to locate a factory, the developing of a long-term strategy for a conurbation, the securing of a planning gain, are all products of a pattern of decision-making which reflects power structures within a particular society. The power structures are culturally determined and the relationship between actors, though shaped by their cultural inheritance, is nevertheless in a constant state of evolution. The procedures that have been developed to handle the decision-making and the plans which are used, sometimes as end products, sometimes as staging posts, are in effect formalizations of this unwritten pattern of power. They are the winks and the nudges to which meaning is ascribed.

The significance of this approach to comparative planning arose from the way in which I formulated my intention to undertake research on French development control. The interest had arisen because, in struggling to understand how the French system of regulatory plans and codified planning law worked in practice, rather as the Oxford–Leiden researchers had done, I found myself asking how flexible the French planning system could be in response to unforeseen change. In using the term—flexibility—which is part of a British professional planner's stock-in-trade, I had at the outset failed to see that what is really involved in flexibility is the issue of discretionary power, a concept which by and large planners have handled much less fluently than they have flexibility in procedures. The fact that discretionary power and discretionary freedom were the issues became clear because of the decentralization of powers to local government that the Mitterrand administration introduced from 1982 onwards. Planning powers were at the heart of the decentralization programme. I found myself, therefore, asking whether the new powers did indeed increase the ability of local government to 'act flexibly', whether in other words there had been a real decentralization of power to mayors of communes, or whether decentralization was more in the mind than in the act.

The purpose of this paper is not to plough over once again the issues that have been tackled elsewhere (Booth 1989a, 1989b, 1989c, 1991a, 1991b). What follows is a presentation of three case studies of rather different sorts. This presentation attempts to expose the cultural dimension in the research carried out, to look at the impact of the cultural dimension on the outcome of events and to explore the impact of a researcher from one culture on a system determined by quite another.

2. Transports Griset: An Exception to the Rule

The first case concerns the extension to a road haulage depot on the outskirts of a semi-rural commune on the southern edge of the Lyon conurbation. In planning terms, the case represents the classic landuse decision-making dilemma. Transports Griset had been set up in 1948 in an outlying hamlet in the commune of Vernaison. By degrees, the tiny family firm had grown to become a major enterprise employing 200 staff and with 150 lorries using the depot. The case examined was the application by the Griset firm to create a further extension to the depots, loading bays and parking areas. Residents of the area immediately adjacent to the depot were vociferously opposed to the extension which they saw as a gross intrusion into an otherwise quiet agricultural and residential district. The research focused on discovering who had effectively taken the decision, how and on what basis it had been taken, what the role of the regulatory zoning plan had been and what impact the residents' complaints had had on the outcome. The application, let it be clear, had been approved and the work had been completed before the investigation took place. The decision was of particular interest because the zoning for this part of Vernaison as contained in the plan was in part for agricultural use and in part for low density residential development. New commercial and industrial development was excluded.

Within the analysis of this case, three separate interactions seemed important. First of all, there were the relationships that existed at the level of the commune. Essentially, this involved the mayor of the commune and his deputy with responsibility for planning who under the decentralization act were in principle responsible for the decisions on development control applications; the residents of the hamlet of Le Pellet; and M. Griset, son of the firm's founder, and now himself managing director. Secondly, there was a whole set of relationships at supra-communal level. The mayor was locked into a network that involved the urban community, the technical officials of the state, the professionals of the *agence d'urbanisme* and the prefect of the *département*. Thirdly, there was the understanding that both of these groups (in which the mayor and his deputy were a common factor) had of the plan and the zoning regulations that applied to the area.

2.1 The Mayor and his Commune

In the first group, the relationships are relatively straightforward. Yet to regard the residents as acting according to the classic principles of an environmental protection lobby or as 'nimbies' (members of the not-in-my-backyard school of thought) obscures some of the subtleties, even though elements of both attitudes are present in their case. Significant in the way they told the story was the fact that "M. Griset does not come from around here", albeit that the firm had been in continuous existence in Vernaison since 1948. Significant, too, that "he had threatened the residents with a gun". The objective reality of these statements is not the issue but an interpretation of them seems to suggest that the depot was not merely an environmental intrusion on the landscape; it was crucially a social one. This was a culture that, for all that it was on the edge of a dynamic urban area, found acceptance of outsiders difficult to handle.

222   *Philip Booth*

Their attitude to the mayor was equally disenchanted, but the disenchantment was not to be equated with, say, the wrath of English ratepayers against their local government. The very closeness of the mayor to his residents would suggest that it could not be. Secondary sources on French local government confirm a generally held attitude to mayors which sees them as above party politics acting with benevolent paternalism in the best interests of their communes. (Birnbaum, 1979; Dupuy, 1985). This conception of communal life deriving from a vision of an idealized rural France is still a powerful force in French thinking; farmers were still, in the 1980s, by far the largest occupational group represented by the mayors of France and, the deputy mayor of Vernaison responsible for planning at the time of the research was himself a farmer. The residents' disenchantment therefore can be seen as a sense of betrayal. They felt betrayed because the mayor, as 'father of his commune', had not protected their immediate interests, and betrayed, too, at a deeper level because he had not upheld the old agricultural order to which he and they belonged and into which the commercially successful Griset had intruded.

But if the deputy mayor and the residents of Le Pellet shared a common culture, why had they not closed ranks voted against the extension of the depot? Part of the explanation must come from the deputy mayor's ambivalent attitude to the Griset firm. In talking about the case, the deputy mayor was dismissive of the residents' objections as *vieilles histoires*, a neighbour dispute without further significance. But there is some reason for doubting that he really believed this. Mayors of small communes in principle have considerable power to govern the activities of their residents, and their status within the commune is high. They are, however, chronically threatened by the lack of resources and the crushing leverage of the administrative structure above them. They are therefore faced with a constant pressure to underwrite their own position and that of their commune. This they may do through development projects—and rationally minded administrators often point to the folly of every commune trying to develop its own swimming pool, when one might suffice for several neighbouring villages—or by attracting or maintaining commercial and industrial development. This latter has a significant advantage of not only being a tangible sign of the commune's prestige, but also an important addition to the tax base. The mayor could not afford to let the Griset company act out its almost certainly idle threat to move elsewhere. Loss of face and loss of finance outweighed environmental considerations, electoral advantage and even traditional values.

2.2 *The Technical Services*

The second set of relationships involves the mayor with other actors—the administrators involved within the Lyon conurbation. The complexity of interactions which impinge on the Vernaison case are, it must be made clear, specific to Lyon. Nevertheless, there is an important sense in which these interactions appeared to be informed by widely held cultural attitudes about the right way of doing things.

Some brief description of administrative structures is necessary if analysis is to make sense. In 1969, 55 communes that make up the major part of the conurbation of Lyon were required to come together to form a supra-communal authority, the *communauté urbaine*, for the provision of services at the level of the conurbation. It was a rare example of compulsion to cooperate which did not, however, seek to replace the communes. Instead, it merely set in place an upper level structure of a council of delegates in charge of a large technical service unit which manages urban transport, school and college buildings, water, sewerage and waste disposal, and forward planning, among other activities. Significantly, the forward planning activities are carried out, not in-house, but by an *agence d'urbanisme* overseen by a separate commission of delegates, which is therefore one removed both from the urban community and the communes. The urban community and the planning agency provide the technical resources not only for the conurbation as a whole, but also for the communes individually. These latter, too small and too poor for the most part even in the Lyon conurbation to develop their own technical services, are thus reliant on this outside expertise.

Within the British context, the relationship between elected representatives and the technical officers that serve them is a difficult one, but in any one case it conforms to some shared understanding of respective roles. This in turn allows some understanding of who controls decisions and at what level they are taken. Bring that kind of understanding to bear on the Griset case, and the process and its outcome look inexplicable. Why did the planning agency or the officers of the urban community not warn the mayor that the depot extension would create a significant environmental impact on the immediate locality? The residents could only understand this lack of action in terms of money having changed hands. There was no evidence whatsoever to support such a view. Conversely, there was a whole series of explanations about how various officials and elected representatives saw their role that would lead to a 'hands-off' approach in the case of Transports Griset without the need for bribery or corruption.

The very existence of an urban community that does not in principle remove any of the rights, privileges or duties of its constituent communes is in itself an enigma from a British cultural context. A major part of the explanation is that the commune's role as a bastion of democracy in the face of the centralizing tendency of the state has long been seen as a central principle of France's constitution. Decentralization by Mitterrand was designed to underwrite that very principle. And yet ... the other great concern of the French constitution is the unity of the state and the desire to constrain unbridled use of political power as much at the local as at the central level. There is therefore a curious, and uneasy, equilibrium that exists between the commune and its mayor, champions of democracy and the commune and its mayor, entirely dependent on state benevolence and outside technical resources. This balance is understood in principle by all those who are part of it (see Dupuy & Thoenig, 1984).

The particular balance in the Griset case is entirely concerned with actors at different levels of the local government system, though elsewhere the pattern would typically include actors from the state services as well. The technical officials of the urban community and the planners in the *agence d'urbanisme* have between them unique control of the expertise needed to determine the case, and therefore could have dictated the outcome. In practice they chose not to; and they, too, chose to describe the objections of the residents as a neighbour dispute. They acted on the general understanding that, provided the larger concerns of the state—or in this case of the urban community—were not affected, there was every reason to allow the mayor his head. They were also motivated by another factor. As agents rather than direct employees, their power as technocrats rests upon the degree to which they hold the mayor's confidence. Since decentralization, the fact that mayors have had the power to choose their source of technical advice has given their relationship an added edge. The *agence d'urbanisme* retains its hold on power by not opposing mayors too often and in other cases by offering positive benefits to retain their credibility. This in effect is the purchase price for power.

2.3 The Plan and its Effects

Finally, we come to the role of the plan in the decision-making process. Given that the decision as presented appears to have been entirely dependent on the perceived roles of the participants, the plan appears to have occupied a negligible role in the outcome. Yet that would also be a false conclusion. The *plan d'occupation des sols* (local landuse plan; POS) in force for Vernaison, like all such plans, consisted of a zoning map with regulations for each zone. The zones of Le Pellet were agricultural or low density residential which precluded new industrial commercial development. But there was a significant exception made for extensions to existing industrial or commercial premises provided that "they do not exacerbate the general conditions of the location of the enterprise in the environment" (Agence d'Urbanisme de la Communauté Urbaine de Lyon, 1982). There was thus an important discretionary power incorporated into the regulations themselves that left scope for *ad hoc* decision-making.

We do, however, need to know something more of how the plans are perceived in order to understand how that power would be used in the event. In reading the commentaries of the French professionals and academics upon their planning system, two things become apparent. The first is that the plan and its regulations are to be understood as the local substitute for the national regulations in the *Code de l'Urbanisme*. The purpose of codified law is to maintain a strong framework within which the state administrators, local officials and elected representatives all must work. In turn, the framework is the guarantee of justice and of the unity of the state. That is why, when the decentralization powers for planning were enacted in 1983, the power of mayors to determine planning applications in the name of the commune was made dependent on there being a local plan in force. The POS is not in the first instance the expression of local will. It is the mechanism by which those with the powers to decide are held to account. The second is that the zoning plan is not so much a constraint on development as an exposition of personal proprietary rights to develop. The right to private property is enshrined in the constitution and that appears to be widely held to include the right to develop, even though an absolute right to property is clearly untenable (Comby, 1989). The plan is the guarantee of what an individual landowner may do. When we compare this insistence on property rights to develop, it becomes clear that the plan has to be understood in a very different light from British local plans which form part of the system where development rights were nationalized in the 1947 Town and Country Planning Act.

The discretionary clause in the POS for Vernaison was not really there to allow a balanced and mature appreciation of the merits of the case. It was there so that a proprietary right to develop land on property used for a purpose established before the plan was approved, could be exercised. Yet the residents of Le Pellet felt outraged. For them the plan was a guarantee of the right not to suffer prejudice from the activities of their neighbour. The belief that the rules had not been respected and that the hint evident in other sources that anarchy is the eventual outcome of such behaviour is inescapable.

3. The MARNU, or When is a Plan not a Plan?

The second case picks up the theme of the plan as an expression of particular political and administrative structure. It concerns the form of agreements that may be reached by the technical services of the state, the *Directions Départementales de l'Équipement* (DDE), and rural communes. Again, some description is necessary to make the interpretation which follows intelligible.

When the decentralization of planning powers to local authorities took place, it was made dependent, as we have seen, on the existence of the POS. There was, apart from the entitlement to exercise new power, a further incentive to local authorities to prepare, or have prepared, a plan, in that where there was no plan in force, development was limited by national regulation to the existing built-up areas (*parties actuellement urbanisées*). There were, however, two difficulties. The first was that the coverage of the country by local plans was distinctly limited, particularly in rural areas. When decentralization came into force only 6443 POS had been published, most of which covered a single commune, for the 36 433 communes of the whole country. The task that confronted the traditional sources of plan-making expertise, the DDE, was enormous and there were those who doubted whether decentralization could ever work (e.g. Wilson, 1988). The second was that small communes believed that the detailed zoning and regulations of the POS were too detailed for their needs, or perhaps too constraining on their activities, or that the new powers were an unwelcome burden of responsibility. At all events, in rural areas there appeared to be some considerable reluctance on the part of the communes to embark on plan-making.

Apart from the general constraint on development outside built-up areas of communes without a plan, the criteria for decision-making are contained in the *Code de l'Urbanisme*. Even before decentralization, there had been the possibility of communes entering agreements with the DDE as to how they as decision-makers would apply the national regulations in the particular commune. In 1986, the law was changed so that these agreements, known as MARNU (Modalités d'Application de la Réglementation Nationale Urbaine; agreement on the application of the national urban regulations), need no longer be contingent on starting work on a POS. As a result, brief written documents itemizing the nature of the agreement and the clauses of the code to be applied, coupled with what looks to all intents and purposes like a simplified form of local landuse plan, began to appear in rural areas. The ministry issued guidance as to how these documents might be prepared.

My interest in these agreements was based on two alternative hypotheses: either that they represented a step backwards from decentralization and the intentions of the Mitterrand reforms, or that they were a pragmatic response to the needs of the smaller communes. The strength of the first hypothesis was fairly clear: no plan, no decentralized powers. A commune would remain dependent on the state services. The strength of the second was that the form of the agreement document was not, unlike the POS, laid down by statute. There therefore appeared to be a real possibility of creating a planning document that genuinely reflected local needs. The attitude of a director of a DDE was encouraging (Brevan, 1988). Given that the overwhelming majority of communes who had decentralized powers or who had thought to acquire them by preparing a POS have relied on DDE for technical help, the simplicity and flexibility of the MARNU appeared a potential gain.

To test these hypothesis, an examination was made of the progress in preparation of these MARNU in two *départements*: one almost the archetype of rural France, the other much more urbanized but with nevertheless a large rural hinterland. The study was too small scale to reach firm conclusions, but the view was presented that the potential for MARNU as a planning document was considerable, although innovation could well be hampered by the dead hand of conservative DDE. This conclusion unfortunately has proved to be based on precisely the view of the primacy of planning documents that, at the outset of this paper, I suggested was flawed. The necessity for reconsidering these conclusions comes from the fact that even though MARNU might contain a plan with simplified zoning, for some French they could not be considered as a planning document.

The reasons for this attitude lie at the heart of what the French regard plans to be for. A MARNU offers no legal guarantees to residents of a commune (it is not, specifically, opposable in law by third parties). Nor does the agreement to apply certain clauses of the national regulations in certain zones preclude their application elsewhere. No legal certainty—therefore not a plan (although this was precisely the kind of open-ended discretionary plan that as a British planner I could understand). My culturally determined attitudes to planning had led me astray. To understand the likely impact of the change in the law, I would have to return to considering the roles of the actors and their interaction with each other. The apparently technical enthusiasm of Brevan for a new planning instrument was without doubt influenced by her own position within the system. The MARNU would ensure a continued, if not a strengthened, role for the DDE.

3

226   *Philip Booth*

4. The Consultation Préalable: Advice and Consent

The third case concerns a procedure which is wholly atypical of normal French planning practice and may even be unique. The commune of Lyon itself has had a devolved power to handle planning applications since before decentralization by virtue of legislation for large towns. The commune had in fact chosen to entrust the processing of applications to a special administrative unit within the urban community technical services, but still nevertheless relied on the DDE, the traditional processing authority, to prepare the paperwork and calculate the taxes payable on development. The commune also relied on the *agence d'urbanisme* for advice on the planning and three-dimensional aspects of planning applications. The procedure that had been adopted by the commune was to hold a monthly consultation meeting for architects and developers of major schemes in the city, which had either been submitted for planning approval or were likely to be submitted some time in the future. Here they could present their schemes to a panel chaired jointly by the deputy mayor responsible for planning and the departmental *architecte-conseil* (consultant architect) and at which members of the planning agency, the urban community and the DDE were all present for comment. The outcome of these meetings was minuted advice on the acceptability of the scheme, and the preferred architectural solutions to given problems. The *architecte-conseil*, it must be explained, is a post which exists in every *département*, and exists to give architectural advice on applications to DDE and to communes which would otherwise lack the necessary expertise.

This monthly *consultation préalable* was of interest because it appeared to be a positive move to make decision-making in planning rather more transparent than it usually is, and to consider the implications of a development proposal albeit primarily in architectural terms. There was also an underlying feeling that these meetings represented a move towards a more British way of doing things, in the interaction between technical officers and the responsible elected representative, and in the debating of merits and demerits rather than legalities. Nevertheless, there were some aspects of the process which seemed inexplicable. It was not clear why the deputy mayor and the consultant architect chaired the meeting jointly and added their joint signatures to the minutes. Nor was it clear why a consultant architect was needed at all given the strength of design expertise in the *agence d'urbanisme*. This in turn raised a question about the relationship between the consultant architect and the technical officers.

Part of the explanation proved to be historical. The *consultation préalable* had grown up from the traditional pattern of architectural advice given on planning applications within all *départements*. The presence of the deputy mayor was in fact a later addition to an already formalized process, a result of a growing political interest in development proposals within the city. The fact that it had subsisted in this form, however, suggested a rather more complex set of factors were at work. In the end, it became clear that assumptions could not be made about relationships between technical officers and elected representatives. Neither the planners of the *agence d'urbanisme* nor the administrators of the urban community, for all their longstanding involvement in development control in Lyon, were there as direct employees of the commune, charged with the carrying out of a statutory duty. They might be more correctly seen as agents or perhaps as consultants whose role was essentially contingent on the commune's grace and favour. From the point of view of the commune, the presence of the consultant architect was a reminder that the commune had access to other sources of advice and expertise if it was no longer satisfied with those that it already employed as agents. From the point of view of the technical services, there appeared an equal desire to be seen as truly independent and neutral in the process even when—or particularly because—employment and credibility depends on the goodwill of the elected representatives. This perhaps explains why it serves everyone's interest that the *agence d'urbanisme* is not fully part of the urban community.

There was, however, another explanation. One of the features of the French administration both at local and central levels is the way in which power is personified in particular actors. The mayor represents his commune and must alone accept responsibility within it for upholding the law to a degree that a leader of a British local authority does not. This appears to lead mayors, where they have the ability to do so, to choose several sources of advice in order to avoid being reliant on, and possibly hoist by, any one. The *consultation préalable* is thus to be understood as a sound investment policy on the part of the deputy mayor as well as a way of keeping the troops in order. The meeting represents in microcosm the equilibrium of forces in French administration, which leads to sometimes unwieldy structures.

What the *consultation préalable* was clearly not about was a decision-making process of the kind undertaken by a British planning committee with its discretionary power to consider other material considerations and its technical support service there to uphold local policy and to act as guardians of the elected representatives' political conciousness. The consensus which appeared to emerge from the meetings was on one level an entirely appropriate attempt to find the best technical solution to the series of development problems being presented. But on another level, it was about the sharing of responsibility and the resolution of tensions between the variety of technical services employed and the political decision-maker. The *consultation préalable* can also be seen as a way of affirming the assertion, frequently made, that all the services worked well together. It is therefore part window dressing, part a genuine means of reducing conflict, part a means of maintaining established positions and part a useful way of finding solutions. If it introduces an important element of openness in decision-making in Lyon, it also serves to bolster entrenched positions.

5. Conclusions

What these case studies do is require us to look at the planning process as not only a technical one but one in which cultural values may be of greater significance. In the case of Transports Griset, that cultural understanding appears to have had a major impact on the outcome. In the case of MARNU, there was a perception of the nature and purpose of plans which would frame the formulation of research. In the case of the *consultation préalable*, we are dealing with a procedure that can only be understood if the perceptions of the participants of their own role and of each other's is seen clearly within a cultural context. In each case, the impact of culturally derived perceptions on outcome is of considerable importance. For planners, observing attitudes, reactions and behaviours is not just some kind of anthropological game, because the attitudes, reactions and behaviours have a direct bearing on matters of landuse allocation and built form.

The methodological consequences of this approach are considerable. First, it implies that the kind of comparative studies which attempt to make direct comparison between types of planning procedures or policies run the risk of foundering on basic misapprehensions. It suggests that the one-way study of a foreign country is likely to be more rewarding, provided that the framework for research is established within the terms of the country to be studied. This does not rule out comparisons at a later stage, but these would probably be concerned with fundamental issues such as accountability or access to decision-making. They would not be about whether zoning is better than indicative planning or whether regulations are preferable to discretionary control.

Secondly, it implies a thorough grounding in the culture of the country being studied. Here, there is a real problem of what this will mean in practice. It could be argued that a knowledge of language is critical to success, because the use of language is often a vital clue to underlying attitudes. Beyond that, however, there is still a problem; the meaning of culture, even as defined in this paper, is so broad that it is difficult to know how to start. But if the questions to which, as planners, we seek answers involve how decisions are taken, or even how

228   *Philip Booth*

solutions to problems are derived, we shall also certainly need to become aware of the attitudes to decision-making, to authority, to political accountability and the relationship between professional and political power. In the French system, these are deeply rooted in history, particularly of the past 200 years. Getting to the heart not just of administrative structures but of their rationale will be of the greatest importance. Dealing with circumstantial evidence from literature, and from newspapers, radio and television would help to develop the more narrowly focused study of culture that a particular project requires. Constant observation is also always important. There is of course no guarantee that in the interpretation of what we observe we shall not be wrong. What is clear, however, is that the process of absorption of culture is not a quick one, even if we become adept at framing the right questions, and we will need to beware of the instant judgement.

The third implication is that we shall need to be acutely aware of our own involvement in the research process. We shall come to cross-national research fully armed with our own culturally determined assumptions and prejudices. There is the real danger of judging another system of planning by criteria derived from, for example, Britain and proving that the other system fails because in effect it is not British. We are engaged, therefore, because our background is not neutral. We are engaged also, however, in the way in which those we observe regard us. In some senses, to be an outsider is to have a salient advantage in not being committed to the pattern of power that involves those we talk to. But there is the real disadvantage of inadvertently becoming aligned with a particular position in the system and of finding information being presented in certain ways or even being withheld because of that perceived alignment. How we, as researchers, are perceived may be an important clue to the interpretation of the information we gather.

Cross-national research at its best is a fascinating and rewarding field because it forces us to enquire not only about other systems of planning but also in the end about our own. We are not only learning about other countries, we are also deepening the understanding of the nature and purpose of landuse planning.

Acknowledgements

The case study research referred to was made possible by grants under the ESRC/CNRS Exchange Award Programme.

References

AGENCE D'URBANISME DE LA COMMUNAUTÉ URBAINE DE LYON (1982) *Plan d'Occupation des sols Secteur sud-ouest.* Lyon: Agence d'urbanisme de la Communauté urbaine de Lyon.

AGNEW, J., MERCER, J. and SOPHER, D. (1985) *The City in Cultural Context.* Boston: Allen and Unwin.

BIRNBAUM, P. (1979) Office holders in the local politics of the French Fifth Republic, in: J. LAGROYE and V. WRIGHT (Eds) *Local Government in Britain and France.* London: Allen and Unwin.

BOOTH, P. (1989a) *Rural development control in France: Agreements between state and commune in Rhône and Haute-Loire.* Working Paper TRP 83, Department of Town and Regional Planning, University of Sheffield.

BOOTH, P. (1989b) *Rules, discretion and local responsibility: Development control case studies in the urban community of Lyon.* PhD thesis, University of Sheffield.

BOOTH, P. (1989c) How effective is zoning in the control of development?, *Environment and Planning B* 16, pp. 401–415.

BOOTH, P. (1991a) The theory and practice of French development control, in: J. DOAK and V. NADIN (Eds) *Town Planning's Response to City Change.* Aldershot: Gower.

BOOTH, P. (1991b) From commune to communauté urbaine: the legacy of the revolution in the planning of Lyon, in: D. WILLIAMS (Ed.) *1989: The Long and the Short of It.* Sheffield: Sheffield Academic Press.

The Cultural Dimension in Planning   229

BREVAN, C. (1988) Environnement et modalités d'application des règles générales d'urbanisme, *La Revue Juridique du Centre-Ouest*, 1, pp. 186–189.

COMBY, J. (1989) L'impossible propriété absolue, in: ASSOCIATION DES ÉTUDES FONCIÈRES (Ed.) *Un Droit Inviolable et Sacré*. Paris: Association des Études Foncières.

DAVIES, H.W.E. (1980) *International transfer and the inner city*. Occasional Paper 5, School of Planning Studies, University of Reading.

DAVIES, H.W.E., EDWARDS, D. HOOPER, A.J. and PUNTER, J.V. (1989) *Planning Control in Western Europe*. London: HMSO.

DUPUY, F. (1985) The politico-administrative system of the département in France, in: Y. MÉNY and V. WRIGHT (Eds) *Centre–Periphery Relations in Western Europe*. London: Allen and Unwin.

DUPUY, F. and THOENIG, J.-C. (1984) *L'Administration en Miettes*. Paris: Fayard.

GEERTZ, C. (1973) *The Interpretation of Cultures*. New York: Basic Books.

GUNSTEREN, H. VAN (1976) *The Quest for Control: A Critique the Rational-Central-Rule Approach in Public Affairs*. London: John Wiley & Sons.

HANTRAIS, L., MANGEN, S. and O'BRIEN, M. (1985) *Doing cross-national research*. Cross-national Research Papers 1, Aston Modern Languages Club, Aston University, Birmingham.

HILDENHEIMER, A. (1986) Politics, policy and policey as concepts in English and continental languages: an attempt to explain divergences, *The Review of Politics*, 48(1), pp. 3–30.

LISLE, E. (1985) Validation in the social sciences by international comparison, in: L. HANTRAIS, S. MANGEN and M. O'BRIEN (Eds) *Doing cross-national research*. Cross-national Research Papers 1, Modern Languages Club, Aston University, Birmingham.

MACRORY, R. and LAFONTAINE, M. (1982) *Public Inquiry and Enquête Publique*. London: Environmental Data Services.

MASSER, I. (1986) Some methodological considerations, in: I. MASSER and R. WILLIAMS (Eds) *Learning from Other Countries*. Norwich: Geo Books.

THOMAS, D., MINETT, J., HOPKINS, S., HAMNETT, S., FALUDI, A. and BARRELL, D. (1983) *Flexibility and Commitment in Planning*. The Hague: Martinus Nijhoff.

WAKEFORD, R. (1990) *American Development Control: Parallels and Paradoxes from an English Perspective*. London: HMSO.

WILLIAMS, R. (1986) Translating theory into practice, in: I. MASSER and R. WILLIAMS (Eds) *Learning from Other Countries*. Norwich: Geo Books.

WILSON, I. (1988) *French land use planning in the Fifth Republic*. Nijmeegse Planologische Cahiers 27, Geografisch en Planologisch Instituut, Katholiek Universiteit Nijmegen.

Originaltext Faludi 1999

3

Patterns of Doctrinal Development

Andreas Faludi

An analysis of the positive Dutch experience in spatial planning—Euro-English for "... setting frameworks and principles to guide the location of development and physical infrastructure" (Healey 1997a, 4)—suggests that there is an underlying planning *doctrine*. In this paper, after presenting the concept of planning doctrine, I discuss predoctrinal situations where a doctrine, even though within reach, cannot be said to exist. At the other end of the scale, I discuss a situation where doctrine is mature. This allows us to infer patterns of doctrinal development. Insight into such patterns gives us an appreciation of the situation of planning, including the potential for change.

ABSTRACT

Planning doctrine refers to a conceptual scheme giving coherence to planning by means of conceptualizing an area's shape, developmental challenges, and ways of handling them. The focus of this article is on the dynamics of such a doctrine. Drawing on Thomas Kuhn, *doctrine* implies a phase model, from a pre-doctrinal to a doctrinal situation and on to one where revolution threatens. Four pre-doctrinal situations are discussed: the Flemish Structure Plan, the Second Outline Structural Plan for the Benelux, European Union planning, and Florida's state growth management. Conceiving of them as *pre-doctrinal* focuses on how, in the fullness of time, they may evolve. However, what about situations where, like in the Netherlands, doctrine is mature? In those situations, a 'Laudanian Model' of evolutionary change (after Larry Laudan, a critic of Kuhn) opens up new perspectives.

■ THE CONCEPT OF PLANNING DOCTRINE

Planning doctrine refers to "... a body of thought concerning (a) spatial arrangements within an area; (b) the development of that area; (c) the way both are to be handled" (Faludi and Valk 1994, 18). The success of Dutch planning is said to be due to such a doctrine. As a concept, planning doctrine has been used before in the North American planning literature to explain examples where planning has experienced unusual exposure and acclaim. The best example is the analysis by Foley of postwar London. He discerns in the London plans "... the influence of certain ideas, the crystallization of a reinforcing web of these ideas into doctrine, and the need to review this doctrine in the light of ongoing experience" (Foley 1963, viii). As is well known, the plan proposed to check the growth of London. "The fashioning of this web ... was creative social intervention of high order. Once formed, it became the basis of social policy—and, in more subtle fashion, the foundation of a highly influential body of doctrine providing a durable nucleus for political and professional consensus in town and regional planning circles and among the members of an informed elite more generally...." (4-5).

Foley's source of inspiration is Selznik's analysis of the experience at the Tennessee Valley Authority, where the latter relates doctrine to leadership. One of the responsibilities of leadership is "... to develop a *Weltanschauung*, a general view of the organization's position and role among its contemporaries" (Selznik 1953, 47).

The concept of doctrine is inherently dynamic. The often vague and contradictory nature of the ideas embodied in it can explain why some achievements, once attained, come under threat. Thus, in his classic analysis of British town planning, Foley (1960, 89ff) focuses on the strain of ideological inconsistency. Friedmann and Weaver focus most clearly on the developmental pattern of what they call regional planning doctrine. They explain that a historical approach "... will help to throw regional planning doctrine into relief by revealing its origins, the options that were received, the influence of circumstantial events, alternative formulations that were neglected, and the new forms of doctrine that are beginning to emerge"

Andreas Faludi is a professor of spatial planning at the University of Nijmegen, the Netherlands; a.faludi@net.HCC.nl.

Journal of Planning Education and Research 18:333-344.

334   *Faludi*

(Friedmann and Weaver 1979, 2). Without making the analogy with the work of Kuhn (1970) explicit, they refer to the emergence of new streams of development doctrine as a "crisis" and a "paradigm shift" (161ff).

All this relates to an idea of good currency in the literature: framing. Framing is "… a way of selecting, organizing, interpreting and making sense of a complex reality so as to provide guideposts for knowing, analyzing, persuading and acting. A frame is a perspective from which an amorphous, ill-defined problematic situation can be made sense of and acted upon" (Rein and Schön 1986, 4). Without such a frame, or doctrine, spatial planning cannot be anything but haphazard.

Spatial planning doctrine must incorporate a spatial organization principle expressed by means of an abstract representation of the plan area, together with directions for its development. Successful images rest on powerful metaphors. Thus, one of the most pronounced concepts invoked in the Greater London Plan, the Green Belt, connotes containment of urban growth threatening a cherished middle-class way of life. The equivalent Dutch concept of a Green Heart in the core of the Dutch metropolis called Randstad, or rim city, rests on an organic metaphor—the country as a body, with its well-being depending on the continuing health of its core. Preserving it continues to be the rationale for much of Dutch planning (Valk and Faludi 1997).

Doctrine presupposes an agency with responsibility for the plan area. In addition, the agency must view its jurisdiction, or part thereof, as the legitimate object of planning. Lastly, doctrine refers to durable concepts, not to the vanity of the day.

The notion thrives on the recognition that planning ideas are the outcomes of discussions, negotiations and conflict resolution involving an element of rhetoric. This bears resemblance to *The Argumentative Turn in Policy Analysis and Planning* (Fischer and Forester 1993). More generally speaking, it relates to the view of planning as communication (Sager 1994), and as a collaborative exercise (Healey 1997b). Putting store in coalition formation, the notion of doctrine also resembles the work of Hajer (1989, 1995) on the formation of discourse coalitions.

The concept of doctrine has been the object of extensive discussions (Alexander 1996). Like that of a paradigm, doctrine implies a phase model, from a pre-doctrinal to a doctrinal situation and beyond, to one where a revolution removes doctrine and its standard bearers from the scene (see Faludi 1987, 130-131; Faludi and Valk 1994, 23; Alexander and Faludi 1996, 36-48). This paper is about this phase model. The paper starts out with pre-doctrinal situations, by far the most prevalent. Identifying them as pre-doctrinal indicates directions in which they could and, at least according to the inherent logic of the phase model, should develop.

■ PRE-DOCTRINAL SITUATIONS

Pre-doctrinal situations inevitably include uncertainty as to the way in which planning should be conceptualized.

Four such situations will be discussed: the Flemish Structure Plan, the Second Outline Structural Plan for Benelux, European Union planning, and Florida's state growth management.

The Structure Plan for Flanders

Flanders is one of the three regions into which Belgium has been "federalized" (Vries and Broeck 1997, 60-61). Being in the core of the European heartland poses challenges and offers opportunities to Flanders.

The area of Flanders is 13,500 square kilometers (5,400 square miles) and the population almost 6 million, making for an unusually high density. The principal language is Dutch, but Flanders shares in the individualistic Belgian culture (Vries and Broeck 1997, 61). Although the region has planning concerns with ribbon development and fragmentation of open space (Albrechts 1998), a new Structure Plan may be indicative of Flanders taking advantage of its position at the crossroads of Latin, Germanic, and Anglo-Saxon cultures.

Federalization has culminated in a 1992 accord on a Flemish Council with its own funding and powers. With the formation of Flanders having reminiscences of struggles for independence, the making of a plan may signify the taking of possession of the land. However, Loeckx (1995, 49) and Albrechts (1997) put more emphasis on the urgent problems that the Structure Plan must address. Hands-off attitudes of the past have allowed the built-up area to expand by no less that 41 percent from 1980 to 1992, the year when work on the Structural Plan started (Albrechts 1995a, 29). Such growth puts pressure on open space. In a comparative analysis with Dutch planning, Hoogerbrugge (1995) explores whether there is a Flemish doctrine in the making.

Two university professors were commissioned to prepare the Structure Plan. One of them, Albrechts (1995a, 30; 1997), describes the philosophy behind it as that of sustainable development. Sustainable development depends on carrying capacity and the quality of space. Carrying capacity is defined as the maximum extent to which locations may be used without damage. The quality of space refers to its potential for various uses. This translates into a vision of "Flanders open and urban."

> Actual spatial reality does not answer to this vision, nor does its present evolution, and this even less so, if we take the terms 'open and urban' to mean two totally different types of space. The spatial distribution of functions and activities is rather such that these two types of space are strongly intertwined (Albrechts 1995a, 30; my translation).

"Flanders open and urban" translates into a policy of deconcentrated clustering. Under this policy of clustering new development near established settlements, development may continue to spread on a macro-scale. However, it ought

3

to be planned so that open space remains, as much as possible, intact. In addition, the plan:

> ... tries to establish structuring principles which will be capable of imposing some order on the existing chaos. There is a strong emphasis on linear infrastructure and mobility ... as structuring frameworks. An advanced network of strengthened urban areas, linked by a group of linear infrastructures ... will ensure that the Flanders nebula once more has clear outlines (Loeckx 1995, 51).

The notion of infrastructure extends to nature areas: "It is river valleys rather than any conceivable regulations that currently provide some articulations, with chaos prevailing in the periphery. The plan aims to make one coherent, structuring framework from the river valleys together with linked open areas and corridors of open space" (Albrechts 1998, 417). Containment is a key to the plan. Albrechts expresses hope that the structuring principles will help also in preventing what he calls *Balkanization*, or the dominance of local perspectives over any overall framework. At the same time, the plan takes on lessons from the strategic planning literature. These lessons point to planning as a collaborative process that engages the power structures, is normative and innovative, and combines top-down and bottom-up approaches (Albrechts 1997, 1998)

An area receiving much attention is the urban network called the Flemish Diamond, the corners of which are marked by the cities of Antwerp, Leuven, the Brussels agglomeration (Brussels itself falling outside the jurisdiction of Flanders, with no effective coordination between the two) and Ghent (Figure 1). The Flemish Diamond contains three metropolitan areas with an overall population of almost 4

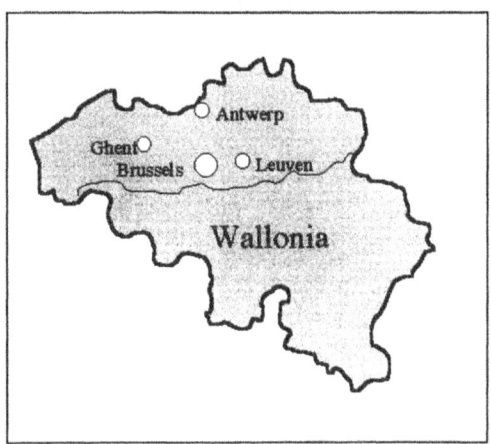

Figure 1. The Flemish Diamond.

million. It has good infrastructure and educational facilities and, above all, two harbor complexes and an international airport that the Structure Plan identifies as gateways.

Albrechts (1998) takes account of the naming-and-framing aspect of designating the area as the *Flemish Diamond*. Writing in terms like those employed in this paper, he claims that this metaphor is capable of decoding its potentials. Elsewhere, he describes the Structure Plan as the first step toward creating a Flemish planning culture (Albrechts 1994, 173).

Commenting on the Flemish plan, Zonneveld (1995) is complimentary toward the Structural Plan's conceptualization of Flemish space, praising the use of metaphors. To him, however, "Flanders open and urban" is less convincing than the concepts: *link* for infrastructure, *engine* for the port functions, and *spine* for the structure formed by nature areas. Zonneveld shows that the Structure Plan invokes a clear spatial organization principle without sacrificing complexity, consisting as it does of overlapping systems, like infrastructure corridors, ecological zones, and open space.

Political acceptance will, of course, be the key issue. In this respect, a controversy surrounding the Nature Policy Plan for Flanders has perhaps been an ominous sign. The Nature Plan incorporated the "Green Outline-Structure" of the Structural Plan. Frustrated by the General Agreement on Tariffs and Trade (GATT) threatening lower prices and by policies to control manure output (Kuijken 1994, 29), agriculturists were incensed by the prospect of more restrictions and opposed both the Nature Policy Plan and the Structure Plan. This was unfortunate, since the makers had put great store in broad support as a precondition of planning effectiveness.

Cabus (1994, 104) expresses doubts over whether there has been genuine dialogue. Loeckx (1995) is more optimistic:

> The whole iterative process of negotiations with the administrative sectors, the countless discussions and information sessions and the twenty preliminary studies have, so to speak, re-established physical planning in Flanders. The documents still contain too many bland compromises and sector dogmas, but that is probably the price that had to be paid for safeguarding the possibility of carrying out a number of radical trend reversals (53; my translation).

The draft plan foresaw 44 core decisions, some of them quite detailed and technical. However, as a result of the furor around the Nature Policy Plan, the government was careful not to make too much of this. The powerful political advisers to the planning minister took some of the sting out of the discussion by swapping the notion of *open areas* with the, in the eyes of agriculturists, less objectionable *outlying areas*. Given that the substance remained unchanged, this may seem trivial. However, it is one of my central assumptions here that rhetoric is important.

Concurrently with the provisional approval of the Structure Plan in mid-1996, new legislation was passed in July 1996. Also, the planning team is being strengthened, bringing the complement up from a pitiful seven to almost 20. Finally, provinces have now embarked on preparing structure plans, and their planning staffs are being expanded, too.

What does this all add up to? Hoogerbrugge (1995, 59-60) doubts whether there is a planning doctrine emerging. However, the intention of key actors is clearly that of formulating a discourse about the new Flemish situation. There are other indications of a pre-doctrinal situation. Unlike Belgium before federalization, Flanders considers planning as an element of that new discourse (Delmartino and Soeters 1994, 268). The actors involved are using rhetorical means. Going by the testimony of one of those commissioned with the work (Albrechts 1997), they are sensitive also to the peculiarities of the Flemish political setup. Meticulous in the care they take to be accessible to all, but subservient to none of the interested parties, they appear to have become a moral force. Around the appointment of permanent staff, they have won a decisive battle. Thus, the planning community, albeit still small, is growing. This much is certain—planning in Flanders will never be the same again.

The Second Outline Structural Plan for Benelux[1]

In 1994, the planning ministers of the Netherlands, Luxembourg, the Belgian regions of Flanders and Wallonia, and the Brussels Capital Region gave the go-ahead for the Second Structural Outline for the Benelux. In May 1996, it came out in draft form (Union économique Benelux 1996).

Figure 2. The Benelux.

On 8 July 1997, the ministers gave their provisional approval, with an action program still pending (Benelux Economische Unie 1997).

The Benelux is comprised of Belgium, the Netherlands, and Luxembourg (Figure 2). They formed a customs union in 1944 and the Benelux Economic Union in 1958. The Second Benelux Governmental Conference (the decision-making body of the Benelux) brought planning under the umbrella of the Benelux in 1969. The Third Conference gave the go-ahead for an Outline Structural Plan, predecessor to the present document. In 1986, this became the first transnational planning document ever to be adopted in Europe and possibly in the world. In the early 1990s, though, it was felt to be out of date and to carry too little commitment, and so a new venture started, with European Union support.

The constraints are only too evident. The differences within the Benelux include those of language, culture, administration, and geography. There are five languages spoken among the three countries; administrative setups diverge, and cultural differences are pronounced. The territory is diverse with many international borders crisscrossing it. Combining sparsely populated and highly dynamic areas, a central issue in Europe—cohesion—is also manifest. To mention another European issue, the sustainability of this area—overall one of the most densely populated in Europe—is equally at stake.

Planning issues are shaped by each Benelux partner's perception of its position. The Netherlands and Flanders view themselves as being on the seaboard. Wallonia and Luxembourg have relatively sparsely populated areas. Wallonia also includes part of the belt of heavy industry stretching south into France and in dire need of restructuring. On top of this diversity, the Netherlands, Flanders, and Wallonia are in themselves highly differentiated, with longstanding memories of regional autonomy.

An added complication is that, as indicated, Belgian regions have full planning powers. In fact, the three regions, and not Belgium as such, participated in the making of the new Structural Outline. This partnership sought to attain prior agreement on how the Second Structural Outline was supposed to work. Memories of the notorious case of a logistic park in an area designated as a cross-border green belt in the original plan led to the demand that the new document have binding force. In the end, this turned out to be a bridge too far. The Benelux Treaty does not cover planning. Renegotiating the treaty was not possible at the time, so what would the adoption of this Outline Structural Plan amount to? The answer was for the plan to generate commitment, so that its key elements, labeled essential decisions, would be taken on voluntarily. In this way, the Outline Structural Plan would form a common frame of reference throughout the Benelux. As a preliminary document puts it:

> The Ministers of Town and Country Planning of
> the countries or regions undertake to transpose

decisions defined as essential in the Second Outline Structural Plan for the Benelux, into their national or regional plans and their national or regional policies. In case of a derogation considered important by one or several Benelux partners, they agree to proceed by means of a partial review of the Outline (Benelux Economische Unie 1994, 65).

Deliberations on the status and scope of the plan took three years, and the work itself two more. Predictably, once the draft was out, there was controversy over which decisions to deem essential. Some reckoned only decisions entailing hard commitments deserved this designation. Others were afraid that this would diminish the weight of the Outline Structural Plan and of the vision of the intended spatial structure that it articulates. The Dutch, for one, were unable to square the essential decisions with their own Planning Key Decisions, which are adopted by parliament after public consultation. Eventually, essential decisions became policy options, representing joint recommendations by the planning ministers to the governments of the Benelux partners.

This relates to the issue of how strategic plans work as what they are intended to be: frameworks for decision making. Whilst this is a general problem, it is particularly acute in transnational planning. The intention was to produce a document that, following the subsidiarity principle (much talked about in Europe these days), had added value, over and above the plans of the various partners. This implied that, rather than being comprehensive, the Outline had to address issues selectively.

The first issue was mobility as it was affected by location policy. Thus, it was not location policy as such that was at issue, but its effect on overall mobility patterns, more specifically car usage. After all, location policy as such was the responsibility of each of the five partners. The second issue was primary infrastructure, on the strength of the argument that the Benelux infrastructure has cross-border effects and is thus a matter of transnational concern. The third issue was rural areas, a field of policy with room for mutual learning and where coordination was expected to mutually enhance the policies of the Benelux partners. In addition, this theme included the main ecological structure with its common-good properties. Above all, the Structural Outline Plan was trying to give " ... a clear, overall concept of the possibilities for reinforcing the position of the Benelux-region in a broader context, and of coherence within the region" (Benelux Economische Unie 1994, 64). As the document explains:

The existence of a clear planning vision means that a well-supported perspective of spatial development for Benelux must be developed. This development perspective lays down the guidelines for the overall ... structure of the Benelux, and the relations with the main regions located on the external borders. "Well supported" means that the development perspective for the Benelux is based to a large extent on spatial perspectives drawn up on national and regional level .... (Benelux Economische Unie 1994, 67).

Will this plan carry weight? Is there a Benelux planning community to carry the idea forward? No more than a handful of planners, working part-time, prepared the plan, with headquarters staff providing secretarial support and a steering group, representing the five partners, providing guidance. Fortuitously, the scores of people on numerous cross-border commissions were kept informed, and the planners were drawn from throughout the Benelux. The National Spatial Planning Commission, made up of high-level Dutch civil servants from virtually every ministry, was also consulted. While there was an effort to generate commitment, awareness of Benelux planning is minimal outside this network. Only the future can tell whether the Structural Plan will have the intended impact.

The planners involved were only too aware of the problem.

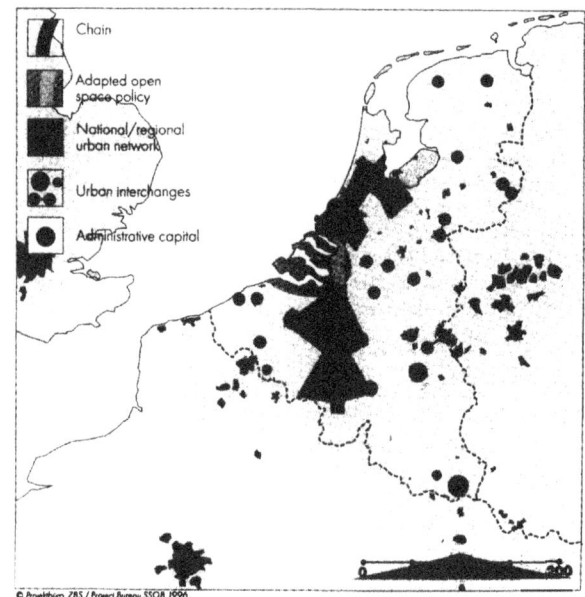

Figure 3. The North-South Chain.

The Structural Plan is designed not to superimpose itself on existing plans, but to supplement them. The idea is for the plan to have added value, over and above national and regional planning documents. This value needs to come from concepts putting Benelux into its wider context, such as a north-south chain of urban networks linking the Dutch so-called Randstad to the Flemish Diamond, discussed above, and stretching further into Wallonia (Figure 3). Unfortunately, alongside the essential decisions, this idea disappeared when it came to adopting the Outline Structural Plan. A concept that was retained was that of a study area called Benelux+, extending beyond the boundaries of Benelux.

So does the Benelux move toward a planning doctrine? In its favor, there is recognition of the role of shared views regarding the positioning of the Benelux in European space, as well as of the inevitable role of rhetoric in exercises such as this. This extended to the use of spatial imagery capable of transcending cultural and linguistic boundaries. Prior to the work on the plan, care was also taken to form a team, the members of which had the ear of the five national and regional planning agencies.

However, we must hold against this the weak setup characteristic of transnational organizations. The appeal of the spatial concepts would have to be great indeed in order to transgress divisions, cultural and otherwise, among the five partners. Of course, in this respect, as in so many others, the future is open. Such difficulties are even greater in the case of the European Union.

European Union Planning[2]

The European Union (EU) with its 15 members is the successor to the European Economic Community established under the Treaty of Rome in 1958. Its policies, such as those for regional and for environmental issues, have spatial impacts. The so-called Common Agricultural Policy has vast spatial implications also. The Trans-European Networks, which include high-speed trains and other services, are perhaps the best way of bringing home to Europeans that their continent is being welded together.

Nevertheless, the EU has been slow in taking on a planning role. European planning is said to be:

> ... implicit, fragmented, uncoordinated and dispersed in many sectoral policies. In this way the technocratic discourse is predominant. All kinds of decisions are cast as mere technical decisions and thereby to a large extent depoliticized.... Important political controversies and substantially controversial developments are reduced to norms, rules, procedures.... This allows existing power groups to make important decisions hiding the question who holds the power. This means that the ordinary citizen is completely absent at important planning decisions (Albrechts 1995b, 13).

The idea of a vast free-trade zone conceives of space merely as an obstacle (Prantl 1994, 17). Even so, it can be argued that European integration is a spatial project (Swyngedouw 1994; Kuklinski 1997; Nijkamp 1995, 33). Indeed, the very notion of, as the saying goes, a level playing field for people, goods, capital, and services is a spatial metaphor.

The shape of the EU (Figure 4) adds to the necessity of spatial planning. In Europe, islands and peninsulas and mountain ranges make differentiation inevitable (Haller 1994, 366). Agglomeration economies accrue in the core, whereas the periphery cannot take full advantage of economies of scale afforded by integration. So integration needs a helping hand in improving spatial relations by overcoming distance and/or physical barriers in the interest of European cohesion.

A European response has been in the air for some time. There were German proposals for a council of European planning ministers in 1964. The Dutch joined the quest, proposing that the then European Economic Community takes on a planning role (Klerx 1995, 46-47). In 1983, the Council of Europe adopted a European Regional/Spatial Planning Charter (Williams 1996, 77-80), and the European Parliament, too, passed resolutions proposing a form of planning.

In the late 1980s, Germany and the Netherlands once again took initiatives. The Dutch minister planted the idea of convening an informal meeting of planning ministers in the mind of his French counterpart, who subsequently hosted such a meeting in 1989 (Klerx 1995, 124-125). When they themselves played host to the ministers, they saw to it that a Committee on Spatial Development was set up with top planners from member states to prepare future meetings.

Meanwhile, the European Commission published *Europe*

Figure 4. *The shape of the European Union.*

2000 (Commission of the European Communities 1991). The overt reason behind *Europe 2000* was to enhance the Commission's ability to assess funding applications. Behind this was a French initiative for the European Commission to engage, as the French do, in spatial planning to underpin its regional policy (Klerx 1995, 122-125).

After *Europe 2000* came *Europe 2000+* (Commission of the European Communities 1995), incorporating the outcomes of a series of trans-regional studies. These efforts notwithstanding, European sector policies continue to dominate. What is still absent is spatial coordination. Thus, there is no plan defining the meaning of European policies for the territory of the European Union, nothing Member States can refer to in positioning themselves in European space.

In terms of European policy-making, the meetings of planning ministers represent an intergovernmental approach: representatives of member states formulating joint policy. This contrasts with the *communautarian* approach taken in matters within the *acquis communautaire*, European regulations transcending national sovereignty. Under certain conditions, the European Commission can assume powers not explicitly mentioned in the treaties. Environmental policy came about in this manner, before it became formalized in 1987 (Majone 1996a, 68-69). The dynamism of the *acquis communautaire* has led Folkers (1995) to exclaim: "The EU is essentially designed for permanent change of its institutions in the direction of more economic and political integration" (88; my translation).

Fearing quite rightly that *Europe 2000* might be the prelude to the European Commission assuming a planning role, the German Parliament proposed that member states (and thus not the European Commission) prepare a joint European Spatial Development Perspective (Krautzberger and Selke 1996, 45-46). Under the German presidency in 1994, the so-called Leipzig principles (Informal Council of Spatial Planning Ministers 1994) were adopted as a starting point. Another German initiative concerned the 1996-1997 revision of the Treaty on European Union. The proposal was to make it mandatory upon the European Commission to coordinate its policies on the basis of such a European Spatial Development Perspective (Faludi 1997).

There are two reasons for these initiatives. Firstly, whilst German support for European integration is an article of faith enshrined in the constitution (Rometsch 1996, 61), federalist Germany is weary of French-style centralism, said to pervade the European Commission. A planning document prepared *bottom-up* suits German predilections. It also gives the German states, sharing as they always do in the preparation of German national policy, a voice in European affairs (Krautzberger and Selke 1996, 68-69). Secondly, like others in northwest Europe, the Germans put great store by a spatial plan conceptualizing the territory as a whole (Krautzberger and Selke 1996, 22). German proposals for Europe are aimed at such a form of planning. The English term used is spatial planning policy,

defined as "... that component of political activity which is both dedicated to creating a condition which conforms with the general orientation and to the sustainable development of the area, and which is intended to resolve spatial problems." Such planning is said to be different from French *aménagement du territoire*. In Germany:

> ... the task is more broadly based. In addition to the idea of the utilization of space, the German term also includes the notion of balance (i.e., coordination) and is not geared solely to the economic function of space. In France, on the other hand, although the term has a long history ..., it describes a somewhat narrower approach dominated by actions to promote economic utilisation of space which almost exclusively characterise the current understanding of the term 'aménagement du territoire' (Malchus et al. 1996, 9).

So, the Germans would not only like to see a bottom-up approach but also a European Spatial Development Perspective going beyond regional considerations to a balanced view of spatial claims.

In 1993, the council of planning ministers called upon the Committee on Spatial Development to prepare, just as the German Parliament had proposed, a European Spatial Development Perspective. Under the authority of the Committee—and with the active support of the European Commission—this project has since taken shape. The first official draft of the *European Spatial Development Perspective* (Commission of the European Communities 1997) was submitted under the Dutch presidency in June 1997, with work to be completed under the German presidency in 1999.

The current draft has four parts. Part I outlines the approach, providing an analysis of urban structures, rural areas, access to infrastructure, access to knowledge, natural heritage, and cultural heritage. Part II focuses on European dimensions of spatial issues and the impact of Community policies on the European territory: the Common Agricultural Policy, the Structural Funds and the Cohesion Fund, the Trans-European Networks, and the environmental policy. On this basis, the document formulates policy aims and options for the entire European territory. It makes reference to the three principles adopted under the German presidency: a more balanced and polycentric system of cities and a new urban-rural relationship; parity of access to infrastructure and knowledge; and prudent management and development of Europe's natural and cultural heritage. All this is supposed to culminate in a framework for integrated spatial policy. Part IV is on "Carrying Out the European Spatial Development Perspective."

The framework in Part III is no more than an agenda for discussion. Although the essence of an integrated spatial approach lies in the combination of selected options into coherent spatial strategies, this geographic integration is left to transnational initiatives, already under way, under the so-called

INTERREG II-C program (each involving at least three states). So the European Spatial Development Perspective by itself does not provide an integrated spatial development perspective. In this respect, it is interesting also that there are no maps in Part III, and that the four maps in the appendix are marked as illustrative only (without actually saying what they illustrate). The absence of even an attempt to arrive at an overall conceptualization of European space—the purpose of the excercise, according to some advocates—is disquieting. Maybe this was, indeed, "a bridge too far" (Faludi 1996a). On the plus side, that the 15 member states have come this far in formulating one joint document seems nothing short of miraculous and augurs well for the future.

Much will depend on the emergent pattern of relations between the EU and member states. This continues to be surrounded by uncertainty. Nobody less than former president of the European Commission is said to have described the European Union as an "unidentified political object" (Schmittner 1996, 37). Could the so-called subsidiarity principle be interpreted as giving the European Union a planning mandate? As embodied in article 3-b of the Treaty on European Union, the principle is designed to ensure that the national identity of each member state receives the respect it deserves. However, the principle also stands "... for the need to take decisions at the appropriate spatial level" (Albrechts 1995b, 12-13). It is conceivable, therefore, that under the subsidiarity principle the EU as a whole might be understood as deserving special attention. Alternatively, the European Commission might eliminate planning powers on the strength of the argument that member states use planning as a barrier to competition or that such powers are essential to achieving the objectives of the treaty.

In all this, the concept of Europe as an entity with common problems and a shape that one can grasp (which is what a European doctrine would require) might in itself become an influential force. In this respect, the European Spatial Development Perspective could play a decisive role in generating a European planning discourse. In the fullness of time, such a discourse might change the very notion throughout Europe of whether or not European planning is needed.

So is European planning in a pre-doctrinal situation? The potential for doctrine seems to be there. Northwest-European planners in particular want to engage in spatial coordination that goes beyond doling out support. This requires images: "The only realistic chance European spatial development policies have if they want to influence spatial development processes at the macro scale is the production of spatial images of a sustainable Europe" (Kunzmann 1996, 163). Clearly, though, those who entertain such hopes must battle the institutional setup. As with the Benelux, the problem lies with the planning principles—notions of what a European-level plan would be, how it should be formulated, how Europe will be ultimately effected. There are no examples to draw upon. To paraphrase Delors, European planning, too, is an "unidentified object."

Florida State Growth Management[3]

Rather than doctrine as in the Netherlands, "controlled growth ideology" is evident in a study of Florida's state growth management program (Evers 1997). The question is, can Florida be described as being in a pre-doctrinal stage?

Since the early 1970s, several U.S. states have enacted growth management to mitigate negative effects of urban growth, with Florida a forerunner. For as long as the term *growth management* stands for policy concerned with the location, intensity, and timing of development (Chinitz 1990, 6), the Netherlands is a European leader in growth management.

Evers points out that the settings of the Netherlands and Florida are also similar. Both are predominantly flat coastal regions with a large share of the approximately 15 million residents wedged between the coastal zone and a protected area. Politically, though, Florida is characterized by a competitive two-party system with frequent shifts between Democratic and Republican control, whereas the Dutch political system is stable. Despite boasting a relatively prosperous economy, Florida ranks among the lowest in the U.S. in terms of spending on education, social welfare, and the like. This is due largely to a low-tax climate in Florida, where the state constitution prohibits the levying of income taxes and sets limits on property taxes. The Netherlands is a welfare state with high taxation.

Given this political climate, Florida's local authorities do not receive much in aid from the state, nor do they receive significant amounts of federal aid. So, boosting development has been a common way of filling municipal coffers. This contrasts with a high level of support for Dutch municipalities. This gives much say to the national government: "He who pays the piper calls the tune," as the saying goes.

In Florida, environmental and other public interest groups have little voice at the state level. The court system often remains their only recourse, whereas the Netherlands has a corporatist system with manifold channels of influence.

Against this backdrop, growth management in Florida originated from 1970s environmental legislation. A series of acts allowed for designation of "areas of critical state concern" and "developments of regional impact," including a State Comprehensive Plan Act that provided the groundwork for Florida's renowned 1985 Omnibus Growth Management Act (GMA). There is a three-tier top-down planning system. Municipalities are required to draw up local comprehensive plans and submit them for regional and state approval. These plans are checked for consistency with the State Comprehensive Plan (SCP) and other rules. The state has been strict, rejecting about half of the originally submitted plans. According to Evers, this, coupled with the considerable sanctions imposed for failure to submit plans on time, has served to secure Florida's top-down epithet.

Evers also points out that the austerity by which plans have been evaluated for consistency has sparked resistance and the creation of unimaginative checklist plans merely seeking to satisfy minimum state standards. Admittedly, some flexibility has now been built into the system, and additional funds have been made available for planning.

Evers' conclusion is that Florida's growth management system meets only two of the three criteria for a planning doctrine. There is an agency responsible for planning, and the whole state has become the object of growth management. However, state growth management cannot, as yet, be said to be a durable phenomenon, like Dutch doctrine is. Evers regards this not just as a question of time. As indicated, in lieu of analyzing it as an example of an emergent planning doctrine, Evers describes current planning thought in Florida with a new concept: controlled growth ideology. He says that the adversarial relationship between planning and development interests has prompted a *regulatory/coercive* response. What this term implies is that the force of entropy is so powerful that, unless checked, growth will always be chaotic. The inability to effectively overcome (or incorporate) the influence of private interests provides a further disincentive for planners to develop ideas about the spatial structure. In any event, shifts in party politics impair implementation so much that, at one point, growth management was in danger of being scrapped.

What is notably absent in Florida is the long Dutch tradition of public intervention in land use decisions and the conscious efforts by planners to educate, and indeed (why not?) *indoctrinate,* the public with planning values. The result is that, while Florida's growth management continues to enjoy a high level of public approval, it is uncertain whether it will grow into the established tradition that it already is in the Netherlands.

Evers illustrates this by referring to a specific policy issue, deteriorating infrastructure. State legislation mandated boldly that no future development could occur unless infrastructure had been financially accounted for or that capacity was "concurrent" with development. Since capacities were often already stretched, and since, equally often, local governments did not have the funds to improve upon the situation, the latter had little recourse but to exact impact fees for development. As a result, the concurrency policy boiled down to a pay-as-you-grow approach.

Both local governments and developers were strongly opposed to the concurrency policy. They feared that it would halt or slow growth, which both sectors considered vital (if for different reasons). In fact, the concurrency policy did slow growth in some regions, but at the expense of driving it out to rural areas where infrastructure capacity was still available and where few impact fees were charged. So Evers concludes that state growth management created, as a perverse consequence, more urban sprawl. In addition, since the concurrency policy did not cover social infrastructure like schools, funds were sometimes diverted from these sources toward meeting concurrency requirements for physical infrastructure.

To Evers, this case shows that development interests were largely successful in avoiding the concurrency policy. Moreover, other political forces were able to limit concurrency to only physical and not social infrastructure. Therefore this case illustrates an inherent weakness of the controlled growth ideology. Faced with similar challenges, the planning systems in the Netherlands and Florida have evolved quite differently. Paradoxically, the Netherlands, famous for its high level of planning, employs more persuasion and coordination tactics, while Florida relies more on regulation and legal coercion.

Perhaps the most important difference between the two is that Dutch planning doctrine has an important *spatial* dimension to it. The National Planning Reports deploy a profusion of maps and visual devices, not merely as references, but as crucial elements in policy. In contrast, Florida's controlled growth ideology is more administrative and legalistic, seeking to establish standards that will assure that development progresses responsibly. For instance, the State Comprehensive Plan contains very few maps.

One of the key assumptions behind the analysis of planning in terms of its doctrine is, indeed, the centrality of spatial metaphors to success in planning. Controlled growth ideology is only defined in relation to its opposite, namely, unrestrained growth. This conclusion comes back in Evers' chief recommendation: to increase the spatial aspect of planning at the state level by producing more map-related documents.

What can we make of this? Clearly, to describe the situation of Florida's state growth management as *pre-doctrinal* might be seen as presumptuous. Is the Dutch model then the only one to follow? Of course, the Dutch model as such cannot be prescribed for the Florida situation. At most, the intention can be to draw lessons from the analysis of Dutch planning in terms of planning doctrine. One important lesson is that the successes Dutch planning seems to enjoy are derived from having a doctrine, as specified in the introduction. This is relevant to other situations only if those concerned define their situation in terms similar to the Dutch. As Evers shows, this is the case at least to the extent that Florida, like the Netherlands, wants to manage urban growth. Going by his analysis, however, there are also shortcomings that need to be rectified—the most important being more attention to the spatial aspects of planning. If followed, these recommendations would bring Florida closer to having a planning doctrine, not because the Dutch model says so, but because this seems to be inherent to the logic of the Florida situation.

For the purpose of this discussion, whether this is likely to happen is not relevant. The point is that, in his analysis of the situation, Evers does indicate that Florida's growth management has the potential for developing a planning doctrine. There is no presumption that Florida will necessarily take this path, only that it represents a possible avenue into the future.

■ THE DYNAMICS OF DOCTRINE

In the first three cases, Flanders, the Benelux, and the European Union, as part of a search for identity, are forming a view of the shape of their area. Such situations can be characterized by conflict, over the meaning of planning and, above all, over the meaning of institutions. The three first cases also show planners attempting to generate order by means of spatial imagery (see also Faludi 1996b). In so doing, planners reconstruct the area as something that can be visualized and understood, and thus acted upon. Where identity is weak and where territorial arrangements are controversial, such conceptualizations of the shape of the territory may exercise a positive influence.

A question arises as to whether all situations without doctrine should be deemed *pre-doctrinal*. Clearly, the answer must be no. Rather, we should reserve the term for situations in which, in the assessment of those in the field, a move in the direction of doctrine is possible. The actors involved must have thought in terms of something akin to doctrine. In other words, pre-doctrinal situations must carry in them a well-understood potential for change. And where this condition exists, the staged model of pre-doctrinal situations followed by situations with mature doctrine provides pointers for the future.

There are contrasting cases, where speaking about a pre-doctrinal situations would be inappropriate. In another comparison with the Dutch situation, Boisvert (1997) wonders whether it is at all possible to achieve consensus in the Montréal Metropolitan Region. Certainly, there have been reform proposals, but so far none has met with approval. The Montréal case, therefore, is an example of situations that are not even pre-doctrinal. Thus, it is possible to discriminate between situations that are pre-doctrinal, and others that are not.

Conceiving of a situation in terms of its doctrinal stage focuses the mind on an inherent dynamic, which in the fullness of time may play itself out. However, what about situations in which mature doctrines are said to exist? According to the staged model, such situations harbor the danger of doctrinal revolutions. In their work on Dutch doctrine, Faludi and Valk (1994, 207-213, 244-245) hint at this possibility. Korthals Altes (1995) takes their story further, thereby addressing whether a doctrinal revolution is in the offing in the Netherlands. An early version of his work is available in English (Korthals Altes 1992), so we can make do with briefly noting his findings.

Korthals Altes writes about the latest Dutch national planning document, the Fourth National Spatial Planning Report, and its supplement, the Fourth Report Extra. At the beginning of the 1970s, the effectiveness of Dutch doctrine was at its peak. However, by the early 1980s, criticism was growing. Growth management was seen as at the end of the "policy life cycle" (Faludi and Korthals Altes 1996). New issues needed to be identified, thus the Fourth Report focused on the Dutch competitive position in a new Europe without borders. This is why a revolution seemed in the offing. In fact, the Fourth Report merely added new themes to the planning agenda, while themes reminiscent of classic Dutch doctrine persisted.

Taking account of the presently less proactive role of gov-

ernment, planning principles, too, have been subject to change. However, this change is not exactly revolutionary either. Against the backdrop of two models of doctrinal change, a Kuhnian Model of "normal planning" punctuated by revolutions, and a Laudanian (after Larry Laudan, a critic of Thomas Kuhn) Model of evolutionary change, Korthals Altes concludes that Dutch doctrine is evolutionary, and thus robust and malleable. The Laudanian Model of evolutionary change opens up new perspectives on the issues raised by the analysis of doctrine. The very term *doctrine* suggests rigid adherence to basic principles. The undertone is that, if and when a crisis hits, it can be handled only by invoking a new doctrine. Under the alternative pattern, in the fullness of time, even the most fundamental principles embodied in doctrine may change. In this sense, the doctrine is open.

For mature doctrines, like the Dutch arguably is, this is important. Revolutions do not follow maturation like night follows day. Even after four rounds of national planning stretching over more than half a century, Korthals Altes still sees a future for Dutch planning doctrine. His is a sophisticated analysis of how, without any major upheaval, even the most cherished principles can make way for new ones. Making practical use of his insights demands an even more sophisticated strategy, and whether even the Dutch, with their highly developed sense for planning, will be able to formulate one in an age of no-nonsense politics remains to be seen. Be that as it may, an analysis in terms of various patterns of doctrinal development assists with appraising the complexities of planning's current situation.

Author's Note: This article is based on a paper presented at the ACSP-AESOP Joint International Congress, Toronto, 24-28 July 1996. I acknowledge help by Louis Albrechts, professor of planning at the Catholic University of Leuven, in supervising Marco Hoogerbrugge, a student at the University of Amsterdam and the author of a master's thesis on the Flemish Structure Plan, and in commenting upon drafts of this paper.

■ NOTES

1. This section draws on a special issue of *Built Environment*. See Zonneveld and Faludi (1997).
2. This section draws on a special issue of *Built Environment*. See Faludi and Zonneveld (1997).
3. Based on a master's thesis submitted at the University of Amsterdam by Evers (1997).

■ REFERENCES

Albrechts, L. 1994. Ruimtelijke ordening tussen droom en werkelijkheid. *Planologisch Nieuws* 14:172-174.
Albrechts, L. 1995a. Bâtir le visage d'une région. *DISP - Dokumente und Informationen zur Schweizerischen Orts-, Regional- und Landesplanung* 122:29-34.
Albrechts, L. 1995b. From agricultural policy towards a policy for rural areas. Paper presented at the symposium *Rural Reconstruction in a*

Market Economy, Mansholt Institute, Wageningen, The Netherlands.

Albrechts, L. 1997. Planning or the (most) difficult art of getting public support. Paper presented at the 11th annual congress of the Association of European Schools of Planning (AESOP), Njimegen, 29-31 May.

Albrechts, L. 1998. The Flemish Diamond: Precious gem and virgin area. *European Planning Studies* 6:411-424.

Alexander, E.R., ed. 1996. Planning doctrine. *Planning Theory* 16:9-101.

Alexander, E.R., and A. Faludi. 1996. Planning doctrine: Its uses and implications. *Planning Theory* 16:11-61.

Benelux Economische Unie. 1994. *Globale Structuurschets: Informatiebrochure Bestuurlijk Akkoord*. Ruimte voor samenwerking. Secretariat-General, Brussels, Belgium.

Benelux Economische Unie. 1997. *Tweede Benelux Structuurschets: Beslisnota*. Ruimte voor samenwerking. Secretariaat-Generaal, Brussels, Belgium.

Boisvert, A. 1997. Montréal et la conurbantion de Hollande: Deux expériences métroplitaines d'aménagement. *Routes et Transports* 26(4):16-26.

Cabus, P. 1994. Het Voorontwerp Ruimtelijk Structuurplan Vlaanderen van 28 maart 1994: een commentaar. *Planologisch Nieuws* 14:102-110.

Chinitz, B. 1990. Growth management: Good for the town, bad for the nation. *Journal of the American Planning Association* 56:3-8.

Commission of the European Communities, Directorate-General for Regional Policy. 1991. *Europe 2000: Outlook for the Development of the Community's Territory*. Luxembourg, Luxembourg: Office for Official Publications of the European Communities.

Commission of the European Communities, Directorate-General for Regional Policy. 1995. *Europe 2000+: Cooperation for European Territorial Development*. Luxembourg, Luxembourg: Office for Official Publications of the European Communities.

Commission of the European Communities. 1997. *European Spatial Development Perspective (E.S.D.P.) First Official Draft*. Luxembourg: Office for Official Publications of the European Communities.

Delmartino, F., and J.M.L.M. Soeters. 1994. Ambtelijke cultuur in Vlaanderen en Nederland. *Bestuurskunde* 3:246-252.

Evers, D. 1997. Growth management strategies in Florida and the Netherlands. Master's thesis. University of Amsterdam, Amsterdam.

Faludi, A. 1987. *A Decision-Centred View of Environmental Planning*. Oxford: Pergamon.

Faludi, A. 1996a. European planning doctrine: A bridge too far? *Journal of Planning Education and Research* 16(1):41-50.

Faludi, A. 1996b. Framing with images. *Environment and Planning B: Planning and Design* 29:93-108.

Faludi, A. 1997. European spatial development policy in "Maastricht II." *European Planning Studies* 5:535-543.

Faludi, A., and W. Korthals Altes. 1996. Marketing planning and its dangers: How the new housing crisis in the Netherlands came about. *Town Planning Review* 67(2):183-202.

Faludi, A., and A. van der Valk. 1994. *Rule and Order: Dutch Planning Doctrine in the Twentieth Century*. Dordrecht: Kluwer Academic Publishers.

Faludi, A., and W. Zonneveld, eds. 1997. *Built Environment* (special issue on European spatial development) 23(4):253-318.

Fischer, F., and J. Forester, eds. 1993. *The Argumentative Turn in Policy Analysis and Planning*. London: Duke University Press/UCL Press Limited.

Folkers, C. 1995. Welches finanzausgleichsystem brauch europa. In *Regionalentwicklung im Prozess der Europäischen Integration*, eds. H. Karl and W. Henrichsmeyer, 87-108. Bonn, Germany: Europa Union Verlag.

Foley, D.L. 1960. British town planning: One ideology or three? *British Journal of Sociology* 11:211-231.

Foley, D.L. 1963. *Controlling London's Growth: Planning the Great Wen 1940-1960*. Berkeley, Calif.: University of California Press.

Friedmann, J., and C. Weaver. 1979. *Territory and Function: The Evolution of Regional Planning*. London, U.K.: Edward Arnold.

Hajer, M.A. 1989. *City Politics: Hegemonic Projects and Discourses*. Aldershot, U.K.: Avebury Gower Publishing Company Ltd.

Hajer, M.A. 1995. *The Politics of Environmental Discourse: Ecological Modernization and the Policy Process*. Oxford, U.K.: Oxford University Press.

Haller, M. 1994. Auf dem weg zu einer "europäischen nation"? In *Europa - wohin? Wirtschaftliche Integration, soziale Gerechtigkeit und Demokratie*,

eds. M. Haller and P. Schachner-Blazizek, 363-385. Graz, Austria: Leykam.

Healey, P. 1997a. The revival of strategic spatial planning in Europe. In *Making Strategic Spatial Plans: Innovation in Europe*, eds. P. Healey, A. Khakee, A. Motte, and B. Needham, 3-19. London, U.K.: University College London Press.

Healey, P. 1997b. *Collaborative Planning: Shaping Places in a Fragmented Society*. London: Macmillan Press, Houndsmills.

Hoogerbrugge, M. 1995. Ruimtelijke planning in Vlaanderen: luchtspiegeling of realiteit? Een verkennende studie vanuit Nederlands perspectief. Master's thesis. University of Amsterdam.

Informal Council of Spatial Planning Ministers. 1994. *European Spatial Planning: Results of the Meeting*. Bonn, Germany: Bundesministerium für Raumordnung, Bauwesen und Städtebau.

Klerx, E. 1995. Plannen voor Europa: Een historische analyse van de Nederlandse pleidooien voor Europees ruimtelijk beleid (stagerapport, De blik gericht op Brussel: Rapport 2). Working paper. Amsterdam study centre for the Metropolitan Environment, Universiteit van Amsterdam, Amsterdam.

Korthals Altes, W.K. 1992. How do planning doctrines function in a changing environment? *Planning Theory Newsletter* 7/8:99-114.

Korthals Altes, W.K. 1995. *De Nederlandse Planning Doctrine in het Fin de Siecle: Ervaringen met Voorbereiding en Doorwerking Van de Vierde Nota Voor de Ruimtelijke Ordening (Extra)*. Assen, The Netherlands: Van Gorcum.

Krautzberger, M., and W. Selke. 1996. *Perspektiven der Bundesstaatlichen Raumplanungspolitik in der Europäischen Union: Das Beispiel Bundesrepublik Deutschland*. Schriftenreihe für Städtebau und Raumplanung, Institut zur Erforschung von Methoden und Auswirkungen der Raumplanung der Ludwig Boltzmann - Gesellschaft, Vienna University of Technology, Vienna.

Kuhn, T.S. 1970. *The Structure of Scientific Revolutions*. 2nd edition. Chicago, Ill.: University of Chicago Press.

Kuijken, E. 1994. De Groene Hoofdstructuur: grondslag voor een nieuw natuurbeleid. *Samenleving en Politiek* 1(8):24-30.

Kuklinski, A. 1997. The New European Space (N.E.S.) - Experiences and prospects: A paper for discussion. In *European Space, Baltic Space, Polish Space - Part One (European 2010 Series)*, ed. A. Kuklinski, 312-330. Warsaw, Poland: European Institute for Regional and Local Development, University of Warsaw.

Kunzmann, K.R. 1996. Euro-megalopolis or Themepark Europe? Scenarios for European spatial development. *International Planning Studies* 1:143-163.

Loeckx, A. 1995. De moeilijke kunst van het tweevoud: Het ruimtelijk structuurplan Vlaanderen. *ARCHIS* 10:48-53.

Majone, G. 1996. The European Commission as a regulator. In *Regulating Europe*, ed. G. Majone, 61-79. London, U.K.: Routledge.

Malchus, Frhr.V. von, C.H. David, U. Höhnberg, R. Klein, P. Klemmer, R. Kunzmann, W. Selke, G. Tönnies. 1996. *Toward Codificating European Spatial Development Policy in the Treaty on European Union*. Dortmund, Akademie für Raumforschung und Landesplanung, Hanover.

Nijkamp, P. 1995. The region and the environment in Europe: Whose concern, whose competence? In *Umwelt und Umweltpolitik in Europa: Zwischen Vielfalt und Uniformität*, eds. K.W. Zimmerman, K.H. Hansmeyer, W. Henrichsmeyer, 33-57. Bonn, Germany: Europa Union Verlag.

Prantl, B. 1994. Die Auswirkungen eines EU-Beitritts auf die österreichische Raumordnung. In *Föderalistische Raumordnung - eine europäische Herausforderung (Schriftenreihe des Instituts für Föderalismusforschung Vol. 47*, ed. P. Pernthaler, 16-29. Vienna, Austria: Wilhelm Braumüller Universitäts-Verlangsbuchhandlung Ges.m.b.H.

Rein, M., and D. Schön. 1986. Frame-reflective policy discourse. *Beleidsanalyse* 15(4):4-18.

Rometsch, D. 1996. The Federal Republic of Germany. In *The European Union and Member States: Towards Institutional Fusion?*, eds. D. Rometsch, W. Wessels, 61-104. Manchester, U.K.: Manchester University Press.

Sager, T. 1994. *Communicative Planning Theory*. Aldershot, U.K.: Avebury.

Schmittner, P.C. 1996. Some alternative futures for the European polity and its implications for European public policy. In *Adjusting to Europe: The Impact of the European Union on National Institutions and Policies*,

344   *Faludi*

eds. Y. Mény, P. Muller, J.-L. Quermonne, 25-40. London, U.K.:
Routledge.

Selznik, P. 1953. *TVA and the Grass Roots: A Study in the Sociology of Formal Organizations.* 2nd edition. Berkeley, Calif.: University of California Press.

Swyngedouw, E.A. 1994. De produktie van de Europese maatschappelijke ruimte. In *Europese ruimtelijke ordening: Impressies en visies vanuit Vlaanderen en Nederland,* eds. W. Zonneveld and F. D'hondt, 47-58. Gent, Den Haag: Vlaamse Federatie voor Planologie, Nederlands Instituut voor Ruimtelijke Ordening en Volkshuisvesting.

Union économique Benelux. 1996. *Espace de Coopération: Deuxième Esquisse de Structure Benelux, project mai 1996.* Secretariat-General, Benelux Economic Union, Brussels.

Valk, A.J. van der, and A. Faludi. 1997. The Green Heart and the dynamics of doctrine. *Netherlands Journal of Housing and the Built Environment* 12: 57-75.

Vries, J. de, and J. van den Broeck. 1997. Benelux: A microcosm of planning cultures. *Built Environment* 23(1):58-69.

Williams, R.H. 1996. *European Union Spatial Policy and Planning.* London, U.K.: Chapman Publishing.

Zonneveld, W. 1995. Het Ruimtelijk Structuurplan Vlaanderen vanuit nationaal en internationaal perspectief: Over het inke luren van witte vlekken. Paper presented at the VFP Studiedag Het Structuurplan Vlaanderen, Autidorium Gemeentekrediet, Brussels, Belgium.

Zonneveld, W., and A. Faludi, eds. 1997. Vanishing borders: The second Benelux structural outline (special issue). *Built Environment* 23(1):5-81.

Originaltext Friedmann 1967

John Friedmann

A Conceptual Model for the Analysis of Planning Behavior

Planning is defined as the guidance of change within a social system. A conceptual model is presented and hypotheses are derived as a means for ordering the data of empirical research into planning processes. Four modes of planning are distinguished: developmental and adaptive, allocative and innovative. In addition, forms of thought relevant to planning, institutions for political guidance and conflict resolution, and types of implementation procedures are discussed in terms of their proper level and position within a comprehensive system.

John Friedmann is director of the Ford Foundation Advisory Program in Urban and Regional Development in Chile.

UNTIL a few years ago, discussions of planning were restricted to consideration of an abstract model of perfect rationality in social decision making.[1] In use, however, this model turned out to be unsatisfactory. As a theoretical model, it failed to lead to fruitful hypotheses and, because of its logical rigidity, it was in-

---

[1] A concise description of this model will be found in Edward Banfield's "Note on Conceptual Scheme" in Martin Meyerson and Edward C. Banfield (eds.), *Politics, Planning and the Public Interest: The Case of Public Housing in Chicago* (Glencoe, Ill.: The Free Press, 1955), pp. 303–330. In a more sophisticated version, this model also underlies much of Jan Tinbergen's influential work, *Economic Policy: Principles and Design* (Amsterdam: North Holland, 1964). For a critique of classical decision theory, see Charles Lindblom, *The Intelligence of Democracy* (New York: The Free Press, 1965).

capable of substantial modification. As a normative model it failed because rationality in real life is always "bounded," so that the recipes for planning that could be drawn from the model were frequently inapplicable.[2]

With the recent upsurge of interest in national planning, however, social scientists have begun to study the actual workings of the planning process. Some students are focusing on the substantive contents of national plans and on the propriety of the strategies adopted; others are doing research on the administrative machinery evolved; still others are curious about how planning got started in particular societies and how the first plans came to be made.[3] But where the earlier theorists erred in ignoring planning practice, the new empiricists are leaning too much in the other direction: they simply look at activities called planning and describe what they see. Although this is leading to the collection of much information, it is also giving rise to unwitting distortions when basic preconceptions have not been made explicit. Simple descriptions of something as ephemeral as national planning is scarcely even of historical value, and certainly does not add significantly to *verifiable* knowledge, which alone is capable of serving as a sound foundation for a theory of planning.[4] The importance of these studies lies primarily in their fresh approach, which has brought the study of planning within the scope of empirical social research.

The present paper is an attempt to create that minimum of conceptual order which is necessary for a scientifically more

[2] On the concept of "bounded rationality," see James G. March and Herbert A. Simon, *Organizations* (New York: John Wiley, 1958), pp. 203–210.

[3] Recent examples include Everett E. Hagen, *Planning Economic Development* (Homewood, Ill.: Richard D. Irwin, 1963); John and Anne-Marie Hackett, *Economic Planning in France* (Cambridge, Mass.: Harvard University, 1963); chapters on the Netherlands, France and Japan in Bert G. Hickman (ed.), *Quantitative Planning of Economic Policy* (Washington: Brookings Institution, 1965); Albert Waterston, *Development Planning: Lessons of Experience* (Baltimore: Johns Hopkins University, 1965); and the several volumes in the National Planning Series published by Syracuse University Press under the general editorship of Bertram M. Gross.

[4] Planning theory was formerly little more than an exercise in the logic of rational decision making; its reformulation on an empirical basis will involve extensive work in the description and explanation of planning phenomena and in generalizations derived from these data.

disciplined study of planning.[5] There are various ways of defining planning.[6] Here planning will be considered as the *guidance of change within a social system*.[7] Specifically, this means a process of self-guidance that may involve *promoting differential growth* of subsystem components (sectors), *activating the transformation of system structures* (political, economic, social), and *maintaining system boundaries* during the course of change.[8] Accordingly, the idea of planning involves a confrontation of expected with intended performance, the application of controls to accomplish the intention when expectations are not met, the observation of possible variances from the prescribed path of change, and the repetition of this cycle each time significant variations are perceived.[9]

To this view of planning as a self-guidance system, a still more general conception may be added that will lead directly into the structure of the model. Planning may be simply regarded as reason acting on a network of ongoing activities through the intervention of certain decision structures and processes. The emphasis here is on intervention and, hence, on *planning for change*. This intervention is made on the basis of an intellectual effort or, more simply, of thought. "Introducing" planning, then, means specifically the introduction of ways and means for using technical intelligence to bring about changes that otherwise would not occur.

[5] A study complementary to the present one and fundamental for any serious research into planning is Bertram M. Gross, "The Managers of National Economic Change," in Roscoe C. Martin (ed.), *Public Administration and Democracy: Essays in Honor of Paul H. Appleby* (Syracuse: Syracuse University, 1965).

[6] For some frequently used definitions of planning, see Yehezkel Dror, The Planning Process: a Facet Design, *International Review of Administrative Sciences*. 24 (1963), 1–13.

[7] This definition is in line with, if somewhat more general than, Bertram M. Gross' definition of planning as the "processes whereby national governments try to carry out responsibilities for the guidance of significant economic change." See his National Planning: Findings and Fallacies, *Public Administration Review*, 25 (1965), 264.

[8] The distinction between self-guidance systems and those in which guidance is imposed by agents external to the system is of theoretical and practical importance, but will not be pursued further in this paper.

[9] This description of the logic of planning coincides with Neil W. Chamberlain's model as developed in his *Private and Public Planning* (New York: McGraw Hill, 1965), especially ch. 7.

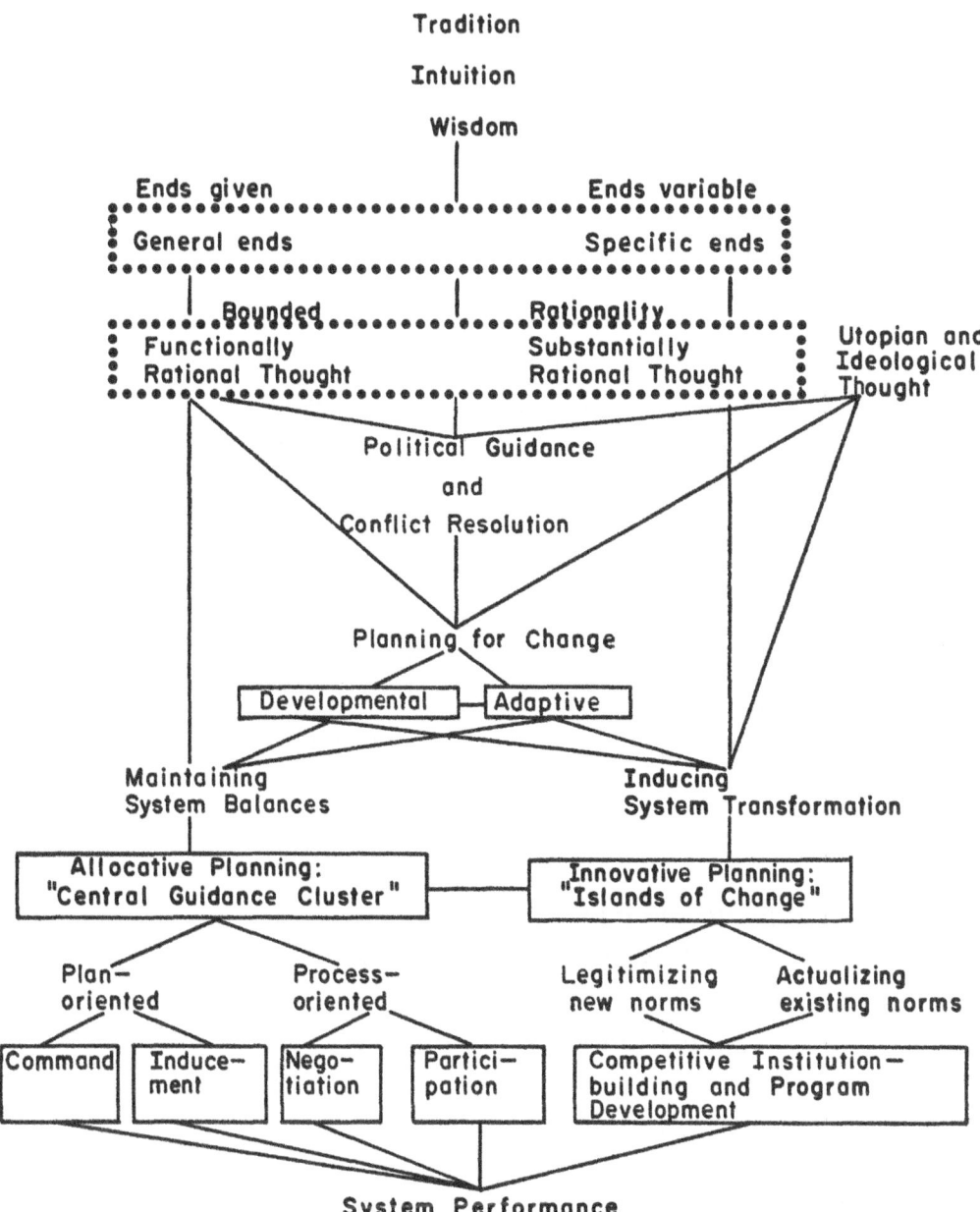

Figure 1. A conceptual model for the analysis of planning behavior.

This is fundamental in my view. Society is a going concern. The ongoing stream of life does not wait for planners to give it direction. Planners must act upon social and economic processes with the fragile instrument of their minds (amplified by whatever practical means they may command) to guide society towards desired objectives. A comprehensive model of planning must, therefore, include forms of thought as an important category for analysis.

THE MODEL

The model proposed here for the analysis of planning behavior has three general characteristics (see Fig. 1): First, it is valid only for what is here called planning for change. Other forms of planning may be identified, such as operations research, but these are not included. Second, it is an attempt to distinguish among different forms of planning for change and to show the relationships among them. Third, the model is intended as an aid to empirical research. On the basis of the actual findings, it will almost certainly need to be refined, modified, and expanded. Specifically absent from the model are the institutional forms of planning and explicit recognition of the time dimension in which planning processes occur.

Forms of planning

A convenient way for entering the model is to consider the two major forms of planning for change. The criterion for distinguishing between them is the relative autonomy of planning units in the making of decisions. Under *developmental planning*, there is a high degree of autonomy with respect to the setting of ends and the choice of means; under *adaptive planning*, most decisions are heavily contingent on the actions of others external to the planning system. In practice, of course, most planning decisions are made along the continuum between complete autonomy and complete dependency, and the behavior of planning systems will differ according to the distribution of decision functions between the two extremes. For instance, planning for urban development at the level of the city will usually be more adaptive than developmental: to a great extent, it will need to respond or adapt to external forces, such as shifts in the locational preferences of national industries, which the municipality cannot significantly influence through its own actions. In planning for national development, on the other hand, the public authorities are able to control a larger number of the variables relevant to its own objectives, so that the nation is much more independent than any of its municipalities. Even among nations, however, there may be differences in the degree of dependency on external forces; and a small, weak nation such as Haiti has to plan more adaptively with respect to international conditions—if it is planning at all—than the city of São Paulo with respect to Brazil.

What are the main differences in the expected behavior between developmental and adaptive planning?

Adaptive Planning.

In adaptive planning, there will be a tendency, to push decisions upward to centers of developmental planning where the parameters for choice at lower levels may be changed. In attempting this, lower-level planning systems will generally rely on political manipulation to achieve their ends. So that negotiations with the central authorities may be conducted with equal technical competence, however, counterplanning may be added to political action. Since, on the government's side, any bargaining in a complex advanced economic system is usually done by qualified technical experts, the contending parties must enter negotiations at least as well prepared.[10]

At the same time, the gradual recognition of interdependence within the system may lead the separate, partisan interests—each engaged in a measure of adaptive planning—to discover a common

[10] A typical example of counterplanning is the large number of national planning agencies that were set up in Latin America when it became known that the Alliance for Progress would make financial aid contingent on the preparation of national plans. In order to deal with the international agencies in Washington, a country had to send economists who could negotiate for aid on the basis of a logical program for development. This program was then compared to the Alliance's own planning for Latin America, whether by the Committee of Nine, the Agency for International Development, or the Interamerican Development Bank. In many Latin American countries today, national planning is primarily a means of obtaining international assistance rather than a means of guiding the use of resources within the country.

or public interest. Such an interest, as the work of the Bureau of Economic Research, the Brookings Institution, the Committee for Economic Development, and the National Planning Association in the United States clearly demonstrates, will lead partisan interests gradually in the direction of quasi-governmental policy planning, although the interests they nominally represent are private. Thus, on the technical side at least, adaptive planning may become fused with developmental planning, that is, subsystem with system planning.

Finally, adaptive planning is typically opportunistic. For instance, one reason for the frequently noted instability of long-term capital improvement programs for municipalities in the United States is that cities cannot afford to lose the federal or state financial aid that frequently plays a decisive part in municipal public works. Since funds from external sources often become available only upon short notice, are tied to specific performance criteria, and normally require matching contributions, significant modifications in the program are frequently made to accommodate the emergence of sudden opportunities for external financing.[11] Similarly, so-called national planning is often related to the availability of funds and the requirements of international sources, as for example, the sudden creation of national planning agencies in Latin America in response to a call for national plans as a basis for Alliance for Progress assistance. These plans closely reflect what each country believes will, at a given time, be the most persuasive program for obtaining funds from the Alliance; they are not necessarily related to the priorities of domestic needs.

Different degrees of autonomy and dependency in decision making tend to be mirrored in a hierarchy of planning authorities which stand in more or less systematic technical and political relation to one another, each level having its appropriate function and decision power. Since each higher level is capable of changing some of the relevant conditions for decisions at all lower levels, and since every change of this sort represents some change of policy, policy planning tends to be emphasized at higher decision levels and programming—the detailed specification of in-

[11] W. H. Brown and C. E. Gilbert, *Planning Municipal Investment. A Case Study of Philadelphia* (Philadelphia: University of Pennsylvania, 1961), ch. 8.

vestments in volume, time, and place—at lower levels. Or, put in another way, developmental planning tends to shade off into policy making, adaptive planning into programming. In fact, however, the two become mixed in varying proportion, according to the point on the autonomy-dependency continuum where planning occurs.

Developmental Planning

In developmental planning, the role of political institutions for guidance and conflict resolution becomes obviously crucial; for it is here that the basic policy decisions are made and that the clashing interests of adaptive planners must be resolved. Developmental planning is not only a technical, but also, and to a large degree, a political function. The relationship between planning and politics is therefore a crucial one.[12]

First, an effective decision process almost always involves both experts and politicians (or policy makers) simultaneously in close interdependency. No politician who values the services of an expert can afford consistently to disregard his judgment, nor will any expert desiring influence, systematically oppose the wishes of his employer. Therefore every decision will be influenced by political interests in varying degrees and, at the same time, it must satisfy some technical criteria. Once a decision is made, it becomes exceedingly difficult to separate the contributions made by each group, for it represents a synthesis of political and expert judgment. Failure to achieve this synthesis will mean that the plans are not carried out or, that policies adopted, being exclusively political, will be inadequate or inappropriate.[13]

Second, successful planning in its more technical aspects must meet certain needs internal to the political process itself. Al-

[12] For excellent discussions of the relation of planning to politics, see Alan Altschuler, *The City Planning Process: A Political Analysis* (Ithaca: Cornell University, 1966). At the national level, one of the best accounts will be found in the forthcoming study by Robert T. Daland, *Brazilian Planning: A Study of Development Politics and Administration* (Chapel Hill: University of North Carolina, February 1966, mimeographed).

[13] Józef Pajestka, Dialogue Between Planning Experts and Policy Makers in the Process of Plan Formulation, paper presented to the International Group for Studies in National Planning, November 15–22, 1966, Caracas, Venezuela.

though these uses of planning are not usually made explicit, they are nevertheless real. They may include: (*1*) symbolic representation of progress, modernity, and so on; (*2*) mobilization of external resources; (*3*) redistribution of the relative influence or weight of participants in a diffused power structure (for example, strengthening the role of the Presidency, of technicians, of the industrial sector, and so forth); (*4*) helping to build a national consensus on fundamental values; (*5*) stimulating an acceptance of development; and (*6*) encouraging counterplanning.[14]

For example, in his recent study on planning in Tanzania, Anthony Rweyemamu writes:

In a new nation like Tanzania, a national plan is a major, albeit incomplete, substitute for the goods which were promised explicitly or implicitly during the struggle for independence. Insofar as it is indicative of a future of abundance, a national plan serves as a unifying agent of an otherwise loose and fragile society. . . . Therefore, even if the economic and social goals are not completely realized, a plan is successful to the extent to which it serves to mobilize the people's energies, bring about national integration and a measure of political consensus.[15]

These varied uses of planning are frequently not only more important than the explicit purposes for which planning is undertaken (more rapid growth, greater efficiency, better coordination) but also inconsistent with these purposes. It is evident that they will also define the respective roles of experts and politicians and help to shape the institutional framework of planning.

In any system, there are large areas of indifference where political behavior is possible without planned intervention. The relative influence of a technical planning function in guiding social and economic change will depend chiefly on five variables: (*1*) the clarity of system objectives, (*2*) the extent of consensus about them, (*3*) the relative importance that politicians attach to them, (*4*) the degree of variance relative to objectives expected in the performance of the system, and (*5*) the extent to which a techni-

---

[14] These and other functions are discussed in J. Friedmann, *Venezuela: From Doctrine to Dialogue* (Syracuse: Syracuse University, 1965). For corroborative evidence, see Robert T. Daland, *op. cit.*, ch. 6.

[15] Quoted in Bertram M. Gross' preface to Fred G. Burke, *Tanganyika: Pre-planning* (Syracuse: Syracuse University, 1965), pp. 19, 20.

cal (as contrasted with a purely political) approach is believed capable of making system performance conform to these objectives. Technical planning, therefore, moves temporarily into the foreground whenever goals are clear, widely held, and deemed to be important; whenever in such a situation system performance is believed to depart significantly from the norm; and whenever, given all of these conditions, expert judgment coupled with a variety of control mechanisms is held to be more effective than political manipulation. Where these conditions do not occur, planning is likely to be reduced to a vestigial function only.

Relation of Kinds of Thinking to Planning

All political and planning activities are in varying degree influenced by different kinds of thinking that may be classified as rational or extra-rational. Rational thought can be further considered as bounded or nonbounded. And bounded rationality may be considered as functionally or substantially rational. Far from being superfluous categories for the analysis of planning, these kinds of thinking are decisive in influencing both the prevaling styles of planning and the actual behavior of planners.

Bounded rationality. This refers to the fact that thought and consequent action intended to be rational are contingent on environmental conditions—the social context of planning—which represent the medium in and through which planning decisions are made.[16] This environment for decision is often discussed in terms of so-called obstacles to planning, but it seems preferable to speak of it simply as the specific set of structural conditions under which planning must occur.[17] In discussing planning in Italy, Joseph LaPalombara underlines the critical importance of the decision environment. He writes:

No one with even the most cursory knowledge of the Italian bureauc-

---

[16] Reference to "adaptive" rationality is made in J. Friedmann, "The Institutional Context," in Bertram M. Gross (ed.), *Action Under Planning* (New York: McGraw-Hill, 1966), ch. 2. The equivalent concept of *bounded* rationality was introduced into the literature by Herbert Simon (see footnote 2). The importance of "context" for planning has also been stressed by Fred W. Riggs, *The Ecology of Development.* Comparative Administration Group, American Society for Public Administration, *Occasional Papers.* (Bloomington, Indiana: 1964).

[17] Albert Waterston, *op. cit.,* ch. 8.

racy could seriously hold that, within its present structure, it is able to support state intervention in the economic sphere, much less to direct and to coordinate economic planning on a national scale.[18]

But, in fact, the limitations of bureaucracy are only one aspect of the decision environment, which is more adequately described in categories such as the following:

1. The number and diversity of organized interest groups and their power to influence decisions.

2. The degree to which political opposition is tolerated or accepted, and the role assigned to it.

3. The dependence of the economic system on private enterprise, and the characteristics of enterprise (size, monopoly, and others) and of entrepreneurial behavior.

4. The efficiency of the relevant information systems: their capacity, load, reliability, promptness, secrecy, etc.

5. The structure of bureaucratic institutions and their performance.

6. The educational level of the population and size of the university-educated elite.

7. The availability of relevant information and its reliability.

8. The predictability of change within the system and of external changes that will affect its performance.

In short, to be "bounded" means that a decision can be no more rational than the conditions under which it is made; the most that planners can hope for is the most rational decision *under the circumstances*. "Until administrative improvements are clearly foreseeable," writes Albert Waterston, "planners must prepare plans which take account of administrative capacity. This means, among other things, that complex forms of planning must be avoided when a country's administration is not ready for them."[19] The author might have broadened this statement to include all conditions that provide the social context for decisions and action.

The concept of bounded rationality suggests the possibility of identifying a number of discrete planning styles which result from

[18] Joseph LaPalombara, *Italy: The Politics of Planning* (Syracuse: Syracuse University, 1966), p. 106.

[19] Albert Waterston, *op. cit.*, p. 292.

the adaptation of the institutional forms and procedures of planning to relatively stable characteristics of their institutional environment. These environments and the forms of planning adapted to them can both be reduced to a few general types. The study of planning styles would therefore be helpful in formulating hypotheses for the comparative study of planning behavior.

It is useful to distinguish two basic forms of bounded rationality. *Functionally rational thought* is rational with respect to the means only; the ends are assumed by the planner to be given and may be more or less rational or even, according to certain criteria, irrational.[20] The ends must remain fairly stable, however, because the decision must appear rational not only before but also *after* implementation. As a rule, the more general the objective, the more stable it will tend to be. It may therefore be concluded that functional rationality in planning is found chiefly in connection with stable, general ends that are applicable to the system as a whole or at least to major portions of it. For example, the general ends of national economic development usually include such values as growth, more equitable income distribution, and full employment. These ends undergo relatively little change through time; but at the same time, they point only in a general direction. Functionally rational thought will try to guide the evolution of the system in this direction.

Substantially rational thought is rational with respect to both the ends and means of action. But this clearly implies the possibility of altering the ends during the action as a result of changing circumstances or new information. One would, therefore, expect to meet with frequent modifications of the ends. Since only specific ends are capable of being modified in this way, it is permissible to posit a strong correlation between substantial rationality and variable ends.

In any planning system, both forms of rationality will usually occur: functional rationality with respect to system ends and substantial rationality with respect to more specific, subsystem ends; that is, one would expect to find stability in the general

[20] This is Karl Mannheim's terminology for describing the two forms of rationality, in *Man and Society in an Age of Reconstruction* (New York: Harcourt, Brace, 1949) pp. 51–60.

direction of the planning effort and adaptability in detailed planning. For example, lengthy conferences operate on what might be called "daily blueprints," that is, frequent revisions of the detailed schedule to accommodate unforeseen events, yet the main purposes of the conference will usually remain the same, although the course towards them may be tortuous and seemingly anarchic.

Nonbounded rationality. The nonbounded form of reason, free from temporal constraints may be called *utopian and ideological thought.* In such thinking there is a picture of an ideal social order, often in considerable detail, and almost always as a final state existing in a perfect equilibrium outside of historical time. The images of perfect communism and perfect capitalism are such utopias, as are the corporate state, national socialism and participant democracy.

Utopian and ideological thought may be considered rational in two senses. Its constructions are not only logical and coherent; they are also concrete representations of abstract social values such as equality, freedom, and social justice, and it is primarily these qualities that make this kind of thinking so often persuasive.[21] Forms of planning are often historical precipitates of utopian and ideological thought. Agricultural planning in the United States, for instance, can be illuminated by analyzing it against a background of Jeffersonian democratic thought; national planning in Spanish-speaking Latin America needs to be viewed in the light of the philosophy of the corporate state; and Indian planning still reflects the influence of Gandhi's social philosophy. What is still more significant is that many of the internal conflicts that rage about specific planning proposals turn precisely on philosophical issues such as these rather than on more pragmatic problems. The outcomes of these conflicts are usually decisive in setting the direction of development for an entire sector or even the whole nation.[22]

[21] Martin Meyerson, Utopian Traditions and the Planning of Cities, *Daedalus* (Winter 1961), 180–193. The influence of utopian thought on economic planning has barely been recognized, however, and merits a full study.

[22] John H. Kautsky (ed.), *Political Change in Underdeveloped Countries: Nationalism and Communism* (New York: John Wiley, 1962). See also J. Friedmann, Intellectuals in Developing Societies, *Kyklos,* 13 (1960), 513–544.

Extra-rational thought. The category of extra-rational thought includes what may be loosely called *tradition, intuition,* and *wisdom.* These forms of thought are not derived from coherent, logical structures, nor based on specific technical expertise. They are, however, the source of most political decisions and, therefore, play an exceptionally large part in planning processes. It must be admitted, however, that recently there has been a steady diminution of the role of extra-rational thought in public decision making. Measurement and calculation are driving intuition and wisdom into more and more exclusive spheres, whereas the rapidity of change renders tradition meaningless as a source of decisions oriented towards the future. The result appears to be a weakening of the political elements in planning accompanied by a strengthening of the role of the technician. How this trend should be assessed is not yet clear.

Allocative Planning

Allocative planning is the assigning of resource increments among competing uses. Typically, this is the task of national planning institutions and, for many people, it is the only task with which planning should be properly concerned. Four characteristics of allocative planning help to define it.

1. *Comprehensiveness.* Allocative planning must be comprehensive with respect to at least the following: (*a*) the interdependence among all of the explicitly stated objectives of the system (or subsystem), (*b*) the interdependence in the use of all available resources of the system (or subsystem), and (*c*) the influence of all external variables on the setting of intermediate targets.[23]

Comprehensiveness has become a preoccupation with allocative planners. They believe that their special contribution to social decision making derives mainly from their ability to manipulate a comprehensive set of variables and objectives and to acquire, as a result, a point of view that necessarily coincides with the interests of the system (or subsystem) as a whole, that is, with the public interest. Their close association with executive power reinforces this conception of themselves. Thus, far from being

[23] Jan Tinbergen, *op. cit., passim.*

mere experts, neutral with respect to values, allocative planners will often defend a set of value propositions as essential to the survival and well-being of the system (or subsystem). Since the concept of a public interest is difficult to maintain, however, especially in pluralistic or in nonintegrated societies, the powers of allocative planning are often resented by groups whose partial concerns are threatened by an insistence on public values arrived at independently of any political process.

2. *System-wide balances.* The optimality criterion, the basic norm for allocative planning, requires a balance among the variable components of the planning system. The model with which allocative planners customarily work is necessarily in equilibrium. Thus, planned investment must not exceed the capacity to invest; total imports must not exceed projected exports; employment gains must not be less than the increase of the labor force; electric energy production must meet projected power consumption. It is a question of determining the right magnitudes as the targets of the economic system. Accompanying a set of carefully worked out quantitative targets, is usually a text suggesting changes in existing policies that are thought necessary for their achievement.

3. *Synthesis.* Neither comprehensiveness nor systematic balance can be obtained without the aid of one or more synthetic models of the economy. These models allow study of the functioning of the system under quasi-experimental conditions as different conditions are considered and their implications are observed. The most common models include national economic accounts, input–output matrices, simulation models, and econometric policy models.[24] These models are abstracted from the institutional and legal framework of the economic system and from the persons through whom the system works.

4. *Functional rationality.* Allocative planning is an attempt to be functionally rational in that the objectives of the system are supposed to be determined externally through a political process that does not significantly involve the planners themselves. Planning, therefore, appears as only a working out of the implications for public policy of norms established independently. As a well-known economist has recently explained it:

[24] See Bert G. Hickmann, *op. cit.*, for a discussion of the current planning models.

The tendency now is to abandon the effort to determine through economic analysis, what is the best form of economic organization or the "best" set of economic policies, and to accept goals established through the political process and stated by governments—full employment, price stability, more rapid economic growth, elimination of pockets of poverty or distressed areas, and the like. For the most part, such goals seem reasonable to economists, but by starting their analysis at the point where government policy is already determined, they avoid value judgments of their own. They may point out inconsistencies among goals, or worry about such new dilemmas as rising cost of living and increasing unemployment side by side, but choice between goals is left to the government, as is the establishment of priorities. Usually some set of measures for achieving goals—once priorities are established—can be suggested, even if there remains doubt as to whether they constitute the best possible set of measures.[25]

The impossibility of remaining uninfluenced by values has already been pointed out. Nevertheless, by shifting the major burden of choice among values to the political process, allocative planning can appear as an activity intended in large measure to be functionally rational and thus, presumably, objective.

Implementation

The institutions charged with implementation of the plans must remain constantly aware of the need to carry out the policies and advance towards the targets of the models. Implementation, however, is not an independent step taken subsequent to plan making; *the kind of implementing mechanism adopted will itself influence the character of the plan and the way it is formulated.* The formulation and implementation of plans are closely interdependent processes, so that the choice of one will in large measure also determine the second. For this reason, allocative planning will be either *plan-oriented* or *process-oriented.*[26]

In Italy, where central planning has not yet advanced very far, the question of the appropriate mechanisms for plan implementa-

[25] Benjamin Higgins, "An Economist's View," in H. M. Phillips (ed.), *Social Aspects of Economic Development in Latin America* (UNESCO, 1963), p. 247.

[26] The best current discussion of problems of planning implementation is that of Bertram M. Gross, "Activating National Plans," in his *Action Under Planning, op. cit.* The terminology adopted in the present paper, however, differs from that of Gross.

3

tion, and consequently for the whole structure of the planning system, is basic to the present controversy there. According to LaPalombara,

> The question of control is a critical one. Is planning to be by "inducement" or "indication," as some Italians claim, or is it to have "compulsive" dimensions? If the latter, will compulsion apply only to the public sector or will it be extended to the private sector as well? If planning is to be compulsory or obligatory for the public sector, what instruments will the government utilize to enforce private-sector adherence to the plan? Will the plan encompass the whole economy or will it be limited to only particularly important sectors? These are merely a few of the questions. Although the 1965 plan provides some tentative answers, many of them remain unclear.[27]

The production of a blueprint and adherence to its basic structure of goals may be viewed so essential by the political leadership of a country that maximum use will be made of the available powers of command and inducement in the endeavor to carry out the plan. In *command planning*, sanctions are applied to compel adherence to clearly formulated objectives and targets. The plan itself may have legal force or may be promulgated in a series of executive decrees in order to obtain specific results.[28] For Jean Maynaud, for instance,

> A plan is not a plan unless it is central and global, involving specific objectives, and directly inserted in the existing socio-economic context even if it anticipates some social change . . . . Public authorities must make a strong effort to assure that the results of planning will be close to what is predicted. This suggests that some compulsion is essential, even in places like Italy, where it appears that the forces of the free market have created rapid economic expansion.[29]

Inducement is a weaker form of activation in that its effects are not, as a rule, experienced as coercion. It arranges the deci-

---

[27] Joseph LaPalombara, *op. cit.*, p. 103.

[28] The concept of command planning has been suggested by Peter Wiles in his brilliant analysis of *The Political Economy of Communism* (Cambridge, Mass.: Harvard University, 1962). See also his "Economic Activation, Economic Planning, and the Social Order," in the aforementioned volume edited by Bertram M. Gross, as well as Zygmunt Bauman, "The Limitations of 'Perfect Planning'" in the same volume.

[29] Joseph LaPalombara, *op. cit.*, p. 104.

sion environment of others in such a way that one kind of decision will tend to be preferred over possible alternative decisions. Typical instruments of inducement are special lines of credit, interest manipulation, subsidies, exchange-rate policies, tax exemptions, and preferential import tariffs.

Both command and inducement are clearly plan-oriented in that the plan will tend to be regarded as a serious long-term commitment in which changes can be made, but not made easily. However, where the performance of subsystems is important for the attainment of system-wide goals (where the carrying out of the plan depends, for instance, on the actions of the private sector), and the imposition of sanctions or indirect controls is impracticable, allocative planners will tend to stress *process* over plan. In this case, the participation of all the principal interests will be enlisted in the formulation of the plan itself. This has come to be known, following the recent French experience, as indicative planning.[30] And, according to LaPalombara, "The very fact that planning procedures will include pluralistic participation undercuts the idea of compulsion."[31]

Both strong and weak forms of process-oriented planning are encountered. The strong form makes extensive use of bargaining; planning appears, therefore, as a process of continuous *negotiation*, with central government agencies among the list of main protagonists. According to Neil W. Chamberlain,

There are few policies which are so "technical," so independent of people's reactions, that they can be instituted without question. Most matters of any consequence involve discussion and compromise. The views of those on whom the functioning of the system depends cannot be wholly ignored unless the system is prepared to part with their services—in which case it must come to terms with their replacements.

A system of bargains among people must be contrived, and it presumably can be contrived more or less efficiently from the viewpoint of achieving the objectives of the system—that is, with varying degrees of sacrifice of system objectives to subsystem (individual) goals.[32]

[30] John and Anne-Marie Hackett, *op. cit.*, for a comprehensive account of French planning.

[31] Joseph LaPalombara, *op. cit.*, p. 104.

[32] Neil W. Chamberlain, *op. cit.*, pp. 7 and 8.

The weaker form of process-oriented planning depends for its implementation on nothing more persuasive than the participation of key actors in the planning process, those who will be charged with implementation. There is but a minimum of negotiation, and the plan document itself will come to be regarded as less important than the possible benefits resulting from a joint consideration of targets, policies, and instruments. These benefits include establishing a dialogue among contending sectors, creating a wider awareness of national problems, providing the main economic actors with a common information base, encouraging socially more responsible decision making, reducing uncertainty in the calculation of sectoral investment programs, and so on.

In the pure case, process-oriented planning would probably not need to have a plan at all except as an informal discussion document which would register the temporary consensus of all the parties involved. But the pure case of process-oriented planning is rare; usually it will be found strongly mixed with a command and inducement, so that a formally adopted plan may have some substance, after all.

Hagen and White in their appreciation of French experience, write:

On approval of the plan, comprehensive and vigorous intervention by the government began to see that the targets were attained. Or rather, the intervention under the preceding plans continued. French planning, M. Pierre Massé, the present director of the *Commissariat au Plan* has said, is "more than indicative and less than imperative." The phrases are correct, but how much more than indicative, even though also much less than imperative, is the actual process! In M. Massé's delicate Gallic phraseology in another place, "the heart of the matter is that French planning is active; it . . . regulates the stimuli and aids at the disposal of the public departments in such a manner that the objectives assigned to the private sector are achieved."[33]

Choice of Kind of Planning

Which kind of planning predominates will depend on the nature of the decision environment and the urgency of the problems to

[33] Everett E. Hagen and Stephanie F. T. White, *Great Britain: Quiet Revolution in Planning* (Syracuse: Syracuse University, 1966), p. 105.

244   ADMINISTRATIVE SCIENCE QUARTERLY

be solved. Allocative planning can occur under either developmental or adaptive planning. In adaptive planning, planners will be chiefly concerned with predicting the behavior of external variables and with adjusting the available policy instruments in order to maintain the system in some sort of equilibrium under the impact of changes which may impinge upon the subsystem (for example, national economic plans are not yet drawn up so as to optimize development potentials concurrently at national and local levels; the local impact of national plans is, therefore, determined largely by chance). The actual scope for allocative planning under these conditions is quite limited, since adaptations are made to external conditions, special opportunities are seized, and decisions are made about what are probably matters of only secondary importance to the community and, hence, are politically more vulnerable.[34]

Under developmental planning, allocative planners perform a quite different role, although they continue to build models or set targets. Their functions in this case can be best understood with reference to innovative planning.

Innovative Planning

This appears as a form of social action intended to produce major changes in an existing social system. According to Neil W. Chamberlain, it creates "wholly new categories of activity, usually large in scale, so that they cannot be reached by increments of present activity, but only by initiating a new line of activity which eventually leads to the conceived result."[35] Unlike allocative planning, it is not *preliminary* to action but, a fusion or synthesis of plan-making and plan-implementing activities within an organizational frame. Four characteristics help to distinguish innovative from allocative planning.[36]

1. Innovative planning seeks to introduce and *legitimize new social objectives*. Its central attention is, therefore, on the main points of leverage that will accomplish this task. By concentrating

[34] W. H. Brown, Jr. and C. E. Gilbert, *op. cit.*
[35] Neil W. Chamberlain, *op. cit.*, p. 175.
[36] This form of planning is discussed more fully in J. Friedmann, Planning for Innovation: the Chilean Case, *Journal of the American Institute of Planners*, 32 (June 1966), 194–203.

on only a few variables, innovative planners inevitably ignore large parts of the total value spectrum of the society into which the innovation is to be introduced. At the same time, only the most general consequences are considered, with attention to those which relate to expected structural changes in the system. The emphasis, therefore, is on the guidance of change through a selective repatterning of the influences on social action rather than on the multiple consequences of alternative allocations.

2. Innovative planning is also concerned with *translating general value propositions into new institutional arrangements and concrete action programs*. This difficult task usually falls upon a creative minority, which is basically dissatisfied with the existing situation. The organization of these groups, their self-articulation, and their functioning—until they themselves become subject to inevitable routinization—may all be thought of as part of the process of innovative planning.

For example, Bertram Gross referring to what he calls an "institutionalized capacity to build other institutions," writes in his introduction to Robert Shafer's treatment of Mexican planning:

A new institutional infrastructure was needed. To build it in small pieces, however disconnected, seemed infinitely superior to the piling up of a vast hierarchical bureaucracy in a small number of ministries. It provided more upward career channels for people with ability and ambition. By placing scarce eggs in many baskets, there was more room for trial and error, more protection against failure. Promotion of new institutions took precedence over their coordination.

This kind of institution building has a pulse rate of its own. The more successful it is in getting things done, the more problems the new institutions create. This leads to increasing pressure to pull things together a little more tightly . . . . But then the effort to get important things done leads once again to new spurts of decentralized institution building. Central promotion of decentralized institutions once again races ahead of central coordination.[37]

3. From this it follows that innovative planners are public entrepreneurs who are likely to have more interest in *mobilizing*

[37] Bertram M. Gross, "The Dynamics of Competitive Planning," preface to Robert J. Shafer, *Mexico: Mutual Adjustment Planning* (National Planning Series No. 4, Syracuse: Syracuse University, 1966), p. xix.

resources than in their optimal allocation among competing uses. They will seek to redirect financial and human resources to those areas which promise to lead to significant changes in the system. In contrast to allocative planners who strive for equal marginal returns, innovative planners seek to obtain the largest amount of resources for their projects, even if this should mean weakening the purposes of competing organizations. Innovative planners are only peripherally concerned with these other purposes; by weakening other parts of the system, they may even gain a temporary advantage for themselves and facilitate the process of transformation.

4. Innovative planners propose to guide the process of change and the consequent adjustments within the system through the feedback of information regarding the actual consequences of innovation, in contrast to allocative planners, whose main endeavor is accurately to predict the chain of consequences resulting from incremental policies and then to adapt these policies to the prospective changes. To state the difference more succinctly: innovative planners are not, as a rule, interested in gradually modifying existing policies to conform to expected results. Innovative planners are more limited in focusing mainly on the immediate and narrowly defined results of the proposed innovation, and more ambitious in advancing a major project and laboring diligently to introduce it into society. Modifications of this project will tend to occur only as a result of political compromise in the course of getting it accepted and the actual consequences of the policy in operation that suggest the desirability of change in its original form. In place of experiments *in vitro* (through the manipulation of econometric models), innovative planners prefer the device of pilot schemes, where the utility of an idea can be observed in action.

Innovative planning is especially prevalent in rapidly changing social systems. It is, in fact, a method for coping with problems that arise under conditions of rapid change, and it will tend to disrupt existing balances. There is much still to be learned about the different ways that major changes are introduced into an established society or how new social systems emerge. But it is certain that equal progress cannot be made on all fronts simul-

3

taneously. Rather, the image that comes to mind is that of successive waves and wavelets of innovation spreading outwards from a number of unrelated focal points, or innovating institutions, from Clarence Thurber's "islands of development."[38] Since it is difficult to sustain innovation at any of these points over prolonged periods, there may be frequent shifts of emphasis in innovation, one wave succeeding another, but in a different direction. It is even more difficult to succeed in establishing effective organizational linkages among institutions engaged in innovative planning, although clearly where a massive effort for change is intended, this is a necessary condition for the successful transformation of the system.

The strategic problem is to identify the critical points for system transformation and to activate innovative planning at these points. But if a system is already undergoing rapid change, the importance of this strategic problem decreases sharply; for the system generates change automatically. It is engaged in what Akzin and Dror call "high pressure" planning. Speaking of Israeli experience, they write:

The high rate of unpredicted change and the central social roles of government activities impose a fast pace of operation upon the civil service. Although nearly all ministries are overloaded with pressing day-to-day problems, energetic senior civil servants continue to launch relatively large numbers of new projects and activities. The constant pressure of issues necessarily lessens systematic long-range thinking and encourages a problem-by-problem manner of decision-making.[39]

But this pragmatic approach, they say, "is frequently the optimal master strategy. For many problems in the economic, social, political, and technological fields, no applicable knowledge is available. Rather than be misled by theories and recommendations based on quite different circumstances, it is wiser to proceed pragmatically."[40]

[38] Clarence E. Thurber, "Islands of Development: A Political and Social Approach to Development Administration in Latin America," paper presented to the National Conference of the Comparative Administration Group, April 17, 1966.

[39] Benjamin Akzin and Yehezkel Dror, *Israel: High Pressure Planning* (Syracuse: Syracuse University, 1966). p. 17.

[40] *Ibid.*, pp. 16–17.

Under high-pressure planning, detailed target achievement is not possible. The general ends will remain fairly stable and give rise to efforts of allocative planners to keep the system in balance and generally moving in the desired direction. But specific ends may be frequently revised in the light of changing conditions and a constant reevaluation of the action. Innovative planning thus appears as a concrete form of substantial rationality.

Innovative planning is typically uncoordinated and competitive, and this is yet another reason why target achievement is, in any functional sense, unattainable. The top leaders of the Israeli government, writes Bertram Gross,

have deliberately nourished the institution-building, empire-constructing, resource-grabbing expansionism of organizations in all sectors of society, including science and education as well as the trade union movement, political parties, and private business. This has meant the promotion of sectoral (or facet) planning. The result has been more and more high-pressure planning and implementation by competitive institutions. Under such circumstances clear-cut coordination by command of central authorities has been neither feasible, essential, nor desirable.[41]

Role of Allocative Planners in Innovative Planning

But not all systems find themselves already engulfed in a process of rapid internal transformation. Allocative planning is sometimes advanced as a means of generating more rapid changes, especially in the economic field. Dissident young engineers and economists, eager to transform traditional stagnation into dynamic industrial systems, regard the creation of central planning agencies as in itself a major act of innovative planning. For them a central planning agency represents a "permanent institutionalized symbol of the Government's sustained commitment" to the goal of rapid economic growth.[42]

But in their enthusiasm, they may forget that their comprehensive econometric models accommodate discontinuous change only with difficulty. The more allocative planning relies on such models, the more conservative is it likely to be. Detailed aware-

[41] In the preface to Akzin and Dror, *op. cit.*, pp. 26–27.
[42] Bertram M. Gross in his preface to Robert J. Shafer, *op. cit.*, p. 13.

ness of interrelations tends to make experts cautious and hesitant to prescribe radical solutions. Allocative planners, then, confront essentially two choices: either to remain satisfied with the *symbol* they have created and to move gradually towards the bureaucratization of the planning function—but, at the same time, to forfeit ambitious goals, or to risk the seeming anarchy of rapid change, consciously using allocative planning for compelling and inducing maximum efforts in key areas and for endeavoring to maintain only a reduced number of strategic balances throughout the system. In the second case, allocative planners will not only resort increasingly to command as a form of implementation, but will encourage large-scale innovative planning to carry out major elements of the plan or to respond to new problems that are generated by rapid change.

In this vitally important interrelationship between allocative and innovative planning, the role of allocative planners is to develop new kinds of leadership, to channel resources to priority areas or points of change, to facilitate communication among the highly competitive innovative organizations, to search for areas of agreement, to help resolve interinstitutional conflicts especially with regard to the use of limited resources, and to encourage organizational links among the many "islands of development." Over time, and as the pace of change slows down, allocative planning tends to replace innovation in the management not only of organizations but also in the social system as a whole.

In general, then, we may conclude, first, that innovative planning is needed to accomplish a major—as compared to only a marginal—reallocation of resources. New institutions and new programs are needed if money is to be spent in radically different ways. The second conclusion concerns the process of translating abstract values into specific projects and programmed activities (goal reduction). Contrary to the belief of some theorists, this is not inherently a logical process but one that requires institutional innovation.

CONCLUSION

The model suggested furnishes a skeleton for the analysis of planning. But why carry out such an analysis in the first place? What may be gained from an analysis of planning behavior?

First, *empirical findings may be incorporated into a positive theory of guided system change.* Many of the elements of such a theory already exist; what has been lacking up to now is a preliminary theoretical framework for ordering the available data and for supplementing them with studies that will ask theoretically relevant questions and begin to test promising hypotheses. In this paper, some of the hypotheses suggested are:

1. Under adaptive planning, there will be a tendency to push decisions upward towards centers of developmental decision making where the conditions for choice at lower levels may be changed.

2. The formulation and implementation of plans are closely interdependent processes, so that the choice of one will in large measure also determine the second.

3. General, system-wide objectives are modified less frequently than more specific subsystem objectives.

4. Innovative planning is typically uncoordinated and competitive.

Second, *empirical findings will permit a systematic analysis of planning pathologies.* What leads to the breakdown of guided system change? Under what conditions and for what reasons does planning cease to be effective? The reasons may include such variables as the failure of planning to adapt itself optimally to its decision environment, the resilience of this environment to change, conflicting relations between experts and policy makers, failure to achieve an optimal distribution of planning functions according to their position on the dependency-autonomy continuum, neglect of either innovative or allocative planning functions, rigidity in planners' attitudes and procedures, and inappropriate mix between plan-oriented and process-oriented forms of implementation.

Third, *empirical findings may serve as a basis for formulating a prescriptive planning theory.* In the light of positive theory and a systematic knowledge of planning pathologies, a normative theory of planning may be formulated that should be superior to existing formulations. Such a theory will have to be expressed as a function of the decision environment of planning.

On the basis of these several purposes, the model raises important questions that can serve as a useful starting point for any research into planning behavior:

1. What is the role of political institutions in goal formulation policy making, and conflict resolution under different planning systems? (The analysis will have to specify not only the social context of planning but also whether planning is developmental or adaptive, and the relations between allocative and innovative planning.)

2. What is the relation of planning institutions and processes to their social context? Can typical planning styles be identified, especially in relation to the mix of implementation procedures? What is the relative importance of allocative and innovative planning under different environmental conditions?

3. What are the political uses served by planning under different systems and how do these uses influence planning behavior?

4. What is the influence of utopian and ideological thought on the formulation, implementation, and substance of planning decisions?

5. What are the dynamic relations between developmental and adaptive planning under different environmental conditions? Under what circumstances does counterplanning appear, and how are the resulting conflicts resolved? Does something like a public or common interest arise from a system in which counterplanning is prevalent? How are planning functions distributed along a centralization-decentralization continuum, and what are their relations horizontally at each level as well as vertically among a hierarchy of ordered centers?

6. What is the relation of policy makers (or politicians) to experts (or technicians) under different planning systems? How does this relationship influence the effectiveness of planning? Does planning lead inevitably to a "depolitization" of major developmental issues?

7. What is the relation of allocative to innovative planning? What are the roles of either type of planning in guiding system change?

8. What are the relations of competitive innovative planning units to each other? What conditions are conducive to greater

coordination among them, and which may be claimed to represent "obstacles to change"?

9. What are the self-images of planners in contrast to the images of planners held by others, and how do these images affect behavior?

Perhaps a philosophical postcript will be permitted. If planning is accepted as the attempted intervention of reason in history, then it is clear that such intervention cannot be immediate and direct, but must be filtered through a series of complex structures and processes to be effective. A definitely anti-heroic picture of reason emerges. It is not the great mind that intervenes, but a multitude of individual actors, each playing his role in a collective process that he does not fully comprehend because he is involved in it himself and lacks perspective. Reason, therefore, to the extent that it operates on society, is a "collective representation" in Durkheim's sense, whose functioning is contingent on structures and forces which are independent of itself.

Originaltext Howe/Kaufman 1979

3

The Ethics of Contemporary American Planners

Elizabeth Howe and Jerome Kaufman

This article, based on a study of a large, randomly selected group of American planners, looks at what planners think is ethical, and why. Although many planners have similar views about what is ethical, sharp differences are also clearly apparent. Chief among the reasons for these differences is role orientation. Consistently, the most politically oriented planners have a more liberal interpretation of what is ethical than the most technically oriented ones, with a third group—high on both the technical and political dimensions—falling in the middle. Other factors such as political views, attitude towards agency, and propensity to express values in the job were also found to be important in explaining why some planners think differently than others about what is ethical. The implications of these findings are drawn for planning theory, practice, and education.

Suppose a planner who works for a high-income suburb recognizes that the community's land development regulations are exclusionary. This makes it quite difficult for poor people or minority group members to live there, even though job opportunities for them exist in the area. The planner, as part of her regular job activities, decides to organize support from local people she knows are in favor of opening up the community so that they will put pressure on the suburban government's officials to change the community's zoning policy.

In acting this way, does this planner behave ethically or unethically? A study conducted by the authors of a random sample of 616 planners who belong to the American Institute of Planners and work for public planning agencies, indicates that opinions differ sharply.[1] Slightly more than half thought the planner's behavior would be ethical. Slightly more than a third thought it would be unethical; and the rest were not sure, one way or the other. Reactions to this short scenario and fourteen others describing the behavior of planners in situations involving ethical

dilemmas are the subject of this study. Figure 1 gives a brief description of each scenario, the tactic the planner used, and the intended beneficiary of the planner's action.

For planners, ethics set the boundaries of acceptable behavior. In theory, a set of commonly held behavioral norms make up the body of professional ethics. Some, but by no means all, of these norms have been codified in the American Institute of Planners' Code of Professional Responsibility. Whether codified or not, these norms ideally represent guidelines for planners to adhere to in conducting themselves as professionals.

Elizabeth Howe is an assistant professor and Jerome Kaufman is a professor in the Department of Urban and Regional Planning at the University of Wisconsin-Madison. Howe, who has a Ph.D. in Social Policy Planning from the University of California-Berkeley, teaches courses in social planning, planning theory, and politics. Kaufman, who has an M.C.P. from the University of Pennsylvania and was Associate Director of A.S.P.O., teaches courses in central city planning, metropolitan planning, and planning implementation strategies.

Planungskultur

Most importantly, they represent the basis for assuring the public who uses planning services that planners will act responsibly in exercising their professional judgment and in applying it.

What makes things difficult for the planner is that some of these norms, or ethical prescriptions, are either so broadly defined that they can be variously interpreted or the norms may conflict when more than one is present in an issue.[2] In the above scenario, for example, two ethical prescriptions are involved—concern for the broader public interest and loyalty to the client community.[3] Partly because of such conflict and ambiguities, all planners may not reach the same conclusion about what is ethical or unethical as the split response to the above scenario indicates.

The primary interest of our study was simply to find out more about how contemporary planners view ethics—what do they agree and disagree about, and what is the nature and extent of their agreement

and disagreement. In other words, our intention is to describe what planners think is ethical, not to judge whether or not planners are ethical by some predetermined standard. To our knowledge, no systematic study of this has ever been done. Yet the subject is important since ethical considerations pervade much of what planners do.

Beyond describing what planners think is ethical, we also wanted to examine factors or characteristics of the planners which explain why some view an action as unethical which others see as acceptable.[4] In essence, we wanted to know what motivates planners in practicing their trade. To get at these causal factors, we asked respondents about their professional role orientations, their attitudes toward a variety of substantive issues in the field, their loyalty to their agency, their propensity to express values in their work, and their political preferences, as well as their background and work situation.

Figure 1. Description of ethics scenarios

Scenario	Tactic	Issue benefiting
1 City planner gives draft recommendations for pollution control plan to environmental group representative who requests them; no agency policy exists about releasing such information.	release draft recommendation on request	environment
2 City planner, who favors low fare to make proposed regional transit system more accessible to the poor, purposely develops estimates showing that system will have high ridership/high revenue yield to counteract low ridership/low revenue yield estimates of regional planners who oppose lower fare.	distort information	mass transit, low income
3 Planning director urges members of important civic group to publicly endorse "park and ride facility" plan, telling them that many neighborhood groups support the plan—director knows, however, that less than half of neighborhood groups consulted so far have agreed to support plan.	distort information	mass transit
4 City planner assigned to work with particular low income neighborhood, without authorization, gives information to head of the neighborhood organization on study being prepared by another planning agency unit which recommends substantial land clearance in this neighborhood.	leak information	low income
5 Regional planning director threatens to use agency's A-95 authority to recommend denial of local projects if local officials do not support regional growth management plan.	threat	environment
6 Regional planner puts several strong recommendations into fair share housing plan which planner feels are expendables that might be later traded off to get commissioners to support central aspects of plan.	use expendables as trade off	low income
7 City planner gives draft recommendations for development plan for largely undeveloped part of city to land developer who requests them; no agency policy exists about releasing such information.	release draft recommendations upon request	development
8 Suburban planner decides to organize support from local people to put pressure on suburb's officials to change community's exclusionary zoning policy.	organize coalition of support to induce pressure	low income

The paper is organized into four sections. We start out with a discussion of the study design. We then consider what planners as a whole think is ethical and unethical behavior, looking particularly at areas of consensus and disagreement, and at the fit between planners' attitudes about ethics and what they say they would actually do. Going beyond the descriptive level, we then consider some important reasons why planners make different ethical choices. Here we focus on the variables mentioned above such as role, political views, and other job related attitudes. Finally we explore some of the implications of our findings for planning theory, practice, and education.

Study design

We sent a mail survey to 1178 members of the AIP, with an overall response rate of 69 percent, and a use-able response from people working in public planning agencies of 616 questionnaires or 53 percent.[5] All responses were completely anonymous. The questionnaire was made up of three parts. The first was made up of a series of 15 short scenarios, each describing how a planner had dealt with a particular ethical dilemma (see Figure 1 for synopses of the scenarios). The respondent was asked if he thought the planner's action was ethical, using a five-point scale; and if he would do the same, using a similar scale.[6]

The issues dealt with in the scenarios were selected to reflect real and difficult ethical dilemmas in the profession. In the interests of keeping the questionnaire to a manageable length, however, we made no attempt to be comprehensive, nor could we provide the respondent with information on the scenario's political and social setting.

Figure 1. (Continued)

Scenario	Tactic	Issue benefiting
9 Regional planner who worked on a wetlands preservation study, without authorization, gives certain findings to an environmental group, because planner feels the agency's director purposely left out those findings, which were objectively documented, from the study draft because they do not support agency policy	leak information	environment
10 Planner who favors increased mass transit use and is preparing a study on need for mass transit decides not to include information from a study done several years ago showing that majority of community's residents opposed expanded mass transit system.	distort information	mass transit
11 Economic planner who initially criticized on technical grounds a proposal by a community development corporation to develop a small industrial park in a ghetto area before the plan commission, later recommends the project to the commission after being told by the director of the director's support for the project.	change technical judgment due to pressure	low income
12 City planner who is a member of Chamber of Commerce, without authorization, gives information to the head of the Chamber of Commerce on an agency study being prepared that will recommend reducing number of on-street parking meters in CBD to lessen traffic congestion.	leak information	development
13 Planning director undertakes a campaign to create a crisis atmosphere about the pollution and health hazards of the city's waterways by holding press conferences next to the city's most polluted waterways to get media coverage.	dramatize problem to overcome apathy	environment
14 City planner gives draft recommendations on scattered site public housing plan to the representative of a white homeowner's group who requests them; no agency policy exists about releasing such information.	release draft recommendations upon request	low income (anti)
15 County planner, without authorization, gives information and advice on own time to a citizen's group which is trying to overturn in court a county rezoning decision which the county planning staff had opposed; the rezoning allows an oil company to build a refinery on a large, tree-covered waterfront property.	assist group overturn an official planning action	environment

The second part of the questionnaire was designed to elicit attitudes about roles in planning, value commitment, orientation toward agency, citizen participation, and about various substantive issues in, or groups affected by, planning: mass transit, the environment, development, and low income and minority groups. The respondent was asked to give his opinion on fifty-three strongly worded statements, using a six-point scale from strongly agree to strongly disagree. The third part of the questionnaire included a variety of demographic questions as well as questions on the respondent's education, political views, and work situation. Of necessity, a mailed questionnaire limited us to information on peoples' *attitudes* about roles, issues, and ethics. We did, however, come closer to actual *behavior* in one respect by asking the respondents not only whether they think the planner's behavior in a given situation is ethical, but whether they would do the same thing if faced with the same situation. Although there have been a number of good in-depth case studies of the behavior of planners on which we could build, none explicitly addressed the question of ethical behavior (Meyerson and Banfield 1955, Altshuler 1965, Rabinovitz 1969, Needleman and Needleman 1974, Jacobs 1978). On the other hand, there has been virtually no research on large numbers of planners—probing into their attitudes on roles, issues, and ethics—so that the tradeoff between behavioral data and larger numbers of respondents seemed well worth making.

What planners think is ethical and unethical

Today's planners do agree strongly about the ethical propriety or impropriety of some kinds of behaviors. On other kinds of issues, there is considerable disagreement; and on any issue there is always a fairly constant ten percent who are uncertain whether any particular behavior is ethically appropriate or not. Differences in the responses seem mainly to be related to the tactic presented in each scenario rather than to the substantive issue or group that benefits. Initially we will discuss the effect of the tactics, and then we will consider the effect of substantive issues. The order of tactics, ranging from most to least acceptable, is given in Table 1. It should be understood that all the scenarios were designed to include political tactics that we thought would be somewhat questionable. Therefore, the rankings reflect planners' opinions of how unacceptable each tactic is.

For eight of the fifteen scenarios, consensus was high; from 66 to 80 percent of the respondents always answered that the behavior described is either ethical or unethical. In these cases, no more than 25 percent of the respondents ever took the opposite position from the majority. Actions to dramatize a problem to overcome apathy, the use of expendables

as a tradeoff, and assistance given on the planner's own time to a group trying to overturn an official action—all of which reflect a fairly activist role in trying to get support for a particular policy—are viewed as the most acceptable tactics. Planners consider threat, distortion of information, and leaking of information as the most unethical tactics, in roughly that order.

It is interesting to note that planners' low toleration for distorting information is consistent with the rule of discipline in AIP's Code of Professional Conduct, which admonishes planners against deceitful conduct. Section (a) of the Rules of Discipline reads: "a planner shall not engage in conduct involving dishonesty, fraud, deceit or misrepresentation." Nevertheless, for the three scenarios in which the planner distorts information (2, 10, and 3), from 13 to 22 percent of the respondents still said that the planner's action was ethical.[7] Strictly interpreted, these planners would be in violation of the AIP rule.

On the other seven scenarios, there was considerably more disagreement. The actions that seemed consistently to provoke such disagreement were releasing the draft recommendations of an unfinished report to a group which requests them, and leaking inside departmental information to an outside group. These issues present particularly clearly the choice between loyalty to one's department and loyalty to some outside group or idea of the public interest.

The scenarios where the planner leaks information to an outside group (4, 8, and 12) actually conflict with Rule of Discipline (d) of the AIP Code, which reads: "Except with the consent of the client or employer . . . a planner shall not reveal . . . information gained in the professional relationship . . . the disclosure of which would likely be detrimental to the client or employer." In each of these scenarios, the planner does in fact reveal information gained in the professional relationship without the employer's consent. Such disclosure might well be interpreted as being "detrimental" to the employer. Although for two of these three scenarios nearly 60 percent thought such an action was unethical, about 33 percent still said the action was ethical. These planners also might be considered to be potential violators of the AIP Code on this point. Clearly, for a minority of planners, the AIP Code of Professional Conduct does not stand up as an irrefutable guide to ethical conduct.

Most planners are generally able to make a choice about the ethical propriety of a particular behavior, even if they do not always agree. But for every one of the scenarios, some always said they were uncertain about the ethical propriety of the behavior described. For all scenarios, the median number of undecideds was 9 percent of the total. The fewest number of undecideds (5 percent) were in the two scenarios where consensus was greatest—scenario 5, where the plan-

Table 1. Rank order of tactics by ethical acceptability

Rank	Tactic	Scenario	Percent total response ethical[a]	Percent total response unethical	Percent total response not sure	Mean[b]
1	Dramatize problem to overcome apathy	13	82	13	5	1.91
2	Use expendables as tradeoff	6	68	21	11	2.33
3	Assist group overturn official action	15	67	24	9	2.31
4	Release draft information on request to environmental group	1	64	27	8	2.46
5	Release draft information on request to white homeowners group	14	54	34	11	2.75
6	Organize coalition of support to induce pressure	8	53	35	11	2.74
7	Release draft information on request to developer	7	47	43	9	2.99
8	Change technical judgment due to pressure	11	42	39	18	3.01
9	Leak information to low income group	4	33	55	12	3.34
10	Leak information to environmental group	9	31	59	10	3.45
11	Distort information	3	22	70	9	3.72
12	Distort information	10	17	74	9	3.87
13	Leak information to Chamber of Commerce	12	16	75	8	3.96
14	Distort information	2	13	81	8	4.05
15	Threaten	5	11	84	5	4.33

[a] Percent for some scenarios is less than 100 percent due to rounding.
[b] The lower the mean score, the higher the number of ethical responses; the higher the mean score, the higher the number of unethical responses. The scale was (1) clearly ethical (2) probably ethical (3) not sure (4) probably unethical (5) clearly unethical.

ner used a threat tactic (the most unacceptable) and scenario 13, where the planner tried to overcome public apathy by dramatizing a problem (the most acceptable). The greatest uncertainty was for scenario 11—where the planner, bowing to pressure from a superior, changed his technical judgment. The split was almost even between those who viewed this behavior as ethical and unethical, but the undecides accounted for nearly 20 percent of the total response.

Previously, we noted that planners react more to the tactic than to the beneficiary in making their ethical choice. This is clear, for example, in the three scenarios (2, 10, and 3) where the planner distorts information to gain more support for a preferred policy. Respondents overwhelmingly consider such an action to be unethical. In each of these scenarios, the issue benefitting would be mass transit, which planners strongly favor.[8] Therefore, the tactic clearly outweighs the issue in this instance.

Yet the benefiting issue, a possible surrogate for values, apparently has some effect on how planners view ethics. Consider Table 2, which shows the responses to three similarly constructed scenarios (1, 14, and 7) where the planner gives the draft recommendations of a plan to someone who asked for them, when no specific policy exists about releasing such information before a plan is completed. Although the tactic is the same, the beneficiary changes in each scenario. While more respondents view this behavior as ethical than unethical in each of these scenarios, a higher proportion apparently feel that giving such information to the environmental group representa-

tive is more acceptable than giving such information to the white homeowners' group representative or to the land developer. Since planners are more pro-environment (70 percent) and pro-low income (57 percent) than pro-development (50 percent), as indicated by their averages on the three attitude items in favor of each issue, values apparently have some effect on ethical choice; though the developer may also be less acceptable because he is also an individual who would benefit financially from the information provided.

The three scenarios (12, 9, and 4) dealing with a planner who leaks information to a representative of a group also suggest the possible effect values might have on ethical choice (see Table 3). Although repondents consider the planner's behavior in these scenarios as much less ethical than that of the planner described in the releasing information on request scenarios (see Table 2), when the leak is to a repre-

Table 2. Effect of values on releasing draft recommendations

Releasing draft recommendations requested	Percent total responses ethical	Percent total responses unethical
to a representative of an environmental group (scenario 1)	64	27
to a representative of a white homeowners' group (scenario 14)	54	34
to a land developer (scenario 7)	47	43

Table 3. Effect of values on leaking information

Leaking information without authorization	Percent total responses ethical	Percent total responses unethical
to a representative of a low income group (scenario 4)	33	55
to a representative of an environmental group (scenario 9)	31	59
to a representative of the Chamber of Commerce (scenario 12)	16	74

sentative of a low income group or to an environmental group, the pro-ethical response is almost twice as high as when the leak is to a representative of the Chamber of Commerce. Again, since planners are less favorable to development than to the environment or to low income groups, values also come into play here.[9]

The relationship between attitudes and behavior

When a respondent says that a planner in one of the scenarios acted ethically, that person is reflecting an attitude about ethics. If, however, that person was faced with the same situation confronting the planner in the scenario, would he behave likewise? To determine whether planners would behave consistently with their ethical perceptions, we not only asked respondents to judge the ethical propriety of the planner's action in each scenario, but we also asked them whether they would behave as that planner did.

We found that there is indeed a strong relation between ethical attitudes and potential behavior. For every scenario, at least 75 percent (in most cases over 80 percent) of the respondents say they would behave consistently.

But this means that for any scenario, from 10 to 25 percent of the respondents would switch—i.e. their attitude does not dovetail with their potential behavior. Two groups of switchers are worth a closer look: those who say the planner's behavior in the scenario is ethical but they would not do it or were not sure they would do it, and those who say the planner's behavior is unethical but they would or might do it anyway. Of the two groups, more judge the behavior as ethical but wouldn't do it than the other way around. Probably the main reason why some say they wouldn't do what they see as ethical is because it's risky. Although threatening someone or going around a superior by leaking information to another group, for example, might be viewed by some as ethical, the possibility of being found out and reprimanded or punished in some way is obviously a deterrent. Not surprisingly, the most acceptable tactics are also the least risky ones. For these scenarios, the number of switches in this direction are lowest. The average percentage of

switches for the three most acceptable tactics (scenarios 13, 6, and 5) is only 13 percent, while the average percentage of switches for the four least acceptable tactics (scenarios 5, 2, 12, and 10) is nearly three times as high or 35 percent.

Far fewer planners would switch the other way; i.e., judging the behavior to be unethical but still saying they would or might do it. For most scenarios, only 5 to 8 percent of those who say the behavior is unethical would switch. But for three scenarios (13, 11, and 6) about 20 percent of those saying the behavior is unethical would switch. Two of these three scenarios (13 and 6) have the highest ethical responses. So, even some planners who saw the behavior as unethical were likely to behave the same way as the great majority who saw the behavior as ethical.

Why planners have different ethical views

What kind of variables seem to explain the differences in ethical choices among planners? Probably the central variable is role. All the scenarios involved situations that were political and actions that were of questionable acceptability. Consistently, the most politically oriented planners found the scenarios more acceptable than did technically oriented ones. In addition to role, there are statistically significant differences on ethics between liberal and conservative planners, between those more and less committed to their agencies—if challenging the agency is an aspect of the scenario—and between value committed and value neutral planners.

We had also hoped to be able to examine the effect of these variables on attitudes about various substantive issues such as the environment, mass transit, development, and redistribution. However, the substantive attitude scales did not effectively measure people's attitudes on these subjects,[10] so that the only way we can get at this effect is by examining the scenarios where the tactic can be held constant and only the issue varied as we did in the last section.

Let us look more closely at the relation between these various independent variables and the dependent variable, acceptable or ethical behavior, starting with the role/ethics relationship.

Roles and ethics

Initially, we thought that an important explanation for differences in ethical perceptions among planners could be traced to a central and continuing conflict in planning, how can planners maintain their technical integrity, yet at the same time, be politically effective? Planning has been struggling over the question of its proper stance as a public profession in a democratic society for many years. The issue has often been posed as a choice between the polar models of the planner as technician (Beckman 1964, Walker

1950, Meyerson 1956) and the planner as a political actor (Rabinovitz 1969, Needleman and Needleman 1974). The former type is supposed to be technically expert, value neutral, and responsible to the public through the political decision-makers he serves. But he is also dependent on those same political decision-makers for the implementation of his "good" advice and plans. The latter, as an ideal type, is more value committed, more responsive to the groups or issues he thinks are particularly related to the public interest, and more willing to work actively through the political system to see that plans are implemented. Only in recent years have there been attempts to think in terms of a role which combines aspects of both technical and political roles to achieve both integrity and effectiveness (Meltsner 1976, Benveniste 1972, Catanese 1974).

For the study, a person's role orientation was determined by his score on two scales, one concerned with technical orientation or attitudes toward analysis, and one concerned with attitudes toward political behavior in planning.[11] For ease in interpreting the results, each scale was dichotomized, and three roles were created (see Table 4):[12] technicians, low on the political scale and high on the technical one; politicians, who had the reverse pattern; and a third group who are high on both scales, who we called "hybrids."[13]

The large number of hybrids is interesting, since those who have developed models which combine both the technical and political dimensions of role seem to indicate that planners who play the combined role are few and rather special.[14] But if our findings are any indication, these polar role types, even combined, are outnumbered by the people who wish to combine aspects of both roles.

As Table 5 indicates, on all scenarios except 9 and 10, the three roles ranked in the same order, with politicians finding the scenarios most ethical, technicians, the least, and hybrids in between; on the two exceptions, politicians and hybrids were virtually identical. On all the scenarios except 10 and 11, the differences between the two extremes, the politicians and the technicians, were statistically significant at the .05 level or better.

The scores of the hybrids are particularly interesting, since they indicate that there seems to be some tension between the two role dimensions. On all but three of the scenarios, the hybrids were not statistically different from the politicians. However, on scenario 5 (on the punitive use of A-95 review power, one of the most unacceptable tactics to all groups) hybrids were not significantly different from technicians. In the remaining scenarios, on expendables (6) and on organizing opposition to the zoning ordinance (8), where the range of opinions was particularly large, the hybrids were significantly different from both other groups. Thus, although hybrids frequently re-

semble politicians, they are almost always more moderate than politicians in their judgments, pulled at least somewhat by their similarity to the technicians as well. In some situations, they may shift to an independent middle ground, while in others they may move all the way to the technical side. With more different kinds of scenarios, this pattern might show up more clearly.

In terms of what planners say they would do on each scenario, the three roles ranked the same way as for ethical judgments. It is interesting, also, that respondents generally are somewhat more conservative about what they say they would do, than about what they think is ethical,[15] and that this does not differ by role.

Political views and ethics

Political views were measured on a scale from 1 to 7, labeled from radical to conservative. The study indicates that 82 percent of the respondents were in the middle three categories—liberal, liberal-moderate, and moderate. The two extremes contained only 23

Table 4. Planner's roles (n = 577)

| | | Political scale | |
		Low	High
Technical scale	Low	(not used as a role) n = 26 % = 4.5	Politicians n = 105 % = 18.2
	High	Technicians n = 153 % = 26.5	Hybrids n = 293 % = 50.8

% is percent of total sample included in table.

Table 5. Means by role for each scenario*

Scenario	Poli-ticians	Hybrids	Tech-nicians
1 release info/env't	2.20	2.37	2.70
2 distort info	3.93	4.02	4.17
3 distort info	3.58	3.64	3.97
4 leak info/low income	3.12	3.20	3.64
5 threat	4.03	4.33	4.45
6 "expendables"	2.07	2.28	2.50
7 release info/developer	2.84	2.94	3.13
8 organize coalition	2.20	2.59	3.31
9 leak info/env't	3.30	3.29	3.76
10 distort info	3.80	3.79	4.02
11 technical judgment	2.88	3.01	3.10
12 leak info/C of C	3.74	3.86	4.26
13 media hype	1.70	1.80	2.17
14 release info/whites	2.50	2.68	2.94
15 assist opponents	2.06	2.20	2.55

* The range for the mean score is from 1 (clearly ethical) to 5 (clearly unethical).

people (3.7 percent of the sample), so that each was combined with the next most liberal or conservative group to make five categories, ranging from conservative to radical.

We thought that political groups might rank the same way roles had, with liberals thinking the scenarios were most ethical, and conservatives the least.[16] In fact, two dominant patterns emerged (see Table 6). The hypothesized pattern did show up for four scenarios (1, 8, 9, and 15) which suggests no obvious logic except that three do deal with environmental issues. Conservatives, however, didn't fit the pattern on five of the scenarios. Surprisingly, on scenarios 3 and 7 they were more like radicals or liberals in their opinions, and on the others they were all to the left of moderate. On 3, 5, and 6, all dealing with tactics, their more liberal stance has a somewhat Machiavellian flavor. On 7 and 14 they were more favorable to the developer and the white homeowners' group, which might have been expected.

In relation to role, we thought that the most political planners might be the most liberal, and technicians the most conservative, with hybrids in between. In actual fact, the single largest group overall is liberals (31 percent of the sample). Politicians were the most liberal, while technicians, though generally more conservative, have a slightly bimodal pattern with high points at both liberal and moderate. Hybrids did have a pattern between the other two, but again it was bimodal like the technicians.

Agency orientation and ethics

A classic dilemma found in any public service profession, including planning, is the possible conflict between what the agency, which presumably serves the public, defines as the public interest, and what the individual professional thinks the public interest is. Whistleblowing (Nader et al. 1972, Finkler 1971),

though risky, is not difficult to justify if the issue involves dishonety on the part of the agency, but what if it is purely a difference of opinion on policy?

In our sample, most planners are quite loyal or committed to their agencies, with 70 percent above the midpoint on the agency orientation scale. On scenarios which involved a challenge to agency policy—all the leaking scenarios, helping the environmental group fight the refinery, and organizing support to challenge zoning policy—these agency-oriented planners thought the action significantly (.05 level) more unethical than the less committed ones. The difference between the two groups was largest on leaking information about the wetlands study (scenario 9), where the challenge to the agency director was most open.

Values and ethics

Overall, 57 percent of the sample had scores above the midpoint scale, measuring propensity to express values in one's work, indicating that despite the idea of the value free planner, the majority of planners think it is acceptable to be open about their values. It is also perhaps not surprising that the values scale has a high correlation ($r_s = .6744$) with the political role scale, indicating that value commitment and a willingness to act on it possibly go together.

Planners who think that they should act on their values in their work are different in their ethical judgments from planners who wish to be value neutral. When the scenarios are grouped together by similar issues, the difference between the value orientations is significant at better than .01 for giving out recommendations, leaking information, and for the most acceptable tactics (scenarios 6, 8, and 13). There is no significant difference, however, between the two groups on the scenarios concerned with the misrepresentation of information.

Table 6. Means for ethics scenarios by political views*

Scenarios	Radicals	Liberals	Lib/Mod	Moderates	Conservatives
1 release info/env't	1.76	2.31	2.59	2.65	2.80
2 distort info	3.86	4.08	3.97	4.27	3.92
3 distort info	3.56	3.72	3.71	3.86	3.47
4 leak info/low income	2.20	3.25	3.61	3.56	3.47
5 threat	3.88	4.28	4.36	4.50	4.37
6 "expendables"	1.69	2.30	2.31	2.57	2.54
7 release info/developer	2.40	3.01	3.06	3.14	3.00
8 organize coalition	2.02	2.71	2.76	2.87	3.22
9 leak info/env't	2.67	3.39	3.50	3.61	3.83
10 distort info	3.88	3.74	3.89	3.94	4.08
11 technical judgment	3.25	2.98	3.04	2.87	3.33
12 leak info/C of C	3.40	3.98	3.94	4.12	4.06
13 media hype	1.60	1.92	1.84	1.98	2.19
14 release info/whites	2.02	2.75	2.83	2.89	2.83
15 assist opponents	1.58	2.12	2.33	2.67	2.77

* The range for the mean score is from 1 (clearly ethical) to 5 (clearly unethical).

Moreover, as we saw earlier, many planners do seem to be influenced, at least to some extent, in what they think is ethical by the intended beneficiaries of their actions. The same tactic used in behalf of different groups is judged differently. We expected this effect to be much stronger for politicians, who approve much more of open value commitment, than for technicians who generally wish to be value neutral. Actually, the results are rather mixed. On the scenarios giving out recommendations (1 and 14), politicians are influenced[17] more by the issue than are technicians; but when it comes to leaking information, there is no difference between the two roles. There is, however, an interesting difference on the leaking scenarios in that the beneficiary is much more important to planners who are not strongly committed to their agencies than for those who are.

Role types

All the variables discussed so far obviously have an effect on what planners think is ethical. But the key variable seems to be role, with the other variables clustering around it in coherent patterns. We will now look more closely at each role type—the technician, the politician, and the hybrid.

The most traditional image of the planner is as a technician. The two major aspects of the technical role are its faith in the efficacy of analysis and its value neutrality; and on both of these, the comparison between the technicians in the sample and the politicians is dramatic. Fully 81 percent of the technicians indicated that they were value neutral compared with only 15 percent for the politicians. Even more strikingly, 94 percent of the technicians said that a planner's effectiveness is based primarily on his reputation for objective, accurate, and in-depth analysis, while only 51 percent of the politicians agreed. Technicians are also much more likely (63 percent) than politicians (12 percent) to argue that plans should stand or fall for their acceptance on their technical quality and internal logic.

In line also with their value neutrality, technicians were more committed to their agencies, and less likely to challenge agency policy. Grouping together all the scenarios which involved leaking information or challenges to agency policy, the technicians on the average thought that those scenarios were more unethical than all the rest of the scenarios grouped together, while the politicians thought those scenarios were more ethical than all the others combined. Technicians were also more committed to their agencies, with 76 percent over 3.5 on the scale, compared with 64 percent of the politicians. On the scenario involving leaks to outside groups, where the tactic could be held constant, planners who were more oriented to their agencies had a narrower range of scores over the three scenarios, indicating that they were more swayed by the impropriety of the tactic itself, and less by any ties to outside groups.

Given these characteristics and views, it is not surprising that technicians would be the most conservative overall in their view of what is ethical, since the scenarios present situations involving challenging the authority of an agency by leaking information or organizing outsiders to change policy, misrepresenting information in analysis, and generally being politically active in trying to get plans adopted. Thus, there were no scenarios where the mean for all technicians was below 2.00, the point on the scale that indicates a judgment of "probably ethical."

Since the quality and integrity of technical analysis is particularly important to them, it is not surprising to find that misrepresenting information (scenarios 2 and 10) is unacceptable; the mean for all technicians on these two is over 4.00, the score for "probably unethical." Moreover, when faced with a conflict of loyalty between agency and analysis, as in scenario 11, analysis continues to exert a strong pull. Agency oriented technicians were more likely to think that the economic planner's capitulation was ethical, but the difference between them and the non-agency oriented technicians was smaller than for any other agency-oriented scenario except one.[18]

Politicians, of course, are at the opposite extreme from technicians in our typology. Overall, they are more interested in influencing policy than the other groups, and are most willing to use a range of openly political tactics to do it. For example, 51 percent of the politicians ranked influencing policy among the three most important aspects of their job, compared with only 30 percent of the technicians. As Table 7 indicates, politicians clearly are much more prone to lobby, mobilize support, and neutralize opposition than are technicians. They are also more interested than technicians in routinely being involved in policy disputes; and a majority, unlike the technicians, think that the quality and depth of analysis done by planners has little to do with their effectiveness.

In general the politicians are more accepting of the political tactics in the scenarios than the technicians are, although this does not mean they think all tactics are ethical. As Table 5 indicates, the mean for all politicians is less than 2.5—in the probably-to-clearly-ethical range—for six scenarios compared to only two for the technicians. Even though politicians are significantly more accepting than technicians of the use of A-95 power as a threat, and of distorting information, the means for these four scenarios (2, 3, 5, 10) range from 3.58 to 4.03, putting them all in the "probably unethical" range.

Politicians are also somewhat less likely to be committed to their agencies and are more likely to have an independent commitment to issues or client groups. They are as value committed (85 percent) as the tech-

Table 7. Support for political tactics by role in percentages (n = 577)

	Politicians	Hybrids	Technicians
lobby actively to defeat harmful proposals	94.3	93.8	62.8
develop groups to support plan	89.6	86.4	42.5
neutralize opposition by mobilizing support	73.3	74.4	26.2
may have to work covertly to gain support	70.1	68.2	50.9
planning should be placed in gov't so it can get involved in disputes	66.7	63.0	22.2
quality and depth of analysis has little to do with effectiveness	57.2	36.8	30.7

nicians are value neutral (81 percent). As indicated previously, they are relatively more willing than technicians to say it is ethical to give recommendations to the environmental group (scenario 1); and although willing to accept the principles as ethical, they are more unwilling actually to give recommendations to the white homeowners' group (scenario 14) or to leak information to the Chamber of Commerce (scenario 13).[19]

Hybrids are the group of planners who are high on both the technical and the political scales. Since they feel the tension between the two role dimensions most, they tend to score between the politicians and the technicians on most variables. Consequently, there are relatively few instances where hybrids are obviously unique by virtue of having scores dramatically higher or lower than the other two groups. Instead, the analysis of this role has to focus on when they seem to be more like politicians, when they are more like technicians, and what kind of overall pattern this produces.

Simply by virtue of the way their role is defined, hybrids are quite close to technicians in their attitudes about analysis, and close to politicians in their attitudes about political tactics. For example, they are similar to technicians in thinking that the effectiveness of planners is based on good analysis, and that plans should stand or fall on their quality and logic. On the political attitudes listed in Table 7, they are virtually identical to the politicians except on the last item on analysis where, again, they are more like the technicians.

In terms of ethics, this means they tend to be more like politicians on tactical scenarios. On the scenarios dealing with leaks and giving out of recommendations, and on the symbolic campaign over pollution, the two groups were statistically indistinguishable. But as the tactics get more varied, and then more questionable,

the differences increase. On using expendables as a bargaining device (scenario 6) and on organizing support (scenario 8), they are statistically different from both other groups; while on the use of A-95 review as a punitive tool (scenario 5), they shift to agreement with the technicians.

It is harder to see clearly the effect of their technical values, since the scenarios posing issues of distorting information (2 and 10) or of giving in on a technical judgment (scenario 11) have a much smaller overall range. This means that even if the difference between technicians and politicians is significant, the hybrids in the middle are difficult to sort out from either.

In terms of who they are, they seem generally to be more like politicians than technicians, though they are still always more moderate. They are liberal, though not as much as politicians, with 42 percent left of center and only 34 percent right of center. Sixty-seven percent are value committed, much closer to the politicians' 85 percent than the technicians' 19 percent. A similar pattern holds for orientation to agency, though the overall range is much smaller. As with politicians, these last two characteristics tend to make them more independent, and possibly more active participants in the planning process. This helps to explain the hybrid's greater acceptance of the scenarios on giving out recommendations, leaking information, and challenging agency policy.

There are also a few interesting and unique qualities of hybrids which give a more rounded image of them. They are somewhat more likely to be over 40, and are disproportionately found in the groups with the least (0 to 2) and the most (21 or more) years of planning experience. This might suggest that the attempt to combine the two aspects of role, while a significant characteristic for all kinds of planners, is somewhat more likely to be true of the inexperienced and possibly idealistic young, and the older, more experienced members of the field.

Also the only instance where they hold views stronger than both other role groups was on the belief that planning should be long range. When taken with their combined technical and political approach to planning, this attitude may indicate a more ambitious idea of what planning should be than is held by either other group.

This image of the hybrids gives some support to Meltsner's (1976, pp. 36–47) idea of the entrepreneurial, technically sophisticated politician: a more active, independent, and skilled actor than either of his other kinds of analysts. Because of its base in attitudes, however, our study tends to emphasize the tension and balance aspects of this role. If, on the other hand, the ethical dilemmas posed in our scenarios are seen as a restraint on the political aspects of the planner's role, these hybrids are generally no more limited than their fellow politicians.

Conclusions

Overall, planners' views of ethics fall into two categories. The first are core professional values, shared by virtually all people in the sample. Two core values that clearly emerge from the study are the un-acceptability of using threats or distorting information. Then there is a large area where values or ethics seem to be relative, with judgments influenced more by such factors as role, political views, and agency orientation. Tactics or means seem to have a more important effect on ethical choices than do the substantive issues or ends the tactics are used to achieve, though substantive issues do have some impact. Since tactics are important, role, which particularly relates to planning strategy and tactics, is an important determinant of planners' ethics. A variety of other variables closely related to role, such as political views, agency orientation, and value orientation, cluster around it in characteristic patterns, having similar effects on what planners think is ethical. It is also significant that planners say that they would act consistently with their ethical views.

Much of our description of planners simply provides greater insight into the already well developed political and technical role models. Our research can enable us to give an indication of how many are in each group, and what they are like, with more statistical precision than previous studies.

But the role we have called the hybrids provides some new insight into the practice of planning, since these people are trying to bridge the gap, combining the characteristics of both roles. Since there are more hybrids in our sample than politicians and technicians combined, it may perhaps be more difficult to maintain one of the polar roles in practice than in theory. But if hybrids have become, and can be expected to remain the dominant group, this poses a number of issues about the adequacy of the ethical standards now in force in the profession.

Our data indicate that ethical standards in planning are relative. It is certainly true that some kinds of behavior, such as distoring information, are considered unacceptable by the vast majority of all planners. But even for these behaviors, there remain some planners who still consider them ethical. And for many political tactics, planners disagree much more about acceptability, depending on such things as their role orientations, political views, value commitment, or sympathies toward the substantive issues at stake.

This variability in ethical judgments means that it is very difficult to establish any single ethical standard that is meaningful to the whole profession. At the extreme, it is not difficult to think of behavior that most people would agree is unethical. Bribery clearly qualifies, and so, probably, does distortion of information; but what about providing information to outside groups so they can fight your agency, or using threats, tradeoffs, or symbolic appeals to get a plan approved?

One might make the argument that it is better for planning to have a restrictive set of ethical standards. Why? Because in a democratic society planners are not decision makers who can be held responsible by the public. They are experts who "attempt to provide public . . . decision makers with the best possible information, analysis, and recommendations to promote the public welfare" (AIP 1977, p. 2). Such a definition would argue that planners should avoid taking upon themselves the right to define the goals that guide the definition and solution of problems, or trying to openly and actively get their own particular views adopted in the political system. They should have a restrictive view of the scope of planning. Our data indicate that the most effective way to ensure that planners have such a restrictive view of planning and of ethical behavior would be to train people as technicians, since, of our three roles, they have not only the narrowest view of the planners' range of discretion, but are even more restrictive in what they are actually willing to do in practice. This is, in effect, the way the present role of ethics of the AIP is drawn; and historically the dominant role model presented to planners had been the technical one.

But according to our study, hybrids are actually the dominant group in the field. In practice, they have a broader image of the proper role of the planner and a less restrictive view of what is ethical. In this they are quite similar to, though more moderate than, the politicians, and between them the two groups make up 70 percent of our sample.

This raises the question of whether it is useful to have a code of ethics, some provisions of which are more honored by many in the breach than in the observance. As Francine Rabinovitz (1969, pp. 133–134) wrote in 1969:

> The profession . . . still officially discourages political roles, having developed no code of conduct that defines what types of political strategies are ethically acceptable and what types are expedient but unprofessional.

Might it be better to reconsider portions of the code in light of the evolving practice of planning? Planners often make ethical judgments implicitly in making strategy and methodological decisions, but they may not give them the sustained and systematic thought that they give to the more practical issues. Moreover, the more political they are in their planning activities, the more they have to face the kind of dilemmas posed in our scenarios, which are not adequately dealt with in the AIP's code of ethics.

Obviously the possibility of making the code less restrictive raises the problem of the tradeoff between technical integrity and political effectiveness. To be too permissive about tactics might lead, as Rabinovitz suggests (1969, Chapter 6), to a loss of legitimacy. On the other hand, to be too restrictive raises the problem, much discussed since the publication of Altshuler's *The City Planning Process*, of being ineffective and unable to carry plans through to fruition.

But the interesting aspect of our study is that it is the hybrids who dominate, not the politicians; and it is the hybrids who make the greatest effort to combine both the technical and the political aspects of role. This would indicate that they are likely to be sensitive to both the problem of loss of legitimacy and the problem of lack of effectiveness. The two aspects of their role do involve a tension which may at times be difficult to balance. As educators, it may be our role to try to develop ways to either reduce the tension or to enable planners to deal more effectively with it. Simply making clear that it exists and that both the political and the technical aspects of role orientation are legitimate and necessary for a balance, may be a step in that direction. Going further, educators could deal head-on in the classroom with the kinds of ethical dilemmas posed in our scenarios, having students grapple with such problems early instead of waiting until they become practitioners to puzzle through the issues these dilemmas pose. But it may also be the role of the professional organization to provide a more carefully thought out and realistic set of guidelines to behavior in this difficult area of practice.

Notes

1. The authors would like to thank all the members of the AIP who participated in the study.

2. Peter Marcuse (1976) demonstrates quite convincingly the divergent ethical prescriptions that underlie professional planning ethics. These include: allegiance, autonomy, knowledge and competence, guild loyalty, concern for the public interest, dissent, loyalty, advancement of knowledge, and statutory responsibility.

3. The public interest concern is most directly reflected in Section 1.1(a) of the Canons of AIP's Code of Professional Responsibility—"A planner serves the public interest primarily." The loyalty to the planner's client concern is reflected in Rules of Discipline (d) of the AIP Code.

4. We also would like to be able to determine what factors are most important in affecting planners' ethical choices. The study was designed with a fairly complex path analysis model in mind. But testing the strength and simultaneous effect of a large number of independent variables (26) on each other and on the dependent variable is a difficult methodological task, especially when, as is the case here, most of the variables are measured only at the ordinal level. This particular paper examines the relationships between six of the twenty-six independent variables and the two dependent variables.

5. Every sixth person on the AIP mailing list was chosen unless that individual's address clearly indicated that he worked for a planning consulting firm, a university, or a private development, or if the code indicated that the person was a student or an affiliate. (The method of choice was count, discard, recount.) The sampling fraction was increased from 12.3 percent to 18.5 percent to allow for the elimination during the sampling procedure of people who obviously did not belong in the sample, and for an additional estimated 27 percent of the final sample who would also turn out to be in one of the categories listed above. We sampled members of the AIP because their mailing list was the only list of professional planners available. We have no way of knowing, however, whether AIP members were typical of all planners.

6. The only two other empirical studies of ethics we could find both use the same kind of scenario format (Carlin 1966, Beard & Horn 1975). One, a study of the ethics of congressman, uses the same five-point scale.

7. These percentages would increase to 20 to 30 percent if those who were uncertain about the ethical propriety of the action were added to those who said the action was ethical.

8. Planners come out strongly in favor of mass transit on the transportation scale in Part 2 of the questionnaire. Seventy percent are pro-mass transit while 30 percent are anti-mass transit.

9. It is interesting to note that, in fact, five of these scenarios form a Guttman scale. A cutpoint of 3.0 was used so that the "not sures" were included along with those who thought the scenarios were unethical. The coefficient of reproducibility for the scale was .9058, and the coefficient of scalability was .7351. The order for the scenarios is the same as in the tables above: scenario 1 is the most acceptable, followed in order by scenarios 14, 7, 4, and 12. This means that the 16 percent of the sample who would leak information to the Chamber of Commerce (scenario 12), would accept providing information to all other kinds of groups; while the 35 percent who would not or were not sure that they would provide information to an environmental group on request, would not provide or leak information to any other kind of group.

10. The variability on substantive issues was much higher in the final sample than in the pretest group. Items used in the construction of Likert scales, which had explained from 30 to 60 percent of the variance in the pretest, explained only from 4 to 20 percent of the variance for the final sample. This makes them weak tools for analyzing substantive attitudes, and we have not relied on such scales in our analysis.

11. We were particularly concerned about the construction of the scales measuring role orientation. Originally, following the drift of much of the literature on planners, we thought of role as a single dimension with the technical and political orientations as its two reverse sides. However, a factor analysis of the pretest results indicated that the items should be separated into at least four scales, a political one, a technical one, one concerned with the planner's attitude toward his agency, and one concerned with attitudes about value commitment. A factor analysis of the final sample showed this pattern even more clearly. The factor analysis used principal factoring with iteration. Both orthogonal (varimax) and oblique rotation were tried, with similar results. For the final construction of the scales, orthogonal (varimax) rotation was used. This grouped the items into six factors. Items with factor loadings of more than .25 were used to create the scales. The political scale included items on political tactics and cynical attitudes toward analysis. The technical scale was made up of two factors, one of which had five items stressing the importance of rationality and technical expertise, and a second, with two items on long-range planning and value neutrality.

Generally, Likert scales of this type have about 20 items in them, but scales with as few as 5, 7, and 10 items are acceptable

(Likert 1967, Hall 1934). The technical and political scales had six and seven items, respectively. The agency orientation and value commitment scales had five items each. Item analysis of the four role scales came out quite strong. Correlations between each item and its scale were not only significant at the .001 level (due in part to the sample size), but also had in each instance, an r^2 explaining at least 23 percent of the variance.

12. The fourth possible role, low on both scales, had only 26 respondents in it, so we did not name it or use it. Meltsner (1976, p. 15) calls the comparable category in his typology the "pretenders."

13. These roles are quite parallel to Meltsner's (1976, Chapter 2) in their underlying dimensions, though not necessarily in their associated characteristics. They were, interestingly enough, developed independently, before either author had read Meltsner's book. They arose logically out of the factor analysis of the pretest results. Meltsner uses the term "entrepreneur" for his high/high category, drawing also on Bardach (1972). The term indicates that these are the most skilled and active analysts of his typology. We wanted a somewhat less value laden term which would indicate that the primary characteristic of our high/high group is that they try to balance the two dimensions of role, and come between the other two roles on a number of our variables. The term "hybrid" denotes "anything of heterogeneous origin or incongruous parts," as animals (especially swine) of mixed parentage.

14. Of Meltsner's group of 116 analysts at the Federal level, only 23 percent were entrepreneurs. However, he was looking at their political and analytic skills, while we have only been able to look at planners' attitudes about what they "should" do. We cannot infer from our data whether the planners interviewed have the skills, opportunity, or even the personal inclination to play the roles they say are best.

15. The measure for action was the mean for each scenario. The means for action were lower than for ethics on 11 out of 15 scenarios for each role, though which 11 scenarios had lower means for action differed from one role to another. Only on one scenario for technicians and on two each for politicians and hybrids, were the action means higher than the means for ethics.

16. Overall, the differences between radicals and conservatives were significant for all scenarios except 2, 3, and 10 (among the least acceptable) and scenario 11 (the most ambivalent). Looking at liberals and moderates reduces the number of scenarios where the difference is significant to seven (1, 2, 4, 5, 6, 9, and 15). Even between the three middle categories, some differences still hold up.

17. Influence is determined by looking at the difference between the means for each scenario. We would expect that if the tactic is held constant, as it is in scenarios 1 and 14, and also in scenarios 4, 9, and 12, the difference in means would be due to the influence of the issue or beneficiary. The size of the difference can be compared for any other variable such as role or agency orientation; the larger the difference is the more the issue or beneficiary matters. Thus, if the difference between scenarios 1 and 14 is larger for politicians than for technicians, then the issue or beneficiary matters more to the politicians.

18. The exception was scenario 12, the leak to the Chamber of Commerce. There was no real difference between the two groups. The beneficiary was less acceptable, and all technicians thought the scenario was unethical.

19. The influence of the environment as an issue was determined in the same way as described in Footnote 17. On unwillingness to act, the mean for action was less than the mean for ethics, and the difference in the means was larger for politicians than for technicians.

References

Alinsky, Saul D. 1971. Of means and ends. In *Rules for radicals*. New York: Random House.

Allor, David. 1970–71. Normative ethics in community planning. *Maxwell Review* 7, 1: 113–137.

Altshuler, Alan. 1965. *The city planning process: a political analysis*. Ithaca, N.Y.: Cornell University Press.

American Institute of Planners. 1962. *Code of professional responsibility and rules of procedure*. Washington, D.C.: AIP.

American Institute of Planners. 1973. *The social responsibility of the planner*. Washington, D.C.: AIP.

American Institute of Planners. 1977. *Planning policies, '77*. Washington, D.C.: AIP.

Bardach, Eugene. 1972. *The skill factor in politics*. Berkeley: University of California Press.

Beard, Edmund, and Horn, Stephen. 1975. *Congressional ethics: the view from the House*. Washington, D.C.: Brookings Institution.

Beckman, Norman. 1964. The planner as a bureaucrat. *Journal of the American Institute of Planners* 30, 4: 323–327.

Benveniste, Guy. 1972. *The politics of expertise*. Berkeley, California: The Glendessary Press.

Bok, Sissela. 1978. *Lying: moral choice in public and private life*. New York: Pantheon Books.

Carlin, Jerome. 1966. *Lawyer's ethics*. New York: Russell Sage Foundation.

Catanese, Anthony. 1974. *Planners and local politics: impossible dreams*. Beverly Hills, California: Sage Publications.

Finkler, Earl. 1971. *Dissent and independent initiative in planning offices*. Chicago: American Society of Planning Officials.

Hall, O. Milton. 1934. Attitudes and unemployment. *Archives of Psychology* 165: 5–65.

Howe, Elizabeth. 1978. Ethical issues and the "publicness" of professions. Paper given at the meetings of the American Society for Public Administration, Phoenix, Arizona.

Jacobs, Alan. 1978. *Making city planning work*. Chicago: American Society of Planning Officials Press.

Kelman, Herbert C. 1965. Manipulation of human behavior: an ethical dilemma for the social scientist. *Journal of Social Issues* 21, 2: 31–46.

Klosterman, Richard. 1978. Foundations for normative planning. *Journal of the American Institute of Planners* 44, 1: 37–46.

Likert, Rensis. 1967. The method of constructing an attitude scale. In *Readings in attitude theory and measurement*, ed. Martin Fishbein. New York: John Wiley & Sons.

Marcuse, Peter. 1976. Professional ethics and beyond: values in planning. *Journal of the American Institute of Planners* 42, 3: 264–274.

Meltsner, Arnold. 1976. *Policy analysts in the bureaucracy*. Berkeley: University of California Press.

Meyerson, Martin, and Banfield, Edward. 1955. *Politics, planning and the public interest*. New York: The Free Press.

Meyerson, Martin. 1956. Building the middle range bridge for comprehensive planning. *Journal of the American Institute of Planners* 22, 2: 58–63.

Nader, Ralph; Petkas, Peter; and Blackwell, Kate, eds. 1972. *Whistle blowing: the report of the conference on professional responsibility*. New York: Grossman Publishers.

Needleman, Martin, and Needleman, Carolyn. 1974. *Guerillas in the bureaucracy: the community planning experiment in the United States*. New York: John Wiley & Sons.

Rabinovitz, Francine. 1969. *City politics and planning*. New York: Atherton Press.

Walker, Robert. 1950. *The planning function in urban government*. Chicago: University of Chicago Press.

Zaltman, Gerald, and Duncan, Robert. 1977. Ethics in social change. In *Strategies for planned change*. New York: John Wiley & Sons.

Originaltext Nadin/Stead 2008

European Spatial Planning Systems, Social Models and Learning

disP 172 · 1/2008 **35**

Vincent Nadin and Dominic Stead

Abstract: An underlying aim of European territorial cooperation initiatives, such as INTERREG, is that they will lead to mutual understanding and learning across national boundaries. However, the effect of mutual learning on national systems and policies of planning is uncertain. After all, spatial planning systems are deeply embedded in their socio-economic, political and cultural context, which can potentially constrain the scope for mutual learning. Moreover, planning systems may have a certain degree of path-dependency, such as the persistence of institutions and cultures. In this paper, we explore the relationship between planning systems and their context and assess the extent to which cooperation and learning might contribute to convergence in styles of planning in Europe, as well as why and how this might be taking place.

We take as our context the prevailing social model or model of society, the collection of common social and cultural values. We examine and compare typologies of planning systems and typologies of social models and find a degree of correspondence between them. The examples of England and the Netherlands are used to illustrate this interdependence. However, they also reveal how planning systems and policies in different contexts are changing in similar ways, and perhaps even demonstrate a measure of convergence. In other words, external factors may be overriding or undermining the influence of the national social model. The implementation of planning reforms may be running ahead of wider changes in the social model.

1. Introduction

The notion of models of society (or ideal types of societies) is used to generalize about the diverse values and practices that shape relationships between the state, the market and citizens in particular places. The closely related concept of the social model is used to generalize about the collections of values that underpin policy positions. Both concepts make use of ideal types to help to explain the more complex reality of models in specific nation-states and regions. The abstract ideal types allow us to compare and

explain approaches to real world problems in particular places, not least, reconciling the competing objectives of economic competitiveness, social cohesion (or social justice) and environmental sustainability. While social models have been employed most extensively in explaining approaches to social welfare policy, we propose that they might also assist in providing explanations for variation in spatial planning systems.

The concept of the spatial planning system has been used as a generic term to describe the ensemble of territorial governance arrangements that seek to shape patterns of spatial development in particular places. Comparative analysis of planning systems in Western Europe has led to a clustering of systems into ideal types that have been applied to the real practices in nations and regions. These might be termed models of spatial planning. Enquiry into models of spatial planning (comparative spatial planning) parallels to a degree the debate about comparative societal models, including questions about how planning systems adjust to external and internal pressures for change through reform, and the result of this in terms of convergence or divergence of systems (Healey, Williams 1993) or the "Europeanization of spatial planning" (Dühr et al. 2007). At the European scale, there has been no serious suggestion of a European planning model, although the concept of territorial cohesion, now dominant in the European spatial planning discourse, has been described as a *spatialization* of the European social model (Davoudi 2007).

The form and operation of planning systems are embedded in their historical context, the socio-economic, political and cultural patterns that have given rise to particular forms of government and law. Underlying the contextual differences is the social model. This is exemplified particularly well in some countries where strong state intervention in spatial development was established as part of the post-war welfare state. However, planning systems and policies are coming under increasing influence from other factors, notably the need to respond to global economic competition, international agreements and European integration (Dühr et al. 2007). Extensive cooperation between planners across national and regional borders has led to

Vincent Nadin is Professor of Spatial Planning and Strategy in the Faculty of Architecture at Delft University of Technology, The Netherlands.

Dominic Stead is a Senior Researcher at the OTB Research Institute for Housing, Mobility and Urban Studies of Delft University of Technology, The Netherlands.

An earlier extended version of this paper was presented at the International Planning Cultures in Europe Conference in Hamburg in June 2007 (where it won the prize for best paper) and at the Association of European Schools of Planning (AESOP) Conference in Naples in July 2007.

a wide exchange of ideas and practices, particularly over the last decade. This learning process has undoubtedly shaped spatial planning systems and policies, and led to a certain degree of convergence or harmonization, which may be in tension with the underlying social model. The European Spatial Development Perspective (ESDP) made reference to this point, cautioning that spatial development policies "must not standardize local and regional identities in the EU, which help enrich the quality of life of its citizens" (CSD 1999: 7).

In this paper, we examine the relationships between social models and models of planning. We summarize how typologies of social models and models of planning have developed over the last 20 years with increasing division into more numerous types, and how the typologies have been applied to particular countries. We demonstrate a strong correspondence in the application of social models and models of planning to particular countries. We then explore the relationship between the dominant social model and the planning system in two countries of northern Europe: the Netherlands and England. This comparison demonstrates the strong relationship between the two. We explain how external forces and learning have combined to shape significant adaptations in the form of planning in the two countries and how this confronts the underlying model of society. This in turn explains why the reform of planning systems has been so challenging and raises questions for future reform.

2. European Social Model(s)

Despite many years of discussion in both academic and political circles, neither the term European Model of Society nor the term European Social Model have been defined with any precision (House of Lords 2004; Jepsen, Serrano 2005; Alber 2006). These two terms are often used interchangeably, although some authors identify distinct differences between them (e.g., Delanty, Rumford 2005). The lack of precision in defining these two concepts has both advantages and disadvantages.[1] Alber (2006) suggests, for example, that the concept of the European Social Model can be seen as "a deliberately ambiguous and elastic political metaphor that aims at fostering an epistemic European policy community with a shared view of social problems" (p. 414).

In EU policy, one of the first references to the European Social Model appeared towards the

end of Delors' 10-year presidency of the European Commission in the 1994 White Paper on Social Policy where it was defined in terms of shared values: democracy, individual rights, free collective bargaining, the market economy, the equality of opportunity for all, social welfare and solidarity (CEC 1994). Since then, the concept of the European Social Model has appeared in various EU policy statements.

The Presidency Conclusions of the Lisbon European Council meeting in 2000, from which the EU's Lisbon Strategy originates, made specific reference to the European Social Model. "Modernizing the European social model, investing in people and combating social exclusion" was identified as one of the main strategies for achieving the Lisbon Agenda's goals (CEC 2000a). What was meant, however, by the term European Social Model was not fully elaborated in the document. Later in 2000, this was provided in an annex of the Presidency Conclusions of the Nice European Council meeting (CEC 2000b). The idea that the European Social Model refers to a common set of values is clearly apparent:

"The European social model, characterized in particular by systems that offer a high level of social protection, by the importance of the social dialogue and by services of general interest covering activities vital for social cohesion, is today based, beyond the diversity of the Member States' social systems, on a common core of values" (CEC 2000b: 4).

In 2002, the Commission's first report on economic and social cohesion made passing reference to the idea of a model of society, stating that "cohesion policy is the guardian of a particular model of society" (CEC 2002: 17). Also around this time, a number of high-level reports were being commissioned by the EU (all of which were searching in various ways for a new impetus for the floundering Lisbon Strategy), which also took up the issue of the European Social Model. These include the Sapir Report (Sapir et al. 2004), the Strauss-Kahn report (Strauss-Kahn 2004), the report of the Michalski group (Biedenkopf et al. 2004) and the Kok report (CEC 2004).[2]

More recently, there have been debates about whether there is not just one model of society in Europe but rather a number of variants or related models. There are, after all, large differences in welfare systems and levels of inequality across European countries (Giddens 2005). Speaking about the future of the European Model at Harvard University in September 2005, the European Commissioner for Eco-

disP 172 · 1/2008 **37**

nomic and Monetary Affairs, Joaquín Almunia, argued that "there is no such thing as a single European social model" (Almunia 2005). The content of his speech suggests that there is the viewpoint that there are a number of different social policy models but that they share a set of common features or underlying aims, notably, reducing poverty and social exclusion, achieving a fairer distribution of income, providing social insurance and promoting equality of opportunity.

Ideal types and classifications of social models

Publication of Esping-Andersen's "Worlds of Welfare" thesis in 1990 (Esping-Andersen 1990) led to an increase in interest in welfare state classification and the subsequent emergence of several competing typologies (Arts, Gelissen 2002; Bambra 2007). Esping-Andersen's typol-ogy was based on the criterion of decommodi-fication (i.e., the degree to which social services are provided as a matter of right and the extent to which individuals can maintain a normal and socially acceptable standard of living without reliance on the market). Esping-Andersen's work has provoked an extensive and ongoing debate about typologies of models of society; the principles or criteria that ought be used for their construction, and the classification of welfare state regimes into types (Bambra 2007).

Criteria that have been used to construct typologies of social models in the welfare state literature have included the decommodification approach (Esping-Andersen 1990), basic income (Leibfried 1992), poverty rates (Ferrera 1996; Korpi, Palme 1998) and social expenditure (Bonoli 1997; Korpi, Palme 1998). The typologies that result from these analyses are summarized in Table 1. In general, the number of different regime types has increased over time

Esping-Anderson 1990	Social-democratic DK, FI, SE, NL	Liberal IE, UK	Conservative AT, BE, FR, DE			
Liebfried 1992	Scandi-navian DK, FI, SE	Anglo-Saxon UK	Bismarck AT, DE	Latin Rim FR, GR, IT, PT, ES		
Ferrara 1996	Scandi-navian DK, FI, SE	Anglo-Saxon IE, UK	Bismarck AT, BE, FR, DE, LU, NL	Southern GR, IT, PT, ES		
Bonoli 1997	Nordic DK, FI, SE	British IE, UK	Continental BE, FR, DE, LU, NL	Southern GR, IT, PT, ES		
Korpi, Palme 1998	Encom-passing FI, SE	Basic Security DK, IE, NL, UK	Corporatist AT, BE, FR, DE, IT			
Sapir 2006	Nordic DK, FI, SE, NL	Anglo-Saxon IE, UK	Continental AT, BE, FR, DE, LU	Mediter-ranean GR, IT, PT, ES		
Aiginger, Guger 2006	Scandina-vian/Nordic DK, FI, SE, NL	Anglo-Saxon/ Liberal IE, UK	Continental/ Corporatist AT, BE, FR, DE, LU, IT	Mediter-ranean GR, PT, ES	Catching-up CZ, HU	
Alber 2006	Nordic DK, FI, SE	Anglo-Saxon IE, UK	Continental AT, BE, FR, DE	Southern GR, IT, PT, ES	New Member States CY, CZ, EE, HU, LV, LT, MT, PL, SK, SI	Other LU, NL

Tab. 1: Welfare state typologies (based in part on Arts, Gelissen 2002)

as a consequence of more sophisticated analyses of welfare systems. Since 1990, the number of regime types in Europe has increased from Esping-Andersen's original three (social-democratic, liberal and conservative) to five or six (Aiginger, Guger 2006; Alber 2006). A number of countries appear in the same position in almost all classifications whereas the position of other countries is quite different for each classification. Finland and Sweden, for example, consistently appear together in the encompassing Nordic/Scandinavian/social democratic category, Ireland and the United Kingdom in the Anglo-Saxon/basic security/liberal category, France and Germany in the Bismarck/conservative/continental/corporatist category and Portugal and Spain in the Latin Rim/ Mediterranean/ southern category. In contrast, countries such as Luxembourg and the Netherlands find themselves together with a different group of countries in almost every classification.

It is important to note here that the various regime types are ideal types that owe their origins to different historical forces (Arts, Gelissen 2002). The allocation of countries to types is not always clear-cut and they may sometimes lie somewhere between types. Contrary to the ideal world of welfare states, the real world exhibits hybrid forms in every country. There may sometimes be quite a lot of variation between welfare systems of countries that appear in the same regime type. Even countries with similar sets of welfare institutions are frequently found to display widely divergent patterns of development (Alber 2006). It is also important to note that the classification of countries into regime types is time-dependent: governments, policies and economic activity can all change over time and directly influence the position of a country in the classification system.

3. European Models of Planning (Planning Systems)

Two main approaches are evident in classifying spatial planning systems. The first starts from other classifications (or families) of the legal and administrative systems within which planning operates, while the second seeks to apply a wider set of criteria but nevertheless produces a similar set of ideal types. Four specific studies of planning systems are discussed below. A summary of the typologies of these studies can be found in Table 2.

Ideal types and classifications of models of spatial planning

Davies et al. (1989) considered planning control in five northern European countries[3] and made a broad distinction between the planning system in England[4] and continental systems. This conclusion followed an earlier comparison of the English and Dutch planning systems (Thomas et al. 1983) which drew attention to the legal certainty provided by systems in continental Europe (at least in the ideal sense) based on the Napoleonic or Scandinavian legal systems, in contrast to the high degree of administrative discretion in the English system created by the legal framework of English common law. Newman and Thornley (1996) also concentrated on a classification of planning systems according to legal and administrative structures, drawing on the five European legal families defined by Zweigert et al. (1987).[5]

Zweigert et al. (1998) explain how all the continental legal systems (Roman, Germanic and Nordic) share a similar legal style: they seek to create a complete set of abstract rules and principles in advance of decision-making. This, they argue, corresponds to particular continental mentality: "The European is given to making plans, to regulating things in advance, and to drawing up rules and systematizing them" (p. 71). In contrast, the English common law system offers far fewer rules. Government does not provide a complete set of legal rules in advance, rather the law has been built up case-by-case as decisions of the courts are recorded. Thus, there is much more emphasis on case law than on enacted law, which provides for more administrative discretion. Faludi (1987) made the same distinction when referring to the continental systems as *imperative* and the English system as *indicative*.

Using legal families and administrative structures to explain differences among planning systems has obvious validity because the legal style and the administrative structure of government provide very strong frameworks for the operation of planning systems. This approach tends to emphasize the differences in the role of plans in the formal regulation of development, such as whether decisions are made through legally binding plans (broadly followed in continental countries) or as and when proposals arise (the main approach in England). We should note here that this approach tends to over-emphasize the formal system of planning in principle as opposed to the reality of its operation in practice.

disP 172 · 1/2008   **39**

The *EU Compendium of Spatial Planning Systems and Policies* (1997) used a wider set of criteria to create four ideal types or traditions of spatial planning (see Table 2). The word tradition was used to emphasize the way that forms of planning are deeply embedded in the complex historical conditions of particular places. The legal family context was used along with six other variables: the scope of the system in terms of policy topics covered; the extent of national and regional planning; the locus of power or relative competences between central and local government; the relative roles of public and private sectors; the maturity of the system or how well it is established in government and public life; and the apparent distance between expressed goals for spatial development and outcomes. As with the models of society, the four traditions of spatial planning are ideal types, that is, a synthesis of the real complex mixture of observable phenomena. They serve as measures against which reality can be compared such that at a point in time and space a planning action might exhibit features of more than one ideal type/tradition. To some extent, the criteria also address the nature of systems in operation, though the idea types still emphasize the formal structure of planning.

The comprehensive integrated approach corresponds quite well to the Scandinavian legal family in the geographical area it covers. The name suggests that the planning system explicitly seeks to provide a measure of horizontal and vertical integration of policies across sectors and jurisdictions. This is in contrast to the land-use planning tradition, which corresponds well to the British legal and administrative family and has the much narrower scope or purpose of regulating land-use change. The other two planning traditions do not correspond so closely to the legal families. The regional economic planning approach cuts across the Napoleonic and Germanic legal families. The urbanism tradition falls within the Napoleonic tradition but for southern Europe only.

The classification of ideal types in the *EU Compendium* (and the limitations this imposes) were used in "a modest update on the movements that took place since" as part of the ESPON program (Farinós Dasí 2006: 112). It gives more emphasis to the distribution of powers relevant to planning among levels of government with a finer analysis of state structures and the decentralization and devolution of competences, especially the varying forms of regional governance and local powers. It concludes, like

Davies et al. 1989*		Common law England		Napoleonic codes DK, DE, FR, NL	
Newman, Thornley 1996	Nordic DK, FI, SE	British IE, UK	Germanic AT, DE	Napoleonic BE, FR, IT, LU, NL, PT, ES	East European
CEC 1997**	Comprehensive integrated AT, DK, FI, DE, NL, SE	Land use regulation IE, UK (and BE)		Regional economic FR, PT (and DE)	Urbanism GR, IT, ES (and PT)
Farinós Dasi 2007***	Comprehensive integrated AT, DK, FI, NL, SE, DE (and BE, FR, IE LU, UK) BG, EE, HU, LV, LT PL, RO, SL, SV	Land use regulation BE, IE, LU, UK (and PT, ES) CY, CZ, MT		Regional economic FR, DE, PT, (and IE, SE, UK) HU, LV, LT, SK	Urbanism GR, IT, ES CY, MT

* Davies et al. (1989) do not give a specific name to the two groups but contrast England and other systems based on their legal frameworks.

** The EU Compendium identifies 'ideal types' of planning traditions. Each country may exhibit combinations of ideal types in different degrees. The ideal types are dominant in the countries indicated here.

*** The ESPON project took the EU Compendium traditions as a starting point and examined how countries, including the transition states of central and eastern Europe, were moving between them.

Tab. 2: Planning system typologies.

the Compendium, that variation is the hallmark of planning systems and that it is difficult to classify, since the member states vary in different aspects of styles or according to which criteria is given prominence. It argues that Belgium, France, Ireland, Luxembourg and the UK are taking up elements of the comprehensive integrated approach. It also asserts that Germany, Ireland, Sweden and the UK are moving towards the regional economic planning style, and that Spain and Portugal are moving towards more land-use regulation.

Learning and convergence?

Comparative analysis of planning systems suggests that continuous adaptation is leading to a general convergence. Some changes are rather detailed institutional and legislative matters, but others reflect a more general "transformation of the style of spatial planning" (Healey et al. 1997: 290). Farinós Dasí (2006) argues that the comprehensive integrated and regional economic planning styles are becoming more common, and, moreover, that this process is producing a "neo-comprehensive integrated planning approach." His argument supports the thesis that the northwestern perspective on planning is becoming more widespread (Rivolin, Faludi 2005). As Healey and Williams noted (in 1993), there are pressures for the convergence of certain aspects of planning systems as "cities and regions become increasingly oriented to competition with European space" (p. 716). Davies (1994: 67) also anticipated such convergence "through the triple effects of *cooperation* between cities and regions in other countries... the competition for investment, tourists and other benefits from the single market, and the *learning process* in working with the [European] Commission" (emphasis added). The limited evidence that we have suggests that Davies was right. There has been a measure of convergence as formal national arrangements for planning have been adapted to address common challenges of global competition and sustainability to which they were unsuited. The debate at the European level has played a part in this (Nordregio 2007) alongside extensive transnational cooperation that has raised awareness and promoted mutual learning. It would be difficult to argue that there is (or could be) a single *European model of spatial planning*, but there does seem to be a strengthening of the common elements, particularly those that are central to the European debate on planning as "a method of securing convergence and coordination between various

sectoral policies" (Report on Community Policies and Spatial Planning 1999).

4. The Evolution of Planning Systems

In this section, we explore the recent trajectories of planning systems, within the wider changes to models of society, in two countries: England and the Netherlands. These countries make for an interesting comparison. The English planning system is very distinctive and corresponds very well to a particular social model while the Netherlands is difficult to classify according to ideal types of social models. Both countries have mature planning and welfare systems but are nevertheless experiencing considerable changes in attitudes and values in society and in the role and guiding principles of the planning system.

England

The ethos of the planning system in England is infused with the dominant and distinctive liberal social model, the pragmatic approach to governance, the common law legal system and the long history of stable national state boundaries. The very idea of spatial planning in the liberal model of society gives rise to fundamental tensions and questions that have characterized planning in England. Though most of the fundamental mechanisms of planning have been stable since the 1950s, its operation and role have fluctuated considerably in parallel with vacillating political ideology and economic conditions. Learning from other countries has also played a part, with a long history of exchange of ideas about planning with other European countries and North America.

The original intellectual arguments for planning came from the social reform movement and its progressive ideological roots, but the notion of planning that arose in practice was dominated by questions of physical form (Ashworth 1954; Sutcliffe 1981; Hall 1992: 49). This was planning with a relatively narrow ambition and, although there was a period of proactive planning in the 1950s and a consistent and firm application of urban containment policy, the formal system of planning in England has been generally reactive and passive, for example, in its emphasis on regulating private sector development (or the public sector acting as private developer). This form of planning fits squarely within the ideal type of land-use management. However, it does address the spatial dimension of tensions that arise in a liberal economy that

disP 172 · 1/2008 **41**

seeks to achieve a more even distribution of social and economic welfare. National government has a dominant position in decision-making, although the system is operated by local authorities. Although formally described as plan-led, there is much negotiation around decisions of any significance and the system offers considerable discretion: decisions on development are made on their merits with no binding zoning instruments. There are extensive opportunities for consultation and objections to policies and development projects.

Shifts in the nature of the dominant liberal social model have been closely interlinked with reforms of the planning system in the past. The rise of neo-liberalism in the 1980s and 1990s with its antagonism toward the welfare state and its adherence to individualism and choice presented a fundamental challenge for planning. It might be argued that the dominant lessons (in the sense of a learning system of planning) at this time were coming from the United States. Evans (1995) argues that planning changed from a "welfare profession" serving the public interest to a skills-based profession selling a service. The ambitions of pre- and post-war grand designs and social and economic goals largely disappeared as planning instead adopted a procedural role in managing the statutory planning – described as little more than "bureaucratic proceduralism" (Evans, Rydin 1997; Tewdwr-Jones 1996).

The 2004 reforms

In this context, attention was concentrated on the efficiency of the system and a government program of Modernizing Planning (DETR 1998) brought forward incremental changes. But contributions from agencies and NGOs to this debate pointed to deep-seated tensions in deregulation in the face of increasing economic competition and sustainable development (TCPA 1999; RTPI 2000; Allmendinger, Tewdwr-Jones 2000). Starting in the late 1990s, transnational cooperation on spatial planning has had a critical influence on the direction of the planning system (DTLR 2001; Nadin 2007). The reform of planning subsequently calls for a stronger role for the planning system in shaping change and a shift in the very culture of planning. Subsequently, the 2004 Planning and Compulsory Purchase Act is at the heart of the most significant reform of the planning system in England since 1968.

The essential features of the system as described above remain the same. The accent has been on reworking the tools of planning to offer planning authorities more opportunities to take the initiative in development, to provide a strategic framework, and to engage stakeholders more effectively. This is summed up in the government's guidance in the notion of "the spatial planning approach" (Nadin 2007). It seeks to meet a desire for more plan-led development and coordination of private investment and sector public policies within the market-driven and fragmented policy environment that arises in a liberal model of society. In some ways, the liberal model is stronger than ever, but at the same time, there are demands for more and more effective public intervention. The solution offered sees planning as a coordinative and collaborative activity injecting a *spatial or territorial dimension* into sectoral strategies and policy; and creating *new policy communities* that reflect the realities of spatial development and its drivers. Planning is being promoted as a learning process. Planning tools have been amended considerably with the objectives of strengthening regional strategic planning capacity and enabling local planning authorities to positively promote appropriate development.

The changes to planning in England have undoubtedly been influenced by transnational learning, particularly ideas coming from the ESDP (Shaw, Sykes 2003, 2004, 2005). Nevertheless, the ESDP exhibits very different assumptions about planning, which arise from continental systems and their roots in different models of society. The key ideas in the European dialogue on planning have been reworked to fit the dominant model in England, but are not proving easy to establish. Reports in 2007 suggest that central aspects of the reformed system are struggling to take root in practice and among the profession (Baker Associates et al. 2006).

The Netherlands

According to Shetter (1988), planning is one of the central cultural institutions in Dutch society. Alexander (1992) describes spatial planning in the Netherlands as relying on a passive regulatory system in which interventions are permitted subject to prior assessment, with stakeholders being consulted at an early stage of the planning procedure. The *EU Compendium* describes the Dutch planning system as one of the most elaborate examples of the comprehensive integrated approach to planning in which "plans are more concerned with the coordination of spatial than economic developments" (CEC 1997). Rather than working with master plans, national planning in the Netherlands works with indicative national policy documents.

Statutory plans are the responsibility of the provinces and municipalities and only the latter have the power to make plans that provide grounds for the refusal of planning permits (Faludi 2005). Much consultation and persuasion therefore takes place to integrate policies of higher levels of government into the plans and policies of lower levels of government (Zonneveld 2006). Central government is the chief source of funding for planning at all levels, and so wields much influence by means of this relationship (Faludi 2005). The Dutch planning system is plan-led: nothing can be developed that is not in accordance with the local land-use plan, since this is legally binding. However, property developers can exert strong influences on the content of a plan, which also gives development a development-led character (EC 1999). Planning practice is strongly influenced by informal ways of using formal rules (administrative pragmatism) (Needham 2005).

The most significant changes in spatial planning in the Netherlands over recent years have been the development of a new national spatial strategy and a new spatial planning act, both adopted in 2006. Zonneveld (2006) characterizes the dominant shifts in Dutch spatial planning that are currently taking place as being from "welfare state spatial planning" to "development planning"[6], which clearly indicates a link between the planning system and the welfare state. The nature of these changes is summarized below (more detail can be found in Needham 2005; Spaans 2006; Vink, van der Burg 2006; Zonneveld 2005, 2006).

The National Spatial Strategy

The National Spatial Strategy (*Nota Ruimte*), approved by the Senate in January 2006, indicates a departure from the restrictive planning discourse (Spaans 2006). It makes a radical break with the centralist tradition in which the national government determines in detail what will be built and where. The National Spatial Strategy's dictum is "decentralize if possible, centralize if necessary", mirroring that of the predecessor of the National Spatial Strategy, the Fifth Memorandum on Spatial Planning. The only difference was that the phrase has been reversed: "centralize if necessary, decentralize if possible." This is perhaps not just a question of semantics but an expression of a deeply

perceived difference in government control (Vink, van der Burg 2006).

According to Zonneveld (2005), the National Spatial Strategy marks a radical departure from traditional Dutch spatial policy in that a new division of responsibilities between the three tiers of government is outlined. Central government takes a step backward in favor of allowing the local authorities, particularly the provinces, to play a key role. The National Spatial Strategy, for example, gives a stronger role to lower levels of government in terms of development control. Vink and van der Burg (2006) contend that there is more focus on development; the Strategy "seeks to tie in with social trends, rather than combating them." The document itself reports that the main difference between this strategy and previous ones "is primarily the method of governance (the how) rather than the policy content (the what)." It asserts that the most important objectives, policy concepts and basic principles from the previous strategies have been retained (Ministry of Housing, Spatial Planning and the Environment 2006). The document also signals a shift in emphasis "from planning to development": more emphasis on development-led planning and less on development control planning (Ministry of Housing, Spatial Planning and the Environment 2006). The Strategy also announces the amendment of the Spatial Planning Act (*Wet ruimtelijke ordening*) to bring it more into line with the philosophy of governance of the new Spatial Strategy (see below). According to Vink and van der Burg (2006), the National Spatial Strategy strengthens the role of the provinces and reduces the number of rules and regulations imposed by central government on others, while creating more scope for local and regional governments, social organizations, private actors and citizens in the planning process. At the same time, however, the National Spatial Strategy also introduces stronger national and provincial powers – national and provincial governments will be able to intervene more forcefully when national or international interests are at stake (e.g., biodiversity, national landscapes).

The Spatial Planning Act

The first comprehensive Dutch spatial planning act dates from 1965, and has since been amended several times. In the 1990s, for example, legislation was amended to speed up large projects (e.g., infrastructure development, river dikes). Most of the changes to the spatial planning act have been marginal, but the overall result disP 172 · 1/2008 43

is a patchwork of different instruments and procedures (Needham 2005). Since 1965, the idea has always been that planning is primarily a coordinating activity (Hajer, Zonneveld 2000).

Work started on a new spatial planning act in 2000, when it was decided not to work within the old 1965 framework, but to develop a new one. Most of the content is uncontroversial or regarded as inevitable (Needham 2005). The new spatial planning act was approved by the Senate in October 2006 and will enter into force in 2008. It has been five years in the making, and many changes to the proposed content have been made in those five years. All have reduced the extent and significance of the proposed changes (Needham 2005). While the new act gives the opportunity for negotiation in planning, the planning system nevertheless retains a strong plan-led orientation.

According to Needham (2005), one of the changes in the planning act that might prove to be very significant is the strengthening of the planning powers of central government at the cost of the planning powers of the municipalities (despite the focus on decentralization in the most recent National Spatial Strategy). The new planning act will, for example, give the national and provincial governments stricter powers for requiring a municipality to follow their policies through the issuing of directives. Vink and van der Burg (2006) report that the new planning act will make it possible for central and provincial government to be able to intervene more forcefully than previously when higher interests are at stake, which can be interpreted as the creation of a more centralized planning system and at odds with the principle of the Netherlands as a decentralized unitary state (Zonneveld 2006). It remains to be seen how centralized the new system will be in practice. Despite new roles for provinces under the new spatial strategy and powers under the new spatial planning act, the Dutch Council of State anticipates "serious consequences" for the position of the provinces in spatial planning and a "strong decline in the importance of the provincial spatial planning policy." Strangely, however, the fate of the provinces in the new Dutch planning system has not generated much debate in the professional domain (Zonneveld 2006).

Reflection

According to Hoekstra (2003: 63), the Dutch system of spatial planning in the 1980s and 1990s was "typical of a social-democratic welfare state", and as such very much in line with the position of the Netherlands within Esping-Anderson's typology (see Table 1). While the 1980s and 1990s marked a period of relative stability for Dutch spatial planning, the beginning of the twenty-first century signaled the start of more wide-ranging changes, primarily in the direction of a more liberal approach to planning. These developments can be interpreted as a movement towards the development-led planning approach that is also evident in England. The recent changes in the Dutch planning system mirror the general direction of welfare reform in the Netherlands over recent decades.

The Netherlands has played an active role in the international spatial planning arena over recent decades, particularly at the European level, which has unquestionably led to learning processes both for the planning community in the Netherlands and, perhaps more importantly, for the planning community outside the Netherlands (who have learnt about the Dutch approach). Indeed, one of the main reasons for the country's active role in the international arena was arguably to try to promote or export concepts and processes from the Dutch spatial planning system, and set the agenda for spatial planning debates. In the 1990s, for example, the Netherlands was one of the driving forces behind the European Spatial Development Perspective (Faludi, Waterhout 2002). More recently, the Netherlands has played an active role in the development of the Territorial Agenda of the EU (Faludi 2007c). Learning has not however just been a one-way process for the Dutch planning community: involvement in the international planning arena has also exposed Dutch policy-makers to a wide variety of spatial planning systems, which has inevitably influenced developments back home.

5. Conclusions

The comparison of models of society and models of spatial planning is a first step in clarifying understanding of the comparative evolution of national forms and policies of planning and the process of convergence in the context of mutual learning. There is first a reasonable correspondence between the ideal types (or categories) of models of society and planning systems (see Tables 1 and 2), which was perhaps to be expected. The planning system is in part an expression of some fundamental values in a society in relation, for example, to the legitimate scope and aspirations of government, the use of land, and the rights of citizens. The definition of types remains much the same even when different criteria are used.

The correspondence of the model of society and the type of planning system is particularly strong for the British/Anglo-Saxon and Nordic models. They are consistently distinguished because of their specific characteristics, demonstrate a very close association between the dominant model of society and the form of the planning system. There is less consistency in the definition of models for continental countries. A few countries are apparently very difficult to classify, including the Netherlands. Countries of central and eastern Europe are most difficult to work into the existing typologies. The classifications of both social models and models of planning are becoming more elaborate and differentiated, recognizing more accurately the real diversity and dynamics in societies.

The examples of England and the Netherlands illustrate how the planning model is embedded in the wider model of society. However, the recent history of change in spatial planning in the two countries has some parallels, despite their different social models and planning systems. The planning systems of both countries are in a process of substantial reform, which seems to be directed by forces that challenge the received models of society. Prominent among these external factors is the learning or the exchange of experience and development of a common reference framework for spatial planning that has arisen from extensive cooperation at European and transnational scales. It is perhaps not surprising, therefore, that recent reforms are creating considerable uncertainty and controversy in each country.

In England, the notion of planning as a tool for spatial policy integration is at the center of the spatial planning approach. This is not a new concern for planning and was perhaps part of the dominant culture in different ways in its earlier history. The aspects of reform that require close partnerships between public, private and civic sectors fits well with the Anglo Saxon liberal model of society, but other changes, such as more long-term strategy and greater policy integration, are more difficult to incorporate. This is recognized in the government's call, alongside new tools and procedures, for a "change in the culture of planning" (Shaw, Lord 2007). Although it is too early to evaluate the reforms, it appears that they are proving difficult to put into practice and the process remains controversial with uncertain outcomes. It is now more difficult to categorize the planning system in England as dominated by the land-use regulation model.

In the Netherlands, the old system of spatial planning (i.e., before the introduction of the new national spatial strategy and spatial planning act in 2006) is very much congruent with a social-democratic welfare model. Furthermore, the new national spatial strategy and spatial planning act signal substantial shifts that are currently taking place in Dutch spatial planning and welfare policy more widely, and a move towards a more liberal approach. Because both the spatial planning strategy and act have only recently been adopted (and the act had not yet come into force at the time of writing), it is too early to say exactly how spatial planning processes and practices may change. The changes in spatial planning closely reflect the trends in various recent welfare reforms (e.g., social security, labor market policy, healthcare and immigration), where the Dutch social model has also undergone some significant changes (see, for example, de Gier et al. 2004). Despite changes in the spatial planning and welfare systems, we believe that the Netherlands still has a planning system that can be categorized into the "comprehensive integrated approach" (CEC 1997) and still has one of the most elaborate examples of spatial planning in Europe.

There are clearly limitations in the use of ideal types or models to classify planning systems and explain their evolution. Any model will be a considerable abstraction of the true variety that the nation states and regions exhibit. Zweigert et al. (1998) note the dangers in reducing the complexity of variation between countries to a few "families". Much depends on the particular criteria employed. In this paper we have explored the potential for explaining the evolution of the formal arrangements for spatial planning with reference to the underlying social model, in the context of transnational learning and Europeanization. The findings suggest that there is potential in continuing such an investigation. European, transnational and cross-border spatial planning initiatives that provide learning opportunities could be usefully informed with a deeper understanding of the dynamic interrelationships of planning reform and social model.

More recent analysis of both social models and planning systems has called for a finer classification of systems that recognizes more diversity and divisions. The examples of England and the Netherlands also show how very similar arrangements can be developed in practice in quite different settings. They suggest that the received models may overemphasize the British-Continental divide. The practice of planning may develop in a similar fashion despite their differing legal and administrative contexts (as has happened in the United Kingdom). External factors including learning through transnational cooperation seem to be leading to a measure of convergence or harmonization of systems, although this creates tensions as changes in administrative systems run ahead of changes in the social model.

Notes

1   Analogous to the debates about spatial planning concepts such as polycentricity (e.g., Richardson & Jensen, 2000; Shaw & Sykes, 2004).

2   See Faludi (2007a, b) for more detailed accounts of these reports.

3   The five countries are Denmark, England, France, The Netherlands and West Germany.

4   The focus of this paper is on England, although the same arguments also apply to the rest of the United Kingdom in matters of principle.

5   Zweigert et al. (1998) also identify three other legal families (South East Asian, Islam and Hinduism), making eight legal families in total.

6   The term development planning in this context implies development-led planning.

References

AIGINGER, K.; GUGER, A. (2006): The Ability to Adapt: Why it differs between the Scandinavian and Continental European Models. *Intereconomics* 14(1): 14–23.

ALBER, J. (2006): The European Social Model and the United States. *European Union Politics* 7(3): 393–419.

ALEXANDER, E. R. (1992): *Approaches to Planning: Introducing Current Planning Theories, Concepts and Issues.* Langhorne PA: Gordon and Breach.

ALLMENDINGER, P.; TEWDWR-JONES, M. (2000): New Labour, new planning? The trajectory of planning in Blair's Britain. *Urban Studies* 37(8): 1379–1402.

ALMUNIA, J. (2005): *The Future of the European Model* (SPEECH/05/560): Speech at Minda de Gunzburh Center for European Studies, Harvard University, 26 September 2005. [http://europa.eu/rapid/searchAction.do]

ARTS, W.; GELISSEN, J. (2002): Three worlds of welfare or more? *Journal of European Social Policy* 12(2): 137–158.

ASHWORTH, W. (1954): *The Genesis of Modern British Town Planning.* London: Routledge and Kegan Paul.

BAKER ASSOCIATES; O'ROURKE, T.; UNIVERSITY OF LIVERPOOL; UNIVERSITY OF MANCHESTER; UNIVERSITY OF THE WEST OF ENGLAND (2006): *Making timely progress and the Integration of Policy. Spatial Plans in Practice: Supporting the reform of local planning.* London: Department for Communities and Local Government.

BAMBRA, C. (2007): Sifting the Wheat from the Chaff: A Two-dimensional Discriminant Analysis of Welfare State Regime Theory. *Social Policy & Administration* 41(1): 1–28.

BIEDENKOPF, K.; GEREMEK, B.; MICHALSKI, K. (eds.) (2004): The Spiritual and Cultural Dimension of Europe. Vienna/Brussels: Institute for Human Sciences/European Commission. [http://ec.europa.eu/dgs/policy_advisers/archives/experts_groups/index_en.htm]

BONOLI, G. (1997): Classifying welfare states: a two-dimension approach. *Journal of Social Policy* 26(3): 351–372.

COMMISSION OF THE EUROPEAN COMMUNITIES – CEC (1994): *European Social Policy – A Way Forward for the Union.* White Paper. COM(94)333 Final. Luxembourg: Office for Official Publications of the European Communities.

COMMISSION OF THE EUROPEAN COMMUNITIES – CEC (1997): *The EU Compendium of Spatial Planning Systems and Policies.* Regional Development Studies. Luxembourg: Office for Official Publications of the European Communities.

COMMISSION OF THE EUROPEAN COMMUNITIES – CEC (1999): *Report on Community Policies and Spatial Planning. Working Document of the Commission Services.* Brussels: European Commission.

COMMISSION OF THE EUROPEAN COMMUNITIES – CEC (2000a): *Presidency Conclusions.* Lisbon European Council Meeting, 23–24 March 2000. SN 100/00 EN. Luxembourg: Office for Official Publications of the European Communities.

COMMISSION OF THE EUROPEAN COMMUNITIES – CEC (2000b): *Presidency Conclusions*, Nice European Council Meeting, 7–9 December 2000. SN 400/00 EN. Luxembourg: Office for Official Publications of the European Communities.

COMMISSION OF THE EUROPEAN COMMUNITIES – CEC (2002): *Commission Communication. First progress report on economic and social cohesion.* COM(2002)46 Final. Luxembourg: Office for Official Publications of the European Communities.

COMMISSION OF THE EUROPEAN COMMUNITIES – CEC (2004): *Facing the Challenge: The Lisbon Strategy for Growth and Employment.* Report from the High-Level Group chaired by Wim Kok. Luxembourg: Office for Official Publications of the European Communities.

COMMITTEE ON SPATIAL DEVELOPMENT – CSD (1999): *European Spatial Development Perspective: Towards Balanced and Sustainable Development of the Territory of the EU.* Luxembourg: Office for Official Publications of the European Communities.

DAVIES, H.W.E. (1994): Towards a European planning system? *Planning, Practice and Research* 9(1:) 63–69.

46 disP 172 · 1/2008

DAVIES, H.W.E.; EDWARDS, D.; HOOPER, A. J.; PUNTER, J.V. (1989): Comparative Study. In: Davies, H.W.E. (ed.) *Planning Control in Western Europe*. London: HMSO, pp. 409–442.

DAVOUDI, S. (2007): Territorial Cohesion, the European Social Model, and Spatial Policy Research. In: Faludi, A. (ed.) *Territorial Cohesion and the European Model of Society*. Cambridge MA: Lincoln Institute of Land Policy, pp. 81–103.

DELANTY, G.; RUMFORD, C. (2005): *Rethinking Europe: Social Theory and the Implications of Europeanization*. London: Routledge.

DEPARTMENT FOR TRANSPORT, LOCAL GOVERNMENT AND THE REGIONS – DTLR (2001): *Planning: Delivering a Fundamental Change*. London: DTLR.

DEPARTMENT OF ENVIRONMENT, TRANSPORT AND THE REGIONS – DETR (1998): *Modernising Planning*. London: DETR.

DÜHR, S.; STEAD, D. & ZONNEVELD, W. (2007): The Europeanisation of Spatial Planning through Territorial Cooperation. *Planning Practice and Research* 22(3): 291–307.

ESPING-ANDERSEN, G. (1990): *The Three Worlds of Welfare Capitalism*. Cambridge: Polity Press.

FARINÓS DASI, J. (ed.) (2007): *Governance of Territorial and Urban Policies from EU to Local Level*. Final Report of ESPON Project 2.3.2. Esch-sur-Alzette: ESPON Coordination Unit.

EVANS, B.; RYDIN, Y. (1997): Planning, professionalism and sustainability. In: Blowers, A.; Evans, B. (eds.) (1997) *Town Planning into the 21st Century*. London: Routledge.

EVANS, B. (1995): *Experts and Environmental Planning*. Aldershot: Avebury.

FALUDI, A. (1987): *A Decision-centred View of Environmental Planning*. Oxford: Pergamon Press.

FALUDI, A. (2005): The Netherlands: A Culture with a Soft Spot for Planning. In: Sanyal, B. (ed.) *Comparative Planning Cultures*. London/New York: Routledge, pp. 285–307.

FALUDI, A. (2007a): The European Model of Society. In: Faludi, A. (ed.) *Territorial Cohesion and the European Model of Society*. Cambridge MA: Lincoln Institute of Land Policy, pp. 1–22.

FALUDI, A. (2007b): Territorial Cohesion Policy and the European Model of Society. *European Planning Studies*, 15(4): 567–583.

FALUDI, A. (2007c): Making Sense of the Territorial Agenda of the European Union. *European Journal of Spatial Development* 25: 1–21 (November 2007).

FALUDI, A.; WATERHOUT, B. (2002): *The Making of the European Spatial Development Perspective. No Masterplan*. RTPI Library Series. London/New York: Routledge.

FERRERA, M. (1996): The Southern Model of Welfare in Social Europe. *Journal of European Social Policy* 6(1): 17–37.

GIDDENS, A. (2005): The world does not owe us a living! *Progressive Politics* 4(3): 6–12.

GIER, E. DE; SWAAN, A. DE; OOIJENS, M. (eds.) (2004): *Dutch Welfare Reform in an Expanding Europe:*

The Neighbours' View. Amsterdam: Uitgeverij Het Spinhuis.

HAJER, M. A.; ZONNEVELD, W. (2000): Spatial Planning in the Network Society – Rethinking the Principles of Planning in the Netherlands. *European Planning Studies* 8(3): 337–355.

HALL, P. (1992): *Urban and Regional Planning*. Third edition. London: Routledge.

HEALEY, P.; WILLIAMS, R. (1993): European Urban Planning Systems: Diversity and Convergence. *Urban Studies* 30(4/5): 701–720.

HEALEY, P.; KHAKEE, A.; MOTTE, A.; NEEDHAM, B. (1997): *Making Strategic Spatial Plans: Innovation in Europe*. London: UCL Press.

HOEKSTRA, J. (2003): Housing and the Welfare State in the Netherlands: an Application of Esping-Andersen's Typology. *Housing, Theory and Society* 20(2): 58–71.

HOUSE OF LORDS (2004): *European Union*. Twenty-Ninth Report. London: HMSO.

JEPSEN, M.; SERRANO P. A. (2005): The European Social Model: An exercise in deconstruction. *Journal of European Social Policy* 15(3): 231–245.

KORPI, W.; PALME, J. (1998): The paradox of redistribution and the strategy of equality: Welfare state institutions, inequality and poverty in the Western countries. *American Sociological Review* 63(5): 661–687.

LEIBFRIED, S. (199): Towards a European Welfare State. In: Ferge, Z.; Kolberg, J. E. (eds.): *Social Policy in a Changing Europe*. Frankfurt: Campus-Verlag, pp. 245–279.

MINISTRY OF HOUSING, SPATIAL PLANNING AND THE ENVIRONMENT (2006): *Nota Ruimte – National Spatial Strategy Summary. Creating space for development*. The Hague: Ministry of Housing, Spatial Planning and the Environment.

NADIN, V. (2007): The emergence of the spatial planning approach in England. *Planning Practice and Research* 22(1): 43–62.

NEEDHAM, B. (2005): The New Dutch Spatial Planning Act: Continuity and Change in the Way in Which the Dutch Regulate the Practice of Spatial Planning. *Planning Practice and Research* 20(3): 327–340.

NEWMAN, P.; THORNLEY, A. (1996): *Urban Planning in Europe: International Competition, National Systems, and Planning Projects*. London: Routledge.

NORDREGIO (2007): *Application and effects of the ESDP in the Member States*. Final Report, ESPON project 2.3.1. Esch-sur-Alzette: ESPON Coordination Unit.

RICHARDSON, T.; JENSEN, O.B. (2000): Discourses of Mobility and Polycentric Development: A Contested View of European Spatial Planning. *European Planning Studies* 8(4): 503–520.

RIVOLIN, U. J.; FALUDI, A. (2005): The Hidden Face of European Spatial Planning: Innovations in Governance. *European Planning Studies* 13(2): 195–216.

ROYAL TOWN PLANNING INSTITUTE – RTPI (2000): *A New Vision for Planning: Delivering Sustainable*

Settlements, Communities and Places, Mediating Space, Creating Place. London: RTPI.

SAPIR, A. (2006): Globalization and the Reform of European Social Models. *Journal of Common Market Studies* 44(2): 369–390.

SAPIR, A.; AGHION, P.; BERTOLA, G.; HELLWIG, M.; PISANI-FERRY, J.; ROSATI, D.; VIÑALS, J.; WALLACE, H. with BUTTI, M.; NAVA, M; SMITH, P. M. (2004): *An Agenda for a Growing Europe: The Sapir Report.* Oxford: Oxford University Press.

SHAW, D.; SYKES, O. (2003): Investigating the application of the European Spatial Development Perspective (ESDP) to regional planning in the United Kingdom. *Town Planning Review* 74(1): 31–50.

SHAW, D.; SYKES, O. (2004): The concept of polycentricity in European spatial planning: Reflections on its interpretation and application in the practice of spatial planning. *International Planning Studies* 9(4: 283–306.

SHAW, D.; LORD, A. (2007): The cultural turn? Culture change and what it means for spatial planning in England. *Planning Practice and Research* 22(1): 63–78.

SHETTER, W.Z. (1987): *The Netherlands in Perspective: The Organizations of Society and Environment.* Leiden: Martinus Nijhoff.

SPAANS, M. (2006): Recent changes in the Dutch planning system. Towards a new governance model? *Town Planning Review* 77(2): 127–146.

STRAUSS-KAHN, D. (2004): *Building a Political Europe – 50 Proposals for Tomorrow's Europe.* Report of the Round Table: A Sustainable Project for Tomorrow's Europe. Brussels: European Commission. [http://ec.europa.eu/dgs/policy_advisers/archives/experts_groups/index_en.htm]

SUTCLIFFE, A. (ed.) (1981): *British Town Planning: The Formative Years.* Leicester: Leicester University Press.

SYKES, O.; SHAW, D. (2005): Tracing the Influence of the European Spatial Development Perspective on Planning in the UK. *Town and Country Planning* 74(3): 108–110.

TEWDWR-JONES, M. (ed.) (1996): *British Planning Policy in Transition: Planning in the 1990s.* London: UCL Press.

THOMAS, H.D.; MINETT, J. M.; HOPKINS. S.; HAMNETT, S. L.; FALUDI, A.; BARRELL, D. (1983): *Flexibility and Commitment in Planning.* The Hague/Boston/London: Martinus Nijhoff Publishers.

TOWN AND COUNTRY PLANNING ASSOCIATION – TCPA (1999): *Your Place or Mine: Reinventing Planning.* London: Town and Country Planning Association.

VINK, B.; BURG, A., VAN DER (2006): New Dutch spatial planning policy creates space for development. disP 164(1): 41–49.

ZONNEVELD, W. (2005): In search of conceptual modernization: The new Dutch National Spatial Strategy. *Journal of Housing and the Built Environment* 20(4): 425–443.

ZONNEVELD, W. (2006): *Planning in retreat: The changing importance of Dutch national spatial planning.* Paper presented at the Conference of the European Group of Public Administration, 6–9 September 2006, Università Bocconi, Milan.

ZWEIGERT, K.; KÖTZ, H.; WEIR, T. (translator) (1987): *An Introduction to Comparative Law.* Second Edition. Oxford: Clarendon Press.

ZWEIGERT, K. (author); Kötz, H. (author) & Weir, T. (translator) (1998): An Introduction to Comparative Law. Third Edition. Oxford University Press, Oxford.

disP 172 · 1/2008 **47**

Vincent Nadin
Faculty of Architecture
Delft University of Technology
P.O. Box 5030
NL–2600 AA Delft
v.nadin@tudelft.nl

Dominic Stead
OTB Research Institute
Delft University of Technology
P.O. Box 5030
NL–2600 AA Delft
d.stead@tudelft.nl

Originaltext Fürst 2007

PNDonline III|2007

3

PNDonline - **eine Plattform des Lehrstuhls für Planungstheorie und Stadtentwicklung mit Texten und
Diskussionen zur Entwicklung von Stadt und Region**

Planungskultur

Auf dem Weg zu einem besseren Verständnis von Planungsprozessen?

Prof. Dr. rer. pol **Dietrich
Fürst**, Universität
Hannover, Lehrstuhl
Landesplanung und
Regionalforschung, seit
2003 im Ruhestand

Die folgende Analyse
basiert auf einer vom
Institut für Landes- und
Stadtentwick-
lungsforschung und Bau-
wesen deds Landes NRW
(ILS NRW), Dortmund
(Prof. Danielzyk) in Auftrag
gegebenen Expertise.

Einführung

„Planungskultur" ist an sich kein wissenschaftlicher Begriff: er ist schlecht definiert,
adressiert insofern ein nicht gut abgegrenztes
Untersuchungsfeld und ist nicht mit spezifischen Theoriehintergründen besetzt. Meist
wird Planungskultur lediglich auf einen Integ-
ral-Begriff reduziert, der ein Ensemble von beobachteten Variablen-Kombinationen belegt.
In der Literatur werden „Planungskultur" und
(vor allem in der angelsächsischen Literatur)
„Planungs-Stile" synonym verwendet, häufig
in sehr allgemeiner Definition, etwa wie bei
George C. Hemmens als *„a general model of
professional practice behavior"* (Hemmens 1988,
85). Im Folgenden werden deshalb Planungs-
stile und Planungskultur nicht unterschieden.

Wann der Begriff „Planungskultur" in die Diskussion eingeführt wurde, ist unklar. Nach
John Friedmann sollen es Klaus Selle et al.
(Keller et al. 1993) mit ihrem hermeneutischen
Zugang zum Wandel von Planerverhalten in
vier Ländern[1] gewesen sein (Friedmann
2005b, 30). Allerdings findet sich der Begriff
schon in früheren Arbeiten, z.B. bei Spath
1985 und offenbar schon bei Bolan 1973[2] (nach
Faludi 2005, 286). Wie dem auch sei: In der

Zwischenzeit hat Planungskultur nicht den
Rang eines wissenschaftlichen Begriffs erlangt, sondern bestimmt lediglich einen
wissenschaftlichen Fokus. Schon der Begriff
der *Kultur* wird sehr uneinheitlich verwendet –
eine Zählung aus dem Jahr 1993 kommt auf
über 160 Definitionen (Faure 1993, zit. nach
Jann 2000, 328).

Planungskultur steht jedoch im Kontext des in
vielen sich mit Verhalten befassenden Disziplinen zu beobachtenden „cultural turn", d.h.
zur Einschätzung, dass Verhalten nicht nur
sach-rational, sondern kulturgeprägt erfolgt
und dass insbesondere Interaktionen von einem kulturgeprägten Interaktionsmodus beeinflusst werden. Im Folgenden soll Planungskultur deshalb zunächst in den Kontext
der sozialwissenschaftlichen Diskussion zu
kulturbestimmten kollektiven Handlungsformen in unterschiedlichen Handlungsberei-
chen gestellt werden (Kap. 2). Diesem folgt eine kurze Übersicht über Vergleichsstudien zu
(institutionellen) Planungssystemen (Kap.3),
aus der bereits hervorgeht, dass ein internationaler Vergleich von Planungskulturen auf besondere Schwierigkeiten stoßen wird (Kap. 4).
Gleichwohl wird in der Literatur mit guten Argumenten auf eine Konvergenz von Rahmenbedingungen für Planungskulturen hingewiesen, woraus allerdings – wegen der Pfadabhängigkeit von Planungskulturen – nicht unbedingt gefolgert werden kann, dass sich Pla-

---

[1] Deutschland, Frankreich, Italien, Schweiz.

[2] Aufsatz im Sammelband von Andreas Faludi: A Reader
in Planning Theory, Oxford: Pergamon Press 1973.

nungskulturen immer mehr angleichen werden, auch wenn dafür vieles spricht (Kap.5). In jedem Falle aber zeigt die Diskussion von Planungskulturen eine sehr enge Beziehung zur „Governance-Diskussion", worauf kurz eingegangen wird (Kap.6). Abschließend werden Hinweise für mögliche „Synergieeffekte" mit vergleichbaren Forschungsrichtungen angeführt (Kap. 7).

„Planungskultur" im Kontext kulturbestimmter kollektiver Handlungsmuster

Das Bewusstsein dafür, dass soziale Prozesse eingebettet sind in regionale Traditionen, Normen, Werthaltungen, Einstellungen und Denkmuster ist in den Sozialwissenschaften seit langem verbreitet und hat eigene Forschungszweige ausgelöst. In den Politikwissenschaften spricht man von „politischer Kultur" (Berg-Schlosser 2005) oder kulturgeprägten „Politikstilen" (vgl. Richardson 1982), in den Verwaltungswissenschaften von „Verwaltungskulturen" (Jann 2000), in den Organisationswissenschaften von *corporate culture* (Sackmann 2006). In der Managementliteratur gibt es den Begriff der *„planning styles"*, womit unterschiedliche Planungs-Verfahren verbunden werden, die sich aus Tradition, Routinen, spezifischen Akteurskonstellationen in einem Unternehmen entwickelt haben. Aber auch in der Wissenschaftstheorie wird von kulturgeprägten „Stilen" gesprochen (vgl. Galtung 1983). Allen gemeinsam ist das Bestreben, Prozesse nicht losgelöst von ihren Kontexten zu erfassen und zu deuten und insbesondere die *„longue durée"* von kulturellen Prägungen kenntlich zu machen.

Dem Fokus „Planungskultur" am nächsten kommen der politikwissenschaftliche, verwaltungswissenschaftliche und organisationswissenschaftliche Ansatz. Vereinfacht geht es hier um Einstellungen, Deutungsmuster und Werthaltungen, die Angehörige eines Handlungskollektivs teilen, das sie einem Kollektiv oder Institutionen entgegenbringen und das durch das Kollektiv und die Institution verstärkt, aber auch konserviert wird. Beispielsweise bezieht sich der Begriff der „Politischen Kultur" *„auf unterschiedliche Bewusstseinslagen, Mentalitäten, „typische" bestimmten Gruppen oder ganzen Gesellschaften zugeschriebenen Denk- und Verhaltensweisen. Sie umfasst alle politisch relevanten individuellen Persönlichkeitsmerkmale, latente in Einstellungen und Werten verankerte Prädispositionen zu politischem Handeln, auch in ihren symbolischen Ausprägungen, und konkretes politisches Verhalten"* (Berg-Schlosser 2005,

743). Jann (2000, 329) formuliert bündiger: *„Politische Kultur meint Muster und Orientierungen, ein mind set gegenüber sozialen und politischen Tatbeständen. ... Etwas überspitzt gesagt: Politische Kultur ist das, was man durch Meinungsumfragen erfassen kann."*

Da Begriffe vom Zweck ihrer Nutzung und damit dem dahinter stehenden Erkenntnisinteresse und den sie leitenden Theorien bestimmt werden, macht es wenig Sinn, sich mit dem Begriff abstrakt auseinander zu setzen. Eine Durchsicht einiger Beiträge zur Planungskultur zeigt, dass der Begriff in unterschiedlichen Kontexten mit unterschiedlicher Intention verwendet wird:

Für die Ebene der Stadtplanung:

- Keller et al. (2006, 279 f.) sehen in der Planungskultur Werthaltungen, Rollenverständnis und Aufgabeninterpretation der Planer im Umgang mit Planungsverfahren und –instrumenten.

- Walter (1995) verbindet mit Planungskultur vor allem einen Wandel von Planungsinhalten, hinter denen sich ein Wandel der gesellschaftlichen Einbindung von Raumplanung vermuten lässt.

- Im Feld der Bauleitplanung wird Planungskultur häufig mit dem Thema der Betroffenenbeteiligung und Interessenvermittlung in Verbindung gebracht (vgl. Sandner 1998, Wentz 1992).

- Das BBR versteht unter Planungskultur die Herstellung der gebauten Umwelt und den alltäglichen Umgang damit (*Baukultur – Planungskultur*, Heft 11/12.2002 der Informationen zur Raumentwicklung).

- Die Vereinigung SRL interpretiert Planungskultur primär als normatives Konzept von Verhaltensnormen für Planer (SRL, Selbstverständnis[3]).

- In den USA ist der Begriff weniger gebräuchlich (vgl. Friedmann 2005); wenn er verwendet wird, dann bezeichnet er meist die gesamte Planungspraxis (vgl. Cullingworth 1993, 7 f.); Eher gebräuchlich ist der Begriff der *planning styles*, hier vor allem ausgelöst durch die Studie von Carl Abbott („The Oregon Planning Style"). Der Bundesstaat Oregon wurde bekannt durch sehr frü-

---

[3] Information unter www.srl.de/srl/selbst/index.php

he Ansätze einer ökologisch orientierten Raumplanung („smart growth") unter aktiver Mitwirkung der Zivilgesellschaft und Teilen der Wirtschaft. Der *„Oregon Planning Style"* wird in der amerikanischen Planungsliteratur zu „smart growth" häufiger zitiert (vgl. Weitz 1999, Wheeler 2000), zumal er Auswirkungen auf die Kommunalebene hatte. Viel Aufmerksamkeit hat das Beispiel von Portland gefunden, wo systematisch eine Art „paradigmatischer Steuerung" genutzt wurde, um durch Planungsprozesse einen Wandel im Bewusstsein der politischen und wirtschaftlichen Eliten zu bewirken (Hovey 2003, der von *the structural power of language* spricht).

Für die Ebene der Raumplanung:

- Faludi (2005, 285 f.) stellt auf ethische und paradigmatische Prädispositionen von Planern gegenüber staatlicher Steuerung und der Einbindung Privater ab. Danach ist Planungskultur *„the collective ethos and dominant attitude of planners regarding the appropriate role of the state, market forces, and civil society in influencing social outcomes".*

- Friedmann (2005a, 183) fasst Planungskultur praktisch als Planung im weitesten Sinne auf, *„as the ways both formal and informal, that spatial planning is conceived, institutionalized, and enacted"*[4]. Friedmann spricht zwar von *„spatial planning"*, bezieht sich aber überwiegend auf die Planung in Großstädten.

- Ludwig (2004) interpretiert Planungskultur als Interaktionsverhalten von Planern und „stakeholders" und sucht nach Veränderungen im Kontext der bayerischen Regionalmanagementpraktiken.

- Ähnlich verfährt Stefanie Sixel (2005) in ihrer Untersuchung zum Wandel der Planungsprozesse bei der Formulierung von Regionalen Entwicklungskonzepten.

- Kühn und Moss vom IRS (Erkner) betrachten die Planungskultur primär unter Aspekten der Steuerungs- und Planungsmodelle (Kühn, M., Moss, T., Hg.: Planungskultur und Nachhaltigkeit, Berlin: Vlg. für Wissenschaft u. Forschung, 3. Aufl., 2001)

Meiner Einsicht ist der Begriff im Kontext der Planungsdiskussion dann besonders sinnvoll, wenn man ihn auf Interaktionsprozesse der Planung und deren „Steuerung" durch kulturelle Prägungen bezieht. Dann könnte sich herausstellen, dass gut funktionierende Planungsprozesse durch „Planungskulturen" geprägt werden, die der Koordination unterschiedlicher Interessen durch die in einer Gesellschaft üblichen Koordinationsmodi besonders gut entsprechen. Oder umgekehrt: dass Planungsprozesse dort schlechter funktionieren, wo Koordinationsmodi verwendet werden, die zu wenig auf die kulturellen Prägungen der Beteiligten abgestellt sind.

Eine auf Interaktionen und deren mentale Prägungen ausgerichtete Konzeption von „Planungskultur" scheint die Mehrheit der sich mit Planungskultur Befassenden in das Zentrum ihrer Überlegungen zu rücken. Deshalb bezieht sich der hier präferierte Begriff auf

Werthaltungen und Einstellungen, die von den Beteiligten eines Planungsprozesses geteilt werden und ihr Interaktionsverhalten bezogen auf (wahrgenommene) Planungs-Aufgaben, Verhalten in der Gemeinschaft oder Gruppe und Durchsetzung von Eigeninteressen bestimmen.

Das Zusammenwirken (Interdependenz) von Akteuren zur Koordination ihrer Belange wird dabei im Wesentlichen von fünf Variablen-Gruppen bestimmt (vgl. Scharpf 2000):

(a) Variablengruppen der Interaktion:

(1) die individuelle *Handlungsorientierung*, worunter Einstellungen (*attitudes*), paradigmatische Deutungsmuster (*beliefs*), Werthaltungen (*values* und ethische Normen), Interessen und Fähigkeiten gehören;

(2) die *Interaktionsorientierung* der Akteure, die im Wesentlichen auf Interaktionsnormen sowie Einstellungen (kompetitiv vs. kooperativ, personenbezogen vs. sachbezogen[5], konsensorientiert vs. ergebnisorientiert[6]) basiert,

(3) die *Akteurskonstellation*, die das Machtverhältnis der Akteure sowie den Verhaltensspiel-

---

[4] In Friedmann 2005b (S.30) sagt er: „local, regional, and national differences in planning institutions and practices – I shall call them cultures"

[5] die französische Interaktionskultur ist stärker personenbezogen, die deutsche stärker sachbezogen, was im interkulturellen Austausch zu Interaktions-Störungen führen kann

[6] die holländische Interaktionskultur gilt als stärker konsensorientiert als die deutsche, während die deutsche für stärker ergebnisorientiert gehalten wird

raum von den für Organisationen tätigen Personen bezeichnet.

(b) Variablengruppen der externen Einflüsse

(4) der *institutionelle Handlungsrahmen*, der sich auf rechtliche, organisatorische und administrative Regelungen (z.B. auch Routinen) bezieht. Hier spielen Unterschiede wie *Common Law* (in Großbritannien)[7] und legalistische Steuerung (in Frankreich) hinein, aber auch unterschiedliche Eigentums-Traditionen, die aus englischem Common Law und römischem Recht stammen[8] (vgl. Booth 2005)

(5) sowie die Handlungs-Situation, die vor allem vom *Problemfeld* und den auf dieses wirkenden Einflussfaktoren geprägt wird, worunter auch ein genereller Wandel von Stimmungslagen und Paradigmen gehört..

Diese Einflussfaktoren müssten dann in der Diskussion der Planungskulturen berücksichtigt werden, wobei allerdings nur die längerfristig konstanten und situationsunabhängigen Variablen relevant sein dürften, also primär die Gruppen (1) bis (4). Aber auch diese wandeln sich (Keller et al. 2006). Keller et al. (2006, 283 ff.) berichten deshalb aufgrund eines mehr als 12-jährigen internationalen Diskurses zum „Wandel der Planungskulturen" nicht nur über den Wandel von Verhaltensweisen (zum „kooperativen Planen" und zur verstärkten Nutzung der „informellen Instrumente"), sondern auch über Veränderungen im institutionellen Ansatz: mehr „Ent-Standardisierung", mehr Projektorientierung im Rahmen der „Renaissance der großen Pläne", Neuorientierung durch „Gestalten ohne Wachstum", Stärkung der privatwirtschaftlichen Raumgestaltung im Zuge von De-Regulierung und Ent-Staatlichung.

Vergleichsstudien zu *institutionellen* Planungssystemen

Internationale Vergleiche von Planungssystemen hat es im EU-Kontext seit den 90er Jah-

ren des letzten Jahrhunderts zahlreiche gegeben, und zwar im Wesentlichen aus vier Gründen: Erstens hat die Diskussion auf EU-Ebene zur Entwicklung des *Europäischen Raumentwicklungskonzepts* (EUREK) die in den Mitgliedstaaten für Raumplanung Zuständigen sensibilisiert, dass die grenzüberschreitende Koordination der Planungen und die Einflussnahme auf EU-Planungs-Ansätze eine gewisse Vereinheitlichung der Planungssysteme erforderlich machen könnte. Zweitens hat die Raumplanung in allen Mitgliedstaaten, zumindest auf regionaler Ebene, ihre Funktion der „regionalen Entwicklungssteuerung" stärker ausgebaut: Sie integriert sich damit in Anstrengungen der Regionen, ihre Wettbewerbsposition gegenüber anderen Regionen zu verbessern. Drittens beobachtet man in vielen Mitgliedstaaten, dass sich auf regionaler Ebene – unterstützt oder sogar angeleitet durch die Regionalplanung – neue Muster der *governance* entwickeln. Viertens beginnt international eine Renaissance der Planungstheorie-Diskussion – primär bezogen auf metropolitane Räume/ Stadt-Regionen im globalen Wettbewerb, aber intensiviert durch Herausforderungen der „nachhaltigen Regionalentwicklung".

Treibende Kräfte waren

- die EU (im Rahmen der Entwicklung des Europäischen Raumentwicklungskonzepts, EUREK). Die EU hat 1996/7 eine Reihe von Länderstudien erstellt[9], die abschließend vergleichend ausgewertet wurden (EU 1997). Nadin et al. haben den Zusammenhang zwischen Regionalplanung und nachhaltiger Regionalentwicklung im Auftrag der EU für Finnland, Griechenland, Irland, Italien und die Niederlande untersucht (Nadin et al. 2001).

- Mitgliedstaaten und Hochschuleinrichtungen, manchmal nur über den Vergleich von zwei oder drei Ländern, manchmal über einem EU-weiten Vergleich:

    - für die Fortentwicklung der französischen Regionalplanung durch Auswertung der regionalen Planungssysteme vergleichbarer EU-Mitgliedstaaten durch Marcou im Auftrag des *Ministère de*

---

[7] Das Common Law führt im englischen Planungssystem zu einem sehr viel ausgeprägteren Abwägungsverhalten, das auf Fallbeispielen und zentralstaatlichen Planning Guidelines beruht.

[8] Das Common Law lässt beispielsweise teilbare Grundstücks-Verfügungsrechte zu, was in den USA zu einem ausgeprägten Markt von property rights führt: Das Eigentum an Boden kann von den Nutzungs- und Verfügungsrechten über den Boden getrennt gehandelt werden. Das römische Recht basiert dagegen auf dem Konzept des „dominiums" und subsumiert alle Rechte am Boden unter einem Rechtstitel, dessen Umfang jedoch durch staatliches Recht eingeschränkt werden kann.

---

[9] Dargestellt wurden: Belgien, Dänemark, Deutschland, Finnland, Frankreich, Griechenland, Groß-Britannien, Irland, Italien, Niederlande, Portugal, Österreich, Spanien, Schweden. Der deutsche Beitrag wurde von Gerd Schmidt-Eichstädt und Paraic Fallon (1996) erarbeitet.

l'Equipement du Logement et des Transports (2004)[10]

- für die Vorbereitung eines Erfahrungsaustauschs mit der spanischen Raumplanung durch Hildenbrand (1996)[11]

- für einen vergleichbaren, primär auf Lokalebene bezogenen Vergleich: Newman/Thornley (1996), Williams (1996)

- für den Wandel der Regionalplanung in der EU zur „strategischen Planung" durch Healey et al. (1997), Salet/Faludi (2000)

- für die nordeuropäischen Staaten (in Reaktion auf EUREK) durch Böhme (2002)[12]

- für die Raumplanung auf nationaler Ebene – auch über die EU hinaus – durch Masser/ Williams (1986), Roberts et al. (2000) sowie Alterman et al. (2001)[13]

- für die Verbesserung der grenzüberschreitenden Planungs-Kooperation u.a. von Kistenmacher/Clev (1994) (Vergleich Deutschland und Frankreich)

- die ARL (im Rahmen ihrer wachsenden Kooperation mit den angrenzenden EU-Mitgliedstaaten sowie im Rahmen der Umsetzung der EU-INTERREG-Programme). Die ARL hat „Raumordnungs-Glossars" für unterschiedliche Nachbarländer erstellt, in denen die Planungssysteme der Nachbarländer skizziert wurden (allerdings wurde keine vergleichende Auswertung vorgenommen).

Darüber hinaus finden sich Länderberichte in Mitgliedstaaten, die lediglich als „Graue Papier" verfügbar sind und die wissenschaftliche Diskussionsebene nicht erreichten.

Kennzeichnend für diese Vergleiche ist:

- die europa-weiten Untersuchungen enden in der Mitte der 90er Jahre des letzten Jahrhunderts und nehmen die neuere europäische Rechtsentwicklung mit Rückwirkungen auf die nationalen Planungssysteme[14] noch nicht auf (Ausnahme: Alterman et al. (2001); Böhme (2002), Marcou (2004));

- ihr Fokus ist auf die systematische Darstellung der Systeme zu einem bestimmten Zeitpunkt gerichtet, wobei meist ein grober historischer Abriss eingebunden ist, aber keine systematische historische Längsschnittanalyse vorgetragen wird; Auch dazu gibt es Ausnahmen. So hat die Studie von Alterman et al. (2001) für die nationalen Darstellungen das folgende Kriterien-Bündel vorgegeben: Ziele, Inhalte, Funktion der nationalen Planungen; Instrumente und Umsetzungsverfahren; Einschätzung der Erfolge und Misserfolge; Böhme geht es vor allem um die Verknüpfung der EUREK-Prozesse mit der nationalen Planungs-Praxis als „diskursiven Prozess" und als netzwerkartige Elite-Verknüpfungen vor neokorporatistischem Hintergrund.

- sie sind institutionell ausgerichtet; dabei interessiert zudem primär nur die Aufbauorganisation, sehr viel weniger die Ablauforganisation;

- ein wirklicher Vergleich findet eher zurückhaltend statt: Im Vordergrund steht die Information über die Planungssysteme anderer Länder, während vergleichende Fragestellungen eher formaler (= institutioneller) Natur sind. Die Studien von Hildenbrand (1996) und Marcou (2004) gehen dabei vielleicht noch am weitesten, weil ihr Untersuchungsauftrag auf *best practices* gerichtet war. Deshalb werden darin nicht nur die Planungssysteme der Länder nach einer einheitlichen Systematik dargestellt, sondern mit großer Akribie die Planungsinstrumente der Pläne auf den unterschiedlichen Ebenen (Staat, Region, Kommune) abgehandelt.

- sofern Vergleiche vorgenommen werden, beziehen sie sich primär auf die Institutionen und Pläne/Programme, nicht/ kaum auf

---

[10] Die Untersuchung konzentrierte sich auf die Kompetenzverteilung zwischen nationaler und regionaler Ebene, auf das Verhältnis von regionaler zu sektoraler Planung, die Planungsinstrumente (mögliche Planaussagen), das Aufstellungsverfahren, die Vollzugsinstrumente sowie planerische Ausbildungs- und Vereinigungs-Strukturen.

[11] Diese Studie entstand im Zusammenhang mit der Erstellung des Regionalen Raumordnungsprogramms für Andalusien und wurde 1992 abgeschlossen, aber kurz vor Veröffentlichung nochmals aktualisiert (Stand: 1995). Untersucht wurden: Deutschland, Frankreich, Groß-Britannien, Italien, Niederlande, Portugal, Schweiz.

[12] Dargestellt wurden: Dänemark, Finnland, Island, Norwegen und Schweden. Böhmes Fragestellung ist darauf gerichtet, wie das EUREK in die nationalen Planungssysteme integriert wird. Er verfolgt einen kognitionstheoretischen Ansatz, dem herrschenden Paradigma des communicative planning oder diskursorientierten Planungsverfahren verpflichtet.

[13] Dargestellt wurden: aus dem EU-Bereich: Dänemark, Deutschland, Frankreich, Großbritannien, Irland, Niederlande, Schweden, außerhalb der EU: Israel, Japan, USA.

[14] Fauna-Flora-Habitat-Richtlinie, Wasserrahmenrichtlinie, UVP-Richtlinie, SUP-Richtlinie.

die Prozesse, Handlungsmodelle oder konkrete Problemlösungs-Verfahren

- es gibt keinen problembezogenen Vergleich: Es wird nicht versucht, international vergleichend die Bearbeitung typischer Planungsaufgaben durchzuführen.

Wenn es folglich analytisch gehaltvolle Ländervergleiche (basierend auf suchenden Forschungsfragen) kaum gibt, so stellt sich die Frage: warum? Offenbar sind sie problematisch, weil Planungssysteme nicht isoliert, sondern nur im Kontext ihres Staatssystems sinnvoll untersucht werden können. Für viele vermuteten „Defizite" gibt es häufig *äquifunktionale Mechanismen* oder Strukturen, die diese „Defizite" letztlich wieder ausgleichen. So muss in einigen Planungssystemen mitgeregelt resp. mit-gesteuert werden, was in anderen Planungssystemen durch andere Institutionen abgefangen wird (das betrifft insbesondere das Verhältnis der Regionalplanung zur Kommunalplanung und zu Fachplanungen). Solche „äquifunktionalen Mechanismen" finden sich

- für das Verhältnis der sog. planerischen „Ordnungsfunktionen" zu den „Entwicklungsfunktionen": Häufig werden Ordnungsfunktionen nur noch projektbezogen, im Rahmen von Entwicklungsfunktionen, wahrgenommen, z.B. über eine Reihe von Prüf- und Verhandlungsverfahren

- für das Verhältnis von nationaler zu regionaler und lokaler Planung: die planerischen Spielräume können auf den unterschiedlichen Ebenen unterschiedlich groß sein

- für die Koordinationsmechanismen: während einige Staaten auf Verhandlungen setzen, arbeiten andere stärker mit staatlichen Regulierungen oder persuasiven Strategien

- für das Verhältnis von öffentlicher und privater Planung (vgl. europäische Tradition vs. us-amerikanische Tradition)

Eine systematische Diskussion äquifunktionaler Strukturen und Mechanismen ist nirgends zu finden – eher neigen die Autoren dazu, die Planungssysteme losgelöst von ihrem institutionellen Kontext zu diskutieren.

Vergleich von Planungskulturen

Eine methodisch anspruchsvollere vergleichende Untersuchung von Planungskulturen

wurde bisher nicht vorgelegt. Studien, die sich mit Interaktionsverhalten in Planungsprozessen befassen, finden sich primär auf nationaler oder sub-nationaler Ebene. Diese nachträglich vergleichend auswerten zu wollen, stößt auf erhebliche Schwierigkeiten, sofern

- sie zu unterschiedlichen Zeiten entstanden sind, d.h. unterschiedliche institutionelle und sonstige Kontextbedingungen berücksichtigen

- zu heterogene Kontexte berücksichtigen müssen,

- nach Kriterien entstanden sind, die der jeweiligen Forschungsfrage entsprechen, die aber nicht einer Vergleichsstudie entsprechen: Infolgedessen sind die ausgewählten Analyse-Variablen, die Gewichtung der Variablen und die Interpretation kaum vergleichbar (z.B. werden vielfach dialogische Verfahren, informelle Vorgehensweisen, die zugrunde liegenden Akteursnetzwerke u.ä. in den Fallstudien zu wenig berücksichtigt)

- sehr unterschiedliche Vorstellungen von „Planungskultur" vertreten.

Sicherlich könnte man durch die Auswahl geeigneter Fallbeispiele einen Teil der Probleme eines internationalen Vergleichs ausräumen („*the prudent comparativist does not choose his countries (hier: Fallbeispiele) by choice: he is guided by pertinent criteria*" (Dogan/Pelassy 1981, 38, zit. nach Nohlen 2005a, 510). Aber selbst dann bleibt das Ergebnis gegenüber einem genuinen Vergleich suboptimal, wenn die jeweiligen Fragestellungen der zu untersuchenden Fallbeispiele zu wenig kompatibel sind. Denn gerade beim Vergleich auf Basis von Fallbeispielen sind methodische Fragen gravierend. Vielmehr ist das Feld der internationalen Vergleichsstudien *„densely populated by non-comparativists, by scholars, who have no interest, no notion, no training in comparing"* (Sartori 1991, 243).[15]

Soweit es bei Planungskulturen vergleichende Studien gibt, beruhen sie methodisch auf hermeneutischen Ansätzen, z.B. auf

---

[15] Als Hauptprobleme international vergleichender Studien auf der Basis qualitativer Indikatoren (Einstellungen, Werthaltungen, Paradigmen) gelten: dass anstelle direkt messbarer Indikatoren „proxies" gewählt werden müssen, dass interpretierte Daten statt Beobachtungen genutzt werden, dass gleiche Fragen in unterschiedlichen kulturellen Kontexten unterschiedlich wahrgenommen und folglich beantwortet werden (vgl. van Deth 2003)

- Experten-Diskursen (vgl. Keller et al. 2006[16]; 1996, Sanyal 2005[17]),

- auf Teilnehmender Beobachtung resp. Expertenerfahrungen (Friedmann 2005a, 2005b)[18]

- Fallstudien (Knieling/During 2006[19] (laufend))

Diese Studien zeichnen sich methodisch meist dadurch aus (das gilt nicht für „Cultplan"),

- dass ein systematischer Vergleich von Ländern nicht vorgenommen wird (kein einheitliches Kriteriensystem),

- dass sie sich tendenziell stärker auf die städtische als regionale Ebene beziehen (auch bedingt durch die häufig fehlende überlokale Planung in vielen Ländern) und

- dass sie meist persönliche Planungserfahrungen wiedergeben: Die Verlässlichkeit der Aussagen ist meist nicht gut nachprüfbar.

- dass nirgends empirisch-quantifizierende Methoden gewählt werden (z.B. Umfrageforschung[20]).

Nicht genutzt wurden bisher die in der Forschung zu politischen Kulturen verwendeten Vergleiche über „kulturelle Indikatoren" (z.B. über Inhaltsanalysen unterschiedlicher Quellen) oder „semiologische Interpretationen" (Auswertung von Ritualen und Symbolen) (Berg-Schlosser 2005, 746 f.). Gerade diese aber spielen beim Verständnis von Handeln in politisch-administrativen Systemen eine große Rolle, nicht zuletzt, weil Images und *cultural semantics* heute die keywords of our time geworden sind (Jay 1998). Ein gutes Beispiel für die konstruktive Nutzung von Inhaltsanalysen und semiologischen Interpretationen bietet ein unter Leitung von Rod A.W. Rhodes auf EU-Ebene durchgeführter internationaler Vergleich politisch-administrativer Handlungssysteme.

Leitfragen waren: 1. ob es Veränderungen in den governance-Mustern gibt, 2.wie die nationalen (sub-nationalen) Eliten diese governance-Veränderungen wahrnehmen und deuten und 3. wie die governance-Muster von den jeweiligen Landes-Traditionen geprägt werden. Rhodes beauftragte in den jeweiligen Ländern „Rapporteurs" mit der Untersuchung und forderte sie auf, einen kognitiv-subjektiven Ansatz zugrundezulegen. Die Länder-Rapporteurs sollten ihre Einschätzungen durch Auswertung von parlamentarischen Debatten, Regierungserklärungen, Beratungs-Papieren für die Regierung, Befragungen, Auswertung von Auto-Biographien einflussreicher Politiker etc. entwickeln.[21]

Konvergenzvermutungen

Die Relevanz von Planungskulturen könnte sich dann als hoch erweisen, wenn sich das institutionelle System der Planungen zwischen den EU-Mitgliedstaaten immer mehr angleichen sollte. Denn dann könnte der schneller institutionelle Wandel mit dem viel langsameren Wandel von kulturellen Prägungen in Konflikt geraten.

---

[16] Donald Keller, Michael Koch und Klaus Selle (1996) berichten von Experten-Diskursen zum Wandel der Raum- (und Stadt-)planung in Deutschland, Frankreich, Italien und der Schweiz. Sie stellen fest, dass in allen Ländern trotz unterschiedlicher institutioneller Rahmenbedingungen, politischer Kulturen und Planungs-Historien Gemeinsamkeiten darin bestehen, dass die Komplexität der Planungsinhalte zunimmt, der Planungsstil kooperativer wird, Projektbezüge immer mehr in den Vordergrund treten und Planer systematischer professionalisiert werden (Anpassung der Ausbildungssysteme). Die von Selle et al. durchgeführten Diskurse zum Wandel der Planungskulturen erheben nicht den Anspruch eines systematischen Vergleichs, sondern sind heuristisch orientiert mit dem Ziel, über Expertenbefragung und Expertendialoge typische Veränderungen in den Planungsprozessen und Planungsansätzen herauszuarbeiten und Erklärungen dafür zu suchen.

[17] Der von Bishwapriya Sanyal (2005) herausgegebene Tagungsband ist das Ergebnis eines Vortragszyklus am MIT im Sommer 2002. Gegenstand waren „Planungskulturen" für die Stadtplanung in fortgeschrittenen Entwicklungsländern (Iran, Hong Kong, Shenzhen, Kalkutta, Mexiko-Stadt) sowie Erfahrungen aus Industrieländern (Japan, England, Frankreich, Niederlande (Andreas Faludi !), Australien, USA). Eingeleitet wurde der Kongress durch einen Beitrag von Sanyal über „hybride Planungskulturen", Friedmann über „Planungskulturen im Übergang", Castells über „Space of flows, space of places".

[18] John Friedmann (2005) berichtet in beiden Artikeln (die inhaltlich sehr ähnlich sind) über den Wandel der Raumplanungskonzepte und Planungssysteme in Japan, China, Indien, Russland, Südafrika, Niederlande, USA, British Columbia.

[19] Das Projekt Cultplan[19] soll kulturbedingte Einflüsse auf Planungsprozesse anhand von ca. 20 Fallstudien in unterschiedlichen INTERREG III-Projekten untersuchen. Dazu wurde ein gemeinsamer analytischer Rahmen entwickelt, der drei Kontext-Systeme umfasst: den Projekt-Kontext, den Kontext der institutionellen Entscheidungsstrukturen und den Kontext der gesellschaftlichen Rahmenbedingungen. Jeder Kontext wird mit eine Reihe von Variablen abgebildet.

[20] Damit wird in der vergleichenden Politikwissenschaft häufiger gearbeitet, z.B. um Einstellungs- und Werthaltungs-Unterschiede herauszuarbeiten (vgl. Hansen/Lauridsen 2004 zum Vergleich der Werthaltungen und Deutungsmuster bei der Übernahme von new public management in der öffentlichen Verwaltung von kommunalen Hauptverwaltungsbeamten in 14 OECD-Staaten).

[21] Die Ergebnisse finden sich in der Zeitschrift *Public Administration*, 81(2003), H.1 für die Länder Deutschland (Werner Jann), Frankreich, Groß-Britannien (Bevir/Rhodes) und die Niederlande (Walter Kickert).

Konvergenzvermutungen speisen sich aus fünf Quellen.

(1) Erstens gibt es eine breite Diskussion über die Aufwertung der regionalen Selbststeuerung und die Stärkung der Regionsebene als gesellschaftliche Handlungsebene im Gefolge von Globalisierung und neo-liberalen Denkmustern. Sie lässt – verbunden mit dem intensivierten Regionen-Wettbewerb – Konvergenzprozesse vermuten. In allen EU-Mitgliedstaaten ist erkennbar, dass (vgl. Keller et al. 1996, Buchmüller 2000, DISP 2002, Friedmann 2005)

- die Regionalisierungsprozesse als Folge der Globalisierung und des Wandels des Staates[22] an Intensität und Steuerungs-Relevanz gewonnen haben (Hein 2002, Kettl 2000, Wolch 1990, kritisch: MacLeod 2001),

- mit der Regionsaufwertung ein wachsender Bedarf nach regionaler Selbststeuerung und Organisation regionaler kollektiver Prozesse der Wirtschaftsentwicklung einhergeht und neue Muster der *regional governance* entstehen (Newman 2000, Martin 1999, Keating 2000). Das gilt um so mehr, als auch die Wirtschaft die Region immer mehr als *supportive system* wahrnimmt (Sternberg 2001, Koschatzky 2002, Moßig 2002, Blotevogel 2000) und sich für die Entwicklung „ihrer" Region engagiert („*corporate citizenship*")

- die Raumplanung sich stärker der Entwicklungsfunktion zuwendet. Sie leitet dann ihre gesellschaftliche Relevanz aus ihrer Stärke ab, Quer-Koordinationsleistungen/ Synergieeffekte zwischen regional relevanten Akteuren über planerische Diskurse herzustellen (Healey et al. 1997). Planung übernimmt insofern immer mehr die Funktion, sich an Prozessen der *regional governance* zu beteiligen, Akteure in der Region zu kollektivem Handeln zu befähigen, Orientierung zu vermitteln und einen Beitrag zur wirtschaftlichen Wettbewerbsfähigkeit der Region zu leisten. Damit verbindet sich auch eine stärkere Hinwendung zur strategischen Planung, d.h. umsetzungsorientierte Entwicklungs-Planung auf der Basis differenzierter SWOT-Analysen

- Raumplanung sich immer häufiger als innovativer Problemlöser denn als „Ordnungshüter" versteht: experimentelles Handeln, Modellvorhaben, organisierte „Problemlösungs-Wettbewerbe" etc. spielen eine immer größere Rolle und Raumplaner müssen das Nachhaltigkeitskonzept ernst nehmen, d.h. immer häufiger auf vier Ebenen gleichermaßen „denken können": räumlich, ökonomisch, ökologisch und sozio-kulturell.

- Raumplaner zwar immer mehr die globalen Handlungszwänge berücksichtigen und beobachten, welche Lösungen in anderen Ländern entwickelt werden. Aber ihr Fokus ist stärker als früher auf die Ausschöpfung der endogenen Potenziale gerichtet, auf die Nutzung der Stärken ihrer Region, auf die Entwicklung der relevanten „Kapitalgüter" (menschliches, soziales, kulturelles und insbesondere: Wissenskapital, Umweltkapital, Urbanitäts-Kapital)

- Raumplaner ihr eigenes Unterstützungs-Potenzial mobilisieren: *stakeholders*, deren Perspektive jedoch im Wege „paradigmatischer Steuerung" stärker auf gemeinsame Belange der Raumentwicklung gerichtet wird. Der sog. „Dritte Sektor" (Priller/ Zimmer 2006, Backhaus-Maul 2006) gewinnt für die Raumplanung eine nicht zu überschätzende Bedeutung.

- Damit einher geht eine engere Verbindung zwischen Planung und *regional governance*: Es geht um die Organisation und Orientierung von kollektiven Prozessen der Raumentwicklung, in denen die einzelnen Akteure nicht borniert ihren eigenen Interessen nachgehen, sondern sich als Teil eines regionalen Gesellschaftssystems verstehen, das kollektive Belange wahrzunehmen hat.

Aufwertung der Region heißt gleichzeitig, dass die Regionalplanung Teil der regionalen systemischen Wettbewerbsfähigkeit wird, woraus sich ein erstes Faktorenbündel herleitet, das Anpassungszwänge ausübt. Offenbar läuft dieses in Richtung strategischer Planung und Regionalmanagement. Zumindest wird europa-weit ein solcher Wandel der Raumplanung konstatiert (Healey et al. 1997).

(2) Ein zweiter Anstoß geht von der EU aus. Mit ihren Strukturfonds, aber auch mit planerischen Ansätzen im Umweltschutz/ Naturschutz (Wasserrahmenrichtlinie, Fauna-Flora-Habitat-Richtlinie, Richtlinie zur Strategischen Umweltprüfung) drängt die EU die Mitglied-

---

[22] In allen Mitgliedstaaten bemüht sich der Staat, unter unterschiedlichen Handlungsphilosphien (enabling state, aktivierender Staat, Abbau des Wohlfahrtsstaates) über „outsourcing" sowie Dezentralisierung/ Kommunalisierung sich auf „Kernfunktionen" zurückzunehmen.

staaten, mehr planerische Anstrengungen zu entwickeln, um Risiken vorzubeugen und kollektive Selbsthilfekräfte zu mobilisieren (vgl. Howe/White 2002, Tewdwr-Jones, Williams 2001, Sommermann 2002 für ein vergleichbares Feld, generell: Sturm/ Pehle 2005). Hinzu kommt eine – von Faludi u.a. angestoßene – Diskussion zur Wirkung des EUREK auf die nationalen Planungssysteme. Da das EUREK nur orientierenden Charakter hat, fehlt ein unmittelbarer nationaler Handlungsdruck. Gleichwohl ist feststellbar, dass die nationalen Planer sich häufig auf das EUREK beziehen und zumindest inhaltlich, wenn nicht sogar verfahrensmäßig Anpassungen vornehmen (vgl. Böhme 2002, 2003, Faludi 2001, 2003, Jensen/Richardson 2001). Empirische Belege scheint es zumindest für die mediterranen Mitgliedstaaten (Frankreich, Griechenland, Italien, Portugal und Spanien) zu geben, wo sogar behauptet wird, die EU-Politik habe dort zur Übernahme neuer Ideen und Praktiken geführt, was man als *„Prozess der kulturellen Innovation in südeuropäischen Planungstraditionen"* bezeichnen könne (Giannakourou 2005).

(3) Ein dritter Anstoß resultiert aus der Planung selbst. Planer und Planungssysteme kommen im Gefolge des Umbaus des Staates (Dezentralisierung, De-Regulierung, Privatisierung) immer stärker unter Druck, einerseits nicht als „Verhinderungsplanung" im Regionenwettbewerb zu erscheinen, andererseits ihre gesellschaftliche Relevanz nachzuweisen. Das erste ist eng mit der Forderung nach „Verschlankung" und Beschleunigung von Verfahren als Standortargument verbunden und hat in Deutschland zum Ausbau informeller Formen der Steuerung geführt (§ 13 ROG). Das zweite wird als Folge der Neuen Steuerungsmodelle und der Fiskalkrise des Staates akut: Regionalplaner müssen begründen können, was sie leisten und warum die öffentliche Hand dafür Geld ausgeben sollte. Dabei scheinen sich europaweit Überzeugungen durchzusetzen, dass Raumplanung ihre gesellschaftliche Relevanz im *Management der Interdependenzen* im Rahmen eines Staatsverständnisses finden muss, das auf Stärkung gesellschaftlicher Selbsthilfekräfte und Abbau paternalistischer Wohlfahrtskonzepte verpflichtet ist. In vielen Ländern formierten sich *„new planning institutions ... in the form of development corporations, rather than planning agencies, because what inspired the moment was entrepreneurship and development, not regulations and planning"* (Sanyal 2005, 9).

(4) Eine große Rolle dürften zudem die Internationalisierung der Planerausbildung und der internationale Austausch in den Planungswissenschaften spielen, wobei insbesondere das Netzwerk AESOP[23] zu nennen ist. Darüber werden Denkmuster und Wertesysteme einander angenähert, aber auch das methodische Vorgehen stärker vereinheitlicht, was über die Ausbildungssysteme direkt in die Planungspraxis hinein wirkt.

(5) Schließlich gibt es begünstigende Politiken. Sie sind Ausdruck davon, dass – als Folge der wachsenden Arbeitsteilung – eine zunehmende gesellschaftliche Nachfrage nach Koordinations- und integrierender Führungsfunktionen entstanden ist. Diese Nachfrage kann zwar von verschiedenen Stellen befriedigt werden – entsprechend erwächst der Regionalplanung auf regionaler Ebene Konkurrenz durch die Regionalisierung der Wirtschaftsstrukturpolitik oder durch Projekte der integrierten ländlichen Entwicklungsplanung. Aber die Regionalplanung hat hier gute Chancen, besonders „wettbewerbsfähig" zu sein. Beispiele dafür sind planerische Bemühungen, die für Stadtzentren massive Störeffekte des großflächigen Einzelhandels am Stadtrand über regionale Einzelhandelskonzepte abzumildern oder die Windenergieanlagen über „Eignungsgebiete" (auch *offshore*) räumlich geordnet aufstellen zu lassen oder Konzepte für die räumliche Verteilung von Groß-Freizeitanlagen zu entwickeln.

In der Praxis erfolgt der Anpassungsprozess allerdings nur langsam und sprungweise, denn Konvergenzpotenziale werden durch die Transaktionskosten der Anpassung erheblich restringiert:

▪ Erstens werden Veränderungen von Systemen nur aufgenommen, wenn der Veränderungsdruck sehr hoch oder Nicht-Anpassung mit Sanktionen verbunden ist. Die Regionalplanungssysteme sind hier im internationalen Vergleich unterschiedlich betroffen. Relativ veränderungs-resistent zeigt sich das stark durch-institutionalisierte deutsche System (Fürst 2005): EU-Richtlinien, die das Planungssystem tangieren, werden tendenziell „assimiliert", indem sie den bestehenden rechtlichen Strukturen subsumiert werden (z.B. integriert in das Abwägungsgebot). *Generell gilt, dass Transaktionskosten um so höher sind, je stärker formali-*

---

[23] Association of European Schools of Planning, z.Zt. c/o Gert de Roo, Universität Groningen, Fakultät für Raumwissenschaften (Faculty of Spatial Sciences).

siert und institutionalisiert Prozesse sind. Verwaltungsverfahren sind deshalb sehr viel veränderungs-resistenter als Planungsprozesse: EU-Richtlinien werden den bestehenden Routinen und Regelsystemen primär subsumiert als dass sie diese reformieren würden (Sommermann 2002)

- Zweitens sind die Transaktionskosten der Systemänderung durch die jeweilige institutionelle Einbindung der Planungssysteme vergleichsweise hoch: Man kann Planungssysteme mitunter nur verändern, indem gleichzeitig auch Änderungen im institutionellen Rahmen vorgenommen werden (vgl. Sommermann 2002 für ein vergleichbares Feld: Verwaltungsverfahren und Verwaltungsprozessrecht). Zudem müssen Änderungen der Planungssysteme von den Planern mitgetragen werden – aber das ist kaum zu erwarten, wenn Änderungen dazu führen, dass die Planung ihre Steuerungskraft verliert, weil die eingeübten *Standard Operation Procedures* durch Verfahren ersetzt werden, die nicht über das Akzeptanzkapital verfügen, das die alten haben.
Deshalb sagt Sanyal (2005, 15), dass „*these planning cultures seem to have evolved with social, political, and economic influences, both internal and external, creating hybrid cultures whose complexity can only be understood through deep historical analysis*".

Der Beitrag der Planungskultur-Diskussion zur governance-Diskussion

Planungskultur-Untersuchungen sind dort sinnvoll, wo sie mit-ursächlich sein könnten für Störungen von Interaktions- und Koordinationsprozessen. Das gilt sicherlich für grenzüberschreitende Planungen, wo Kulturdifferenzen offensichtlich sind, aber auch für den Wandel institutioneller Koordinationsmechanismen im Verhältnis zu den kulturellen Prägungen der sie tragenden Akteure. Koordinationsmechanismen sind immer eine Mischung aus normativer Steuerung (über Handlungsnormen wie Solidarität und Altruismus, Traditionen, paradigmatische Voreinstellungen), tauschfömiger Steuerung (über Markt, Verhandlungen sowie Anreizsysteme) und zwingenden Formen der Steuerung (z.B. über Mehrheitsentscheidungen, hierarchische Steuerung, Konditionalprogramme der Steuerung[24]). Bei Governance kommt es auf die

konkrete Mischung der Koordinationsmechanismen an (Regime).

Insofern hat Planungskultur einen engen Bezug zur *regional governance-Diskussion*. Das ist von einigen Autoren bereits angemerkt worden (vgl. Healey 2006, Albrechts et al. 2003). Denn auch bei *regional governance* lassen sich unterschiedliche Regime ausmachen, eingebunden in jeweils spezifische Kontexte. Auch hier könnte man von „*governance*-Kulturen" oder „*governance*-Stilen" sprechen. Unter *regional governance* wird hier verstanden: *regionale Formen der partnerschaftlichen Selbststeuerung, welche kollektives Handeln zwischen Akteuren unterschiedlicher gesellschaftlicher Subsysteme erlauben und auf die Entwicklung der Region ausgerichtet sind.* Die Besonderheit der neuen Formen liegt darin, dass netzwerkartige Steuerungsformen dominieren, um organisierte Akteure der unterschiedlichen Handlungslogiken (Markt, Politik, Gruppenhandeln) integrieren zu können.

Regional governance befasst sich praktisch mit dem, was auch in der „Planungskultur-Diskussion" verhandelt werden müsste. Das Konzept:

- adressiert primär ein mentales Phänomen, das sich auf Werthaltungen, Einstellungen und Deutungsmuster, und zwar sowohl im Verhältnis Individuum←→Aufgabe (Handlungsorientierung) als auch Individuum←→Gruppe bezieht (Interaktionsorientierung),

- es ist eingebettet in die Interaktions-Kultur einer Gesellschaft, die sich u.a. in der politischen Kultur, im Institutionensystem und im Handeln öffentlicher Akteure niederschlägt,

- es richtet sich auf Steuerungsfragen und folglich den Umgang mit gesellschaftlicher Interessenvielfalt und Informationsprozessen.

Mit *regional governance* verbinden sich zudem ähnliche Phänomene wie mit Planungskultur:

- neue Akteurs- und Einfluss-Konstellationen bestimmen die Prozesse (z.B. vermehrte Integration von Wirtschaftsakteuren als „*stakeholders*"),

---

[24] Unter „Konditionalprogrammen der Steuerung" versteht man Regelsysteme, die dem Adressaten der Steue-

rung nur eine wenn-dann-Entscheidung offenlassen: Wenn die Kriterien der Regelung zutreffen, müssen Entscheidungen gemäß den Vorgaben der Regelung getroffen werden.

- neuartige Themen und Lösungen müssen bearbeitet werden (v.a. bezogen auf „nachhaltige Raumentwicklung"),

- die Formen der Interaktionen wandeln sich (z.B. Zunahme von Verhandlungen, Verträge),

- regions-spezifische Formen der *governance* müssen in die bestehenden Institutionen integriert werden, weil diese letztlich die Ergebnisse umsetzen müssen.

Ähnlich wie in Deutschland, wo noch parallel neue Formen der Selbststeuerung von der regionalen Strukturpolitik (Gerlach/Ziegler 2002; Eckstein 2001), von der Agrarstrukturpolitik (RegionenAktiv, LEADER) und von der Umweltpolitik („Lokale Agenda-Prozesse") initiiert werden, haben sich auch in anderen EU-Mitgliedstaaten noch keine durchgehend-konsistenten neuen *regional governance*-Muster herausgebildet. Aber die Integration der unterschiedlichen Impulse (insbesondere zwischen Regionaler Strukturpolitik und Regionalplanung) wird intensiver werden, je mehr sich der Staat auf „Kernfunktionen" im Konzept des *„enabling states"* zurückzieht und die gesellschaftlichen Selbststeuerungskräfte unterstützt (Bevir/ Rhodes 2003; Jann 2003). Solche Verbindungen werden zur Zeit noch wenig diskutiert (vgl. aber Östhol/ Svensson 2002). Die Relevanz dürfte jedoch hoch sein:

- Erstens, weil Regionalplanung einen wichtigen Beitrag dazu leisten kann, insbesondere unter der Leitvision des *sustainable development*;

- Zweitens, weil Regionalplanung sich immer intensiver auch in die Förderung der regionalen Wettbewerbsfähigkeit einbindet – was insbesondere die Interaktionen zur Wirtschaft intensiviert und sich heute bereits in Ansätzen der Raumplanung niederschlägt, gemeinsam mit der Wirtschaft regionale Raum-Strukturkonzepte zu erarbeiten („*Regionale Einzelhandelskonzepte*", *Regionale Eignungsgebiete für Windenergie, Regionale Freizeitwirtschaftskonzepte*);

- Drittens wird – insbesondere die Regionalplanung – mit dem Ausbau der sog. *Neuen Steuerungsmodelle* in der Verwaltung ihre arbeitsteiligen Beziehungen zu anderen „Anbietern" von Leistungen intensivieren müssen.

Hier bietet sich ein Analyse-Rahmen für Planungskulturen in Beziehung zur Governance an, der sich auf die Steuerungsfähigkeit von *governance* resp. Regionalplanung richtet und in den Vergleichsstudien bei Östhol/Svensson (2002) erfolgreich angewendet wurde: Die Steuerungs-Muster werden über den Umfang und die Art der Beteiligten (*inclusiveness*), die Rückkopplung zu den regionalen Akteuren und Institutionen (*accountability*) sowie die kollektive Handlungsfähigkeit (*coherence*) erfasst.

Literatur

Abbott, C.: The Oregon planning style, in: C.Abbott, D.Howe, S.Adler, Hg.: Planning the Oregon way. A twenty-year evaluation, Corvallis/Or.: Oregon State University Press 1994

Adam, F.: Social capital across Europe. Findings, trends and methodological shortcomings or cross-national surveys, Berlin: Wissenschaftszentrum 2006 (Discussion Papers P 2006-010)

Albrechts, L., Healey, P., Kunzmann, K.: Strategic spatial planning and regional governance in Europe, in: Journal of the American Planning Ass. 69(2003), 113-29

Alterman, R., Hg.: National level planning in democratic countries. An international comparison of city and regional policy-making, Liverpool: Liverpool U.P. 2001 (Town Planning Review Special Study Nr. 4)

Backhaus-Maul, H.: Gesellschaftliche Verantwortung von Unternehmen, in: Aus Politik und Zeitgeschichte 12/2006, 32-38

BBR, Hg.: Baukultur – Planungskultur, in: Informationen zur Raumentwicklung, Heft 11/12.2002

Berg-Schlosser, D.: Politische Kultur/Kulturforschung, in: D.Nohlen, R.-O.Schultze, Hg., Lexikon der Politikwissenschaft, Bd.2, München 2005 (3.Aufl.), 743-48

Bevir, M., Rhodes, R.A.W.: Searching for civil society: Changing patterns of governance in Britain, in: Public Administration 81(2003), 41-62

Bevir, M., Rhodes, R.A.W., Weller, P.: Traditions of governance: Interpreting the changing role of the public sector, in: Public Administration 81(2003), 1-17

Blotevogel, H.H.: Zur Konjunktur der Regionsdiskurse, in: Informationen zur Raumentwicklung 9./10.2000, 491-506

Böhme, K.: Discursive European integration. The case of Nordic spatial planning, in: Town Planning Review 74(2003), 11-29

Böhme, K.: Nordic echoes of European spatial planning: Discursive integration in practice, Stockholm 2002 (Nordregio Report Nr. 8)

Booth, Ph.: The nature of difference: traditions of law and government and their effects on planning in Britain and France, in: Sanyal, B., Hg.: Comparative planning cultures, New York: Routledge 2005, 259-84

Bossaert, D. u.a.: Der öffentliche Dienst im Europa der Fünfzehn: Trends und neue Entwicklungen, Maastricht: European Institute of Public Administration 2001

Buchmüller, L. et al.: Planen, Projekte, Stadt? Weitere Verständigungen über den Wandel in der Planung, in: DISP 141(2000), 55-59

Bundesministerium des Innern: Moderner Staat – Moderne Verwaltung, Berlin 1999 (BMI-Publikation, auch: www.staatmodern.de)

Culling worth, J.B.: The political culture of planning: American land use planning in comparative perspective, New York: Routledge 1993

Cultplan: Status report Cultplan (working package 21), Wageningen Sept. 2006

Derlien, H.U.: Observations on the state of comparative administration research in Europe – rather comparable than comparative? in: Governance 5(1992), 270-311

DISP 148 (2002): Special Issue über "Zukunft der Raumplanung"

DISP 115(1993): Special Issue über „Planungskulturen in Europa"

van Deth, J.: Measuring social capital: orthodoxies and continuing controversies, in: International Journal of Social Research Methodology 6(2003), 79-92

Eckstein, G.: Regionale Strukturpolitik als europäischer Kooperations- und Entscheidungsprozeß, Frankfurt/M.: 2001 (Europäische Hochschulschriften, Reihe XXXI Politikwissenschaft, Bd.440)

Eatwell, R., Hg.: European political cultures. Conflict or convergence? London: Routledge 1997

EU: The EU compendium of spatial planning systems and policies, Brüssel 1997 (CX-03-97-879-C) (Luxemburg: Office of Official Publications, 1997)

Faludi, A.: The Netherlands: A culture with a soft spot for planning, in: Sanyal, B., Hg.: Comparative planning cultures, New York: Routledge 2005, 285-308

Faludi, A.: The application of the European Spatial Development Perspective. Introduction to the special issue, in: Town Planning Review 74(2003), 1-9

Faludi, A.: The application of the European Spatial Development Perspective: Evidence from the north-west metropolitan area, in: European Planning Studies 9(2001), 663-75

Friedmann, J. (2005a): Globalization and the emerging culture of planning, in: Progress in Planning 64(2005), 183-234

Friedmann, J. (2005b): Planning cultures in transition, in: Sanyal, B., Hg.: Comparative planning cultures, New York: Routledge 2005, 29-44

Fürst, D.: Entwicklung und Stand des Steuerungsverständnisses in der Raumplanung, in: DISP 163(2005), 16-27

Fürst, D.: Steuerung auf regionaler Ebene vs. regional governance, in: Informationen zur Raumentwicklung 8./9.2003, 441-50

Fürst, D.: Regional governance – ein neues Paradigma der Regionalwissenschaften? in: Raumforschung und Raumordnung 59(2001), 370-80

Galtung, J.: Struktur, Kultur und intellektueller Stil. Ein vergleichender Essay über sachsonische, teutonische, gallische und nipponische Wissenschaft, in: Leviathan 11(1983), 304-38

Gerlach, F., Ziegler, A.: Mit Innovation und Kooperation zu mehr Beschäftigung? Zur Diskussion neuerer Beschäftigungsförderungsansätze in Regionen am Beispiel der Territorialen Beschäftigungspakte und des Inno-Regio-

Programms, in: Mitteilungen aus der Arbeitsmarkt- und Berufsforschung H.3/2002, 429-39

Giannakourou, G.: Transforming spatial planning policy in Mediterranean countries: Europeanization and domestic change, in: European Planning Studies 13(2005), 319-31

Hansen, M.B., Lauridsen, J.: The institutional context of market ideology: A comparative analysis of the values and perceptions of local government CEOs in 14 OECD countries, in: Public Administration 82(2004), 491-524

Healey, P.: Relational complexity and the imaginative power of strategic spatial planning, in: European Planning Studies 14(2006), 525-45

Healey, P., Khakee, A., Motte, A., Needham, B.: Making strategic spatial plans: Innovation in Europe, London: UCL-Press 1997

Hein, W.: Globalisierung und Regionalisierung. Neue theoretische Ansätze und die Chancen des Empowerment durch Global Governance, in: Nord-Süd aktuell 2002, 215-34

Hemmens, G.C.: Thirty years of planning education, in: Journal of Planning Education and Research 7(1988), 85-91

Hesse, J.J., Hg.: Local government and urban affairs in international perspective: Analyses of twenty Western industrialised countries,, Baden-Baden: Nomos 1991

Hildenbrand, A.: Política de ordenación del territorio en Europa, Sevilla 1996 (Universidad de Sevilla)

Hofstede, G.: Culture's consequences: comparing values, behaviors, institutions, and organizations across nations, 2.Aufl., Thousand Oaks/Cal.: Sage 2001

Hofstede, G., Hofstede G.J.: Cultures and organisations software of the mind, New York: McGraw-Hill 2005

Hovey, B.: Making the Portland way of planning: The structural power of language, in: Journal of Planning History 2(2003), 140-74

Howe, J., White, I.: The potential implications of the European Union Water Framework Directive on domestic planning systems: A UK case study, in: European Planning Studies 10(2002), 1027-38

Innes, J., Gruber, J.E.: Planning styles in conflict at the San Francisco Bay Area's metropolitan transportation commission, Berkeley 2001 (Working Paper des Institutes of Urban & Regional Development of the University of California, WP-2001-09)

Jann, W.: State, administration and governance in Germany: Competing traditions and dominant narratives, in: Public Administration 81(2003), 95-118

Jann, W.: Verwaltungskulturen im internationalen Vergleich. Ein Überblick über den Stand der empirischen Forschung, in: Die Verwaltung 33(2000), 325-50

Jay, M.: Cultural semantics: Keywords of our time, London: Athlone Press 1998

Jensen, O.B., Richardson, T.: Nested visions: New rationalities of space in European spatial planning, in: Regional Studies 35(2001), 703-17

Keating, M.: The new regionalism of Western Europe, London: Edward Elgar 2000

Keller, D.A., Koch, M., Selle, K.: Verständigungsversuche zum Wandel der Planungskulturen. Ein Langzeit-Projekt, in: K.Selle, Hg.: Zur räumlichen Entwicklung beitragen. Konzepte, Theorien, Impluse, Dortmund: Rohn-Verlag 2006 (Reihe „Planung neu denken,", Bd.1), 279-91

Keller, D.A., Koch, M., Selle, K.: „Either/or" and „and": first impressions of a journey into the planning cultures of four countries, in: Planning Perspectives 11(1996), 41-54

Keller, D.A., Koch, M., Selle, K., Hg.: Planungskulturen in Europa. Erkundungen in Deutschland, Frankreich, Italien und in der Schweiz, Darmstadt: Verl. f. Wiss. Publ. 1993 (= DISP 115(1993))

Kettl, D.F.: The transformation of governance: Globalization, devolution, and the role of government, n: Public Administration Review 60(2000), 488-97

Kistenmacher, H., Clev, H.G.: Raumordnung und raumbezogene Politik in Deutschland und Frankreich, Hannover 1994 (ARL-Beiträge Bd. 129)

Knodt, M.: Tiefenwirkung europaischer Politik : Eigesinn oder Anpassung regionalen Regierens?, Baden-Baden: Nomos 1998

Knodt, M.: Die Prägekraft regionaler Politikstile, in: B.Kohler-Koch u.a., Interaktive Politik in Europa: Regionen im Netzwerk der Integration, Opladen: Leske + Budrich 1998, 97-152

Koschatzky, K.: The "New Economic Geography": Tatsächlich eine neue Wirtschaftsgeographie? in: Geographische Zeitschrift 90(2002), 5-19

Kühn, M., Moss, T., Hg.: Planungskultur und Nachhaltigkeit, Berlin: Vlg. für Wissenschaft u. Forschung, 3. Aufl., 2001

Ludwig, J.: Neue Planungskultur in der Regionalentwicklung: Untersuchung anhand der Erarbeitungsverfahren und Strategie Regionaler Entwicklungskonzepte in Bayern, Diss. Bayreuth 2004

MacLeod, G.: New regionalism reconsidered: Globalization and the remaking of political economic space, in: International Journal of Urban and Regional Research 25(2001), 804-29

Marcou, G.: La planification à l'echelle des grands territoires. Etude comparative (Allgemagne, Expagne, Italie, Pay-Bas, Royaume-Uni), Paris 2004

Martin, R.: The new "geographical turn" in economics: Some critical reflections, in: Cambridge Journal of Economics, 23(1999), 65-91

Masser, I., Williams, R.H., Hg.: Learning from other countries: The cross-national dimension in urban policymaking, Norwich/UK: Geo Books 1986

Moßig, I.: Konzeptioneller Überblick zur Erklärung der Existenz geographischer Cluster, Evolution, Institutionen und die Bedeutung des Faktors Wissen, in: Jahrbuch für Regionalwissenschaft 22(2002), 143-62

Nadin, V., Brown, C., Duhr, St.: Sustainability, development and spatial planning in Europe, London, New York: Routledge 2001

Newman, P.: Changing patterns of regional governance in the EU, in: Urban Studies 37(2000), 895-908

Newman, P. and Thornley, A.: Urban Planning in Europe. International competition, national systems and planning projects. London: Routledge 1996

Nohlen, D.: Vergleichende Regierungslehre/ Vergleichende Politische Systemlehre, in: D.Nohlen, R.-O.Schultze, Hg., Lexikon der Politikwissenschaft, Bd.2, München 2005, 1099-95

Nohlen, D. (1995): Vergleichende Methoden, in: D.Nohlen, R.-O.Schultze, Hg., Lexikon der Politik, Bd.2: Politikwissenschaftliche Methoden, München 1995, 507-17

Priller, E., Zimmer, A.: Dritter Sektor: Arbeit als Engagement, in: Aus Politik und Zeitgeschichte 12/2006, 17-24

Priller, E., Zimmer, A., Hg.: Der Dritte Sektor international. Mehr Markt – weniger Staat? Berlin: sigma 2001

Reichel, P., Hg.: Politische Kultur in Westeuropa. Bürgerschaft und Staat in der Europäischen Gemeinschaft, Frankfurt/New York: Campus 1983

Richardson, J.J., Gustaffson, G., Jordan, G.: The concept of policy style, in: Richardson, J.J., Hg.: Policy styles in Western Europe, London: Allen & Unwin 1982, 1-16

Roberts, P., Shaw, D., Walsh, J.A. (Hg.) Regional Planning and Development in Europe, London: Ashgate Press 2000

Sackmann, S.: Success Factor: Corporate Culture. Developing a Corporate Culture for High Performance and Long-term Competitiveness, Gütersloh: Bertelsmann 2006

Salet, W., Faludi, A., Hg.: The revival of strategic planning, Amsterdam: Royal Netherlands Academy of Arts and Sciences 2000

Sandner, R.: Stadtteilforen in Berlin. Ein Beispiel zur neuen Planungskultur? in: DISP 134(1998), 20-23

Sanyal, B., Hybrid planning cultures: The search for the global cultural commons, in: Sanyal, B., Hg.: Comparative planning cultures, New York: Routledge 2005, 3-28

Sartori, G.: Comparing and mis-comparing, in: Journal of Theoretical Politics, 3(1991), 243-57

Scharpf, F.W.: Interaktionsformen. Akteurzentrierter Institutionalismus in der Politikforschung, Opladen: Leske + Budrich 2000

PNDonline III|2007   15|15

Schmidt-Eichstädt, G., Fallon, P.: The EU compendium of spatial planning systems and policies. Germany. Final draft prepared for the European Commission, DG XVI, Berlin 1996 (TU-Berlin)

Schnapp, K.-U.: Ministerialbürokratien in westlichen Demokratien. Eine vergleichende Analyse, Opladen: Leske + Budrich 2004 (Interdisziplinäre Organisations- und Verwaltungsforschung, Bd.11)

Seligson, M.: The renaissance of political culture or renaissance of the ecological fallacy? in: Comparative Politics 34(2002), 273-91

Selle, K.: Neue Planungskultur – Raumplanung auf dem Weg zum kooperativen Handeln? in: K.M. Schmals, Hg.: Was ist Raumplanung? Dortmund 1999 (Dortmunder Beiträge zur Raumplanung, Bd.89), 210-26

Sixel, St.: Regionale Entwicklungskonzepte in Mecklenburg-Vorpommern: ein Beitrag zur Weiterentwicklung der Planungskultur im ländlichen Raum, Diss. Rostock 2005

Sharpe, L.J., Hg.: The ris of the meso government in Europe, London: Sage 1993

Sommermann, K.-P.: Konvergenz im Verwaltungsverfahrens- und Verwaltungsprozessrecht europäischer Staaten, in: Die öffentliche Verwaltung 55(2002), 133-43

Spath, Chr.: Bürgerbeteiligung im Baugenehmigungsverfahren? Ein Vergleich der Baugenehmigungspraxis in der Bundesrepublik Deutschland und Großbritannien, in: Archiv für Kommunalwissenschaften 24(1985), H.2, 242-60

Sternberg, R.: New Economic Geography, in: Zs. f. Wirtschaftsgeographie 45(2001), 159-80

Sturm, R.: Die Politikstilanalyse. Zur Konkretisierung des Konzeptes der Politischen Kultur in der Policy-Analyse. In: Hartwich, H. H. (Hg.): Policy-Forschung in der Bundesrepublik Deutschland. Opladen: Leske + Budrich 1985, S. 111-116

Sturm, R., Pehle, H.: Das neue deutsche Regierungssystem. Die Europäisierung von Institutionen, Entscheidungsprozessen und Politikfeldern in der Bundesrepublik Deutschland, 2. Aufl., Wiesbaden: VS-Verlag 2005

Tewdwr-Jones, M., Williams, D.: The European dimensions of British planning, London: Spon 2001

Van Waarden, F.: Persistence of national policy styles: A study of their institutional foundations, in: B. Unger, F.van Waarden, Hg.: Convergence or diversity? Aldershot: Avebury 1995, 333-73

Walter, R.: 50 Jahre Planungskultur in den Niederlanden: veränderte Leitthemen der Raumordnung seit 1945, in: DISP 122(1995), 24-28

Weitz, J.: From quiet revolution to smart growth: State growth management programs, 1960 to 1999, in: Journal of Planning Literature 14(1999), 266-337

Wentz, M.: Sozialer Wandel und Planungskultur, in: Wentz, M., Hg.: Planungskulturen, Frankfurt/ Main: Campus 1992 (=Die Zukunft des Städtischen 3), 10-19

Wheeler, St.M.: Planning for metropolitan sustainability, in: Journal of Planning Education and Research 20(2000), 133-45

Williams R. H.: European Union spatial policy and planning, London: Paul Chapman Publishing 1996.

Williams, R.H., Hg.: Planning in Europe, London: Allen & Unwin 1984

Wolch, J.R.: The shadow state: Government and voluntary sector in transition, New York: Foundation Center 1990

Originaltext Taylor 2013

TPR, 84 (6) 2013 doi:10.3828/tpr.2013.36

Zack Taylor

Rethinking planning culture: a new institutionalist approach

Scholars of planning have long grappled with the dilemma of how to explain variation among places' traditions, modes or styles of planning practice and the legal and institutional frameworks that govern spatial development and implement planning policies. In a related effort, historians have explored the international diffusion of planning ideas and practices, the study of which has gained contemporary relevance in the context of European integration and globalisation. At the core of these enterprises is an attempt to understand change – to specify how and why planning practices are changing and why distinct patterns of planning practice have evolved in different places and at different times. Recent work has embraced the concept of 'planning culture' as the basis for explanation, yet this work has lacked focus. This article argues that historical institutionalism as developed in the social sciences provides a more precise explanatory framework for comparative planning research.

Keywords: planning culture, new institutionalism, historical institutionalism, planning history, comparative planning systems

Scholars of planning have long grappled with the dilemma of how to characterise differences among places' traditions, modes or styles of planning practice and the legal and institutional frameworks that govern spatial development and implement planning policies (Larsson, 2006; Masser and Williams, 1986; Newman and Thornley, 1996). In a related effort, historians of planning have explored the international diffusion of planning ideas and practices (Healey and Upton, 2010; Home, 1997; Ward, 2000). Once of largely historical interest, studies of cross-national influence and exchange have gained contemporary relevance in the context of European integration and globalisation (Tewdwr-Jones and Williams, 2001). At the core of these enterprises is an attempt to understand change – to specify how and why planning practices have changed and are still changing today and why distinct patterns of change have occurred in different places and at different times. Recent work on these topics has embraced culture as the basis for explanation (Knieling and Othengrafen, 2009; Sanyal, 2005), yet for the most part this work has been vague and unfocused. Quite different social phenomena have been conflated under the rubric of 'planning culture' with the effect of undermining its analytic traction.

In this article I argue that the new institutionalist paradigm – particularly the genre of historical institutionalism as developed in political science, sociology and political economy – can provide a more precise explanatory framework for compara-

Zack Taylor is Assistant Professor in the City Studies Program, Department of Human Geography, University of Toronto, Scarborough; email: zack.taylor@utsc.utoronto.ca
Paper submitted January 2012; revised paper accepted March 2013.

3

tive research on planning systems and practices. The new institutionalism refers to a diverse family of approaches to understanding stability, change and causal processes in social and economic systems. Historical institutionalism emerged in political science and political sociology in the 1970s as a rejection of several themes that had dominated the social sciences since the early 1950s: the atomistic rationalism and society-centrism of behaviouralism (Campbell et al., 1960), the system-level generalisations of structural-functionalists (Almond and Powell, 1966; Parsons and Smelser, 1956), the pluralist presumptions of system self-equilibration and therefore stability (Dahl, 1961) and the ethnocentrism of linear stage theories of modernisation and development (see Gilman, 2007). While divergent in their methods and interests, new institutionalists conceptualise institutions neither as epiphenomenal to economic structure or culture, nor as passive background conditions or arenas in which social relations occur. Rather, institutions are seen as causal variables that structure the opportunities and constraints faced by individual and collective actors and therefore favour some outcomes or patterns of activity over others.

The article begins by describing the emergence of cultural analysis as a means of explaining the international diffusion of planning ideas and as the basis for comparing national planning frameworks and processes. It goes on to discuss imprecision in the literature regarding the definition of planning culture – imprecision that compromises its explanatory power and which the application of new institutionalist concepts may remedy. The article continues with an overview of how the new institutionalist paradigm can be applied to research on historical and contemporary change in planning systems and practices and concludes with a brief discussion of why planning scholarship has avoided a comprehensive engagement with historical institutionalism.

Cultural explanation in the study of planning

Albeit for different reasons, planning historians and practitioners have long been interested in the diffusion of ideas. Planning historians have seen diffusion across national borders as productive of technical innovation and, through encounters with local organisational structures and practices, hybridisation of practices (Nasr and Volait, 2003; Ward, 2002). For their part, planning practitioners have often looked to other places for 'best practices'. The literature is replete with studies of the transnational adaptation of the zoning principle (Logan, 1976), the garden city (Ward, 1992) and the new town concept (Masser and Williams, 1986, Chapters 10–14); the replication of techniques and physical layouts in different places through the activities of globetrotting consulting firms (Ward, 1999); and the imposition of colonial powers' planning ideas for their present and former possessions (Home, 1997). For the most part, these have eschewed the explicit building and application of theory. An exception is Ward's (1999; 2000; 2010) typology of planning diffusion mechanisms, which distinguishes between different forms of borrowing and imposition based on the relative power of the sending and receiving nation or system. While these are useful aids to description, there is little to guide systematic assessment of the relative power of relevant actors, especially in the middle categories where the power differential is small. Still, Ward's typology hints at the role domestic institutions and practices may play in shaping processes of international policy transfer.

The notion that planning policies must be compatible with the structures and logics of the institutional frameworks that carry them out necessarily leads to the related but more encompassing comparative literature on planning systems. Perhaps because planning is so fragmented in the United States and Canada, with distinct implementations of comprehensive planning and development control instruments in operation in each state and province, the notion of planning constituting a coherent system has found the greatest purchase in Britain and continental Europe. British scholar J. B. Cullingworth's (1987; 1993; 1994) surveys of American and Canadian planning are the exceptions that prove the rule. Over the past half-century, European integration has spurred interest in comparing distinct national approaches to regulating the present and future use of land and property. One of the most extensive efforts was Masser and Williams' edited collection *Learning from Other Countries* (1986), which summarised the proceedings of a series of conferences on the theory and method of cross-national planning research (see also Masser, 1984). This project synthesised numerous themes and issues, not least the methodological problem of avoiding miscomparisons stemming from the observer uncritically viewing a foreign context through the lens of his or her home system's norms and assumptions. Although several chapters examine how the new town concept diffused internationally earlier in the century, the focus of Masser and Williams' collection is squarely on probing the potential for policy learning and transfer, not on illuminating the historical processes through which policies evolved. There is little discussion of exactly why national planning systems differ in structure, aim and effect, only that they do. An exception is Sutcliffe's introductory chapter, which hints at the determinants of systemic difference. He approvingly cites Friedmann's hypothesis that '[d]istinctive styles of national planning are associated with different combinations of *system variables*, including the level of economic development attained, the form of political organisations, and historical tradition' (Friedmann, 1967, 33 [emphasis added]). Albeit vague, this statement can be generalised as recognition that macrostructural conditions ('level of economic development'), formal institutional structures ('the form of political organisations') and professional and societal norms ('historical tradition') are independent variables that somehow shape the evolution of national planning systems.

More recently, scholarly attention has focused on how globalisation and, in Europe, the expansion of European Union involvement in heretofore national or local planning processes are changing domestic planning practices (Albrechts et al.,

3

2001; Salet et al., 2003; Tewdwr-Jones and Williams, 2001). Much of this work is descriptive, rather than directed at theory-building. Juxtaposing city or country case studies usefully establishes a common ground for comparison, but provides little in the way of systematic evaluation of the historical question regarding why national planning systems differ and what role these differences may play in mediating external influences. An exception is Newman and Thornley's *Urban Planning in Europe* (1996). They identify five European legal and administrative planning 'families'– the British, Napoleonic, Germanic, Nordic and East European – defined as such by the degree to which development control is discretionary as opposed to regulatory; authority is centralised or decentralised; the public–private relationship is conflictual or coopera-tive; and, borrowing from Healey and Williams (1993), plan-making, land use regula-tion and land assembly and servicing are integrated or handled separately. Although he does not create a typology, Larsson (2006) similarly examines the planning systems of fourteen Western European countries with respect to seven subjects: consistency and completeness across the national territory, the balance between responsibilities and capacities at each level of government, the degree to which higher-level govern-ments bind lower-level activities, flexibility, integration and coordination across sectors of activity, private–sector participation and compensation to private actors for public actions. Stead and Nadin (2009) outline a two-level model in which the legal and administrative characteristics and intellectual traditions of the planning system are related to what they call the 'social model' – the distinct national pattern of linkages between economy, society and the state – which they relate to Esping-Andersen's (1990) seminal typology of welfare capitalisms.

Although the body of work on diffusion and system comparison has largely been focused on the formal institutions, instruments and practices of planning, cultural explanations have never been far from the surface. The appeal to culture has become more explicit as scholars have become increasingly attuned to the seeming incom-mensurability of terminology and patterns of practice in different national–linguistic contexts. By this account, a myopic focus on formal legal and organisational structures and regulatory instruments may obscure the informal norms and values that guide their application, lead analysts to perceive similarities where none exist and thereby overstate the potential for the successful transplantation of policies into new contexts (Booth, 1993; 2011; Healey and Upton, 2010). Assimilating more general characteri-sations of local 'planning styles' (Abbott, 1994; Wheeler, 2004), these concerns have crystallised into a new literature on 'planning culture' (see especially Knieling and Othengrafen, 2009; Sanyal, 2005).

The limitations of planning culture as an explanatory concept

While the growing body of work on planning culture has produced many useful insights, it suffers from two deficiencies. The first is imprecision. 'Planning culture' has been used to describe quite different phenomena, thereby undermining its analytic leverage. This can be illustrated by viewing planning culture along two dimensions. To begin, there is the question of whether planning culture is seen as something that causes change – in other words, as an independent variable – or, alternatively, as something that is an object of influence, a dependent variable. Then there is the question of whether the 'culture' in question is construed as equivalent to norms shared by professional planners and other actors directly involved in spatial development or, alternatively, as the norms and values of broader society. We may think of the former as 'organisational culture' and the latter as 'societal culture'. Juxtaposing the two dimensions produces three viable combinations, each of which is discussed with examples (see Table 1).

Table 1 Two dimensions of planning culture

	Analytic focus	
Planning culture as:	Organisational culture	Societal culture
Independent variable	(1) National planning traditions influence professional planning and spatial development practices and outcomes (Healey, 1997) and, in Europe, supranational spatial policymaking (Faludi, 2004).	(2) Planning institutions and professional practices are embedded in a societal culture that organises meanings and public and private sector relationships to land, property and the state (Booth, 2011; Cullingworth, 1993; Healey, 2010; Knieling and Othengrafen, 2009; Sanyal, 2005; Waterhouse, 1979).
Dependent variable	(3) Planning culture is being restructured by macrostructural forces (Friedmann, 2005a) or European integration (Keller et al., 1996; Sanyal, 2005; Tewdwr-Jones and Williams, 2001).	

Note: The lower right quadrant is empty, because no one has argued that planning culture exerts a meaningful independent influence on broader political culture.

First, let us consider planning culture as an organisational culture – a set of norms and values that is internal to the planning system and practice. This view is consistent with Healey's discussion in *Collaborative Planning* (1997) of national 'traditions' of practice that define the boundaries of what planners do. Her most recent articulation of planning culture borrows from policy studies in describing planning as at once a policy community made up of actors implicated in land use change (Rhodes, 2008),

3

an epistemic community made up of practitioners drawing on a common knowledge base (Haas, 1992) and a community of practice defined by the 'doing' of planning (Wenger, 1998) – all of which are embedded within formal institutions (Healey, 2010, Footnote 6). Sanyal (2005, 3) similarly sees planning culture as internal to the profession: 'the collective ethos and dominant attitude of professional planners in different nations toward the appropriate roles of the state, market forces, and civil society in urban, regional, and national development'. While Healey is concerned with the impact of national professional norms on domestic spatial planning and development processes, Faludi (2004) shows how they shaped national bargaining positions in the creation of the European Spatial Development Perspective. The planning traditions of the three key national actors – France, Germany and the Netherlands – each embodied a different balance between indicative versus binding instruments, national versus local authority and state control versus private initiative, each a product of historically evolved norms and values. Conflicts between them led to an ill-defined compromise at the European level. Others reverse the causal arrow, positioning planning culture as a dependent variable. Friedmann (2005a; 2005b), for example, is interested in how neoliberal globalisation exerts homogenising pressure on national or local planning cultures. Similarly, Keller et al. (1996) and Tewdwr-Jones and Williams (2001) are interested in how the expansion of European Union policymaking is altering domestic planning practices.

Whether positioned as an independent or dependent variable, the works described so far define planning culture in terms of the professional norms and practices implicated in spatial development. An alternative perspective shifts the analytic focus to broader society. In this view, planning culture is seen as determined by, or as a subset of, the broader political or societal culture. As mentioned earlier, Booth (1993; 2011) argues against a narrow focus on deracinated planning instruments and techniques, calling instead for a full accounting of planning's embeddedness in national cultural and linguistic norms and structures. Similarly, Cullingworth (1993) situates the exceptional nature of America's decentralised and fragmented planning system in the individualistic and localist DNA of American political culture, while Waterhouse (1979) concludes that despite similarities in the formal structures of planning in Toronto and Munich, their 'application differs significantly according to specific traditions, specific forms of land ownership and specific interpretations of the concept of social justice' (322). Knieling and Othengrafen (2009) call for a 'culturised planning model' that relates the formal and informal characteristics of the planning system to their national institutional contexts, legal and administrative traditions, patterns of market relations and ways of life, which are seen as ultimately determined by culture. In their view, 'each national or regional context is characterised by particularities of history, by attitudes, beliefs and values, political and legal traditions, different socio-economic patterns and concepts of justice, interpretations of planning tasks and responsibilities, and different structures of governance – in other [words]: by its specific cultural characteristics' (xxiv). Sanyal (2005) also characterises planning culture as a reflection of the 'larger social culture in which it is embedded' (22), stating that 'planning culture is not an independent variable' (20). Friedmann, too, sees planning culture as derivative of societal culture, defining it as 'the ways, both formal and informal, that spatial planning in a given multi-national region, country or city is conceived, institutionalised, and enacted' (2005a, 184). The point of discriminating along these dimensions is not to say that only one of these conceptualisations of planning culture is correct. Rather, it is to say that identifying these different processes with the same label undermines the concept's utility in causal explanations. The broader political culture should be kept analytically distinct from professional attitudes, norms and discourses and also from the more specific realm of policy ideas.

The imprecise definition of planning culture leads to a second deficiency – the difficulty of accounting for cultural stability and change. Societal cultures are implicitly viewed as fixed and durable, while (at least in Quadrant 3) organisational cultures are sometimes seen as more changeable. Lacking precision, however, is an account of why this may be so and why planning cultures, however defined, may be more susceptible or resistant to change in some circumstances than in others.

Historical institutionalism: explaining stability and change

To add precision to causal explanations and to account for stability and change in planning systems, norms and practices, I suggest that planning scholars interested in these questions shift their focus from planning cultures to the formal institutions in which norms and traditions are embedded. Viewing formal institutions as sites of cultural production and reproduction can enable clearer understandings of how stability and change may be patterned by their structural features. To account for these processes, planning scholars and especially historians of planning have much to gain from explicit engagement with historical institutionalism as it has emerged in the social sciences and policy studies. The following overview calls attention to concepts used in historical institutionalist analysis that are relevant to the study of stability and change in planning systems, norms and practices.

At the core of new institutional analysis is an analytic distinction between formal and informal institutions, ideas and structures. For new institutionalists, formal organisational structures, laws and decision rules are distinguished from informal norms and conventions, and both are distinguished from ideas and macrostructural variables (Hall and Taylor, 1996; Hay, 2008; Pontusson, 1995). Historical institutionalism in political science and political economy tends to be polity-centred; the focus is on governing institutions and the formalised rules and informal conventions associated with them. By recognising the relative autonomy of the state apparatus from society, these analyses illustrate how the configuration of institutions privilege some interests and courses of action over others (Nordlinger, 1988; Skocpol, 1985). This is not to say that the state is necessarily monolithic, intentional or efficient in its actions (March and Olsen, 1984). Rather, historical institutionalists are interested in disaggregating the state to discover how its components possess distinct and often contradictory interests, preferences and logics (Clemens, 2006).

The focus on the formal and informal institutions of governance does not mean that the influence of ideas, discourses and macrostructural variables is ignored. Ideas regarding means and ends are central to processes of institutional creation, while formal institutions such as training and credentialing regimes reproduce and propagate policy ideas. While the issue continues to be debated, many analysts insist that ideas are not in themselves institutions and that they are best considered separately as motivating frames for the mobilisation of action, including action directed towards institutional change (see Schmidt, 2008; 2010; Smith, 2006). Historical institutionalists also do not ignore the impacts of changes to broader social and economic structures. Indeed, scholars working at the seam of international relations and domestic politics see the configuration of state and non-state institutions and actors as determinants of globalisation's domestic impact (Keohane and Milner, 1996; Schmidt, 1995) – a research agenda with clear parallels to the aforementioned work on the Europeanisation of domestic planning practices. At the same time, Pierson (2004, Chapter 3) draws attention to long-term and cumulative processes that induce institutional change – for example, demographic shifts, technological advances and climate change – and the importance of timing and sequence in shaping outcomes, noting that different paths may stem from similar initial conditions if the external context differs (Pierson, 2000b).

Applying these concepts to the historical and contemporary study of planning may aid in disentangling the different social phenomena that have been subsumed under the rubric of planning culture. A focus on the legal and organisational dimensions of the planning system – such as statutes and regulations, professional organisations and schools and bureaucratic organisations – reveals the mechanisms by which broader societal norms and power relations are produced and reproduced. It avoids the positioning of societal culture as a fixed and all-determining independent variable that is prior to all other social phenomena. Students of political behaviour have long recognised that members of societies are socialised into superordinate orientations toward political and social objects that are independent of individual preferences and interests (Almond and Verba, 1963; Campbell et al., 1960; Pye and Verba, 1965). Yet to view the configuration and activities of the state and private institutions simply as expressions of national culture reductively denies the potential for culture to change, ignores the institutional mechanisms through which it is reproduced and transformed through time and downplays political contestation.

The focus on governing institutions leads to a second concern: the problem of how to explain stability and change. A core objective of historical institutionalism is to delineate and explain the determinants of historical paths or trajectories (or their absence) – in other words, to explain stability and change without presuming either to be natural (Thelen, 2003). Earlier formulations retained the pluralist presumption of homeostasis, the precept that social and economic systems tend toward equilibrium and change only when subjected to an external stimulus (Krasner, 1984). These so-called 'punctuated equilibrium' models were criticised for their bias toward stability (Streeck and Thelen, 2005). Still, there remains a conceptual distinction between 'critical junctures' or episodes of change and the periods of apparent stability that lie between them. Generally, new institutionalists have identified four sources of stability and change: transaction costs, policy feedback, the interaction of political context and endogenous discretion and enforcement regimes and institutional complementarities. Each is discussed in turn.

Transaction costs

Borrowing from the new institutional economics (NIE), which emphasises the roles played by laws, regulations, property rights and other formal and informal institutions in lowering the cost of economic exchange and collective action (Coase, 1998; North, 1990; Ostrom, 1990; Williamson, 1975), some historical institutionalists conceptualise 'path-dependence' – stability through time – in relation to the high transaction costs associated with change. This may be conceptualised in terms of sunk costs: once organisational structures have been established and standard operating procedures defined, it is costly to dismantle them or direct them to new purposes. A more nuanced perspective emphasises the coordinating role of institutions and practices. Coordination reduces the costs for implicated actors, because they can make commitments and investments with reasonable expectations of future outcomes. Institutional change, even if it would ultimately result in greater aggregate efficiency, imposes costs on actors, because it alters the allocation of costs and benefits in unpredictable ways. This produces resistance to change.

The latter approach has been applied to planning by Alexander (1992), who stresses the stabilising effect of public land use regulation in a complex market populated by a variegated array of actors. In more recent work, Alexander (2005) moves from the descriptive to the normative, suggesting that if planning in a complex society is, above all, a problem of coordination of social and economic resources, then the primary role of the planner is to design coordinating institutions that stack the deck in favour of desired planning goals. Alexander hints at potential sources of institutional change – in particular, the planner as an instigator of change – but does not extend the analysis to explain why particular planning institutions and practices emerged in the first place or why they differ from one place or country to the next.

3

Policy feedback

A related source of stability is what Pierson (2000a; 2006) characterises as positive policy feedback. Adapting the concept of 'increasing returns' from economics, Pierson shows how institutional and policy choices allocate resources, thereby shaping the composition of interest groups, benefiting some actors over others, altering actor preferences and influencing public opinion and political behaviour. A prime example is the provision of social benefits. The institutionalisation of the welfare state was not simply the product of changing political attitudes or cultures. The beneficiaries of welfare state policies have not only defended particular institutions and policies at the ballot box and in the streets, but they also represent a reorientation of societal norms and attitudes in relation to state intervention.

Planning systems may generate positive feedback in several ways. In market economies, property development largely occurs through the investment of private capital. Private interests that are party to the urbanisation process – including developers, builders, lenders, public- and private-sector professionals and home owners – are likely to support the continuation of institutions and practices that increase the predictability of investments in property. In this way, we can understand the market appeal of privately regulated common-interest developments, such as condominium apartments and gated communities (Deng et al., 2007), as well as the durability of sometimes exclusionary zoning practices (Nelson, 1980). As Moore (1979) notes, zoning and other forms of development control emerged in the first instance not as means of implementing comprehensive planning visions, but as mechanisms of protecting property values and neighbourhood character in established areas; as stabilising mechanisms that property owners routinely support. Another example is the emergence of interest groups and social movements dedicated to defending policies that protect peri-urban rural landscapes for their amenity and productive value. While the London Green Belt is perhaps the highest profile example (Edwards, 2000; Elson, 1986), the emergence of regional coalitions of urban environmentalists, suburbanites and farmers in favour of urban containment is well documented (Abbott and Howe, 1993; Hanna and Webber, 2010).

As hinted at by Healey (2010) in her discussion of planning traditions and norms (what I have called organisational culture), it is important to recognise the cognitive or intersubjective dimension of institution-bound relations. Through dense interaction, actors operating within a common institutional framework – what Marsh and Rhodes (1992) characterise as integrated policy communities – may come to develop common values, beliefs and understandings of policy problems and their resolution. Shared norms and practices reduce coordination costs, thereby creating an incentive to maintain and continue to operate within the established institutional framework. At the same time, the very structure of the institutional framework shapes these meanings. Planning in a market economy consists of ongoing close interactions among planning professionals, community residents, property developers and interest groups. These interactions are regulated by statutory rules, oversight mechanisms, codes and practices inculcated through professional education and so on. To the extent that these script and routinise interaction and regulate conflict, they generate increasing returns by lowering the transaction costs of coordination.

A useful example of this is found in Vancouver, Canada, where the creation in the early twentieth century of metropolitan water and sewerage service boards institutionalised a framework for voluntary intermunicipal discussion and collaboration among local politicians and professionals. When the post-war population boom occurred, this governance model was extended to regional planning with the creation in 1948 of the Lower Mainland Regional Planning Board. This, in combination with the contemporaneous establishment of the University of British Columbia planning school, which instilled a particular set of regional planning ideas in the nascent profession, helped entrench a professional and political consensus on the need for regional planning to protect agricultural land and promote compact and self-sufficient urban centres. Without the prior existence of a widely accepted regional governance model that facilitated collaborative interactions and the early establishment of a local centre for planning education, regional planning ideas would not have taken root. Through these institutions, planning ideas and practices were produced and reproduced, resulting in remarkable continuity in policy content and outcomes (Artibise et al., 2004; Chadwick, 2002; Oberlander and Smith, 1993).

Interaction between political context and internal discretion regimes

Thelen and her collaborators explain how different patterns of often gradual change occur as a result of two variables: political context or the degree to which external decision-makers possess the authority to block change – in other words, act as 'veto players' (Tsebelis, 2002) – and the degree to which actors internal to the institution exercise discretion over the implementation or enforcement of mandated policies (Mahoney and Thelen, 2010b; Streeck and Thelen, 2005). If decision-makers possess the constitutional or legal resources to block changes to an institution *and* internal actors cannot instigate change by interpreting or enforcing rules differently, then opponents may pursue their objectives by creating new institutions in parallel to existing ones ('layering'). If they do not, opponents are able to reform, replace or eliminate the existing institutions ('displacement'). A prime example of layering is the creation in the United States during the 1970s of environmental impact assessment regimes, which were largely independent of existing subnational and local land-use planning and governance systems. In New South Wales and other Australian states, however, displacement occurred: traditional regulatory planning and environmental instruments were unified under a single framework (Gurran, 2011, Chapters 7–8). The

3

fragmented nature of authority in the United States multiplies potential veto points, inhibiting change, whereas the centralisation of executive authority in Australia permitted the imposition of a new system.

More interesting is what may occur when actors can exercise discretion over the interpretation and enforcement of existing rules, thereby altering their purpose and effects ('conversion'). Over the past century in North America, Britain and elsewhere, the same planning instruments have been used to facilitate or restrain urbanisation, conserve rural and wilderness lands for their productive uses or their own sake and preserve privileged social spaces or promote social equity. These transformations represent the adaptation or conversion of institutions to new purposes. A more ambiguous outcome is what Hacker (2004) and Mahoney and Thelen (2010a) characterise as 'drift': when broader social, economic and environmental transformations undermine the effectiveness of an existing institution, discretion is insufficient to bring about necessary adaptation and it is in the interest of decision-makers to abstain from or actively oppose reform. The oft-repeated failure to expand the territorial scope of metropolitan planning authorities when their boundaries are overtaken by urbanisation is an example of drift.

Institutional complementarities

The varieties of capitalism (Hall and Soskice, 2001) and worlds of welfare capitalism (Esping-Andersen, 1990) literatures in comparative political economy identify mutually supportive 'complementarities' between institutionalised subsystems of domestic political economies. These works have spawned numerous refinements and syntheses that contrast historically evolved national configurations of state regulation, corporate governance, labour relations, skills development, lending and social assistance regimes that condition both the productivity and innovativeness of firms and domestic responses to transnational competitive pressures (Amable, 2000; Kitschelt, 1999; Pontusson, 2005; Thelen, 2004). The notion of complementarity provides a useful lens for understanding policy diffusion and institutional transplantation: by virtue of interlocking dependencies, not all forms and practices are possible in all milieux. At the same time, the emergence of disjunctures between complementary subsystems may be a source of change. Orren and Skowronek (2004), for example, describe how political and institutional change can be driven by conflict between different 'orders' – institutional structures that embody different and often contradictory purposes, logics and resources. Change in one part of the political, economic and social system may, by virtue of complementary linkages across domains, generate change in other parts.

As aforementioned, while the recent comparative planning systems literature has explored the notion of patterns of interlocking institutions at the national or subnational scale, deeper engagement with comparative political economy promises to more systematically elucidate historically evolved relationships between those subsystems implicated in spatial development – land-use planning, the property rights regime, real estate finance, infrastructure planning and finance and so on – and others, not least the organisation and regulation of business firms and welfare state programs. Booth et al.'s (2007) comparison of the British and French planning systems points in this direction. The authors supplement discussion of the legal and organisational structures of spatial planning with analysis of the two countries' different property right regimes and land tenure systems, the structure of land and property markets and the financing of development public–private partnership arrangements, public transportation governance and policy and the influence of EU policies on domestic planning activity.

An example: regional planning and governance in Portland, Oregon

These four concepts do not operate in a vacuum. Indeed, it is worth undertaking a more detailed examination of a case that is often viewed in cultural terms, yet whose historical development is intimately bound up in the interaction of institutional logics. Portland, Oregon, is celebrated in the American context for the high quality and coherence of its planning (Abbott, 2000; Abbott et al., 1994). Its ostensible success is attributed to distinct local public attitudes favourable to compact development and protection of the natural environment – in essence, a local planning culture. This telling makes the emergence of effective regional planning appear inevitable, when in fact the creation of Portland's regional planning regime was fiercely contested and institutionalised only through a series of contingent episodes of institutional change and adaptation.

Perhaps more than most parts of the United States, Oregon possesses a localist political culture that values local autonomy and the sanctity of property rights and participatory democracy. Oregon was an early adopter of constitutional 'home rule' for municipal governments (whereby limits are placed on the ability of the state government to unilaterally legislate on municipal affairs) and policymaking by public initiative or referendum. Seen through the lens of institutional complementarity, we can see how the early establishment of a highly decentralised and autonomous system of municipal government was compatible with and reinforced a discourse of unconstrained property rights and how both, in tandem with the initiative system, limited the scope of legitimate action by state and local governments in the field of planning. The first attempts to create a regional planning capacity for the region, the Metropolitan Planning Commission (1957–67) and the Columbia Region Association of Governments (CRAG) (1967–79), were ineffective. These institutions were established as clearinghouses for federal transportation and planning grants, but had a little support from area municipalities. Ultimately, CRAG's planning activities were seen as violations of local autonomy and individual property rights. By the late 1970s, its planning agenda had stalled.

3

It was in this institutional context that a group of reformist local elites secured private funding to lobby for the rationalisation and consolidation of the myriad local governments and special-purpose bodies in the Portland region. Their objective was not primarily to institute regional planning, but to create a more unified and accountable system of service delivery. They argued for a single agency to operate infrastructure and provide major services in the three-county area. The outcome of this process was the creation of Portland's elected regional government, Metro, in 1978 (Abbott and Abbott, 1991). To secure popular ratification in a referendum (in essence, to circumvent the multiple veto points in the fragmented and decentralised political system), reform advocates engaged in subterfuge. The referendum question was deliberately worded to appeal to the majority of voters who desired the abolition of CRAG – in fact, most voters likely believed that they were voting against the creation of a new regional government – and the boundaries of the new entity were defined so as to exclude as much of the rural fringe as possible. Contrary to its advocates' intentions, opposition from municipalities and counties prevented Metro from assuming new functions. Only late in the process did it acquire the function for which it is now best known: the administration of the state-mandated urban growth boundary. Had it not become a planning agency, it would have remained an almost empty shell, managing solid waste disposal and a zoo.

Metro's planning activities have been under almost continuous attack in the legislature and at the ballot box from property rights and anti-government activists. Metro has insulated itself from challenges through avoidance of conflict and the assiduous cultivation of public support. An 'inside–outside' coalition of peri-urban farmers and urban environmentalists has emerged to defend Metro and its urban containment strategy. As the land use program has matured, actors implicated in the planning and urban development process have come to share a set of organising principles and standard operating procedures. Landowners and developers who would otherwise be natural opponents of land use regulation have made their peace with planning, because it provides market certainty. Metro has also protected itself from challenge through a programme of institutional redesign. Playing into localist ideology, Metro acquired a 'home rule' charter in 1993. This effectively insulated Metro from both unilateral interference from the Oregon state legislature and also challenges from other local governments and interest groups.

In this example, we see how the making of Portland's distinctive 'planning style' (Abbott, 1994) or local planning culture is a story not only of institutional design (albeit with unintended outcomes), but also of institutional constraints on action and institution-bound conflict over the production and reproduction of norms and practices. From an unlikely beginning, Portland Metro has carefully administered state-level planning policies in a way that has generated positive feedback through the nurturing of support coalitions. Working from a cultural approach would not reveal the contingency of Portland Metro's creation, its institutionalised autonomy from opponents, its generative role in the creation and propagation of influential professional planning ideas and its capacity to nurture support coalitions. A cultural perspective would also obscure the complex and competitive relationships between the state, Portland Metro, the counties and local governments.

Conclusion: a role for new institutionalism amid normative concerns

This overview suggests that scholars of planning practice and history and also of comparative planning systems may gain useful insights from institutional theories as they have developed in the social sciences. Why, then, has historical institutional analysis found so little purchase in planning studies? After all, the influence of formal and informal institutions on action is well recognised in planning scholarship. One possible explanation is the deep-rootedness of modern Anglo-American planning in a meliorative and sometimes utopian project of social transformation (Friedmann, 1987; Hall, 2002; Teitz, 2007). This has steered contemporary scholars' and practitioners' engagement with institutions in a normative direction; in other words, toward a focus on institutional design. As Beauregard (2005, 204) puts it:

> The current interest in 'institutional transformation' is an attempt to […] recapture the spirit of reform that so enthused planning in the late 19th and early 20th century, and to resurrect planning's long-lost utopian tradition. The objective is not to do planning better but to change the structures within which planning takes place. The implicit assumption is that an improved institutional environment will be more hospitable to planning and enable it to be more effective.

Although rooted in different philosophical grounds, the normative focus on designing an ideal institutional context for deliberation in planning is visible in Forester (1989), Friedmann (1973, Chapter 8) and Healey's (1997) influential theorisations of communicative action and 'transactive' or 'collaborative' planning. Given many planning scholars' theoretical, empirical and ultimately normative focus on institutional design, it is no surprise that Teitz (2007, 31–32) concludes that there is no obvious 'explicit role' for social-scientific theories of institutional dynamics. This is unfortunate. In this article I have sought to demonstrate that historical institutionalism may remedy some of the shortcomings of cultural analysis, not least its conflation of organisational and societal culture. It does so in a way that does not deny the importance of societal culture, professional norms, ideas and discourses. Rather, it emphasises the role that institutions play in producing and reproducing them. A renewed and systematic focus on institutional dynamics promises to reveal the determinants of historical trajectories and pathways of institutional and policy development, explain variation across space and time, account for stability and change and, in so doing, equip scholars and practitioners of planning to better understand opportunities and constraints as they contemplate contemporary challenges.

3

References

ABBOTT, C. (1994), 'The Oregon planning style', in Abbott et al. (eds), 205–26.

ABBOTT, C. (2000), 'The capital of good planning: metropolitan Portland since 1970', in R. Fishman (ed.), *The American Planning Tradition: Culture and Policy*, Washington, DC, Woodrow Wilson Center Press, 241–62.

ABBOTT, C. and ABBOTT, M. P. (1991), *Historical Development of the Metropolitan Service District*, Portland, OR, Metropolitan Service District.

ABBOTT, C. and HOWE, D. A. (1993), 'The politics of land use law in Oregon: senate bill 100 twenty years after', *Oregon Historical Quarterly*, **94**, 5–39.

ABBOTT, C., HOWE, D. A. and ADLER, S. (eds) (1994), *Planning the Oregon Way: A Twenty-Year Evaluation*, Corvallis, OR, Oregon State University Press.

ALBRECHTS, L., ALDEN, J. and DE ROSA PIRES, A. (eds) (2001), *The Changing Institutional Landscape of Planning*, Aldershot, Ashgate.

ALEXANDER, E. R. (1992), 'A transaction cost theory of planning', *Journal of the American Planning Association*, **58**, 190–200.

ALEXANDER, E. R. (2005), 'Institutional transformation and planning: from institutionalization theory to institutional design', *Planning Theory*, **4**, 209–23.

ALMOND, G. A. and POWELL, G. B. (1966), *Comparative Politics: A Developmental Approach, The Little, Brown Series in Comparative Politics. An Analytic Study*, Boston, MA, Little, Brown, & Company.

ALMOND, G. A. and VERBA, S. (1963), *The Civic Culture: Political Attitudes and Democracy in Five Nations*, Boston, MA, Little, Brown, & Company.

AMABLE, B. (2000), 'Institutional complementarity and diversity of social systems of innovation and production', *Review of International Political Economy*, **7**, 645–87.

ARTIBISE, A., CAMERON, K. and SEELIG, J. (2004), 'Metropolitan organization in greater Vancouver: "do it yourself" regional government', in D. Phares (eds), *Metropolitan Governance without Metropolitan Government?*, Aldershot, Ashgate, 195–211

BEAUREGARD, R. A. (2005), 'Introduction: institutional transformations', *Planning Theory*, **4**, 203–7.

BOOTH, P. (1993), 'The cultural dimension in comparative research: making sense of development control in France', *European Planning Studies*, **1**, 217–29.

BOOTH, P. (2011), 'Culture, planning and path dependence: some reflections on the problems of comparison', *Town Planning Review*, **82**, 13–28.

BOOTH, P., BREUILLARD, M., FRASER, C. and PARIS, D. (eds) (2007), *Spatial Planning Systems of Britain and France*, London, Routledge.

CAMPBELL, A., CONVERSE, P., MILLER, W. and STOKES, D. (1960), *The American Voter*, New York, NY, Wiley/University of Michigan Survey Research Center.

CHADWICK, N. (2002), *Regional Planning in British Columbia: 50 Years of Vision, Process and Practice*, Vancouver, School of Community and Regional Planning, University of British Columbia.

CLEMENS, E. S. (2006), 'Lineages of the Rube Goldberg state: building and blurring public programs, 1900–1940', in I. Shapiro, S. Skowronek and D. Galvin (eds), *Rethinking Political Institutions: The Art of the State*, New York, NY, New York University Press, 380–443.

COASE, R. H. (1998), 'The new institutional economics', *American Economic Review*, **88**, 72–74.

CULLINGWORTH, J. B. (1987), *Urban and Regional Planning in Canada*, Oxford, Transaction Press.

CULLINGWORTH, J. B. (1993), *The Political Culture of Planning: American Land Use Planning in Comparative Perspective*, New York, NY, Routledge.

CULLINGWORTH, J. B. (1994), 'Alternate planning systems: is there anything to learn from abroad?', *Journal of the American Planning Association*, **60**, 162–72.

DAHL, R. (1961), *Who Governs? Democracy and Power in an American City*, New Haven, CT, Yale University Press.

DENG, F., GORDON, P. and RICHARDSON, H. W. (2007), 'Private communities, market institutions, and planning', in N. Verma (ed.), *Institutions and Planning*, Oxford, Elsevier, 187–205.

EDWARDS, M. (2000), 'Sacred cow or sacrificial lamb? Will London's Green Belt have to go?', *City*, **4**, 106–12.

ELSON, M. (1986), *Green Belts: Conflict Mediation in the Urban Fringe*, London, Heinemann.

ESPING-ANDERSEN, G. (1990), *The Three Worlds of Welfare Capitalism*, Cambridge, MA, Polity Press.

FALUDI, A. (2004), 'Spatial planning traditions in Europe: their role in the ESDP process', *International Planning Studies*, **9**, 155–72.

FORESTER, J. (1989), *Planning in the Face of Power*, Berkeley, CA, University of California Press.

FRIEDMANN, J. (1967), 'The institutional context', in B. M. Gross (ed.), *Action under Planning: The Guidance of Economic Development*, New York, NY, McGraw Hill, 31–67.

FRIEDMANN, J. (1973), *Retracking America: A Theory of Transactive Planning*, Garden City, NY, Anchor Press/Doubleday.

FRIEDMANN, J. (1987), *Planning in the Public Domain: From Knowledge to Action*, Princeton, NJ, Princeton University Press.

FRIEDMANN, J. (2005a), 'Globalization and the emerging culture of planning', *Progress in Planning*, **64**, 183–234.

FRIEDMANN, J. (2005b), 'Planning cultures in transition', in B. Sanyal (ed.), *Comparative Planning Cultures*, New York, NY, Routledge, 29–44.

GILMAN, N. (2007), *Mandarins of the Future: Modernization Theory in Cold War America*, Baltimore, NJ, Johns Hopkins University Press.

GURRAN, N. (2011), *Australian Urban Land Use Planning: Principles, Systems and Practice* (2nd edn), Sydney, Sydney University Press.

HAAS, P. M. (1992), 'Introduction: epistemic communities and international policy coordination', *International Organization*, **46**, 1–35.

HACKER, J. (2004), 'Privatizing risk without privatizing the welfare state: the hidden politics of social policy retrenchment in the United States', *American Political Science Review*, **98**, 243–60.

HALL, P. G. (2002), *Cities of Tomorrow: An Intellectual History of Urban Planning and Design in Twentieth Century* (3rd edn), Oxford, Blackwell.

HALL, P. V. and SOSKICE, D. (eds) (2001), *Varieties of Capitalism*, Oxford, Oxford University Press.

HALL, P. V. and TAYLOR, R. (1996), 'Political science and the three new institutionalisms', *Political Studies*, **44**, 936–57.

HANNA, K. S. and WEBBER, S. M. (2010), 'Incremental planning and land-use conflict in the Toronto region's Oak Ridges Moraine', *Local Environment*, **15**, 169–83.

HAY, C. (2008), 'Constructivist institutionalism', in R. A. W. Rhodes, S. A. Binder and B. A. Rockman (eds), *The Oxford Handbook of Political Institutions*, New York, NY, Oxford University Press, 56–74.

3

HEALEY, P. (1997), *Collaborative Planning: Shaping Places in Fragmented Societies*, Vancouver, UBC Press.

HEALEY, P. (2010), 'Introduction: the transnational flow of knowledge and expertise in the planning field', in P. Healey and R. Upton (eds), *Crossing Borders: International Exchange and Planning Practices*, London, Routledge, 1–26.

HEALEY, P. and UPTON, R. (2010), *Crossing Borders: International Exchange and Planning Practices*, *Rtpi Library Series*, London, Routledge.

HEALEY, P. and WILLIAMS, R. (1993), 'European urban planning systems: diversity and convergence', *Urban Studies*, **30**, 701–20.

HOME, R. K. (1997), *Of Planting and Planning: The Making of British Colonial Cities (1st edn)*, *Studies in History, Planning, and the Environment*, London, Spon.

KELLER, D. A., KOCH, M. and SELLE, K. (1996), '"Either/or" and "and": first impressions of a journey into the planning cultures of four countries', *Planning Perspectives*, **11**, 41–54.

KEOHANE, R. and MILNER, H. (eds) (1996), *Internalization and Domestic Politics*, Cambridge, Cambridge University Press.

KITSCHELT, H. (1999), *Continuity and Change in Contemporary Capitalism*, Cambridge, Cambridge University Press.

KNIELING, J. and OTHENGRAFEN, F. (eds) (2009), *Planning Cultures in Europe: Decoding Cultural Phenomena in Urban and Regional Planning*, Aldershot, Ashgate.

KRASNER, S. (1984), 'Review article: approaches to the state: alternative conceptions and historical dynamics', *Comparative Politics*, **16**, 223–46.

LARSSON, G. (2006), *Spatial Planning Systems in Western Europe: An Overview*, Amsterdam, IOS Press.

LOGAN, T. H. (1976), 'The Americanization of German zoning', *Journal of the American Institute of Planners*, **42**, 377–85.

MAHONEY, J. and THELEN, K. (2010a), 'A theory of gradual institutional change', in J. Mahoney and K. Thelen (eds), 1–37.

MAHONEY, J. and THELEN, K. (eds) (2010b), *Explaining Institutional Change: Ambiguity, Agency, and Power*, Cambridge, Cambridge University Press.

MARCH, J. G. and OLSEN, J. P. (1984), 'The new institutionalism: organizational factors in political life', *American Political Science Review*, **78**, 734–49.

MARSH, D. and RHODES, R. A. W. (eds) (1992), *Policy Networks in British Government*, Oxford, Clarendon Press.

MASSER, I. (1984), 'Cross national comparative planning studies: a review', *Town Planning Review*, **55**, 137–49.

MASSER, I. and WILLIAMS, R. H. (1986), *Learning from Other Countries: The Cross-National Dimension in Urban Policy-Making*, Norwich, Geo Books.

MOORE, P. W. (1979), 'Zoning and planning: the Toronto experience, 1904–1970', in A. F. J. Artibise and G. A. Stelter (eds), *The Usable Urban Past: Planning and Politics in the Modern Canadian City*, Toronto, Macmillan, 316–41.

NASR, J. and VOLAIT, M. (eds) (2003), *Urbanism Imported or Exported? Native Aspirations and Foreign Plans*, Chichester, John Wiley & Sons.

NELSON, R. H. (1980), *Zoning and Property Rights: An Analysis of the American System of Land Use Regulation*, Cambridge, MA, MIT Press.

NEWMAN, P. and THORNLEY, A. (1996), *Urban Planning in Europe: International Competition, National Systems, and Planning Projects*, London, Routledge.

NORDLINGER, E. A. (1988), 'The return to the state: critiques', *American Political Science Review*, **82**, 875–901.

NORTH, D. C. (1990), *Institutions, Institutional Change and Economic Performance*, New York, NY, Cambridge.

OBERLANDER, H. P. and SMITH, P. J. (1993), 'Governing metropolitan Vancouver: regional intergovernmental relations in British Columbia', in D. N. Rothblatt and A. Sancton (eds), *Metropolitan Governance: American/Canadian Intergovernmental Perspectives*, Berkeley, CA, Institute of Governmental Studies Press, University of California, 329–73.

ORREN, K. and SKOWRONEK, S. (2004), *The Search for American Political Development*, Cambridge, Cambridge University Press.

OSTROM, E. (1990), *Governing the Commons: The Evolution of Institutions for Collective Action*, Cambridge, Cambridge University Press.

PARSONS, T. and SMELSER, N. J. (1956), *Economy and Society: A Study in the Integration of Economic and Social Theory*, Glencoe, IL, Free Press.

PIERSON, P. (2000a), 'Increasing returns, path dependence, and the study of politics', *American Political Science Review*, **94**, 251–67.

PIERSON, P. (2000b), 'Not just what, but when: timing and sequence in political processes', *Studies in American Political Development*, **14**, 72–92.

PIERSON, P. (2004), *Politics in Time: History, Institutions, and Social Analysis*, Princeton, NJ, Princeton University Press.

PIERSON, P. (2006), 'Public policies as institutions', in I. Shapiro, S. Skowronek and D. Galvin (eds), *Rethinking Political Institutions: The Art of the State*, New York, NY, New York University Press, 114–31.

PONTUSSON, J. (1995), 'Putting political institutions in their place and taking interests seriously', *Comparative Political Studies*, **28**, 117–48.

PONTUSSON, J. (2005), *Inequality and Prosperity: Social Europe vs. Liberal America*, Ithaca, NY, Cornell University Press.

PYE, L. W. and VERBA, S. (eds) (1965), *Political Culture and Political Development*, Princeton, NJ, Princeton University Press.

RHODES, R. A. W. (2008), 'Policy network analysis', in M. Moran, M. Rein and R. E. Goodin (eds), *The Oxford Handbook of Public Policy*, New York, NY, Oxford University Press, 425–47.

SALET, W. G. M., THORNLEY, A. and KREUKELS, A. (eds) (2003), *Metropolitan Governance and Spatial Planning: Comparative Case Studies of European City-Regions*, London, Spon.

SANYAL, B. (ed.) (2005), *Comparative Planning Cultures*, New York, NY, Routledge.

SCHMIDT, V. A. (1995), 'The new world order, incorporated: the rise of business and the decline of the nation-state', *Daedalus*, **124**, 75–106.

SCHMIDT, V. A. (2008), 'Discursive institutionalism: the explanatory power of ideas and discourse', *Annual Review of Political Science*, **11**, 303–26.

SCHMIDT, V. A. (2010), 'Taking ideas and discourse seriously: explaining change through discursive institutionalism as the fourth "New Institutionalism"', *European Political Science Review*, **2**, 1–25.

3

SKOCPOL, T. (1985), 'Bringing the state back in: strategies of analysis in current research', in P. Evans, D. Rueschemeyer and T. Skocpol (eds), *Bringing the State Back In*, Cambridge, Cambridge University Press, 3–37.

SMITH, R. M. (2006), 'Which came first, the ideas or the institutions?', in I. Shapiro, S. Skowronek and D. Galvin (eds), *Rethinking Political Institutions: The Art of the State*, New York, NY, New York University Press, 91–113.

STEAD, D. and NADIN, V. (2009), 'Planning cultures between models of society and planning systems', in J. Knieling and F. Othengrafen (eds), *Planning Cultures in Europe: Decoding Cultural Phenomena in Urban and Regional Planning*, Aldershot, Ashgate, 283–300.

STREECK, W. and THELEN, K. (2005), 'Introduction: institutional change in advanced political economies', in W. Streeck and K. Thelen (eds), *Beyond Continuity: Institutional Change in Advanced Political Economies*, New York, Oxford University Press, 1–39.

TEITZ, M. (2007), 'Planning and the new institutionalisms', in N. Verma (ed.), *Institutions and Planning*, Amsterdam, Elsevier, 17–35.

TEWDWR-JONES, M. and WILLIAMS, R. H. (2001), *The European Dimension of British Planning*, New York, NY, Spon Press.

THELEN, K. (1999), 'Historical institutionalism in comparative politics', *Annual Review of Political Science*, **2**, 369–404.

THELEN, K. (2003), 'How institutions evolve: insights from comparative historical analysis', in J. Mahoney and D. Rueschemeyer (eds), *Comparative Historical Analysis in the Social Sciences*, New York, NY, Cambridge University Press, 208–40.

THELEN, K. (2004), *How Institutions Evolve: The Political Economy of Skills in Germany, Britain, the United States, and Japan*, Cambridge, Cambridge University Press.

TSEBELIS, G. (2002), *Veto Players: How Political Institutions Work*, Princeton, NJ, Princeton University Press.

WARD, S. V. (1999), 'The international diffusion of planning: a review and a Canadian case study', *International Planning Studies*, **4**, 53–77.

WARD, S. V. (2000), 'Re-examining the international diffusion of planning', in R. Freestone (ed.), *Urban Planning in a Changing World: The Twentieth Century Experience*, London, Routledge, 40–60.

WARD, S. V. (2002), *Planning the Twentieth-Century City: The Advanced Capitalist World*, New York, NY, Wiley.

WARD, S. V. (2010), 'Transnational planners in a postcolonial world', in P. Healey and R. Upton (eds), 47–72.

WARD, S. V. (ed.) (1992), *The Garden City: Past, Present and Future*, Oxford, Taylor & Francis.

WATERHOUSE, A. (1979), 'The advent of localism in two planning cultures: Munich and Toronto', *Town Planning Review*, **50**, 313–24.

WENGER, E. (1998), *Communities of Practice: Learning, Meaning, and Identity*, Cambridge, Cambridge University Press.

WHEELER, S. M. (2004), *Planning for Sustainability Creating Livable, Equitable and Ecological Communities*, New York, NY, Routledge.

WILLIAMSON, O. E. (1975), *Markets and Hierarchies, Analysis and Antitrust Implications: A Study in the Economics of Internal Organization*, New York, NY, Free Press.

Serviceteil

© Springer-Verlag GmbH Deutschland, ein Teil von Springer Nature 2019
T. Wiechmann (Hrsg.), *ARL Reader Planungstheorie Band 2*, https://doi.org/10.1007/978-3-662-57624-3

Artikelverzeichnis

Albrechts, Louis (2004): Strategic (Spatial) Planning Reexamined. In: Environment and Planning B: Planning and Design 31 (5): 743–758.© 2004 by Albrechts, Louis, Reprinted by Permission of SAGE Publications, Ltd.

Healey, Patsy (2009): In Search of the "Strategic" in Spatial Strategy Making. In: Planning Theory & Practice, Vol. 10 (4), 439–457.Reprinted by permission of the publisher Taylor & Francis Ltd, ▸ http://www.tandfonline.com

Mastop, Hans; Faludi, Andreas (1997): Evaluation of Strategic Plans: The Performance Principle. In: Environment and Planning B, Vol. 24 (6), 815–832.© 1997 by Mastop, Hans; Faludi, Andreas, Reprinted by Permission of SAGE Publications, Ltd.

Kühn, Manfred (2008): Strategische Stadt-und Regionalplanung. In: Raumforschung und Raumordnung, Vol. 66, No. 3, 230–243.© Akademie für Raumforschung und Landesplanung (ARL) und Bundesamt für Bauwesen und Raumordnung (BBR) 2008

Fürst, Dietrich (2012): Internationales Verständnis von „Strategischer Regionalplanung". In: Vallée, D. (Hrsg.): Strategische Regionalplanung. Hannover, 18–30. Verl. d. ARL, 2012 (Forschungs- und Sitzungsberichte der ARL 237). – ISBN 978-3-88838-066-2, 18–30. ▸ https://nbn-resolving.org/urn:nbn:de:0168-ssoar-337095© Akademie für Raumforschung und Landesplanung (ARL) und Bundesamt für Bauwesen und Raumordnung (BBR) 2008

Ritter, Ernst-Hasso (2007): Strategieentwicklung heute. Zum integrativen Management konzeptioneller Politik (am Beispiel der Stadtentwicklungsplanung), in: PNDonline, Vol. 1, 1–12.© Verlag Dorothea Rohn

Mintzberg, Henry (1987): The Strategy Concept I: Five Ps For Strategy. In: California Management Review, Vol. 30, No. 1, 11–24.© 1987 by Mintzberg, Henry, Reprinted by Permission of SAGE Publications, Ltd.

Weick, Karl E. (1987): Substitutes for strategy, In: Weick, Karl E. (2000): Making Sense of the Organization, Wiley-Blackwell, Chichester, West Sussex, 345–356.© 2001 by Weick, Karl E., Reprinted by permission of the publisher Blackwell-Wiley, Oxford, ▸ www.blackwellpublishing.com

Bryson, John M. (1988): A Strategic Planning Process for Public and Non-profit Organizations. In: Long Range Planning, Vol. 21, No. 1, 73–81.Reprinted from Long Range Planning, Vol. 21, No. 1, Bryson, John M., A Strategic Planning Process for Public and Non-profit Organizations, 73–81, Copyright (1988), with permission from Elsevier

Sanyal, Bishwapriya (2005): Hybrid Planning Cultures: The Search for the Global Cultural Commons, In: Sanyal, B. (Hrsg.) Comparative Planning Cultures, New York, Routledge, 3–25.Copyright (2005) From Comparative Planning Cultures by Sanyal, Bishwapriya. Reproduced by permission of Taylor and Francis Group, LLC, a division of Informa plc

Booth, Philip (1993): The Cultural Dimension in Comparative Research: Making Sense of Development Control in France. In: European Planning Studies 1 (2): 217–229.Reprinted by permission of the publisher Taylor & Francis Ltd, ▸ http://www.tandfonline.com

Faludi, Andreas (1999): Patterns of Doctrinal Development. In: Journal of Planning Education and Research 18 (4): 333–344.© 1999 by Faludi, Andreas, Reprinted by Permission of SAGE Publications, Ltd.

Friedmann, John (1967): A Conceptual Model for the Analysis of Planning Behavior. In: Administrative Science Quarterly, 12 (2): 225–252.© 1967 by Friedmann, John, Reprinted by Permission of SAGE Publications, Ltd.

Howe, Elizabeth; Kaufman, Jerome (1979): The Ethics of Contemporary American Planners. In: Journal of the American Planning Association 45 (3): 243–255.© The American Planning Association, ▸ www.planning.org, reprinted by permission of Taylor & Francis Ltd, ▸ http://www.tandfonline.com on behalf of The American Planning Association.

Nadin, Vincent; Stead, Dominic (2008): European Spatial Planning Systems, Social Models and Learning. In: disP – The Planning Review 44 (172): 35–47.© ETH – Eidgenössische Technische Hochschule Zurich, reprinted by permission of Taylor & Francis Ltd, ▸ http://www.tandfonline.com on behalf of ETH – Eidgenössische Technische Hochschule Zürich.

Fürst, Dietrich (2007): Planungskultur- Auf dem Weg zu einem besseren Verständnis von Planungsprozessen? pnd online.© PND Online, RWTH Achen

Taylor, Zack (2013): Rethinking Planning Culture: A New Institutionalist Approach. In: The Town Planning Review 84 (6): 683–702.© Liverpool University Press

Sachverzeichnis

The manufacturer's authorised representative in the EU is Springer
Nature Customer Service Centre GmbH, Europaplatz 3, 69115 Heidelberg,
Germany. If you have any concerns regarding our products, please
contact ProductSafety@springernature.com

Printed and bound by CPI Group (UK) Ltd, Croydon, CR0 4YY

27/04/2026

02097675-0001